BIOLOGY

LIFE FEATURES

THIRD EDITION

Tom M. Graham
University of Alabama

Pearson
Custom
Publishing

Cover Photo: "A white barn owl swoops down to a snowy landing," by © Manfred Danneger/Tony Stone Images.

Printed in the United States of America

10 9 8 7 6 5 4 3 2

Please visit our web site at www.pearsoncustom.com.

ISBN 0–536–02421–9

BA 990948

PEARSON CUSTOM PUBLISHING
160 Gould Street/Needham Heights, MA 02494
A Pearson Education Company

Photo Credits

BLACK & WHITE

6 © 1997 NCEAS; **33** © David Newman/Visuals Unlimited; **36** © John D. Cunningham/ Visuals Unlimited; **38** © Science VU/Visuals Unlimited; **42** © Jim Merli/Visuals Unlimited; **82** (A): © R. Kessel–G. Shih/Visuals Unlimited; **112** (A) Centers for Disease Control and Prevention; **112** (B) Centers for Disease Control and Prevention; **112** (C) Dr. Gerald Goldstein/Licking County Archaeological and Landmarks Society; **114** © Veronika Burmeister/Visuals Unlimited; **115** © T.E. Adams/Visuals Unlimited; **116** © John D. Cunningham/Visuals Unlimited; **121** Sponge: © Don W. Fawcett/Visuals Unlimited; **121** Jellyfish: © John D. Cunningham/Visuals Unlimited; **121** Flatworm: © John D. Cunningham/Visuals Unlimited; **121** Eels: © John D. Cunningham/Visuals Unlimited; **122** Octopus: © John D. Cunningham/Visuals Unlimited; **122** Clamworms: © John D. Cunningham/Visuals Unlimited; **122** Praying mantis: © Joel Arrington/ Visuals Unlimited; **122** Echinoderms: © Science VU/Visuals Unlimited; **130** Agnathan: © John D. Cunningham/Visuals Unlimited; **130** Sunfish: © Cleveland P. Hickman, Jr./ Visuals Unlimited; **130** African Clawed Frog: © Cleveland P. Hickman, Jr./Visuals Unlimited; **130** African Crocodile: © Leonard Lee Rue III/Visuals Unlimited;**131** Hummingbird: © Len Rue, Jr./Visuals Unlimited; **131** Armadillo: © Leonard Lee Rue III/ Visuals Unlimited; **133** © Tom J. Ulrich/Visuals Unlimited; **134** © James L. Amos /Photo Researchers, Inc.; **153** Left: © C. Gerald Van Dyke/Visuals Unlimited; **153** Right: © G. Shih–R. Kessel/Visuals Unlimited; **168** Desert Biome: © Jonathan Turk/Visuals Unlimited; **168** Deciduous Forest: © Science VU/Visuals Unlimited; **169** © Manfred Kage/ Peter Arnold, Inc.; **183** © G. Prance/Visuals Unlimited; **189** © E. Webber/Visuals Unlimited; **219** Yellow Pitcher Plant: © Howard A. Miller, Sr./Visuals Unlimited; **219** Venus Fly Trap: © John D. Cunningham/Visuals Unlimited; **234** © David M. Phillips/Photo Researchers, Inc.; **293** © Len Clifford/Visuals Unlimited; **294** © Cleveland P. Hickman, Jr./Visuals Unlimited; **297** Male Baboon: © Barbara Gerlach/Visuals Unlimited; **297** Male Toad: © Joe McDonald/Visuals Unlimited; **297** Mocking Bird: © John D. Cunningham/Visuals Unlimited; **307** Cell Division (A): © T. J. Beveridge/Visuals Unlimited; **307** Cell Division (B): © David M. Phillips/Visuals Unlimited; **321** Normal RBC: © Science VU/Visuals Unlimited; **321** Sickle Cells: © Science VU/Visuals Unlimited; **322** © John D. Cunningham/Visuals Unlimited; **383** Wildflowers: © Steve McCutcheon/Visuals Unlimited; **383** Wildebeests: © Len Rue, Jr./Visuals Unlimited; **383** Bison: © Preston J. Garrison/Visuals Unlimited; **383** Bull Elk: © Leonard Lee Rue III/Visuals Unlimited; **383** Maine Forest: © John D. Cunningham/Visuals Unlimited; **383** Arctic Fox: © Irene Vandermolen/Visuals Unlimited; **392** © Renee Lynn/Photo Researchers, Inc.; **398** Dr. Arnold Tamarin; **410** (A–C): GeoScience/PH; **410** (D): A.I. Friedman–IVOR; **413** English Heritage; **428** Dr. N.T. Boaz; **429** Paris Pavlakis, Ph.D.

Contents

Part Seven

Part Nine

Preface

Biology: Life Features is intended for introductory nonscience majors biology courses at the college and university levels. The book is formatted around seven life features—organization, energy transactions, sensation and response, homeostasis, reproduction, growth and development, and evolutionary adaptation which constitute the framework for study in each section. The emphasis on fundamentals reflects the needs of the audience for whom the book was written. For virtually all of the students enrolled, a nonmajors biology course is a one-time, terminal experience. These students, who will become lawyers, artists, actors, journalists, and business graduates, will not be required to understand or employ complex biological principles in their vocations. Accordingly, this is *not* a text for biology majors, minors, premeds, predents, and the like, nor is it a compendium of the classical biology text for majors. As such, it is *not* an attempt to "cover" all the known fields, theories, attitudes, and viewpoints associated with biology.

Stated positively, the intent of the text is to highlight some of the major areas of biology that may evoke in a student an appreciation of science in general and biology in particular, and to provide an opportunity to learn how to think rationally and analytically. Perhaps the crux of a nonmajors science course is to offer students the opportunity to become more aware, better informed citizens, and to prevent them from being fooled, misinformed, or overwhelmed by partial truths, hype, and sensationalism—without drowning them in a sea of scientific minutiae.

The author thanks Pearson Custom Publishing, and especially Terry Brennan, for the opportunity to present the book.

PART ONE

The Spectrum of Science

Chapter One

Science: What It Is—And Isn't

Science: Foundation and Philosophy

This is a book about science; more exactly, it is about a particular science known as *biology*. As such, it is also necessarily a book about philosophy. Philosophical inquiry ultimately is concerned with two fundamental questions: "What is the nature of the universe?" and "How do we discover it?" Science, as we will see, rests on a philosophical premise, and it is critical to your orientation to biology that you understand what science is—and what it is not. When we refer to biology—or any discipline—as a "science," what exactly do we mean? What is science and how do we distinguish it from nonscience?

Pre-Greek civilizations—Sumerian, Babylonian, Egyptian, and others—made attempts at science, but it would be difficult to distinguish their attempts from theology. Their world-view centered on superstition and magic, intertwined with belief in omnipotent and ever-present gods. From this outlook there developed elaborate and pervasive mythologies that persisted for centuries. But in Greece, 2400 years ago, a new spirit began to awaken: the Greeks discovered the supremacy of the mind. In describing the state of human thought before the rise of Greece, the Greek scholar Edith Hamilton wrote: "The human mind played no part in the whole business. It was enslaved by terror. . . . There was no dependable relation anywhere between cause and effect." With the Greeks came a love of knowledge and ultimately the birth of science.

The story of the birth of science began in Greece over 25 centuries ago with a philosopher named Thales, a renowned geometrician, astronomer and engineer. He wrote no book, but is considered the father of philosophy and even the father of science. Unfortunately, we know little else about him. And yet, he is credited with a single statement that marks the beginning of the most extraordinary intellectual revolution in recorded history. "Water," said Thales, "is the cause of all things." This seemingly simplistic statement represented a radical shift from the mythological and superstitious world-view of pre-Greek civilizations. The statement does not reflect a superstitious approach or an appeal to the will of gods, but is an independent *hypothesis,* a critical step in the scientific process. To state that water is the cause of all things is to conceive the world in *natural* terms. From this base, then, it is possible to advance knowledge by debating the statement. Most importantly, the assertion *opens up the possibility of verification.* Investigation can reveal if the cause of all things is, in fact, water, or if it is not.

Today, the statement may seem naive, even childish, but its appeal is to the reasoning mind.

The term "science" is from the Latin *scientia,* meaning "knowledge," and, in part, science pertains to an organized body of knowledge acquired through observation, experimentation, and reason. Science is also a process, an exploratory attempt to understand the universe and to systematize the knowledge acquired from this exploration. A philosophic underpinning of science is the proposition that all natural phenomena have causes for which there are logical and verifiable explanations. The historian Will Durant wrote that "to observe processes and construct *means* is science; to criticize and coordinate *ends* is philosophy." The scientific process, then, is a means to a philosophic end—the end being verifiable knowledge.

It is important to understand that verifiable knowledge does not imply infallibility or omniscience. Science is concerned with acquiring the best available evidence that explains a given phenomenon. In many instances, this evidence, although compelling, may still be incomplete. Consequently, scientific pronouncements are couched as "to the best of our knowledge . . . ," or "the evidence indicates that. . . ." Science, by definition, is neither faultless nor dogmatic.

A characteristic of science is the precise definition of terms. All knowledge, if it is to be communicated, is subject to the accuracy of definitions. In time, terms come to mean what people want them to mean. This is acceptable, providing the terms mean the same thing to everyone; otherwise, rational human communication is impossible. As an example, consider the often quoted phrase "Humans have an instinct for self-preservation." The pivotal term here is "instinct." In common usage, the words "desire" or "urge" could be substituted and the statement would still mean what virtually everyone thinks it means. Instinct, however, refers to innate, genetically programmed behavior, an inborn, automatic form of knowledge that enables an organism to know what to do without the benefit of prior experience. Some spiders, for example, spin elaborate webs the first time they try. While instinctive behavior can become more skillful over time through learning, the crucial point is that instinct itself does not require an opportunity to learn. Applied to self-preservation in humans, instinct is clearly misleading; there is no evidence that humans know instinctively what is good or bad for them. A baby would drink poison, play with fire, or fall into a hole if left unattended. Humans must acquire the knowledge of self-preservation through learning.

It is crucial to our understanding of what science is and is not that we apply the exacting rule of definition to the term "science." The essential precept underlying science is that the universe can be explained and understood solely in terms of physical and chemical laws. If we discover and understand these laws, they are sufficient to explain all natural phenomena. This philosophic view is known as **mechanism.** While this is admittedly a belief system, it is established on certain objective principles, as we will see.

A competing philosophic view is called **vitalism,** according to which there is an extranatural or supernatural "vital force" at work in the universe in addition to the natural laws of physics and chemistry. Vitalistic explanations, to the best of our knowledge, are scientifically dead end propositions because they have not been supported by verifiable evidence. We cannot construct a controlled experiment comparing results when the vital force is present and when it is not. A controlled experiment uses a *control* treatment in which all known conditions affecting the experiment are strictly regulated. The *experimental* treatment, into which variables are introduced, is compared with the control. In comparing the growth of oak trees, for example, there is no known way to determine the effects that the presence or absence of the vital force might have on their growth. On the other hand, determining the effects of water (the variable) on growth can be approached scientifically by watering one tree (the control) and not another. While scientific experiments are not always this simple, the principle of verification still applies. Science, then, is limited to the sphere of the natural universe. Science, by definition, cannot investigate alleged non-natural or supernatural phenomena.

We can illustrate this concept by examining an ongoing controversy. In recent years, science, particularly evolutionary science, has been confronted by the vitalistic philosophy known as *creationism,* also referred to as "intelligent design." Basically, the creationist position is that all life forms were created instantly out of a void by a supernatural power, and that the evolution of more complex life forms from simpler forms has not occurred on Earth. From a scientific standpoint, there are several problems with this position. First, *which* creation scenario is correct or acceptable? There are a number of these, many of which are not in agreement. Second, a strong belief in creationism is not, in itself, sufficient reason to award it equal status with a scientifically established theory. Evolutionary theory is supported by anatomical, genetic, ecological, paleontological, embryological, and biochemical evidence. Creationists still argue, however, that this evidence is incomplete and therefore inconclusive. This argument is irrelevant; evidence is evidence. Science is fraught with theories for which there is incomplete evidence. Third, what exactly would be the scientific explanations relating to creationism? If a supernatural force created all living organisms as they are out of nothingness, what rules of evidence and experimental procedure will be employed in an attempt to verify it? If it is determined that such rules *cannot*, apply, then creationism cannot be discussed scientifically. If the answer is that they *need not* apply, we destroy science.

Reason, as mentioned, is an integral factor in science. Basically, reasoning can be either **deductive** or **inductive.** Deductive reasoning begins with one or more general premises from which are drawn logical conclusions. If our premises are "All dogs have four legs" and "Fido is a dog," we may validly conclude that "Fido has four legs." Notice here that the conclusion is only as sound as the correctness of the premises. Some dogs, through injury or disease, may have fewer than four legs; thus, the conclusion may not be valid. Deductive reasoning always proceeds from the general to the specific.

In contrast, inductive reasoning begins with specifics and proceeds toward generalizations. From the experience of observing that "Every dog we have seen has four legs," we induce the generalization that "All dogs have four legs." In this case, the evidence is not conclusive. The next dog we see may *not* have four legs. Consequently, inductive reasoning indicates what *is probable,* not certain. It was primarily the application of inductive reasoning to natural phenomena over 300 years ago that gave science its greatest impetus. In practice, however, science employs both deductive and inductive reasoning.

The task of science is to discover and demonstrate the most probable explanation for a natural process or event. To accomplish this, science often employs an experimental system known as the *scientific method,* which reflects the application of deductive and inductive reasoning (Fig. 1.1).

The Method and Limitations of Science

Suppose that one morning you have just settled into your biology class when a student bursts through the door, wild-eyed with arms flailing and proclaims in a frenzied screech:

"I just saw a leprechaun run across the campus!" There is dead silence. You blink, glance at the professor and realize you are holding your breath. You wait. More silence. You again glance at the professor. No clue. Overcome with frustration, the student wheels and bolts from the room.

Inherent in this little scenario are two questions fundamental to the scientific approach: Is the observation scientifically acceptable or unacceptable? (What is the nature of the universe?) and How do we determine the answer? (How do we discover it?) Discussing the incident with your bewildered classmates, you might phrase these questions as: "Do you believe this guy or not?" and "How do we know if he's telling the truth?"

The alleged observation of the leprechaun is immediately restricting; that is, an "either-or" alternative. Either the little creature was evident physically or it was not. In science an observation must be accessible to everyone; that is, it must be a phenomenon detectable directly by the senses or indirectly using an instrument. The possibility that the student imagined he saw one of the little people only accentuates this point. Until a leprechaun is demonstrated to exist, or until *something* is known about it, no verification is possible and no scientific conclusions can be drawn.

This is why science cannot attempt to answer such philosophic inquiries as: "What is the meaning of life?" or "Does God exist?" or "What is justice?" Their premises lie outside the realm available to science, and the possible answers, unsupported by evidence, cannot be verified. Science makes neither value judgments nor moral judgments.

At the core of the scientific method—its obligatory first step—is the unbiased observation of natural phenomena. The toughest aspect of observation is objectivity, which is seeing only what is evident rather than what we want or expect to see or what someone tells us we *should* see. The observation may provoke questions we want answered or present problems we wish to investigate. To do so, the question or problem has to be subject to some form of testing. "Why does grass appear green to us?" is a scientifically legitimate question because it is based on objective observation and is subject to certain testing procedures.

Next, various proposals or hypotheses are offered as to the possible solution to the problem. A **hypothesis** is simply a suggested proposition or educated guess. In developing hypotheses, scientists use inductive reasoning; that is, they observe *specific* events from which they induce *general* hypothetical conclusions. Then, by employing deductive reasoning, specific conclusions or expectations are proposed based on the general assumptions of the hypotheses. Ultimately, all scientific hypotheses must be subjected to experimentation, which provides evidence for or against the hypotheses. Let us interrupt here to make a critical point: It is with experimentation that science and nonscience part company. Alleged observations backed with hypotheses based on opinions, intuition, or feelings are legion throughout history. Unsupported by evidence, such hypotheses remain inconclusive and unknowable in the scientific sense. Regardless of the appeal of its claim, a hypothesis remains a hypothesis until the cardinal question is answered: "What is the evidence?"

Through accumulated evidence, some hypotheses are retained and some are discarded. When, despite all efforts to discredit it, a hypothesis is supported by demonstrable evidence, it becomes a **theory**. Scientific theories have predictive value and imply a certain degree of probability. This means that from our observations of A, we can predict with reasonable assurance that our conclusion will be B. There is, then, a certain degree of probability that B will be concluded from what we know about A. Theories having a very high degree of probability and predictive value are called **natural laws.** The laws of gravity, thermodynamics, and planetary motion are well-documented examples.

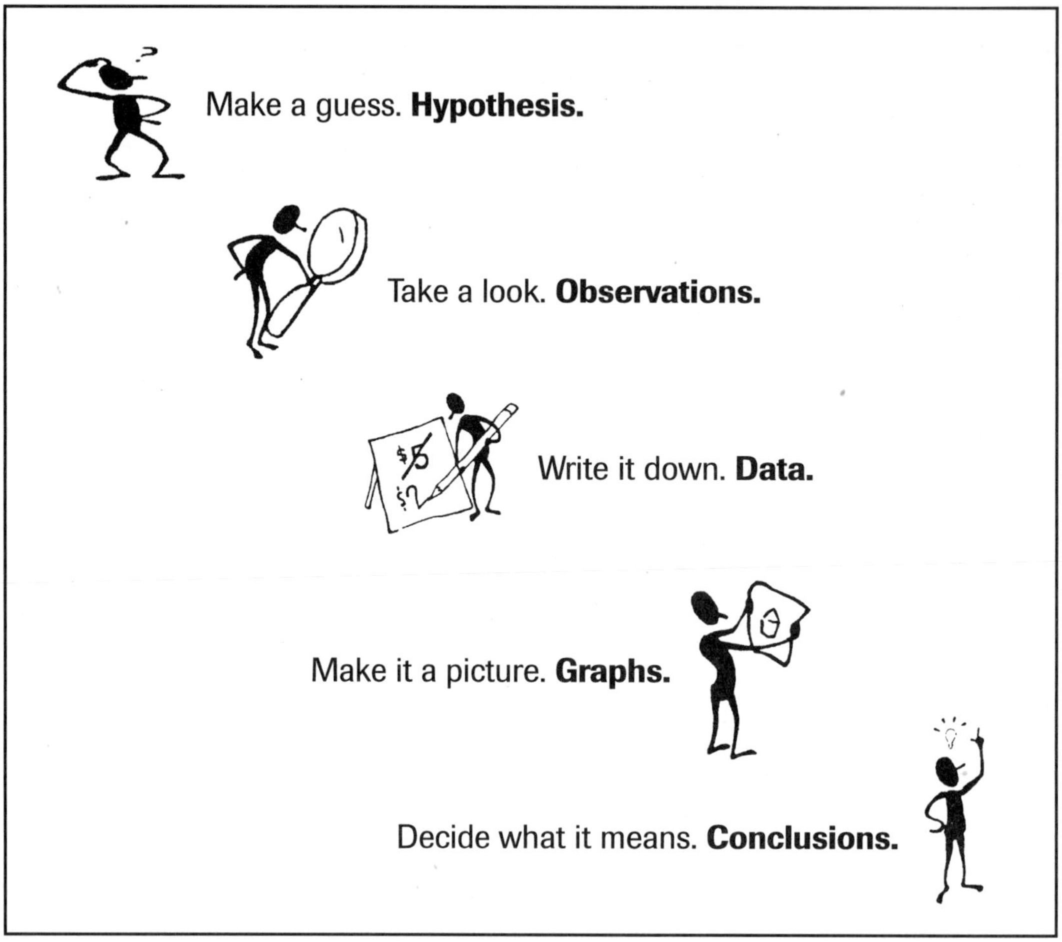

Figure 1.1 The basic steps of the scientific method. Although these steps are important in the analysis of data, scientific inquiry does not always proceed in this lock-step manner.

It is important to your appreciation of science to understand that many scientific discoveries, both great and small, quite often, or probably usually, do not necessarily follow the lock-step sequence outlined in the scientific method. Science really isn't a tightly structured activity. There are numerous instances of accidental finds, such as the discovery in 1889 of the hormone insulin and the discovery of the antibiotic aureomycin in 1944. Many discoveries also result from intuition and from "flashes" of insight that come after wrestling with a problem for a time. This is not to minimize creativity or intellectual discipline, however. As the French chemist Louis Pasteur noted, "Chance favors the prepared mind."

Contrary to popular opinion, the purpose of science is not to discover the ultimate "truth," nor to offer its findings as absolute and final. Scientific theories are always subject to modification if new evidence becomes available. Science, as we have said, attempts to accumulate data based on the *available* evidence. It is often not known if or when additional evidence may be forthcoming. Moreover, scientific findings reflect knowledge of the immediately known universe. We have not been everywhere in the universe nor have we experienced other possible conditions different from those we know at present. Thus, in keeping with the principle of objectivity, science must deal with probability, not certainty and thus cannot present its findings as the absolute and final truth. There is an extremely high probability that the sun will rise tomorrow morning, but no one really *knows* for certain.

Scientific theories become visible through their application in industry, medicine, commerce, engineering, agriculture and other fields. The term for applied science is *technology;* that is, the application of knowledge for practical ends. Obviously, technology has affected virtually every realm of human experience. Along with its supporters, technology also has its critics who point a condemning finger at what they see as excess and waste—too many weapons, too many cars, too many "gadgets"—and blame science. Scientific findings themselves, however, have no intrinsic value; theories cannot be evaluated as "good" or "bad" in a moral sense. One may be tempted to argue that the theories behind the development of the combustion engine or the production of antibiotics are "good." This evaluation, however, refers to their *application* and not to any scientific theory that led to their application. We can detail this concept with a piece of history.

By the first quarter of the twentieth century, science had a fairly comprehensive view of atomic structure. By 1939, the heavy uranium atom had been split into two lighter atoms, a process known as *fission*. In late 1942, a group of scientists succeeded in producing the first manmade chain reaction (a series of fissions). From there, the United States undertook the Manhattan Project that resulted in the world's first atomic bomb. In 1945 we witnessed for the first time its awesome power and destructive force. A terrifying weapon thus gave rise to the nuclear age. But atomic energy also has provided heretofore undreamed of benefits. The generation of electrical power; its use in diagnosing, treating, and curing diseases; supplying power for ships and space flights; and studying photosynthesis are only a few.

The story behind splitting the atom began nearly 24 centuries ago in ancient Greece. The philosopher Democritus proposed that matter consisted of tiny, "uncuttable" particles he called *atoms.* Over 2000 years later, the English chemist John Dalton, capitalizing on this ancient idea, proposed the atomic theory. In later years, researchers discovered subatomic particles and eventually the nuclear age was born. Thus, the world was presented with the atomic theory, which subsequently was applied in the harnessing of nuclear energy. Existing alone, the atomic theory is just that—a theory; in and of itself it is neither "good" nor "bad." No moral judgment is possible; it is only the application of the theory, its use in technology, that can be evaluated.

Science operates within a limited sphere. We have mentioned that controlled experimentation is the one determinant defining this sphere. If a hypothesis cannot be verified by experimental analyses, it is discarded. This makes science a self-correcting, self-protecting process. Attempts to misuse it are self-defeating because the laws of nature cannot be violated. Nonetheless, instances of such attempts have appeared throughout history, some with disastrous consequences.

One of the most disastrous, and tragic, episodes occurred in the Soviet Union during a 30-year period from the mid-1930s to the mid-1960s. T.D. Lysenko, an alleged geneticist, claimed that by soaking winter wheat seeds in ice water, winter wheat could be converted to spring wheat. He based this on the unsupported hypothesis that a cold environment (ice water) could alter the genetic makeup of an organism (wheat). We will consider this inadmissable idea of altering the genetic makeup of an organism through environmental manipulation later in this book. This preposterous notion would have died a deserving death had the Soviet dictator Stalin not endorsed it. Stalin's interest and approval stemmed from his Marxist political philosophy, which emphasized the role of the environment in shaping human beings. Lysenko had rejected the known principles of genetics—that *genes* determine an organism's makeup—in favor of environmental determination of heredity. Stalin embraced this concept in his fanciful attempt to create the "new Soviet man." If only the "correct" environment were provided, he reasoned, it would result in producing the new elite. Similarly, if the desired environment could be provided, Lysenko proposed that the genetic makeup of wheat could be altered. The ignorance and charlatanism behind such proposals were responsible not only for the devastation of Soviet agriculture for many years, but also for the deportation of dissenting, eminent geneticists to concentration camps.

Science ultimately attempts to accumulate experimental evidence to support its hypotheses. The "pseudoscience" just presented declared the correctness of the hypothesis—the genetic transformation of wheat through exposure to the environment—without providing experimental evidence to verify it.

In a different context, another example of the misuse of science is the attempt to apply scientific theories politically or socially. One of the most notable instances of such an attempt arose in the nineteenth century. Other centuries had their mathematics and physical sciences, but the 1800s was the century of the Industrial Revolution, Darwin and evolution. A foremost proponent of evolution was the philosopher Herbert Spencer (1820–1903), who believed evolutionary theory could be applied to virtually any situation. He coined the famous phrase "survival of the fittest," a phrase that seeded the sorry spectacle of *social Darwinism.* Social Darwinists in Europe, and later in America, applied Darwin's theory of biological evolution in a sociopolitical context. They held that certain races and nationalities were superior and thereby "fit" to survive and rule. Spencer advocated that "the ill-fitted must suffer the evils of unfitness, and the well-fitted profit by their fitness." This led not only to racism, anti-Semitism and disdain for the less fortunate, but provided "justification" for such obnoxious sentiments. This misuse of science dramatizes a cardinal concept: The application of scientific theories in a nonscientific context is totally unwarranted and can be damaging or even dangerous. There actually is no such thing as "social" Darwinism any more than there is "Republican" chemistry or "Lutheran" physics.

Biology: The Life Science

History of Biological Thought

Biology, a term derived from the Greek *bios,* "life" + *logos,* "study," is the natural science that is concerned with the structural, functional, and behavioral features of every form of life that has ever existed. Biology seeks to answer the questions "What is the nature of the universe?" and "How do we discover it?" as they apply to the living world.

The influence of biology on the course of history has been profound and dramatic. Its significance as a distinct, organized study emerged from the work of the Greek philosopher Aristotle, the founder of biology and the world's first great biologist. He was the first to classify living organisms according to their common characteristics. Some of these he discerned as unifying characteristics underlying the rich diversity of life. While his concept of unity and diversity was a prelude to the theory of evolution, Aristotle did not recognize it as such. He did recognize, however, that life progressed from simpler to more complex forms and that, among the animals, increasing intelligence often accompanied this progression.

He observed the relation between diet and lifestyle and contributed to the fields of experimental genetics, animal behavior, and marine biology. His writings on comparative anatomy held forth for centuries and he also founded the science of embryology. Important to the philosophy of science and, indeed, to all philosophy was his pronouncement that any entity is what we perceive it to be and that it cannot act in contradiction to its nature. Further, Aristotle presented to the world a view of the human animal. Man, he said, is the rational animal and a happy life is one governed by reason. The French historian Ernest Renan said of Aristotle that he "gave science to mankind," and the poet Dante lauded him as the "master of those who know." These are no exaggerations for the man whose accomplishments laid the foundation for 2000 years of scientific progress.

How far has biology come in over 23 centuries? A very long way indeed. In historical perspective, the advance of biology has been no less than amazing. Following the fall of the Western Roman Empire in A.D. 476, Europe began to sink into the stagnation of the Middle Ages. Science, including biology, sank with it. For over 1000 years, scientific inquiry lay dormant; its rebirth came with the Italian Renaissance in the sixteenth century. In the succeeding 400 years, biological research accelerated at a dizzying pace, spawning discoveries that transformed the world. From the ground-breaking analysis of the human body by the Belgian physician Andreas Vesalius in the 1500s to the development of the theory of biological evolution by Charles Darwin in the 1800s to mapping the location of human genes today, the influence of biology has left no person or institution untouched.

As with all scientists, biologists are concerned with formulating and using theories. Biologists seek to be objective about the living world. Objectivity is a philosophical viewpoint and we have commented that science rests on a philosophical base. The law of identity states that an entity is what it is, that it has a specific nature. It cannot be what it is and something else at the same time.

The law of causality states what an entity can or will do in a given context. As Aristotle declared, an entity cannot act in contradiction to its nature. As biological entities, plants, animals, and other organisms act and interact with each other and with their surroundings within a mutual framework we call *life*. Life is the central focus of biology and of this book. As a science, biology seeks to investigate and understand any phenomenon attributable to living organisms. For centuries, humans have probed the meaning of life, while at the same time being unable to define exactly what "life" is. The question "What is life?" still cannot be answered specifically. It has been suggested that life is the absence of death, or life is that which only living organisms experience. These, however, are circular arguments, virtually devoid of meaning. It is thus more productive to ask, "How do we characterize life?" or "How do we distinguish the living from the nonliving?" As one example, the French scientist Jacques Monod, a 1965 Nobel Prize winner, listed three essential characteristics of living organisms: goal-directed behavior, self-constructing action, and reproduction. These and other life features are the subject of this book.

Aristotle noted that to have knowledge of phenomena in the true sense, we must know their causes, the principles behind them, and how they fit into the scheme of nature. We may be *aware* that such-and-such exists, but we may *know* virtually nothing about it. Moreover, much of our awareness stems from memorization; true knowledge is *understanding*. This is the theme of science. There are those who must know "What?" and "Why?" and "How?" They carry civilization forward, ease the burdens of humanity, and bring clarity out of confusion. They have taught us how to think and, though we praise the brilliance of the concept, have demonstrated that water is not the cause of all things.

Chapter Summary

1. The world-view of pre-Greek civilizations was based on superstition and magic. The birth of science came with the rise of ancient Greece, beginning with the influence of the philosopher Thales.
2. Science is an objective attempt to understand the universe and depends, in part, on the precise definition of terms and reasoning. Two competing philosophic views are mechanism and vitalism.

   Key Terms: *mechanism, p. 3*
   vitalism, p. 3
   deductive reasoning, p. 4
   inductive reasoning, p. 4

3. Science often employs the scientific method in solving problems. Scientific theories are not presented as the ultimate truth; science deals with probabilities that reflect the best evidence for explaining a certain phenomenon.

   Key Terms: *hypothesis, p. 5*
   theory, p. 5
   natural laws, p. 5

4. Biology is the science of life and emerged as an organized study from the work of the Greek philosopher Aristotle.

PART TWO

The Panorama of Life

Chapter Two

Earth: The Habitable Planet

Earth: Its Place in the Solar System

The turbulence was thunderous and violent. Roaring through the empty blackness, the massive whirling disk of cosmic gas and dust grew larger and larger. Reaching a critical size, most of its mass began to contract, creating enormous pressure, and growing ever hotter. Intense heat and pressure set off thermonuclear explosions, hurling dust and gas outward trillions of kilometers into the black void. As the great disk began to cool, specks of solid, carbon-containing crystals solidified within the disk. Collisions between these tiny particles formed numerous chunks of rocky bodies known as **planetesimals**. In the cold, outer regions of the disk, water froze into ice, forming icy clumps that grew into more planetesimals. Collisions between the planetesimals created larger and larger bodies called **protoplanets**. These masses of rock and ice eventually became the nine known planets of our solar system and, for over 4.5 billion years, have orbited the great gaseous disk we call the Sun (Fig. 2.1a). Thus, according to the most widely held theory, known as the big bang theory, the solar system was born. The small rocky planets orbit closer to the Sun; the icy, colder planets lie in the outer reaches of the solar system. Today, tucked in among some 250 billion other stars of the Milky Way galaxy, the Sun spreads its light outward to the nine planets. Of these nine planets, only Earth possessed the unique characteristics that allowed development of life forms capable of using this light as an energy source.

Earth and Other Planets: A General Comparison

Earth is the third planet from the Sun, located at a distance of about 150 million kilometers (km) (93 million miles) from it, and Earth ranks fifth in size of the nine planets in our solar system. The planet closest to the Sun is Mercury, followed by Venus, and, in turn, Earth, Mars, Jupiter, Saturn, Uranus, Neptune, and, at the outlying border of our solar system, Pluto. The relative sizes and of the planets and their orientation to the Sun are shown in Figure 2.1b. The closest planetary neighbor to Earth is Venus, which comes within 42 million km (26 million miles) at its nearest approach. Although similar to Earth in size and mass, Venus is shrouded in heavy clouds and possesses an atmospheric pressure many times that of Earth. With a surface temperature of 450° C (842° F), water on Venus cannot remain in the liquid state, but exists as a vapor. At the other extreme, in the farthest reaches of the solar system, the planet Pluto lies some 4.5

billion to over 7 billion km (2.8–4.5 billion miles) from the Sun. This smallest planet in our solar system has a surface of frozen methane and ice at a temperature of –251° C (–484° F). Its atmosphere is extremely thin; the atmospheric pressure is but a minute fraction of that on Earth.

Figure 2.1A Solar system format reaching a critical size and density, (a) a massive cloud of interstellar gas and dust begins to contract, (b) forming a protostar, and (c) smaller bodies called planetesimals. The protostar eventually contracts into a star (our sun) and the planetesimals clump together, (d) forming first protoplanets and then planets and moons. The smaller planetesimals form asteroids, meteors, and comets. This whole process took place probably between 4.6 and 5 billion years ago.

The planet Earth is almost spherical and has an outer shell, or **crust,** that averages about 65 km (40 miles) in thickness. Over 70 percent of the crust is covered with water. The crust rests on the underlying, molten **mantle,** a layer of rock extending over 2800 km (1750 miles) downward to what is thought to be an iron-rich **core** at the center of Earth. Surrounding Earth is an atmospheric sea of gases rising over 480 km (300 miles) above the surface (Fig. 2.2). The atmosphere is a mixture of gases, over 99.9 percent of which is nitrogen, oxygen, and argon, with traces of carbon dioxide. Specifically, the atmosphere is composed of about 78 percent (by volume) nitrogen and almost 21 percent oxygen, with other gases constituting the remaining 1 percent. Gaseous water vapor is also present in the atmosphere—its amount varying with climate and location. The distance from the Sun and the presence of water and such a large amount of oxygen make Earth unique in the solar system and form the backdrop for the capacity of Earth to produce and support life.

Life as we know it could not exist on either Venus or Pluto. There is, in fact, no evidence that life exists anywhere in the solar system except on Earth. Still, the search for life and the hope for its appearance elsewhere continue. In the United States, for example, the National Aeronautics and Space Administration (NASA) continues its Search-for-Extraterrestrial-Intelligence (SETI) program, which relentlessly scans the skies for signals from other-world civilizations.

For years, unmanned space flights have searched the solar system in vain for any form of extraterrestrial life. One of the more promising sites for life elsewhere, however, is the planet Mars. Although the Viking mission over 25 years ago found no evidence of life on the surface of Mars, some scientists believe life may have arisen there in times past. If so, fossils (imprints or remains of dead organisms) or even living organisms could be secluded somewhere deeper in the terrain. Mars has a temperature range of –125° C to 25° C (–257° F to 77° F), which is within the limits suitable for life as we know it. Its atmosphere is thin and contains carbon dioxide. Pictures taken of Mars

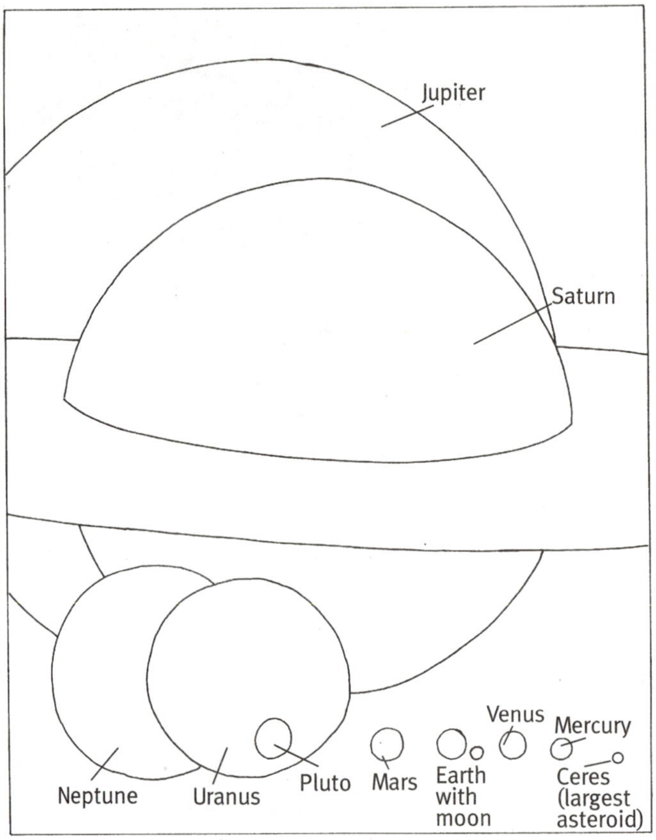

Figure 2.1B Diagram of the relative sizes of the nine planets of our solar system.

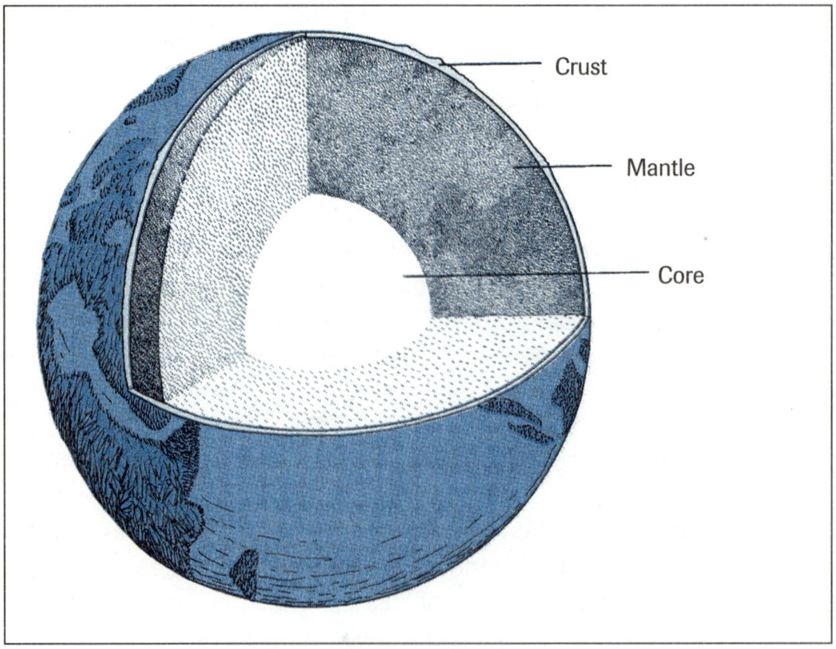

Figure 2.2 The major layers of Earth.
Living things exist only on the crust and the atmosphere above it.

from space have revealed what appear to be dry riverbeds or valleys, but no liquid water is evident. Scientists have suggested that lakes and geysers may have dotted the Martian surface more than 2 billion years ago. Channels cutting across the Martian landscape suggest that numerous "marsquakes," triggered by asteroid collisions billions of years ago, may have cracked open the frozen crust and released huge amounts of underground water deposits. Prominent polar caps of frozen carbon dioxide, or dry ice, have been observed as well; interestingly, these frozen caps vary with the Martian seasons, almost vanishing completely during the Martian summer.

Another possible life-producing candidate is Europa, one of at least sixteen moons of the planet Jupiter. Europa has a gravitational pull that may generate enough heat to melt the deep layer of thick ice that envelops it; if so, the presence of liquid water offers at least the possibility that life as we know it could evolve there.

Requirements for Life on Earth

Perhaps there is very little that so intrigues the mind as the thought of life on other worlds. The bewitching appeal of extraterrestrial life is timeless. For centuries there have been visions and dreams of finding life "out there." How realistic are such dreams? We can frame the answer only within the context of life as we know it on Earth. This assumes a certain bias because we have not encountered life elsewhere and are not familiar with a living system that is structured differently from any found on Earth. Life as we know it, then, could arise only under conditions similar to those on Earth.

It is agreed, therefore, that the origin and evolution of life are dependent on certain basic substances and conditions. Accordingly, if these substances and conditions are present, the origin and evolution of life are probable; in their absence, there is no evidence that life is possible. We will take a look at each of these requirements in turn; they include organic compounds, water, an appropriate temperature, oxygen, and a source of energy.

Organic Compounds

Part of our introduction related to the requirements for life on Earth involves a brief discussion of various chemical substances. It is not necessary at this point to understand their more involved functional significance. We will consider these substances in a very general sense here and then study them more thoroughly in Chapter 4.

The first requirement for life on Earth concerns the basic *structural* components of life: carbohydrates (sugars and substances derived from them), lipids (various fatty substances and steroids), proteins (structural components of hair, bone, and muscle; functional components of our body's defense system), and the self-copying carriers of genetic information called genes. All of these components are based on the carbon atom; that is, their structures are built around a backbone of linked carbon atoms. We call these components **organic compounds** because of the presence of carbon atoms, which are the smallest particles of the element carbon that still retain its characteristic qualities or properties. An element, then, is a substance consisting of certain types of atoms; an element cannot be broken down chemically into a simpler substance. Carbon is a fundamental substance that is made up of carbon atoms. Oxygen, gold, nitrogen, and copper are examples of other elements. A compound is a substance that contains two or more different elements. Organic compounds, therefore, contain the element carbon linked to one or more other elements. No known element other than carbon can form the extended and often complex chains and rings that serve as the framework for the millions of different organic compounds necessary for life as we know it on Earth (Fig. 2.3). Notice that the *arrangement* of the carbon atoms in the compounds shown in Figure 2.3 is different for each compound. This is essentially why each compound is different from any of the others.

Water

A second essential substance for life as we know it is water. All the chemical reactions that take place in living organisms do so in a watery environment. Moreover, living cells dehydrate if not bathed continually in water. About 70 to 90 percent of most living organisms is water, and more substances will dissolve in water than in any other common liquid. The unique properties of water, to be discussed in Chapter 4, make it the only known liquid suitable for the evolution, maintenance, and perpetuation of life on Earth. As mentioned earlier, water exists as a vapor on the planet Venus and as solid ice on Mars. This is one of the major reasons these planets cannot support life as we know it on Earth.

Temperature

You are familiar with the physical effects of temperature—the wilting heat of summer and the chilling blasts of winter winds, for example. Although organisms vary considerably in their temperature requirements, a limited temperature range is necessary for life to exist and flourish on Earth. Living organisms exist mostly within a temperature range of 0° C to 55° C (32° F to 131° F). Exceptions, however, do exist. Bacteria have been discovered in deep ocean vents thriving in a water temperature of 306° C (583° F). Ordinarily, excessively high temperatures inactivate most of the structural components of an organism. Proteins, for example, coagulate when heated above a certain temperature. You probably have observed this phenomenon in a fried egg. When heated sufficiently, the runny protein contents of an egg coagulate, solidify, and turn white. The various proteins comprising a living organism also coagulate at sufficiently high temperatures.

At very cold temperatures, chemical reactions may take place too slowly to sustain life. Ice crystals may form within cells, and blood, tissue fluids, and other body liquids may become sluggish or frozen. A controlled reduction in body temperature, however, has been used at times by physicians as a substitute for or in conjunction with anesthesia during surgical procedures such as amputations and heart surgery. Lowering of body temperature reduces the risk of shock and decreases the need for oxygen by slowing the rate of body metabolism. Metabolism is the sum of the chemical processes that occur within a living organism.

Oxygen

Today all living organisms—with the important exception of some bacteria and yeasts—require an uninterrupted supply of oxygen. Surprisingly, though, we will discover later in this chapter that the *lack* of oxygen in Earth's early atmosphere seems to have been essential for the *origin* of life. Without oxygen chemical reactions within most eukaryotic cells cannot manufacture adequate amounts of the high-energy compound ATP (adenosine triphosphate). Various bacteria and yeasts, however, are capable of manufacturing small amounts of ATP by using organic molecules instead of oxygen. As we will see in Chapter 4, ATP is the universal energy carrier, the one major chemical form in which energy from the breakdown of food is stored in living cells. When this com-

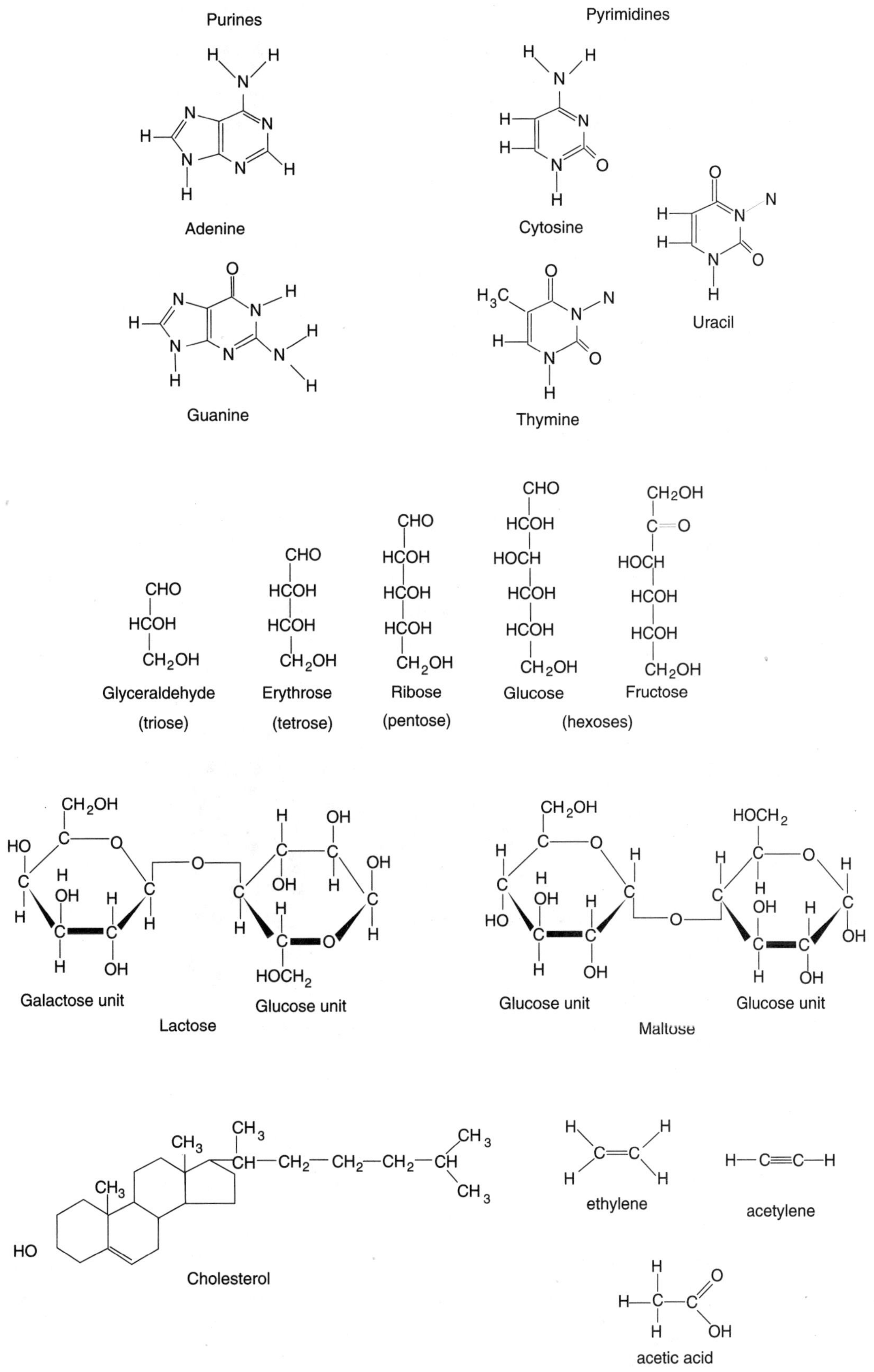

Figure 2.3 A few examples of the numerous compounds formed from the bonding of carbon atoms with other elements.

pound is dissociated, or split apart, energy is released and is used to drive the chemical reactions that sustain life. These reactions constitute part of body metabolism, mentioned above.

On Earth today, oxygen also is essential in another capacity. Oxygen atoms high in the atmosphere unite to form ozone (O_3) molecules that create a protective **ozone layer,** a virtual shield around the planet. This ozone layer protects life on Earth from full exposure to the potentially lethal X-rays, gamma rays, and ultraviolet (UV) radiation emanating from the Sun; from 95 to 99 percent of this radiation is absorbed by ozone in the upper atmosphere (Fig. 2.4). A significant increase in the amount of UV radiation striking Earth could have several serious effects. It has been estimated, for example, that a 5 percent depletion of ozone concentration in the atmosphere could result in millions of additional cases of skin cancer in the United States in the coming decades.

Energy

The fifth and major determining factor related to the existence of life is a constant, usable, external source of energy. **Energy** is defined in a general sense as the capacity to do work; that is, to bring about some action or effect. Living systems are maintained by a steady contribution of energy and energy is required for the many metabolic reactions carried out by living cells. Energy exists in several different forms, such as mechanical, atomic, heat, and light. For Earth, the ultimate energy source is light from the Sun. Earth is an appropriate distance from the Sun to utilize this light energy. In contrast, the planets Mercury and Venus, which are closer to the Sun, have very high surface temperatures that would vaporize water and destroy living tissue. Planets outside

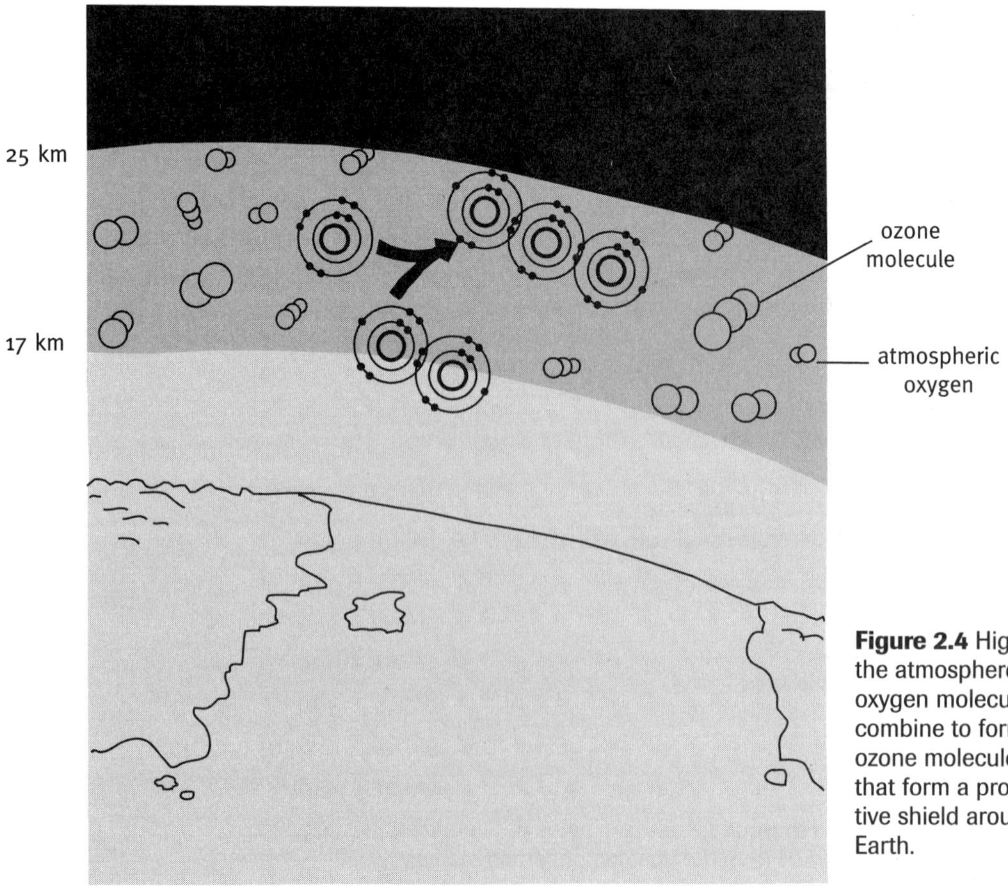

Figure 2.4 High in the atmosphere, oxygen molecules combine to form ozone molecules that form a protective shield around Earth.

Earth's orbit, such as Saturn and Uranus, have very low surface temperatures. Water and tissues on these planets would remain frozen. The temperatures of Earth, however, are not usually harmful to living organisms.

On Earth, green plants, various algae, and some bacteria use the process of photosynthesis to convert solar energy into other forms of energy that they and other living organisms can use. Photosynthesis, which produces sugars and other organic compounds, is the critical process for introducing sources of food energy into the living world. In turn, virtually all organisms use these food sources as fuel for their existence. Without an appropriate energy source, neither you nor any living organism could sustain life.

Is Life Possible Elsewhere?

The presence of carbon, water, oxygen, a suitable temperature range, and a usable energy source are only some of the major essentials required for the continued existence of life as we know it. In addition, Earth also has a sufficient gravitational force to hold an atmosphere around it, but not a force that exerts a crushing atmospheric pressure. Locating another celestial body with all the life sustaining requirements we have mentioned may seem a little too much even to imagine. But since there may be 100 billion billion planets to consider, the existence of life as we know it elsewhere becomes a little more probable. There is, of course, the probability that life on other worlds may exist in forms different from those on Earth. It could be that life on these worlds would have to be defined by criteria different from those we have enumerated here. For example, life based on a fundamental unit other than the carbon atom would have very different requirements for its existence.

There is speculation that some stars may have associated planetary systems where life could evolve. Our nearest neighbor outside our own solar system is a star over 40 trillion km (25 trillion miles) distant. Whether this nearest neighbor has such an associated system of planets is unknown.

It seems, then, that life—as we know it—elsewhere in the solar system is a remote possibility. It is difficult even to comment on the probability of life existing on stars trillions of kilometers away. The probability may seem favorable, but there is no available evidence. As mentioned, what about the possibility of life existing in forms different from those on Earth? This, too, has been the subject of much debate. Hypotheses suggesting ammonia as a substitute for water and the element silicon for carbon have been proposed; but, again, there is no supporting experimental evidence. Silicon, for example, usually exhibits only a weak tendency to bond with itself; consequently, silicon cannot form the long chains and rings characteristic of the compounds of carbon. It is tempting to conclude that life is an isolated, unique occurrence in the universe, but temptation is a fickle verdict. As yet, we are mired in ignorance, and there is so much "out there" waiting to be discovered and understood.

The Theme of Evolution

Physical Evolution: Pangaea and Plate Tectonics

On our planet where life abounds, there is a long and engrossing tale of the parallel between its development and the beginning of life. The evolutionary history of Earth itself is one of dramatic change, sometimes violent, with drastic climatic shifts and geologic upheavals. As the original masses of cosmic gas and dust cooled and the plan-

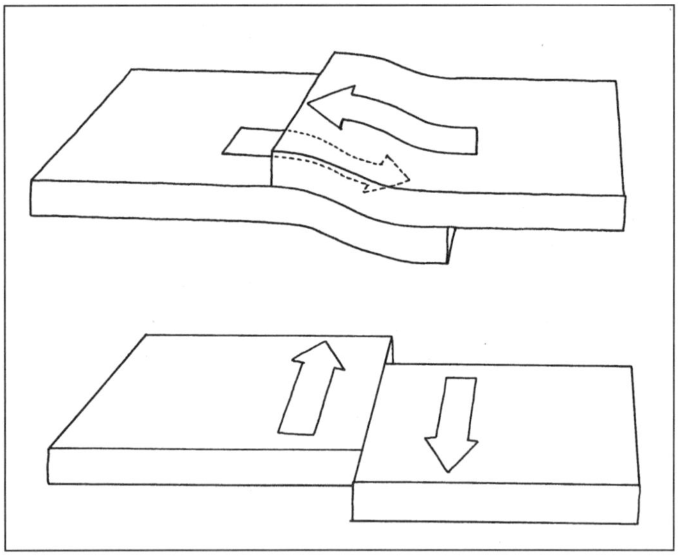
Figure 2.5 Representation of movements of Earth's crustal plates.

ets formed, the young Earth became increasingly hotter. The intense heat maintained atmospheric water vapor in a constant gaseous state. But later, as Earth slowly cooled, the water vapor condensed and began to fall in torrential downpours. After untold centuries of relentless drenching, Earth cooled sufficiently to allow water to remain in the crust in liquid form. During the cooling process, the crust buckled and cracked, forming deep depressions, chasms, and mountainous ridges.

Earth's crust—the outermost layer of our planet—is formed from several enormous rocky plates resting on—or floating on—the underlying, molten mantle (Fig. 2.5). These **crustal plates** move several centimeters a year on the average; they may push or crash against each other, causing mountainous upheavals or deep fissures; they also may slide along each other, producing earthquakes. In California, earthquakes occur in the area of the San Andreas Fault as a result of two plates sliding past each other. The study of the geological processes, such as earthquakes and volcanic activity, that result from movements of the crustal plates is known as *plate tectonics.* **Plate tectonics** is used to explain the theory of **continental drift,** which proposes that the formation of modern-day continents is a result of the breaking apart of a single, huge land mass—called Pangaea—over time. The breaking of Pangaea into several smaller land masses helps explain the geographic distribution of organisms on Earth today. By referring to Figure 2.5, you can visualize the effects that even a tiny movement of the crustal plates could have on altering Earth's surface.

More than 350 years ago, it was noticed that the eastern coast of South America and the western boundary of Africa were nearly complementary. If moved next to each other, the two continental boundaries would interlock like pieces of a puzzle. At first, this was considered merely coincidental and certainly of no geologic significance. Those who advocated the possible connection between the two continents were ridiculed and, for decades, there was little investigation of the observation. Serious research into the theory of continental drift began in the early 1900s and, since the 1960s, there has been overwhelming experimental evidence supporting the theory. It was discovered, for example, that fossils of the same group of reptiles over 200 million years old were found in both West Africa and Brazil, South America—two continents that now are separated by 3000 km (1875 miles) of the Atlantic Ocean. Continued research concluded that, about 200 million years ago, all the present-day continents were actually one gigantic land mass some 50 million km^2 (80 million mi^2) in area. This "supercontinent" was given the name **Pangaea,** meaning "all lands." Studies of the ocean floor indicate that the major land masses began to shift during this period. About ten million years later, Pangaea is thought to have split into two huge land masses called **Laurasia** and **Gondwanaland.** Laurasia, the northern mass, eventually fragmented to form what is now North America, Europe, Asia, and Greenland. Gondwanaland, the southern part, gave rise to Africa, South America, India, Arabia, Australia, and Antarctica (Fig. 2.6). This figure illustrates the wide-ranging geographical effects of continental drift.

The split between South America and Africa began about 135 million years ago, with South America sliding first to the west and later northward. Africa, and then Australia and

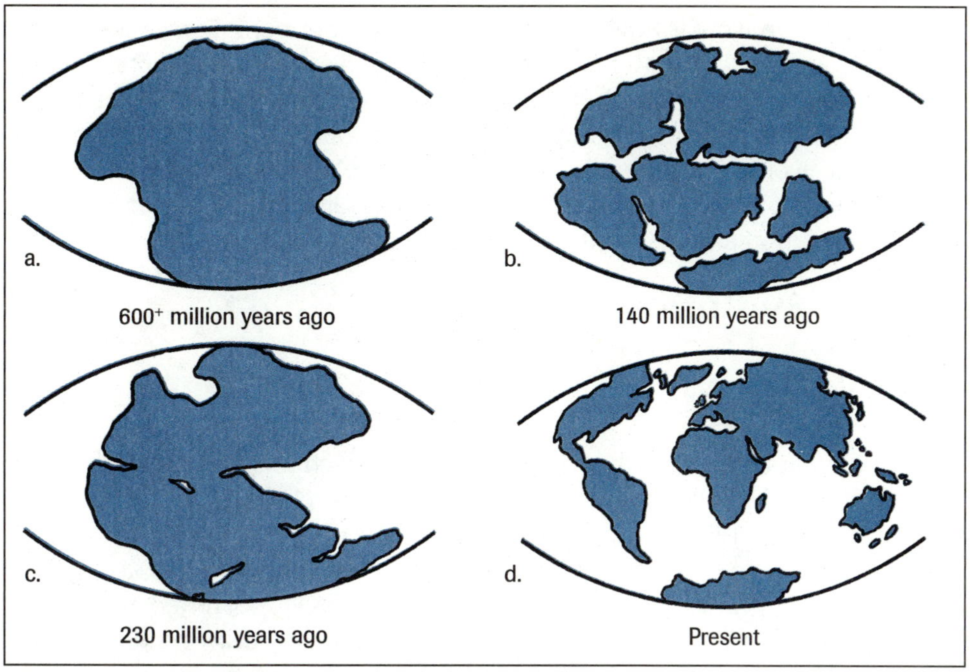

Figure 2.6 (a) Original earth land mass. (b) Pangaea. The continents form a single land mass with practically no points of separation between them. (c) Some of the continents, though widely separated, still are connected by land bridges. (d) The continents as they appear today. Continental drift is ongoing today as North America and South America continue to move away from each other, as do Europe and Africa.

India, separated from Antarctica. India drifted thousands of miles north, crashing into southern Asia. This violent collision produced an upheaval that formed the Himalayan Mountains. Moving ever northward, South America eventually became connected to North America by the Isthmus of Panama. Over 80 million years ago, North America separated from Europe. Less than 500 million years ago, the South Pole was in the area of the Sahara Desert. All of these separations, movements, and new connections of the land masses drastically altered Earth's crust. As continents drifted apart, new oceans, seas, and other waterways were created between and within the newly formed land masses. Climatic conditions changed dramatically, producing great ice sheets, new deserts, and water currents, all of which affected the face of Earth. The continents continue to shift today and will do so in the future. The evolutionary history of Earth is still being written.

What has been the effect of continental drift on the biological world? The answers are framed within the context of biological evolution on the various continents. In a general sense, **evolution** refers to *the change in the genetic makeup of a population over a certain period of time.* In general, a population consists of organisms of the same kind, each of which contributes its genes, or units of hereditary information, to the population. Evolution, then, involves successive changes in a population from one generation to the next as a result of such processes as genetic mutations. Continental drift has been an indispensable factor in this change.

The formation and breakup of Pangaea resulted in the extinction of untold numbers of aquatic invertebrates (animals without a backbone), and initiated radical geographic shifts in different plant and animal populations. Surviving organisms of these ancient cataclysms founded new populations in greatly altered environments. Many populations eventually became geographically isolated on islands and other land masses. Marsupials, for instance, which are currently restricted primarily to Australia, are thought to have originated on what is now the North American continent. Formation of connections between continents allowed migration and wide dispersion of plants and animals that for thousands of years had been restricted in movement by expanses of water. These migrations and dispersions altered the genetic makeup of many populations and redirected their evolutionary future. Undeniably, the **physical evolution** of Earth is a major factor in the planet's biological history and an indispensable factor in our understanding the evolution and dispersion of life on this planet.

Chemical Evolution

The Early Atmosphere

What of the origin of life on Earth? To attempt to approach this question, we must look more closely at Earth itself. One of the apparent factors related to the evolution of life was the composition of Earth's early atmosphere. It is generally agreed that there was virtually no free oxygen in the early atmosphere. Opinions vary, though, as to the exact composition of the primitive atmosphere. Original theories assumed that ammonia and methane were present, but this is not certain. Also, it was long accepted that a high concentration of free hydrogen was present; this, too, has been widely debated. The most abundant **atmospheric gas** then, as now, is thought to have been nitrogen. Other possibilities include ethane, a colorless, odorless gas found in crude petroleum; hydrogen cyanide, a colorless, extremely poisonous gas; carbon dioxide; and perhaps some additional hydrogen-containing gases. Volcanic emissions deep from within the planet are thought to have contributed to the primitive atmosphere by spewing forth gases such as carbon monoxide, hydrogen, and water vapor.

Early and modern theories on the origin of life attribute the formation of some of the basic components of life to chemical reactions between various atmospheric gases. While these reactions and the structures of the gases may not be familiar to you, we will include them here for illustrative purposes and leave the specifics for later chapters.

Hydrogen gas (H_2) and carbon dioxide (CO_2) react with each other to form methane gas, a gas composed of one carbon atom and four hydrogen atoms (CH_4). The O_2 is released as a by-product of the reaction. This reaction illustrates that organic compounds could have been formed long before living organisms appeared. Nitrogen gas (N_2) reacts with hydrogen to form ammonia (NH_3), while the bonding of carbon (C), hydrogen (H), and oxygen (O) atoms produces simple organic compounds, such as acetic acid ($C_2H_4O_2$) and formaldehyde (CH_2O). In turn, reactions between these simple compounds could have produced more complex organic compounds. These include a variety of amino acids (the building blocks of proteins), fatty acids (the basic components of some lipids), certain nitrogen compounds (the components of proteins and nucleic acids), and several other important biological compounds. Thus, over great spans of time, according to the theory, interactions between simple chemicals led to the formation of increasingly complex organic compounds—a process known as **chemical evolution.** Once formed, the products of chemical evolution, along with various inorganic substances, such as minerals and salts, could have been washed into the warm waters of the ancient lakes and seas (Fig. 2.7). When heated, amino acid molecules link together to form long chains, and when exposed to water, these chains cluster together to form microspheres. These microspheres are crudely similar to living cells and are capable of reproducing other microspheres. Figure 2.7, then, illustrates one theory of how chemical evolution could have led to the origin of a living system.

This theory would necessitate the spontaneous formation of complex organic compounds from simple chemicals. Evidence for just such a spontaneous occurrence was first provided in 1953 by Stanley Miller. Miller, then at the University of Chicago, was a graduate student working with Harold Urey, who won the Nobel Prize in chemistry in 1934 for his discovery of deuterium, a form of hydrogen. Miller and Urey designed a glass apparatus containing a mixture of water vapor, hydrogen gas, methane, and ammonia, which are the component gases thought to have been most prevalent in Earth's early atmosphere. As part of the apparatus, a flask of warm water was used to simulate an ancient sea. Using electric sparks to simulate lightning, they repeatedly discharged the electric sparks into the gaseous mixture as it circulated through the apparatus (Fig. 2.8). A week later, analysis of the contents of the flask revealed the presence of several organic compounds, including urea, amino acids, lactic acid, and other organic acids. Thus, assuming the correct conditions, the evidence was conclusive: organic compounds could indeed have formed spontaneously from inorganic substances in the ancient seas. It therefore became more possible to imagine that, through the eventual assembly of these organic compounds, life could have arisen from nonlife. The Miller-Urey experiment has been repeated many times over the years, each experiment yielding essentially the same results as the original.

Biological Evolution

The Origin of Life

All of this leads us to the intriguing question of just how a discrete life form first arose in the waters of our planet and there are some rather engaging and ingenious theories that

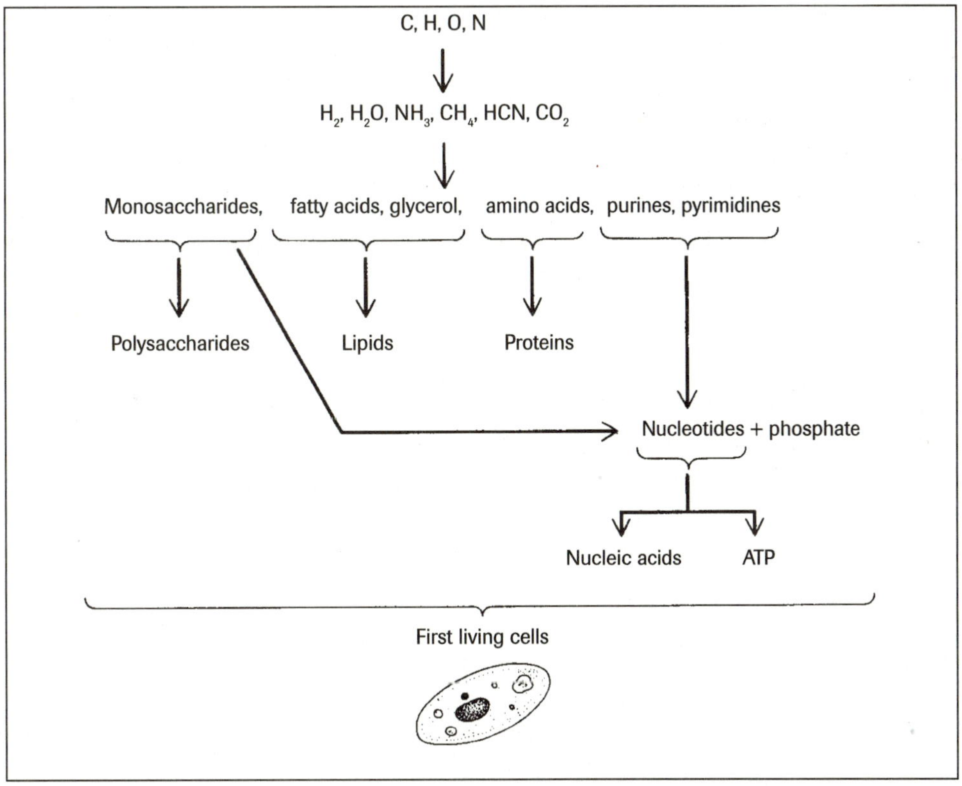

Figure 2.7 An abbreviated sequence of the steps of chemical evolution. Supplied with energy from heat, lightning, ultraviolet radiation, and radioactivity, a few basic elements on the primitive Earth interacted to form a variety of simple compounds, which led to the formation of the fundamental organic compounds of life.

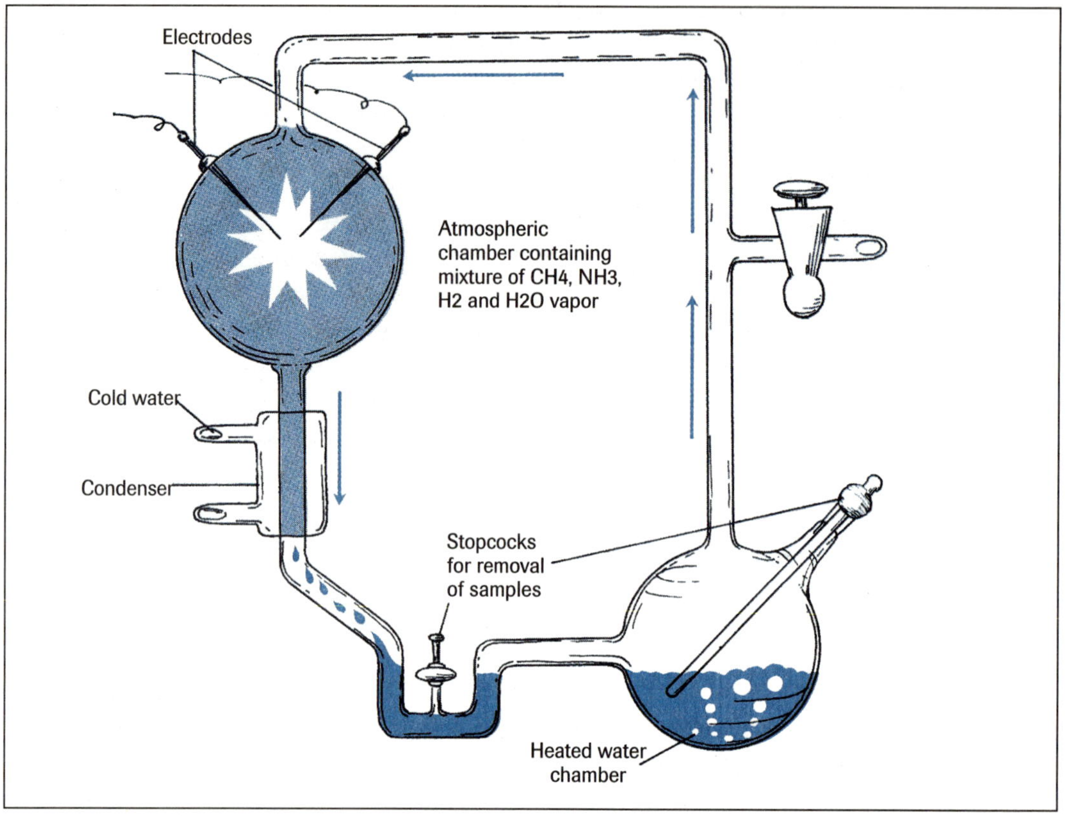

Figure 2.8 Diagram of vacuum-sealed reaction chamber used in the Miller-Urey experiments. The upper, or atmospheric, chamber contained a mixture of methane (CH_4), ammonia (NH_3), hydrogen gas (H_2), and water vapor (H_2O). These gases are considered to have been present in the primitive Earth's atmosphere. Electrodes discharged sparks into the chamber, simulating lightning, a probable source of energy on the primitive Earth. The gas mixture then was cooled by passing it through a condenser. This resulted in the formation of water droplets that were passed into the heated water chamber, which represented the ocean on the primitive earth. If any organic molecules were dissolved in the water droplets, they would pass into the water chamber. Analysis of samples withdrawn from the chamber indeed showed that several amino acids, urea, lactic acid, and a few other organic molecules were present.

propose to answer the question. There is widespread, often heated, disagreement concerning the most likely theory, but there is consensus that life arose from a collection of simple chemicals. From these chemicals, as we have described, more complex organic substances were formed. This sequence of chemical evolution assumes that the appropriate grouping of compounds eventually led to a crude form of living system. The controversy mainly is concerned with how and where this development took place. One of the many problems hovering around the origin of life question concerns energy. For even simple chemicals to join and form more complex substances, there must be an input of energy to drive the reactions. The sources are assumed to be the energy released from lightning, volcanic emissions, solar energy from the Sun, and radioactivity from decaying elements, all of which apparently were abundant when Earth was young (Fig. 2.9).

Oxygen, and the lack of oxygen, have played a central role in **biological evolution.** Without sufficient oxygen to form the ozone layer above the young Earth, deadly UV radiation beamed relentlessly onto its surface. Any potentially evolving land organism would have been destroyed immediately. This suggests, of course, that life did not originate on land; indeed, almost all of the evidence supports a water-based origin, including the observation that UV radiation cannot penetrate water below a certain depth. As the physical and chemical evolution of Earth continued, oxygen formed a layer of ozone around the planet and, importantly, allowed life to emerge from the sea and invade land. Strange as it may seem, although oxygen is required for most organisms today, the *lack* of oxygen appears to have been a prerequisite for the *origin* of life. On early Earth, the basic molecules of life, such as nitrogen compounds, could not have assembled in the presence of free oxygen. These and other organic substances react slowly with free oxygen and are chemically altered and thus destroyed. In the absence of oxygen, chemical substances could have accumulated for hundreds of millions of years in the ancient seas. If you ever have wondered why life could not arise spontaneously today as it must have done on primitive Earth, the basic answer is that its component molecules would be destroyed by reacting with oxygen.

Theories on the Origin of Life

According to one classical theory, life originated in shallow ponds or tidal pools. There are, however, drawbacks to this concept. One is that bombardment of shallow water by UV radiation would have destroyed any newly forming molecule. Another objection is that early Earth may have been tattooed constantly by comets and meteorites, wreaking havoc on anything on or near its surface. A more recent theory proposes that life arose in deep ocean vents, or hot springs, using heat as an original energy source. This theory, too, has its problems. For one thing, the temperature of the hot springs often exceeds 300° C (572° F). At this temperature, water would be boiling and chemical bonding between elements would be quite difficult or impossible. Still, as we have noted, bacteria exist today in deep ocean vents at just about the same temperature.

There also is considerable discussion as to the specific **substrate,** or supporting substance, from which life emerged. Some of the candidates include clay, muddy ooze, and an iron and sulfur-containing compound called pyrite, or "fool's gold." The chemical components of these substances, in conjunction with the energy available, such as heat or radioactive decay, may have produced amino acids and other organic compounds.

One theory proposes that the products of chemical evolution, amino acids and lactic acid, for example, stuck to a substrate, causing the products to be arranged in a certain order.

Figure 2.9 Lightning, solar energy, and volcanic eruptions may have been the sources of energy necessary for chemical evolution when Earth was very young.

If the substrates contained various metals, such as iron, they may have acted to enhance certain chemical reactions. Given the arrangement of the products and their association with the metals, it is possible to link the products to produce larger, more complex organic molecules. Experiments have shown that, in water, these molecules form droplets or microspheres that function in part as primitive "precells." They absorb organic molecules from their surroundings, are the sites of some simple chemical reactions, and sometimes reproduce by splitting to form new spheres when they reach a critical size.

From this beginning, it is assumed that some primitive life form could have evolved. It is a long way, however, from a collection of organic chemicals to even a crude living system such as a functioning cell. And yet, step-by-step accumulated changes is the essence of accepted evolutionary theory. Apparently, living organisms came into existence by virtually imperceptible, step-by-step changes involving the increasingly complex combinations of inorganic substances to form organic compounds. With the eventual establishment of a functioning living system, subsequent evolution gave rise to increasingly complex systems. The changes involved, according to the theory, usually require an extremely long period of time to occur. Opinions differ on the definition of "long," but the point here is that the evolution of life is not considered to have progressed through a series of great "leaps" from primitive structures to highly complex ones.

There still is no specific theory explaining the origin of life and the ideas presented here are, for now, largely speculation. Experimentation has provided limited acceptance of some of the theories, but nothing conclusive as yet. This, however, reflects the essence of the scientific method: proposed theories must be continually revised as new data become available. We indicated with the Miller-Urey experiment that certain organic compounds could be formed from a mixture of gases thought to be present in the primitive atmosphere. Even so, the laboratory conditions employed to perform the experiment may not have been the conditions on Earth 3.5 billion years ago. But for all of our speculation, we have evidence that life has existed on Earth for these 3.5 billion years; further, these early life forms are the ancestors of later forms. The evidence also leaves little doubt that evolution is the process by which all life forms came to be.

So where does all this leave us? Obviously, with a multitude of questions, a few answers, and a gaping void between the two. This, however, often is the case in science. Theories, research, discussions, and more research will follow until, if possible, the solution unfolds. The quest for the answer to the origin of life is too fascinating, too consuming to imagine anything less than a continuing focused and dedicated effort. Why has the answer been so elusive? According to Stanley Miller: "I think we just haven't learned the right tricks yet."

Evolution: The Integrating Theme

The evolutionary history of Earth itself encompasses more than 4.5 billion years of turbulent change. It is important to remember that the events outlined in this chapter took place at a time and in an environment radically different from today. Earth was nearly a billion years old before a life form appeared. The first known fossils of a living organism are strings of cells resembling bacteria: tiny, relatively simple organisms that are nearly 3.5 billion years old. Admittedly, it is difficult for us to think in terms of millions and billions of years, but evidently this time was necessary for the evolution of Earth and of life. Indeed, the evolution of life accompanied and was influenced by the evolution of our planet. On a single isolated sphere among trillions of others, the right combination of water, carbon, and sunlight came together to create an environment unique in the solar system, perhaps in the universe. Given these conditions, life became possible; some say inevitable. As Earth continued to change and the concentration of oxygen in the atmosphere increased, life also changed. Eukaryotes, capable of utilizing oxygen in cellular metabolism, evolved and life progressed to the stage that we know it today.

Evolution is the integrating theme in biology and the principle that correlates the relationship between unity and diversity in the living world. All organisms are thought to be related through descent from a common ancestor **(unity)** and, as descendants of this ancestor radiated into various environments over millions of years, they evolved different adaptive characteristics **(diversity)** that allowed them to survive in those environments. Overwhelming evidence indicates that the development of life can be understood within the context of evolutionary theory. The change that typifies evolution is not change with a purpose or goal in view, but change brought about by natural, impersonal forces. This is not to suggest that the change in question is random or brought about by pure chance; indeed, the development of life, properly understood, stands in direct opposition to chance. We will pursue this concept later. For now, it is sufficient to understand that life and Earth have trod a common path, a path that opened millions of years ago, and is still being traveled into tomorrow.

Chapter Summary

1. According to the most widely held theory, the solar system was formed from a mass of cosmic gas and dust. Earth, by virtue of its atmosphere, temperature, and the presence of oxygen, is the only planet in the solar system known to support life. The basic requirements necessary to support life on Earth include the element carbon, water, a suitable temperature, oxygen, and an outside source of energy.

   Key Terms: *planetesimal, p. 12*; *protoplanet, p. 12*; *crust, p. 13*; *mantle, p. 13*; *core, p. 13*; *organic compound, p. 15*; *ozone layer, p. 18*; *energy, p. 18*

   Concept Question: You are a planetary scientist responsible for evaluating the probability of finding life as we know it on Venus, Pluto, and Mars. What positive and negative factors will affect your evaluation? Select one of these planets and construct an organism that might be able to survive there.

2. Cooling of primitive Earth resulted in extreme geologic upheavals, movement of the rocky crustal plates, and retention of water in the crust. According to the theory of continental drift, Pangaea formed about 200 million years ago and then split into Laurasia and Gondwanaland. As a result, Earth underwent dramatic geologic and climatic changes. Continental drift resulted in the extinction and massive redistribution of plants and animals. Accordingly, biological evolution was affected significantly.

   Key Terms: *crustal plates, p. 20*; *plate tectonics, p. 20*; *continental drift, p. 20*; *Pangaea, p. 21*; *evolution, p. 22*; *physical evolution, p. 22*

   Concept Question: Assume you are standing in the middle of Pangaea some 190 million years ago when it split apart. You are at the top of a tall mountain where later you witness the effects of continental drift. What are some of the significant geologic and biological changes you are privileged to observe?

3. The most widely held theory on the origin of life asserts that life arose from inorganic chemicals present in Earth's primitive atmosphere. Life on Earth is thought to have originated in water, apparently utilizing the energy of lightning, volcanic emissions, the sun, radioactivity, and perhaps other energy sources.

   Key Terms: *atmospheric gas, p. 22*
   chemical evolution, p. 22
   Miller-Urey experiment, p. 23
   biological evolution, p. 25
   substrate, p. 26

   Concept Question: What is the relationship between the characteristics of primitive Earth and chemical evolution? How does your answer correlate with the general theories on the origin of life?

4. Earth is approximately 4.5 billion years old and the earliest known fossils of life are of bacteria about 3.5 billion years old.

   Key Terms: *unity, p. 27*
   diversity, p. 27

   Concept Question: Why do you think there was such a great span of time between the formation of Earth and the appearance of the first known life forms?

Life: Unity in Diversity

Chapter Three

Life: Diversity and Distinguishing Characteristics

Scientists estimate that anywhere from 3 million to 30 million different species, or kinds, of organisms inhabit our planet. Members of a species have similar structural characteristics and usually interbreed. Actually no one really knows how many species there are; only about 2 million different species have been identified. Nevertheless, this number reflects an enormous degree of diversity, or difference, which can become rather overwhelming when biologists wish to compare and analyze different groups of organisms. This problem has been lessened somewhat by formally classifying different species based on a variety of distinguishing characteristics and evolutionary relationships. The scientific discipline concerned with naming, describing, and classifying species is called **taxonomy**. This discipline is encompassed within the field of **systematics,** the study of species diversity and interrelationships within the context of evolutionary theory. Organizing species into distinct groups based on evolutionary relationships helps us view the great diversity of life in a more comprehensible manner. The broadest and most inclusive taxonomic category is the **kingdom**. The current scheme recognized by most biologists places organisms into one of five kingdoms: *Monera, Protista, Plantae, Fungi,* and *Animalia*.

The kingdom Monera includes all bacteria: single-celled or colonial, mostly microscopic organisms that exist virtually everywhere—even on your own skin and in your large intestine. Actually, members of the Monera can be separated into two evolutionary branches. Most are "true" bacteria, or **eubacteria** (G. *eus* true); the second group, the **archaebacteria** (Gr. *archaio* ancient), are less numerous and most inhabit extreme environments, such as waters with very high salt content, low and high acidity or alkalinity, or high temperatures. One species of archaebacteria, for example, exists deep in the oceans at a temperature of 110° C (230° F). The unique structure of bacterial cells is the fundamental characteristic that distinguishes these organisms from members of the other four kingdoms. All other organisms have a more complex cellular structure and are placed in one of the other four kingdoms based on additional identifying characteristics.

The kingdom Protista includes more than 60,000 diverse species of complex, mostly single-celled organisms, such as algae. Protists possess a higher level of organization within their cells than do the simpler bacteria.

All terrestrial and aquatic plants, over 265,000 identified species, are placed in the kingdom Plantae. By general definition, the members of this kingdom are multicelled, complex organisms capable of producing their own food through the process of photosynthesis.

Members of the next kingdom, Fungi, number over 100,000 known species and are found in virtually all environments. They cannot synthesize their own food using the sun's energy as plants do but rather break down and then absorb organic matter from their own environment. Most fungi are multicellular ("many celled"), have a filamentous or threadlike body, and produce a special structure for reproduction. A mushroom is an example of a fungal reproductive structure. The kingdom Animalia includes all terrestrial and aquatic animals. This group cannot produce their own food and must rely on eating other organisms for survival.

Organisms in each kingdom are separated into less inclusive but more definitive categories that include *phylum* (pl. *phyla*) or *division, class, order, family, genus* (pl. *genera*), and *species*. As an example, the taxonomic classification of human beings and roses is shown in Fig. 3.1.

Figure 3.1 The taxonomic classification of roses and humans.

Roses	*Humans*
Kingdom: Plantae	Kingdom: Animalia
Division: Anthophyta	Phylum: Chordata
Class: Angiospermae	Class: Mammalia
Order: Rosales	Order: Primates
Family: Rosaceae	Family: Hominidae
Genus: *Rosa*	Genus: *Homo*
Species: *Rosa virginiana, Rosa multiflora,* and others	Species: *Homo sapiens*

There are at least a million different species of the animal kingdom, about 80 percent of which are insects—and well over one-third of the insects are beetles. The mollusks—oysters, clams, octopuses, squids, snails, slugs, and the like—boast over 100,000 living species. There are 10,000 species of roundworms and about 500 species of leeches. Diversity in the plant kingdom is no less wondrous. The flowering plants alone number over 235,000 species including, for example, 35,000 species of orchids and 3000 species of roses. Within the rose family diversity abounds, claiming in its group strawberries, blackberries, raspberries, cherries, peaches, plums, apricots, apples, pears, and the American Beauty. The point here, of course, is that life exists in so many and varied forms. Despite this obvious abundance of diversity, are there features that are common and basic to the existence of life on Earth? To answer this question, we will explore what is meant by the term "life." There is no comprehensive, all inclusive definition of life because life does not have dimensions, color, or texture. What, then, is a practical way to separate or characterize living organisms—members of the five kingdoms—from nonliving matter? A reasonable approach is to provide a framework of features or principles that distinguish and are shared by all known forms of life. This text considers seven features with which to explore the variety of life and to discover the fundamental unity that underlies this diversity; these will serve as the general framework for the remainder of the book. Other more detailed principles could be added to these seven, and a degree of overlap and similarity between features is inevitable at times. Used as an underlying framework, however, these seven features will aid in presenting the fundamentals of biology in an integrated and comprehensible manner.

Seven Life Features: A General Framework

Organization

A basic feature shared by all living organisms and evident at many different levels is **organization** (Fig. 3.2). Each organism is constructed of one or more microscopic units called cells. An individual **cell** is considered the structural and functional unit of life and is the smallest unit considered to be alive. Based on their particular organization of components and general structure, cells are distinguished either as *prokaryotic* or *eukaryotic*. Only members of the kingdom Monera—the bacteria—are classified as **prokaryotes**. All other organisms are structurally more complex and possess a greater level of internal organization and thus are classified as **eukaryotes** (Fig. 3.3).

As organisms increase in complexity, the amount of organization increases as well. A human being requires much more organization to function properly than does a single-celled bacterium. For example, a group of the same type of eukaryotic cells can be organized to form a **tissue,** which would have a specific role to play in the organism. Muscle tissues and reproductive tissues are examples of organized groups of similar cells that perform a particular function in our bodies. Single-celled organisms cannot

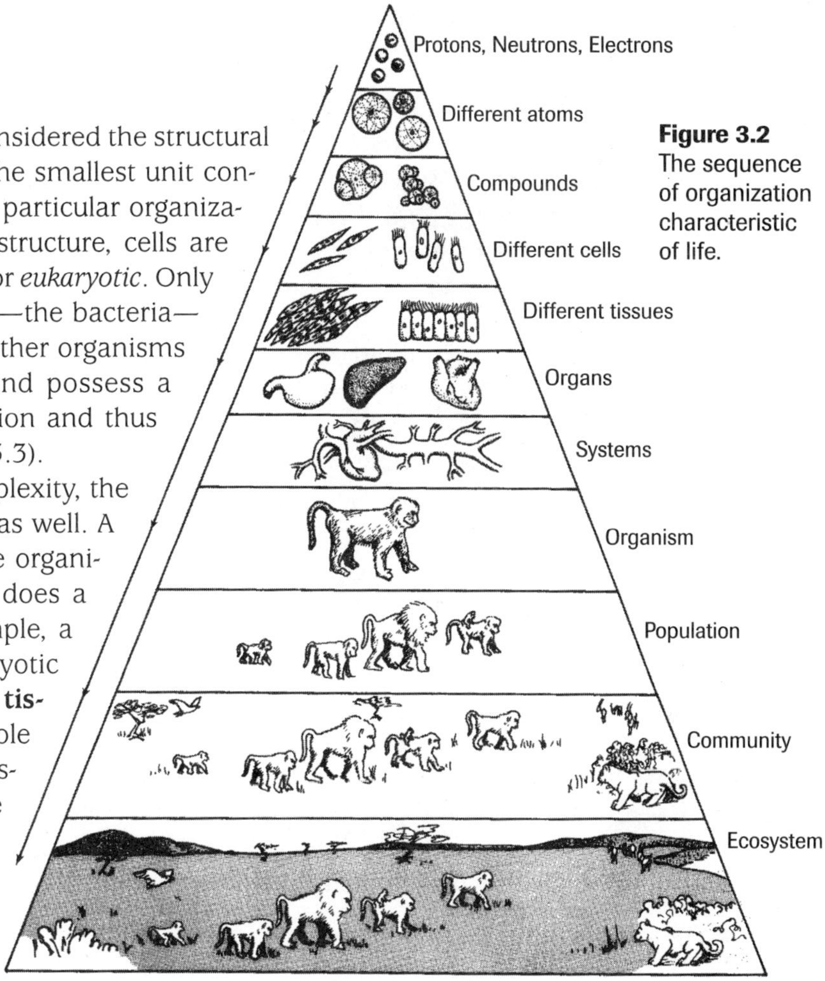

Figure 3.2 The sequence of organization characteristic of life.

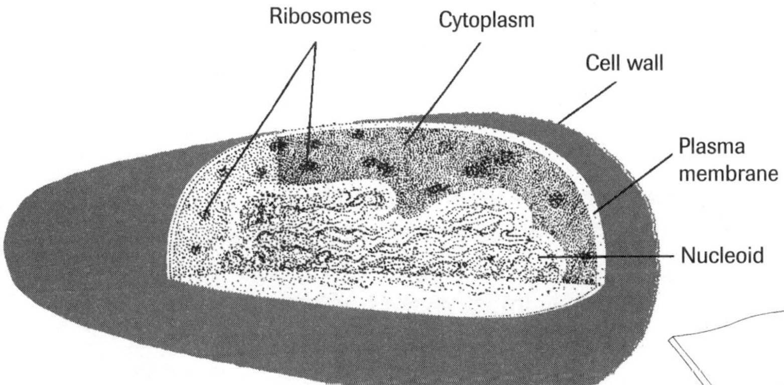

Figure 3.3A A generalized prokaryotic cell. When compared to a eukaryotic cell, the most distinguishing structural characteristics of a prokaryotic cell are the absence of a nuclear membrane and all cytoplasmic organelles except ribosomes.

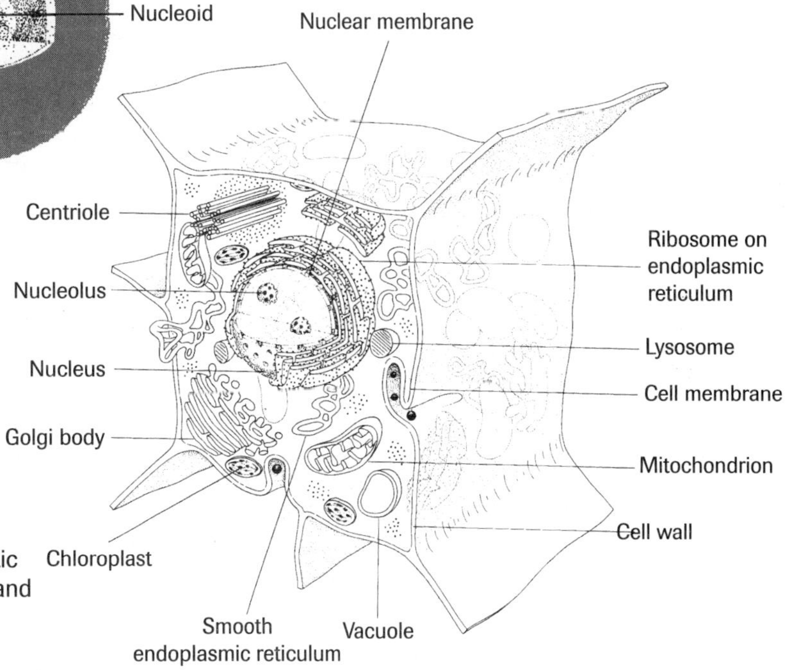

Figure 3.3B Diagram of a eukaryotic cell, showing a composite of plant and animal cell organelles.

exhibit this level of organization. In some organisms, a greater degree of organization is displayed as different types of tissues are integrated to form an **organ,** such as a brain, a plant root, or an eye. An organ is a specialized body structure that carries out a specific function. In turn, different organs make up a **system,** circulatory, respiratory, digestive, and so on, that is integrated with other systems to form an individual **organism.**

A group of the same type of organisms, or a species, living in a particular geographical location constitute a **population**. A population may be humans living in a city or sparrows flocked in a field or dandelions growing on a lawn. Populations of different interacting species inhabiting a certain area make up a **community.** If we consider the community and its interactions with the surrounding nonliving physical components—air, soil, water, and so on—associated with a community, we call this an **ecosystem**. The **biosphere**—the collective area on Earth where life is found—actually is a composite of different ecosystems that forms one large global ecosystem.

Just as we can study the hierarchy of organization extending from an individual cell to an ecosystem, we also can explore levels of organization coincidental with life *below* the level of a cell. Cells are composed of a variety of organic chemical substances called *compounds,* primarily carbohydrates, lipids, proteins, and nucleic acids. The principal inorganic compound of cells is water. These compounds are composed of several *elements,* including carbon, hydrogen, oxygen, and, depending on the compound, other elements as well. Each element is composed of certain types of *atoms,* which in turn consist of fundamental particles that include *protons, electrons,* and *neutrons*. These subcellular levels of organization, along with some related chemical principles, will be considered further in Chapter 4.

Energy Transactions

The second basic feature shared by all life forms and which also is found at each organizational level we will discuss is termed **energy transactions**. Organisms require a continuous source of energy to maintain their existence. They convert energy into other forms within their cells, store energy, and release energy. Energy describes the capacity to do work, to bring about some action or effect, and underlies virtually every activity and biological process engaged in by living systems. It is needed to drive the cellular processes that are ultimately responsible for these activities. The types of cellular processes that combine smaller or simpler substances to yield larger, more complex ones *require* energy and are referred to collectively as anabolic reactions or **anabolism**. Catabolic processes, or **catabolism,** on the other hand, work oppositely by breaking down larger substances into smaller ones, resulting in the *release* of energy. There is a critical relationship between anabolism and catabolism in that the energy released by catabolic processes can be used to drive the anabolic processes. The sum of both processes in an organism is known as **metabolism**. Looking at energy transactions on a more global level of organization, the ultimate source of energy for almost all organisms is the sun. Using solar energy, green plants and other specialized organisms produce their own food, which, in turn, provides food sources for most of the living world.

Sensation and Response

We have considered how energy fuels an organism's activities. One of these activities, shared by all organisms, is the ability to sense and respond to their environment (Fig. 3.4). **Sensation and response** is the third feature we will consider in our general framework of life principles. Through the use of special cell types or structures, or various chemicals, all organisms react to stimuli. A **stimulus** is any environmental factor capable of causing a response by an organism. Heat, cold, pressure, light, odor, sound, and many other factors are examples of stimuli. You blink if a bright light is flashed in your eyes, you quickly withdraw your finger from a hot object, and certain plants will bend in the direction of the sun. Sensation-response mechanisms are part of an organism's total behavior and, for the most part, tend to be protective and to promote the organism's general welfare and survival.

Figure 3.4 Soybeans bending toward the sun, an example of a response to a stimulus.

At the organizational level of the cell, sensation and response are evident, for example, when your white blood cells attack an invading microorganism, such as a bacterium or virus. The chemical toxins released by the invader are sensed by the white blood cells, which respond by migrating to the infected area and destroying the microorganism. The activities of the white blood cells are translated into further sensation-response mechanisms at the organismal level. For instance, you may experience a lessening of pain or discomfort and a subsequent relaxation of the body after the microorganism has been destroyed. Keep in mind that the sensation-response mechanisms of the white blood cells, the microorganism, and your own body are dependent on a continuous input of energy. These mechanisms are part of the regulatory functions of homeostasis, which is discussed next.

Homeostasis

The capacity to sense and respond appropriately to an environmental stimulus is made possible by the ability of an organism to regulate certain biological processes. All organisms tend to maintain a relatively stable internal equilibrium and this tendency is called **homeostasis**. Homeostasis, the fourth life feature, essentially means "staying the same." Even if the external environment changes, various regulatory mechanisms act to maintain an organism's internal stability and allow it to function within a tolerable range. On hot days, many animals perspire or pant as a means to cool the body and maintain a normal body temperature. Internal regulatory mechanisms control the secretion of perspiration and release of heat so that the body temperature normally does not rise to dangerous levels. Regulation of blood sugar level, blood cell production, and control of body fluid balance are other examples of homeostatic mechanisms at work in many animals. Plants, too, must regulate their internal environment. The regulation of water content in the plant, for example, depends on a balance between the uptake of water by the roots and evaporation of water from the leaves.

Homeostasis does not necessarily imply a strictly *constant* internal stability. Blood sugar levels or body temperature, for example, may remain fairly constant over the short term, but may fluctuate within normal limits over the long term. When homeostasis is upset over a long period of time, an organism may experience some malfunction that often results in discomfort or disease.

Before introducing the next fundamental feature shared by all living systems, let's review our investigation thus far. We have seen that all living systems are organized and that this organization extends from subcellular levels up to an entire ecosystem. Energy transactions occur at all of these organizational levels—energy consumption, energy release, energy conversions—fueling chemical reactions and processes, and making and breaking down compounds. We have discussed how all life systems possess some means

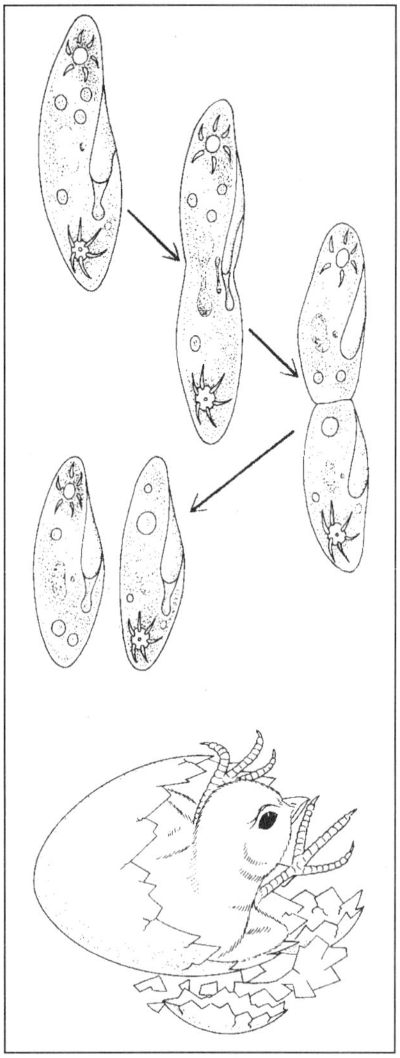

Figure 3.5 Reproduction results in the production of organisms that are similar copies of the parent or parents.

of sensing and subsequently responding to changing environmental conditions and how these actions are one part of an individual's homeostatic mechanism, which works to maintain a relatively constant internal stability. These fundamental features underlie all known living systems on this planet. Yet, for these processes to continue, each organism must possess the ability to reproduce itself. Reproduction—the principle that life arises from life—is the fifth feature we will discuss in our construction of a general framework within which to explore the fundamentals of biology.

Reproduction

Organisms have the capacity to reproduce—asexually, sexually, or both—and pass their hereditary material to their offspring. A reproductive process may be as relatively simple as a bacterial cell splitting in two in a matter of minutes or as complex as the maturation, division, and growth of a fertilized egg cell to produce a multicellular organism (Fig. 3.5). Through reproduction, the perpetuation of life is made possible. As with other life processes, reproduction requires an input of energy and is subject to homeostatic controls. Some organisms, particularly those that live in stable environments, are able to expend more energy in reproducing than are organisms inhabiting unstable environments. In an unstable environment, such as one in which there is extreme seasonal temperature fluctuations, much of an organism's energy is channeled into maintaining homeostasis. In a stable environment, less energy is required to maintain homeostasis and more energy can be expended on other processes, including reproduction.

Growth and Development

The remarkable feat of reproduction—initiated with a single cell and eventually producing an entire living organism—requires both growth of cells and, at times, the development of certain cells into specialized cell types. **Growth and development** can occur in a cell, in an individual, even in an ecosystem, and this will be our sixth feature. Growth in the biological sense occurs through an increase in the number and size of cells. Body cells increase in number by mitotic cell division, which is carried out in essentially the same way by almost all forms of life. The process of mitosis refers specifically to division of the cell nucleus; in most organisms, mitosis is followed by actual division of the cell into two cells. Through mitotic cell division, a single cell splits to become two, each of which in turn may divide to yield four cells, and the process continues as needed. Growth also may occur through an increase in cell size, as in plant root cells that elongate without undergoing cell division. The size of an organism or body structure is determined primarily by the number of body cells and not by the size of the cells. Larger organisms have more cells, not necessarily larger cells, than do smaller organisms (Fig. 3.6).

Development in multicellular organisms is a broad term that refers generally to the growth and formation of specialized *types* of cells, such as skin cells or root cells. At a higher organizational level, development may refer to the formation of organs and other body parts capable of performing a specific function. In most cases, normal growth and development ultimately result in producing a fully formed structure or mature organism.

Certain aspects of growth and development can occur at almost any time in the life of the organism or, indeed, throughout the life of the organism.

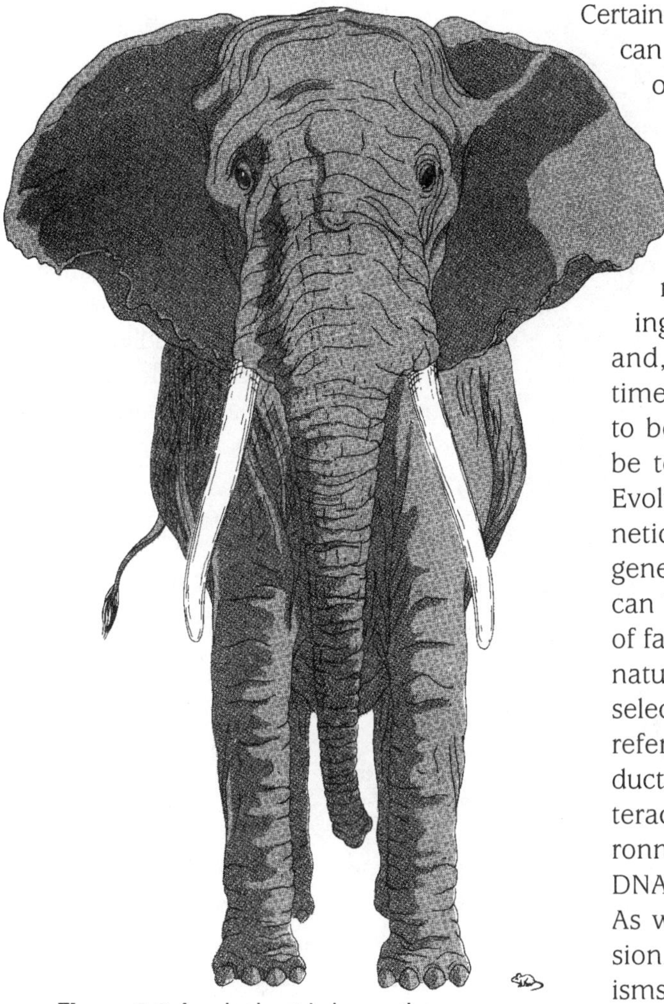

Figure 3.6 An elephant is larger than a mouse because the elephant has more cells, not larger cells.

Evolutionary Adaptation

Underlying all of these fundamental yet striking features of living organisms is the concept of time and, more specifically, change over time. The seventh and final life feature to be woven into our framework will be termed **evolutionary adaptation.** Evolution refers to a change in the genetic makeup of populations from one generation to the next. Such changes can be the consequence of a number of factors, including processes such as natural selection or mutation. Natural selection, which will be discussed later, refers to the different degrees of reproductive success that result from the interaction of organisms with their environment; a mutation is a change in the DNA, or hereditary material, of genes. As we have discovered in our discussion of sensation and response, organisms must sense and respond to changing environmental conditions such as temperature to promote their own welfare. So, too, must populations adapt successfully to their specific environment to ensure their continued survival. This does not mean that populations consciously decide or choose to adapt; in evolutionary terms, adaptation is a consequence of natural selection. An **adaptation** may refer to any structural, functional, or behavioral characteristic that increases an organism's ability to survive in a given environment. The color of some birds, for example, blends with the color of the grasses in which the birds nest (Fig. 3.7). This adaptive camouflage lessens the chance of detection by enemies and increases the likelihood of surviving and producing offspring.

The term may also refer to the short-term or long-term ability of an organism to adjust to environmental demands. Short-term adaptation, as we have examined with sensation-response and homeostatic mechanisms, involves the more immediate adjustments an organism makes when there are relatively abrupt changes in its environment. If the air temperature drops, an animal might respond by tightening its muscles and shivering. Short-term adaptations, however, are consequences of a long evolutionary history. To survive, populations must make adjustments over extended periods of time through long-term adaptation, which is a consequence of evolution. As environmental conditions and requirements change over thousands and millions of years, the continued existence of a population is dependent on the variation (diversity) in traits

Figure 3.7 Evolutionary adaptation: A ptarmigan, all but hidden by its surroundings.

that normally exists in the individuals that make up the population. At least some of the individuals in the population must have characteristics that are particularly suitable to new environmental conditions if the population is to survive. The inability of an individual organism to make short-term adjustments to its environment can result in injury or sometimes death; for populations, failure to adapt can eventually lead to extinction.

These seven general features are used as a basis on which to characterize and unify all known species of organisms. We will begin our exploration with a closer look at the unity and diversity of life on Earth.

Of Snakes and Snails: Exploring Diversity

To illustrate our theme of unity in diversity, let's first consider the differences between two organisms—one familiar to you and one perhaps not—and then examine their underlying unity as revealed by the seven life features. We will discover how both organisms exhibit the basic features of life—but each in its own way.

Our first example involves snakes, members of the kingdom Animalia and representatives of the vertebrates (animals that possess a backbone). We will consider snakes from a comparative and evolutionary standpoint. Snakes are reptiles that evolved from lizard ancestors. Reptiles appeared on the scene about 300 million years ago and gradually replaced the amphibians as the dominant land animals. Reptiles were the first true land vertebrates (many species of amphibians must return to water to reproduce) and remained dominant about 180 million years before mammals began to evolve. Some type of snake may have appeared about 100 million years ago, although the first known snakelike fossil is only 60 million years old. Today there are about 2300 species of snakes.

Snakes, having descended from lizards, once had legs. During their evolutionary history, however, snakes became adapted to their specific environments and this has resulted in the loss of various physical features that were present in their lizard ancestors. Of course they evolved other physical traits that allowed them to fit into their particular environments successfully. Keep in mind that neither snakes, nor any other organism, evolve particular structures because they "need" to or because there is some "goal" toward which they are striving. Evolution has no purpose. The evolutionary exploitation of land by reptiles resulted in adaptations particular to terrestrial environments. Snakes gradually lost their legs, yet there is still evidence of these now unnecessary limbs in the presence of certain bones near the base of the tail in some species, such as boa constrictors (Fig. 3.8). Snakes now "walk on their ribs" as it is sometimes described.

The teeth of snakes point backward toward the throat and are virtually useless for chewing or grinding. These inverted teeth, however, provide the snake a grip that is better adapted to their needs, instead of the gripping power of legs. Being unable to chew, snakes must swallow their food whole. This is accommodated by a jaw that becomes unhinged, allowing the mouth of some snakes to open wide enough to swallow a hundred-pound animal. Egg-eating snakes have a specialized tooth used to break the shell of the egg.

Although they can be voracious eaters, snakes, if necessary, can go for a year or more without food. Snakes cannot hear vibrations through the air but receive vibrations from solid objects through their skin, which, incidentally, is waterproof. The flicking tongue is used for touch and to pick up small particles and tuck them into little cavities in the roof of the mouth. The cavities are associated with small organs linked to the sense of smell. By picking up the particles, the tongue assists in detecting odors. The eggs of snakes have a tough, leathery shell that prevents the fluid contents surrounding the embryo from drying out. This adaptation to land has been instrumental in the evolutionary success of snakes and, indeed, of all reptiles. Some snakes, however, such as the common garter snake, give birth to live offspring. Snakes grow rather quickly, reaching maturity in about 2 years.

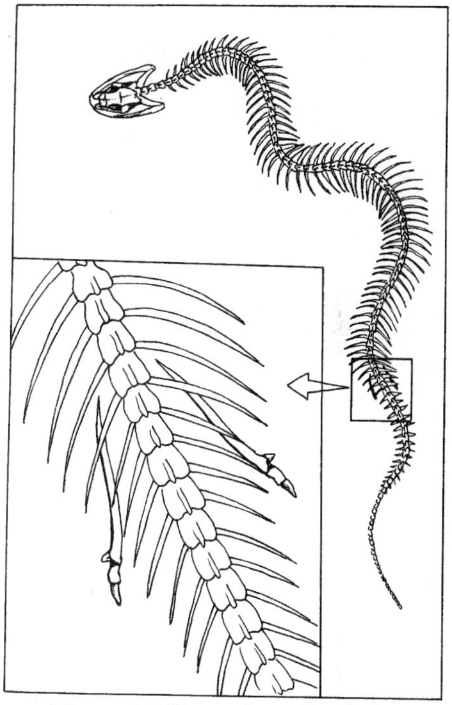

Figure 3.8 The rudimentary limb bones shown in this boa constrictor skeleton are evidence that the ancestors of snakes possessed legs.

These reptiles actually have a variable body temperature higher than that of their surrounding environment. Whereas birds and mammals regulate body temperature using heat released from food metabolism, snakes rapidly raise their body temperature by lying in the sun. If the body temperature becomes too warm, they move out of the sun. In colder weather, snakes become sluggish and hibernate or seek sufficiently warm shelter. Several times a year, snakes shed their waterproof skin and grow a completely new skin.

Our second organism to illustrate unity in diversity belongs to a group of shell-less mollusks called *nudibranchs* (meaning " naked gills "), or sea slugs. They are a group of aquatic invertebrates (animals that lack a backbone) numbering more than 2000 species. These tiny ocean snails usually are beautifully arrayed in colors of red, orange, yellow, green, blue, or some combination of these colors (Fig. 3.9). Some species can change color as their background environment changes. In the course of evolutionary adaptation, nudibranchs no longer possess a shell; some nudibranchs, however, have acquired a protective coloration that strikingly resembles that of shells found on the ocean floor. Nudibranchs feed mainly on tiny "sea flowers" called anemones, which are not flowers at all, but animals. Sea anemones, jellyfish, corals, and other similar aquatic animals are equipped with microscopic stinging organs used in defense and for capturing prey. These stinging organs are housed within tiny oval cells and contain poisonous chemicals. When touched, the anemones release the stinging organs with explosive force. In what has to be one of nature's more remarkable adaptations, the little nudibranchs have acquired the ability to swallow the oval cells without digesting them or causing them to explode. From the snails' digestive tract, the oval

Figure 3.9 Nudibranch, or sea slug, is a mollusk that lacks a shell.

cells migrate to appendages on the outside of their bodies where they remain. When disturbed, the nudibranchs, in turn, launch the stinging organs in much the same way as do the anemones.

Nudibranchs possess a pair of bladelike jaws used to cut prey into small pieces. Food passes from the mouth into a simple saclike stomach, which leads into a system of branching digestive ducts. A few nudibranchs graze on corals or on anemones that contain colonies of algae. After a time, the nudibranch may not return to its normal mode of feeding but may rely solely on food provided by the algae.

Nudibranchs are *hermaphrodites* (from *Hermaphroditus,* a son of the Greek god *Hermes* and the goddess *Aphrodite*), which means that both male and female reproductive structures are present on the same organism. The reproductive process occurs sexually through the mutual exchange of sperm between two organisms. Fertilized eggs are released into the surrounding water, where they grow and develop into mature adults.

Of Snakes and Snails: Exploring Unity

There is clearly an enormous physical difference between a snake and a nudibranch. In terms of the seven life features, however, they exhibit a remarkable underlying unity, as do all forms of life. They simply express this unity in their own particular way; hence, unity in diversity. The evidence that all members of the five kingdoms share these fundamental life features becomes more plausible when we consider the origin and evolution of life. Evidence suggests that all organisms have descended from a common form of life—presumably some type of single cell. The unity, or sameness, evident in organisms can be explained, therefore, in terms of a common ancestry. If this is so, how do we account for the great diversity among living things? Basically, diversity has resulted from the innate capacity of organisms to adapt to Earth's many and varied environments. Environmental adaptation is a consequence of evolution, and it is evolution that provides the unifying link between unity and diversity in all organisms.

We have explored the remarkable diversity between snakes and nudibranchs, but in what ways do they exhibit the unity of life as characterized by our seven life features? As living organisms, both are organized on a cellular basis, being composed of complex eukaryotic cells that are able to divide, grow, and metabolize in a similar manner. Their cells, of course, are comprised of the same basic elements and compounds. The two organisms further exhibit a hierarchy of organization evident in many multicellular organisms by forming specialized tissues, organs, and body systems capable of specific functions. Yet these two organisms have expressed this underlying unity of organization each in its individual way. A snake does not physically resemble a nudibranch in any way, yet both are composed of cells possessing similar compounds organized at a similar level of complexity. A seemingly disparate pair of organisms has revealed a perhaps surprising fundamental unity underneath two such different exteriors.

The snake and the snail both require a source of energy to maintain their well-being. Cells use energy to build larger substances from smaller, simpler ones. All cells, and ultimately organisms, are able to maintain their level of organization and proper function only by receiving a continuing supply of energy. Yet this fundamental unity is displayed very differently for each organism. The snail feeds on coral or algae-containing anemones, cutting its prey into small pieces and converting this food into a usable source of energy. The snake swallows its prey in its entirety, then gradually converts this food into a usable form.

Snakes and nudibranchs interact with other species within their respective communities through feeding and avoiding predators. They also react with the nonliving components of their environments by utilizing oxygen and water. Such interactions, as well as many others, attest to their innate capacity to sense and respond to environmental stimuli.

Snakes mainly inhabit dry deserts, the tropics, and aquatic environments, while nudibranchs swim in ocean waters. To survive, control of body temperature, water balance, excretion of metabolic wastes, and other instances of homeostasis must be equally effective in both organisms. These mechanisms allow both species to make short-term adaptations if their environments are altered. Both snakes and nudibranchs illustrate the principle of biogenesis—life arising from life—in reproducing their own kind. The offspring of both snakes and nudibranchs grow by mitotic cell division, and develop through formation of different types of body cells. While this is evidence for the underlying unity of snakes and nudibranchs, cell division produces two organisms that are structurally and functionally quite diverse.

The physical and functional characteristics of snakes and nudibranchs have resulted from interactions between these organisms and their environments. As a consequence of these interactions, populations of both organisms have evolved and have become adapted to their surroundings. In descending from a common ancestor, the forebears of snakes, nudibranchs, and all other known forms of life radiated outward to inhabit diverse environments. As many life forms successfully adapted to these specific environments over millions of years, great organismal diversity was superimposed on the unity that arose from a common ancestry.

Finding Unity in Diversity: Characteristics of Living Organisms

We have briefly discussed the ways in which scientists try to make sense of and study the overwhelming numbers of organisms inhabiting Earth. This tremendous project is facilitated by grouping similar organisms together—at first into five broad categories or kingdoms, and then further by dividing and grouping them into increasingly specific subdivisions of these kingdoms. Let's now take a closer look at some of the general characteristics biologists use to identify and categorize organisms in different groups.

Cells

The basic unit of life is the cell, and we have found that organisms may have either a prokaryotic or eukaryotic cell structure. In addition, organisms may be either *unicellular* or *multicellular.* Some organisms, including species of algae and protists, form groups of cells known as *colonies.* Most bacteria and protists, various algae, and a few fungi are examples of unicellular forms. It is important to remember that a single cell can perform all life functions, such as metabolism, response, and reproduction. As a result of development, the cells of some colonies and most multicellular organisms become differentiated to form specialized cell types having specific functions. A cell will differentiate by producing a specialized cell structure or component, or a particular shape or size that allows it to perform its specific function. In higher plants, for example, some cells transport water and are constructed in a way that allows them to accomplish this task; some cells are involved in reproduction, others carry out photosynthesis and thus contain the apparatus necessary for this process. In your own body, skin cells form a protective covering for the body, red blood cells transport oxygen, and muscles cells allow for movement. The specialization of cells in a multicellular organism results in **division of labor** among the various cell types and an increase in the degree of organizational complexity of the individual.

Animals

Animals are classified for convenience into invertebrates, nudibranchs are examples, which lack a backbone, and vertebrates, such as snakes, which possess a backbone. Invertebrates are immensely diverse and numerous, constituting over 95 percent of all known animal species. The first animals are thought to have originated approximately 700 million years ago, with most invertebrate phyla arising about 150 million years later. According to some experts, all modern animals evolved from a group of protists, but the fossil record of animals, especially invertebrates, is highly fragmented and often difficult to interpret. Consequently, tracing the evolutionary relationships among the animals is something of a knotty problem.

Invertebrates

The simplest invertebrates include the *sponges,* followed closely by various jellyfish, anemones, corals, and related types called *cnidarians.* Sponges do not possess a tissue or organ level of organization. The body of a sponge is little more than a hollow, double-layered sac of cells with holes punched in it. Cnidarians are more complex structurally and functionally than sponges, and possess simple muscles and nerves.

Somewhat more advanced than the sponges and cnidarians are the various *flatworms* and *roundworms*. Both of these groups exhibit evolutionary advancement over the cnidarians through development of a rudimentary brain and bilateral symmetry (having a body with two equal halves). Flatworms include the parasitic flukes and tapeworms, along with some free-living (nonparasitic) forms.

The major roundworms are the cylindrical nematodes, which include many free-living species as well as various parasitic species, such as pinworms and hookworms. The roundworms possess a complete digestive tract having a mouth and anus. They also have well-developed muscles extending the length of the body.

The *mollusks* include clams, oysters, octopuses, squids, nudibranchs, and others. Adaptations among the mollusks are quite diverse. It may be difficult to imagine that a sluggish, burrowing clam is in the same group of organisms as the darting, jet-propelled squid with its complex brain and highly-developed eyes. The features of both organisms, however, are adaptations to their particular modes of existence. Being specialized for a quiet, sedentary life, clams feed on fine food particles suspended in the surrounding water and use their muscular foot to burrow into sand or mud. Squids, on the other hand, are carnivores that feed on fish, crustaceans, and worms. Thus, squids must locate and pursue their prey. The squid then uses its beaklike jaws and a poisonous secretion to incapacitate and kill its prey.

The segmented worms, or *annelids,* are composed of body segments that appear as little rings. The rings are an adaptation to a burrowing, creeping type of movement. Included in the annelids are the familiar earthworm and some related aquatic forms. Other annelids include a group of mostly marine worms and the leeches.

An enormous group of joint-footed animals—the *arthropods*—include crayfish, lobsters, spiders, scorpions, centipedes, millipedes, and all the insects. Adaptations among the arthropods are fabulously diverse. These animals have developed claws, fangs, poison glands, stink glands, swimming appendages, and a variety of sensory receptors, to name but a few.

The final major group of invertebrates includes spiny-skinned marine animals, such as starfish, brittle stars, sand dollars, and sea urchins, collectively known as *echinoderms.* Their bodies are usually constructed on a five-part plan, with each part radiating like spokes of a wheel from a center. Echinoderms possess a unique network of internal canals called the *water vascular system.* The canals branch into tiny tube feet that function in feeding, anchoring, and gas exchange. Some echinoderms are beautifully colored, with bodies of intricate design.

Vertebrates

The first vertebrates appear in the fossil record about 500 million years ago and are thought to have evolved from some type of invertebrate. Modern vertebrates are far less numerous than invertebrates, but are nonetheless quite diverse and occupy a wide range of environments. The most primitive living vertebrates are the jawless, eel-like agnathans. Some agnathans are parasites on fish. The *fishes* include two major groups—those with an internal skeleton (endoskeleton) of bone, such as trout, catfish, and bass, and those with an endoskeleton of cartilage, such as sharks and rays.

The earliest known land vertebrates are the *amphibians,* a term meaning "two lives", which refers to those amphibians that carry out their life cycles on land and in water. Representative amphibians include frogs, toads, and salamanders. Snakes, lizards, turtles, alligators, and crocodiles are *reptiles,* which are descendants of an ancient amphibian stock. Included in the reptiles are the extinct dinosaurs, nearly all of which disappeared about 65 million years ago. Reptiles have a waterproof skin that prevents dehydration, and they lay leathery, fluid-filled eggs (Fig. 3.10A & B). These two major developments have allowed reptiles to adapt successfully to a completely terrestrial existence. Evolutionary descendants of reptiles are the *birds,* which are vertebrates with feathers. Most birds are capable of

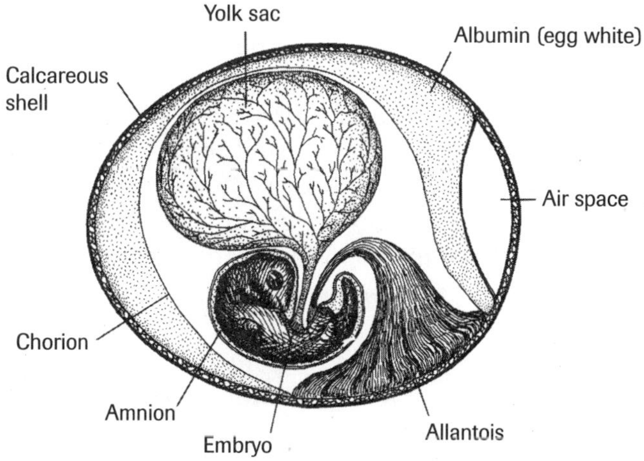

Figure 3.10A The amniotic, egg of reptiles, birds and egg-laying mammals. The evolution of the egg was crucial to the biological success of these vertebrates in that the egg provides protection (the shell), nourishment (yolk and albumin), and a source of water (fluid in the amnion) for the developing young.

Figure 3.10B A female ringnecked snake with her eggs.

flight, but a few, such as the ostrich and penguin, are flightless. The most distinguishing physical characteristics of *mammals* are the presence of hair and the milk-producing mammary glands of females. Mammals evolved from a group of mammal-like reptiles. The duck-billed platypus and spiny anteater are the only egg-laying mammals. The young of marsupial mammals, such as opossums and kangaroos, are born alive and complete their embryonic development in the mother's abdominal pouch, while placental mammals, including humans, develop within the mother's uterus.

Plants

It is thought that plants evolved from a type of green algae about 425 million years ago. All plants are photosynthetic, meaning they are capable of using light energy to synthesize organic molecules; simply put, they can make their own food. Almost all plants reproduce sexually and some employ asexual methods as well. Development of the sex cells, or gametes, occurs within a jacket of protective cells, and the plant embryo develops within the female organ. The life cycles of plants involve what is known as *alternation of generations* in which a gamete-producing, or gametophyte, generation gives rise to a spore-producing, or sporophyte, generation, which, in turn, gives rise to a new gametophyte generation. A plant spore is a reproductive cell that can give rise to a multicellular plant without fusing with another cell. In mosses and related plants, the gametophyte generation is more dominant or conspicuous, while in all other plants the sporophyte generation is dominant. For example, the green mat of moss growing on a tree trunk consists of gametophytes, while the tree on which the moss is growing is a sporophyte.

The great majority of plant species are terrestrial. Members of the plant kingdom can be separated generally into *nonvascular* plants and *vascular* plants. These terms refer to the absence or presence of internal supporting and conducting tissues that transport water, minerals, and food. Nonvascular plants, such as certain mosses, lack these tissues and thus usually are small and found most frequently near a source of water. Vascular plants include the ferns and seed plants (conebearing and flowering plants), along with a few seedless forms. Having a tough, outer coat, seeds provide protection for the developing embryo and also provide a ready supply of food as the embryo develops. In addition, the seed embryo has a dormant phase during which it does not germinate. Add to this their ability to take up and retain water and it is evident why vascular plants have adapted to such a wide range of environments.

This very brief sketch of the diverse characteristics of life will serve as an introduction to our more detailed survey of the kingdoms of life in Chapter 6.

Diversity: The Impact of Evolution

One of several keys to understanding evolution is the effect of the environment on the survival of organisms. The **environment** includes nonliving factors such as temperature, light, and moisture, as well as living factors, which are other organisms. Survival of the individual and survival of the population of which the individual is a part obviously are interrelated. Although an individual organism may survive into old age, it is of little significance to the survival of the population unless that organism produces offspring. Thus, survival of a population depends on some of its members surviving *at least until reproductive age.* "Survival of the fittest," an expression coined by the British philosopher Herbert Spencer, most accurately refers to organisms that produce fertile offspring. Survival, of course, assumes that an organism is adapted to the demands of a given environment. An organism's characteristics allow it to survive and reproduce under the existing environmental conditions. Cacti survive in an arid environment unsuited to apple trees, and our snakes and nudibranchs certainly are ill-suited to each other's environment. Feeding, reproducing, excreting, and other activities of snakes and nudibranchs are considerably different because of the demands or requirements of their contrasting environments. Organisms either cope with these demands or they perish. Organisms possessing the most favorable structural and functional characteristics in a given environment tend to survive in greater numbers than organisms lacking those characteristics. This means that, as the environment changes, populations must either adapt, decline in number, or die. This does not mean that the environment *creates* characteristics in organisms; an organism's characteristics are determined primarily by genetic (inherited) factors.

Genes, which we will examine more closely in Part VII, are the basic units of heredity and are the functional units of chromosomes, which are located in virtually every cell. People, roses, snakes, and nudibranchs are what they are and look the way they look because of the particular genes they possess. These genes contain all of the information needed and ensure that each of the seven life features normally is carried out in an organism; they determine the traits we possess, everything from eye color to height. Genes are organizational units of information, and this information is enclosed on sequences of molecules called *nucleotides.* Nucleotides are the building blocks of genes and the information certain nucleotide sequences contain allows a cell to reproduce, to synthesize the appropriate compounds to carry out metabolism, to construct homeostatic mechanisms, to sense an environmental change and respond to it, and to differentiate and carry out a specific function.

Modern evolutionary theory is thus based on the interaction of genetically determined traits and environmental factors. It is traits that are governed by the influence of many different genes that are most important in the evolution of diverse kinds of organisms. We have seen that snakes and nudibranchs have evolved a number of adaptive features that reflect the influence of their environments. Such is the rule with all forms of life.

Adaptations can assume rather intriguing forms. One classic example involves certain species of orchids having petals that resemble a species of female flies. The flowers attract male flies that become sprinkled with pollen when they land on the petals. The males may carry the pollen to another orchid displaying flylike petals. In this unusual adaptation for pollination, the fly-flower relationship enhances the reproductive potential of the orchids. In the animal kingdom, species such as frogs and flounders, like our nudibranchs, can change color to match their background. This protective coloration serves as a defense against potential enemies. Some animals, such as certain species of butterflies, emit offensive odors or taste unappetizing, and are avoided by predators following an unfortunate encounter. Various harmless animal species, such as certain beetles that closely resemble poisonous wasps, are avoided by predators that cannot distinguish between the two.

Natural Selection

These four interrelated factors—the environment, variation, genetics, and adaptation—are fundamental in understanding diversity and its relation to the theory of evolution. Modern evolutionary theory centers on the concept of **natural selection,** a process whereby individuals with more favorable adaptive characteristics tend to produce more surviving offspring than do the less favorably adapted members of the population. This differential reproductive success results from the effects of the environment on organisms. Being better equipped to exploit their environment, organisms with favorable adaptive traits tend to be naturally selected by that environment. Individuals less successfully adapted are at a disadvantage, and tend to appear less frequently, or die out, through the generations. They are, it is said, at a selective disadvantage. Over extended periods of time, this selective process can result in changes in populations, leading to new species.

Natural selection acts on physical traits, such as size, shape, and color, which, of course, are expressions of genes. All forms of life are considered to have had a common ancestor, so it follows that even snakes and nudibranchs are related to a certain extent. We also said that all forms of life probably arose from some type of ancestral cell. What we are contemplating here is *change*—dramatic change involving development of one form of life from a drastically different form. How could a single cell lead to such diversity evident in the living world? The straightforward answer is that biological change occurs by step-by-step genetic alterations in succeeding organisms. This, according to some evolutionists, normally allows *very small* changes to accumulate gradually over time until a particular structure is formed. A competing view holds that evolution proceeds in relatively rapid spurts rather than gradually. According to this theory, known as *punctuated equilibrium,* a new species is modified during its development from the parent species, but then changes very little for a long period thereafter.

Each of the more primitive (or less developed) intermediate structures formed along the way may be as important to the organisms that possess them as the more developed structure is to its possessor. This applies to eyes, hearing organs, legs, or virtually any structure. It is critical to understand that every change in the sequence of development is built on a previous change. In other words, the changes are cumulative. A "new" or dramatic change rarely if ever arises from scratch. Some changes may be favorable; that is, of adaptive advantage in a given environment. If so, they tend to be selected by the environment. Other traits may be of no adaptive value. These traits, then, tend to be eliminated over time. Remember that natural selection does not act directly on the genes, but on the products or effects of the genes as revealed in the physical, functional, and behavioral makeup of the organism.

Evolution: A Unifying Biological Theme

Biological evolution is the unifying principle, the common thread, that ties together all forms of life. It is an ongoing process that affects the entire genetic makeup of a population. Evolution can occur within a population without new species being formed or more advanced organisms being developed. All that is necessary is that some genetic change occur and that, by definition, is evolution. When we consider the complexity of the living world—the diversity of species and their distribution in the biosphere, the changes observed in old and new populations, the relationships between body structures in different species, and the development of life as revealed in the fossil record—it is evolution that makes it understandable. Throughout its biological history, life has confronted the challenge of new and changing environments. For a few groups, the challenge has been met through evolutionary adaptation. Evolution through natural selection is responsible for the great divergence of life from simpler to more complex forms. But regardless of the diversity this process may create, it does so within the framework of a common unity.

Chapter Summary

1. The five kingdoms of life are Monera, Protista, Plantae, Fungi, and Animalia. Each kingdom is divided into decreasingly smaller, more specific categories, allowing for more detailed classification. Life on Earth exhibits extreme diversity. Life can be characterized and identified by seven basic features: organization, energy transactions, sensation and response, homeostasis, reproduction, growth and development, and evolutionary adaptation. All living organisms exhibit these common features—this unity of life, each organism in its own diverse way.

   Key Terms: *cell. p. 30*, *prokaryotic, p. 31*, *eukaryotic, p. 31*, *metabolism, p. 32*, *stimulus, p. 32*, *energy transaction, p. 32*

   Concept Question: Explain how the seven life features apply to a single cell as well as to a multicellular organism.

2. Evidence indicates that all living things are descendants of a single, original life form. The unity of life, therefore, is understood in terms of a common ancestry. The diversity of life results from the capacity of organisms to adapt to different environments. The evolution of organisms is the unifying link between unity and diversity. Some fundamental characteristics used to describe the diversity of life include unicellular or multicellular structure, invertebrate and vertebrate animals, and nonvascular and vascular plants.

   Key Terms: *adaptation, p. 35*, *division of labor, p. 40*

   Concept Question: In general, how does comparing plants and animals illustrate the unity and diversity of life?

3. Organisms must be adapted to environmental conditions in order to survive and reproduce. Structural, functional, and behavioral characteristics result from genetic factors. Favorable characteristics tend to be perpetuated in a population. The environment may contribute to variations of these characteristics, but does not cause them.

   Key Terms: *environment, p. 42*, *gene, p. 43*

   Concept Question: What is the role of the environment in the development of diversity in living organisms?

4. Within the biosphere, organisms are subjected to specific requirements, such as food, water, and temperature. The fitness of an organism is reflected in its adaptation to its specific environment. Adaptation is especially critical until reproductive age.

   Evolution acts through natural selection. Organisms with more favorable adaptations tend to produce more surviving offspring than do less favorably adapted organisms. This phenomenon can lead to changes in populations, resulting in new species. Evolution basically refers to genetic change in a population over time, and is the unifying principle that explains the relationship between unity and diversity in the living world.

Key Terms: *natural selection, p. 44*

Concept Question: Why is natural selection considered the driving force behind biological evolution? In your own words, explain the phrase: "Evolution is the link between the unity and diversity of life."

PART THREE

Life Feature 1: Organization

Chapter Four

Matter: Structure, Properties, and Interactions

Matter and the Elements of Life

Chapter Three pointed out that all organisms are characterized by a fundamental unity, or sameness, while at the same time exhibiting these common features in extremely diverse ways. The unity of life through a common ancestry suggests that organisms may also exhibit a common structural organization. Indeed, such unity was evident in our brief discussion of cells as the basic units of organisms. This continuity extends well below the cellular level to the chemical level in all living matter. **Matter** is the stuff of which anything—living or nonliving—is made; the physical substance of an object. In biology, our major interest is living matter, but it turns out that the basic kinds of matter—the elements—comprise both living organisms and nonliving matter (Fig. 4.1). An **element** is a pure substance composed of a certain type of atom, and it cannot be broken down into a simpler pure substance by ordinary chemical means. It isn't possible to break down the element gold into two or more simpler substances. It is the different combinations and organization of basic elements that make possible the immense diversity we observe on our planet. The element carbon in your pencil, diamond ring, cheeseburger, or your toenails is the same element. It is the arrangement of the carbon atoms in each of these that makes them different.

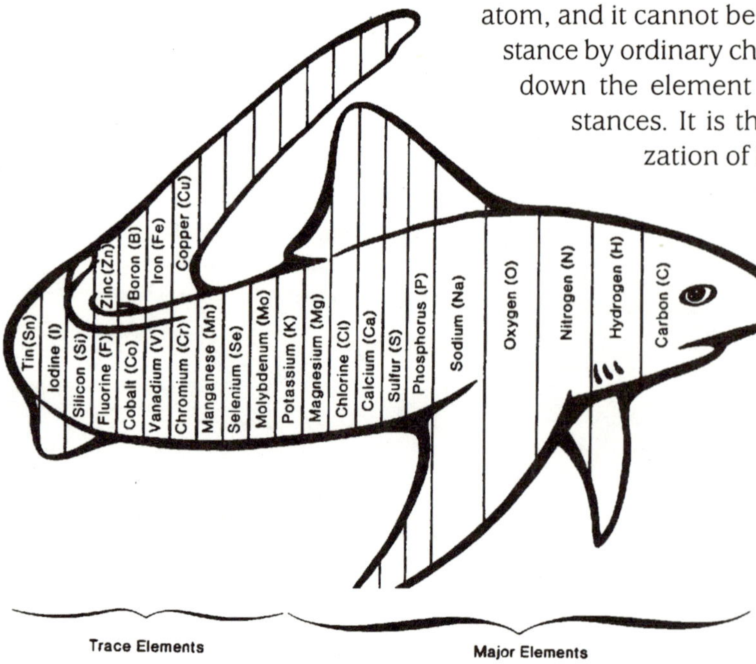

Figure 4.1 Elements found in living things. Some organisms may require many of these elements, but not necessarily all of them.

Through their interactions, the chemicals in a living organism make possible the functioning of a certain structure or part. You are able to read this page because of chemical reactions in your eyes, brain, and associated nerves. Moreover, organisms do not exist solely on their own. The nutrients required for life are chemical substances, as are the nonliving components of the surroundings in which organisms exist. Interactions between living organisms and their environment ultimately are possible because of the nature of chemicals. Air, water, the sun, and all the other factors required for life have a chemical makeup. Life arose from and is constructed of nonliving chemicals which is why we need to know a little about these chemicals, their structures and properties and how they interact.

There are 92 naturally occurring elements, which differ from each other by the structure of their atoms. The six elements carbon, hydrogen, nitrogen, oxygen, phosphorus, and sulfur make up about 98 percent of all living matter. These elements can be designated in symbolic chemical shorthand as CHNOPS, the initial letters in the name of each (Fig. 4.2). Some elements, such as chlorine (Cl), sodium (Na), calcium (Ca), and iron (Fe), are designated by two letters. These four elements, along with magnesium (Mg), potassium (K), and other elements present in trace amounts, constitute the remaining 2 percent of living matter. If you have noticed that the chemical symbols for some elements do not correspond to the elements' names, it is because these symbols are derived from the Latin names of these elements. Iron, for instance, has the symbol Fe, the first two letters in *ferrum,* which is the Latin term for iron. The chemical symbol for potassium is K, the first letter in the Latin term *kalium.*

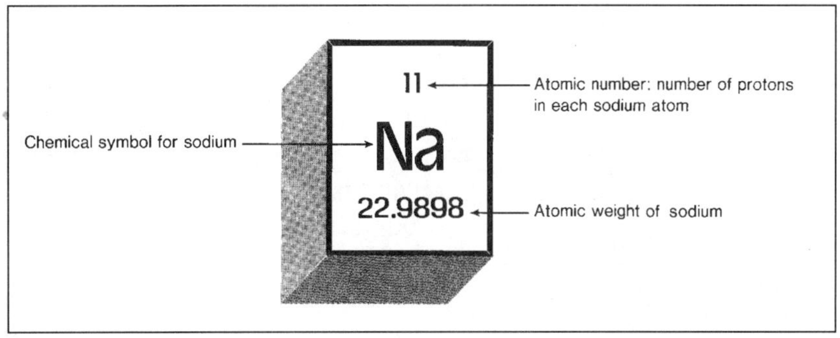

Figure 4.2
A box from periodic table of elements, showing the chemical symbol, the atomic number, and the atomic weight of the element sodium.

We have already stated that all matter is comprised of elements in various combinations. Further, matter is a constant in the universe; that is, matter cannot be created or destroyed. This is known as the **law of conservation of mass.** Another way to express this law is that there is no detectable change in the amount of matter in ordinary chemical reactions. The total mass of the reactants at the start of a chemical reaction equals the total mass of all the products at the completion of the reaction. The atoms participating in a chemical reaction may be rearranged, but they cannot be destroyed and new atoms cannot be created. Matter, however, can change form. Matter can release energy and energy can be stored in different types of matter. Gasoline, for example, stores energy in chemical bonds, which, when broken, release energy to run your automobile. We discovered in Chapter Three how energy transactions occur continuously. We use energy constantly and, in fact, matter and energy are very closely related. Just as we have found is the case with matter, energy cannot be created or destroyed. Energy, like matter, can change form—plants, for example, can absorb light energy from the sun and convert it to chemical energy in the form of starch.

There are three states of matter—solids, liquids, and gases. One way these states of matter can be converted from one state to the other is with a sufficient rise or fall in temperature. Water is in the liquid state between 0° C (32° F) and 100° C (212° F), but at 0° C it freezes to solid ice, and at 100° C it evaporates, becoming gaseous steam, or water vapor (Fig. 4.3). These conversions, however, do not create new matter or destroy

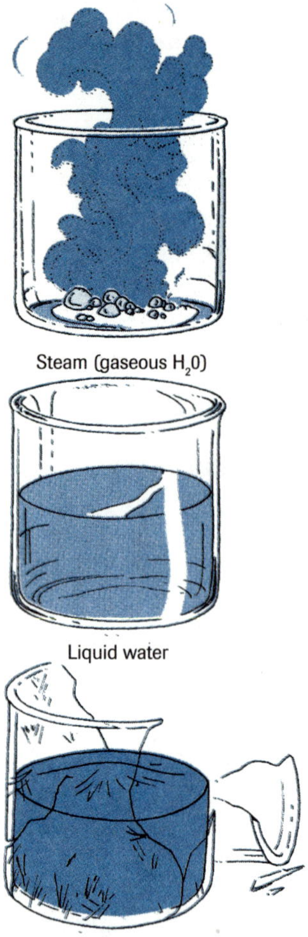

Figure 4.3 The three states of matter (solids, liquids, and gases) may be converted in form by raising or lowering their temperatures.

existing matter. Conservation of matter is a natural law, about which no exceptions are known. We have said that life, with a very few exceptions, exists mostly within a temperature range of about 0° C (32° F) to 55° C (131° F). This restriction is closely tied to the nature and properties of chemicals. If the temperature rises far above or drops far below this range, certain chemicals undergo rearrangements that cannot be tolerated by a living organism. Each chemical possesses certain properties unique to that chemical. A **property** of any substance is simply a characteristic, attribute, or quality it possesses. Hardness, the temperature at which a liquid boils or a solid liquefies, density, color, in fact any characteristic that does not involve a chemical change is *a physical* property. Your physical properties include your height, weight, hair color or waist measurement. Matter exhibits *chemical* properties when it is involved and subsequently changed in a chemical reaction. Iron in your bicycle handlebars, for example, combines with oxygen to produce rust. In this reaction, the arrangement of the atoms in rust is quite different from those of iron or oxygen. A chemical property, then, is a description of changes a substance can undergo when it participates in a chemical reaction. Because organisms are constructed of elements, the physical and chemical properties of these elements are reflected in the structural and functional characteristics of the organisms.

Our primary concern in biology is with the six major elements mentioned, plus a few of the minor, but quite essential, ones. These elements, each with its unique structure and properties, are the basic units that comprise and maintain organisms, and they do so by interacting with each other through chemical reactions. You will recall that during evolution on the early Earth, reactions between chemicals gave rise to several of the basic constituents, such as amino acids, found in living organisms. Our first task, then, is to consider the chemical level of the organization of life—how and why these reactions occur. To do so, we need a little background in atomic theory.

Atoms, Energy, and Chemical Bonds

Every element consists of at least a single **atom**, the smallest possible part of an element that retains the properties of that element. The smallest part of the element gold is a single gold atom. The same concept applies to the other 91 naturally occurring elements. Atoms can be split into two or more fragments, as happens in nuclear fission, and thereby no longer have the same elemental properties. We, however, are not concerned with splitting atoms, but we still need to know about their basic structure.

An atom has two distinct regions, an inner dense region called the *nucleus* and an outer region that is mostly empty space. The center, or nucleus, of an atom is comprised of a certain number of particles called **protons** and **neutrons.** Protons have an electrical charge of +1, while neutrons do not possess an electrical charge.

The number of protons in the nucleus of an atom is called the **atomic number** and this number is unique for each element. In other words, each element is composed of a type of atom that possesses a characteristic number of protons and thus has a unique

Figure 4.4 The periodic table of elements with a list of the elements with their symbols and atomic numbers.

Active metals: 1A, 2A — Transition metals: 3B–2B — Nonmetals: 3A–8A — Inert gases: 7A, 8A

Key: Atomic number / Chemical symbol / Atomic weight (mass)

1A	2A	3B	4B	5B	6B	7B	8B	8B	8B	1B	2B	3A	4A	5A	6A	7A	8A
																1 H 1.00797	2 He 4.0026
3 Li 6.939	4 Be 9.0122											5 B 10.811	6 C 12.01115	7 N 14.0067	8 O 15.9994	9 F 18.9984	10 Ne 20.183
11 Na 22.9898	12 Mg 24.312											13 Al 26.9815	14 Si 28.086	15 P 30.9738	16 S 32.064	17 Cl 35.453	18 Ar 39.948
19 K 39.098	20 Ca 40.08	21 Sc 44.956	22 Ti 47.90	23 V 50.942	24 Cr 51.996	25 Mn 54.9380	26 Fe 55.847	27 Co 58.9332	28 Ni 58.71	29 Cu 63.54	30 Zn 65.37	31 Ga 69.72	32 Ge 72.59	33 As 74.9216	34 Se 78.96	35 Br 79.909	36 Kr 83.80
37 Rb 85.47	38 Sr 87.62	39 Y 88.905	40 Zr 91.22	41 Nb 92.906	42 Mo 95.94	43 Tc (99)	44 Ru 101.07	45 Rh 102.905	46 Pd 106.4	47 Ag 107.870	48 Cd 112.41	49 In 114.82	50 Sn 118.69	51 Sb 121.75	52 Te 127.60	53 I 126.9044	54 Xe 131.30
55 Cs 132.905	56 Ba 137.33	57 La* 138.91	72 Hf 178.49	73 Ta 180.948	74 W 183.85	75 Re 186.2	76 Os 190.2	77 Ir 192.2	78 Pt 195.09	79 Au 196.967	80 Hg 200.59	81 Tl 204.37	82 Pb 207.19	83 Bi 208.980	84 Po (210)	85 At (210)	86 Rn (222)
87 Fr (223)	88 Ra (226)	89 Ac† (227)	104 Rf (257)	105 Ha (260)	106 Unh (263)												

*Lanthanide series (Rare earth metals)	58 Ce 140.12	59 Pr 140.907	60 Nd 144.24	61 Pm (147)	62 Sm 150.35	63 Eu 151.96	64 Gd 157.25	65 Tb 158.924	66 Dy 162.50	67 Ho 164.930	68 Er 167.26	69 Tm 168.934	70 Yb 173.04	71 Lu 174.97
†Actinide series (Uranium metals)	90 Th 232.038	91 Pa (231)	92 U 238.03	93 Np (237)	94 Pu (242)	95 Am (243)	96 Cm (247)	97 Bk (247)	98 Cf (249)	99 Es (254)	100 Fm (253)	101 Md (256)	102 No (253)	103 Lr (257)

NOTE: For elements all of whose isotopes are radioactive, parentheses enclose the atomic weight of the isotope with the longest half-life.

Element	Symbol	Atomic number
Actinium	Ac	89
Aluminum	Al	13
Americium	Am	95
Antimony	Sb	51
Argon	Ar	18
Arsenic	As	33
Astatine	At	85
Barium	Ba	56
Berkelium	Bk	97
Beryllium	Be	4
Bismuth	Bi	83
Boron	B	5
Bromine	Br	35
Cadmium	Cd	48
Calcium	Ca	20
Californium	Cf	98
Carbon	C	6
Cerium	Ce	58
Cesium	Cs	55
Chlorine	Cl	17
Chromium	Cr	24
Cobalt	Co	27
Copper	Cu	29
Curium	Cm	96
Dysprosium	Dy	66
Finsteinium	Es	99
Erbium	Er	68
Europium	Eu	63
Fermium	Fm	100
Fluorine	F	9
Francium	Fr	87
Gadolinium	Gd	64
Gadolinium	Gd	64
Gallium	Ga	31
Germanium	Ge	32
Gold	Au	79

Element	Symbol	Atomic number
Hafnium	Hf	72
Hahnium	Ha	105
Helium	He	2
Holmium	Ho	67
Hydrogen	H	1
Indium	In	49
Iodine	I	53
Iridium	Ir	77
Iron	Fe	26
Krypton	Kr	36
Lanthanum	La	57
Lawrencium	Lr	103
Lead	Pb	82
Lithium	Li	3
Lutetium	Lu	71
Magnesium	Mg	12
Manganese	Mn	25
Mendelevium	Md	101
Mercury	Hg	80
Molybdenum	Mo	42
Neodymium	Nd	60
Neon	Ne	10
Neptunium	NP	93
Nickel	Ni	28
Niobium	Nb	41
Nitrogen	N	7
Nobelium	No	102
Osmium	Os	76
Oxygen	O	8
Palladium	Pd	46
Phosphorus	P	15
Platinum	Pt	78
Platinum	Pt	78
Plutonium	Pu	94
Polonium	Po	84
Potassium	K	19

Element	Symbol	Atomic number
Praseodymium	Pr	59
Promethium	Pm	61
Protactinium	Pa	91
Radium	Ra	88
Radon	Rn	86
Rhenium	Re	75
Rhodium	Rh	45
Rubidium	Rb	37
Ruthenium	Ru	44
Rutherfordium	Rf	104
Samarium	Sm	62
Scandium	Sc	21
Selenium	Se	34
Silicon	Si	14
Silver	Ag	47
Sodium	Na	11
Strontium	Sr	38
Sulfur	S	16
Tantalum	Ta	73
Technetium	Tc	43
Tellurium	Te	52
Terbium	Tb	65
Thalium	Tl	81
Thorium	Th	90
Thulium	Tm	69
Tin	Sn	50
Titanium	Ti	22
Tungsten	W	74
Unnilhexium	Unh	106
Uranium	U	92
Vanadium	V	23
Xenon	Xe	54
Xenon	Xe	54
Ytterbium	Yb	70
Yttrium	Y	39
Zinc	Zn	30
Zirconium	Zr	40

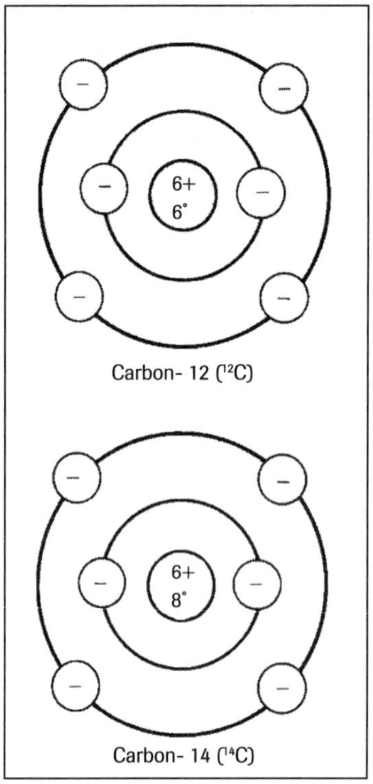

Figure 4.5 (Upper) Diagram of a normal carbon atom having a mass number of 12. (Lower) An isotope of carbon having a mass number of 14. Note that the only structural difference between the two atoms is the two additional neutrons in the atomic nucleus of ^{14}C.

atomic number. The carbon atom has six protons in its atomic nucleus, oxygen has eight, and hydrogen only one. Scientists have used this discovery to organize the chemical elements in order of increasing atomic number on a chart called the Periodic Table (Fig. 4.4).

In the region surrounding the atomic nucleus there are small, negatively charged particles called **electrons.** Electrons have a charge of −1. The number of electrons in a neutral atom normally equals the number of protons. Thus, all neutral atoms, by definition, have a net electrical charge of zero.

Atoms of the same element can have different numbers of neutrons. Such atoms are called **isotopes** (Gk. *isos* same + *topos* place) of that element. Different isotopes of the same element have the same atomic number but may differ in physical properties. The carbon atom, for example, can occur as different isotopes (Fig. 4.5). The most common and abundant form of carbon is carbon-12, an isotope having six neutrons. Another isotope of carbon is carbon-14, which has eight neutrons and differs physically from carbon-12 in that it is *radioactive.* A radioactive isotope has an unstable nucleus and will spontaneously change to become a more stable isotope. During this process, called radioactive decay, energy is emitted in the form of particles or radiation, or both. Radioactive decay occurs at a characteristic rate for each element and can be detected and measured, making radioactive isotopes of invaluable use in the study of biology.

When introduced appropriately into an organism, radioactive atoms can be used as "tracers" to study chemical reactions or even diagnose disorders in certain body systems or specific tissues. The element iodine, for example, is necessary for normal functioning of the thyroid gland of many animals. The thyroid gland produces certain hormones that affect metabolism, growth, and development. By introducing a radioactive isotope of iodine into the body, the utilization of iodine by the thyroid gland during hormone production can be monitored. In another example, water containing a heavy isotope of oxygen has been used to follow the path of oxygen during photosynthesis in plants or algae. At one time, scientists did not know the origin of the oxygen released as oxygen gas (O_2) during photosynthesis. Experiments demonstrated that this oxygen initially entered the photosynthetic process attached to two hydrogen atoms as water. Radioactive isotopes of cobalt, radium, and other elements are used in radiation therapy to destroy cancerous tissues. Finally, radioactivity in rocks and in the fossil remains of organisms is used to determine the age of Earth and its ancient inhabitants.

We have learned that different numbers of neutrons in the nucleus of a given type of atom result in isotopes of that atom. The electrons that encircle the nucleus of an atom also play quite an important role. In a chemical reaction between atoms, it is only their electrons that are directly involved in the reaction. These negatively charged particles move rapidly around the nucleus and, by virtue of their position relative to the atomic nucleus, have different amounts or **levels of energy.** Electrons nearest the atomic nucleus have the least energy and are said to be at a low energy level. Electrons found farther from the nucleus have more energy and thus are at a higher energy level. Helium, for example, with an atomic number of two must also have two electrons since the number of electrons always equals the number of protons in a given neutral atom. Both of these electrons are found in the lowest energy level surrounding the nucleus. In fact, two is the maximum number of electrons that can occupy this energy level. Carbon, with an atomic number of six, has two electrons in the lowest, or first, energy level and the remaining four occupy the second energy level. An electron can move to a higher energy level through an input of energy from an outside source; if an electron *loses* sufficient energy, it will move to a lower energy level.

How do these energy levels relate to chemical reactions and, further, to the phenomena of life? In the cells of green plants, for instance, electrons are raised to higher energy levels by an input of light energy from the sun. These electrons become a source of **potential,** or **stored energy** for the plant cells. When the electrons return to lower energy levels, they release this stored energy to fuel cellular processes. To illustrate this principle, visualize a boulder lying at the foot of a hill (Fig. 4.6). An input of energy, such as that provided by your muscles, can boost the boulder up the hill. The energy expended to move the boulder uphill is transferred to the boulder as stored or potential energy. As the boulder moves up the hill to higher and higher energy levels, it gains increasing amounts of potential energy. At the top of the hill, the boulder has stored more potential energy than it had at any of the lower levels. If the energy input or

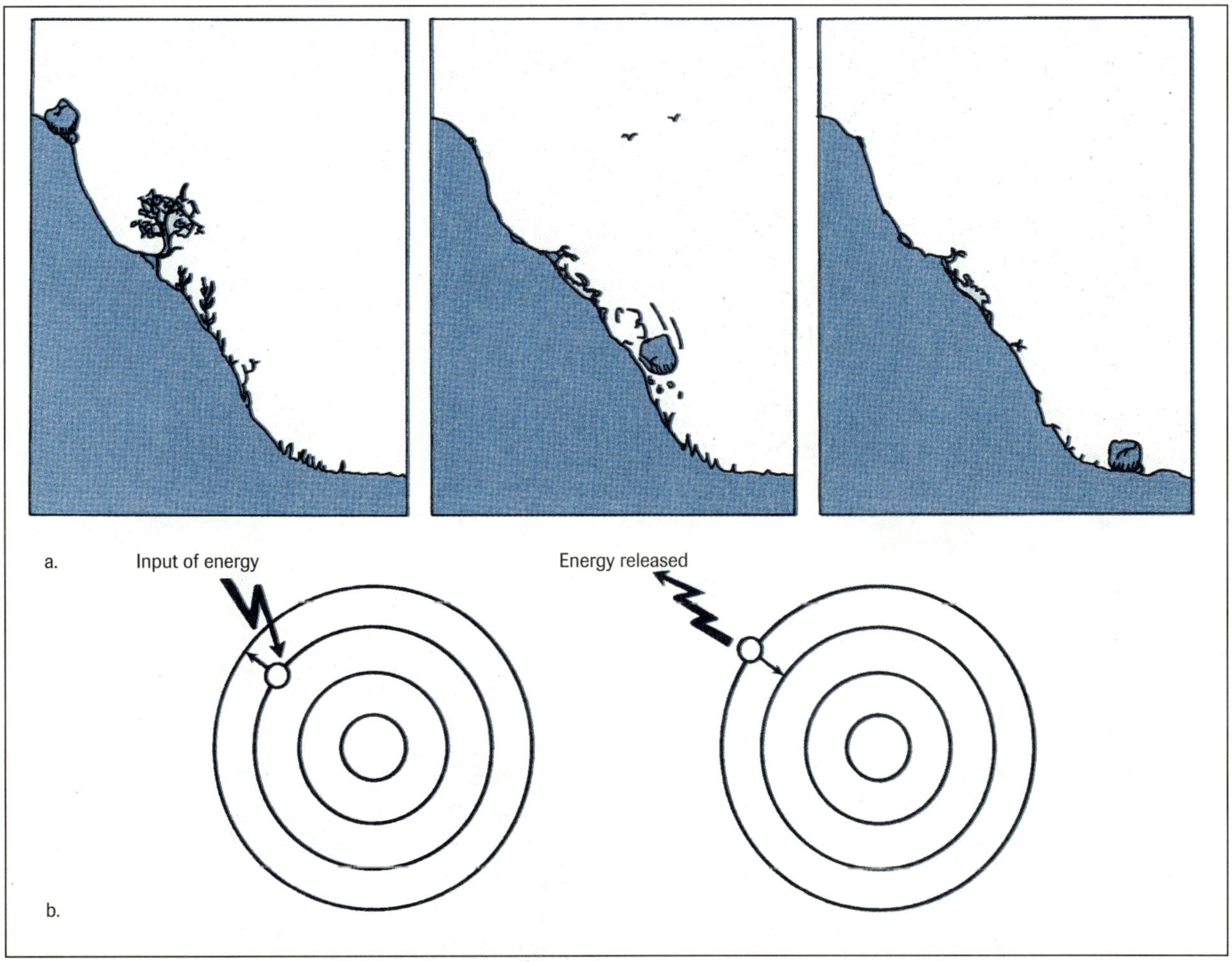

Figure 4.6 A boulder high on the side of a hill has high potential energy. As the boulder rolls down the hill, its potential energy is converted into kinetic energy. At the bottom of the hill, the potential energy of the boulder is lower than when it is at the top of the hill. Similarly, an electron raised to a higher energy level gains potential energy. As it returns to a lower energy level, the electron releases its potential energy in the form of kinetic energy.

energy source is removed, the boulder will roll back, releasing its potential energy as it returns to lower levels of the hill. As the boulder rolls from the top of the hill to the bottom, it releases more potential energy than if it began rolling only a short distance from the bottom. This could be verified to one's satisfaction by standing in its path. This potential energy stored in the boulder is converted into kinetic energy as the boulder begins to roll down the hill. **Kinetic energy** is the energy of motion—the energy that actually does work. Clearly, more potential energy would be released if the boulder began rolling from the top of the hill than if it started rolling from some lower point. This means that the boulder could perform more work because more potential energy would have been released and converted into kinetic energy. Electrons in the atoms of living organisms behave much like the boulder. Provided there is a sufficient input of energy from an outside source, electrons gain potential energy as they are raised to higher energy levels, and release this energy as they return to lower levels. Notice that the energy involved has been converted from one form of energy into another. In our example of green plant cells, light energy is converted into electrical energy; that is, the energy of moving electrons.

We have discovered that energy transactions, one of our features shared by all living organisms, occur even at the subatomic level of life's organization. The arrangement of an atom's electrons in different energy levels not only allows us to visualize their different amounts of energy, but it is also crucial in understanding the attraction, or chemical

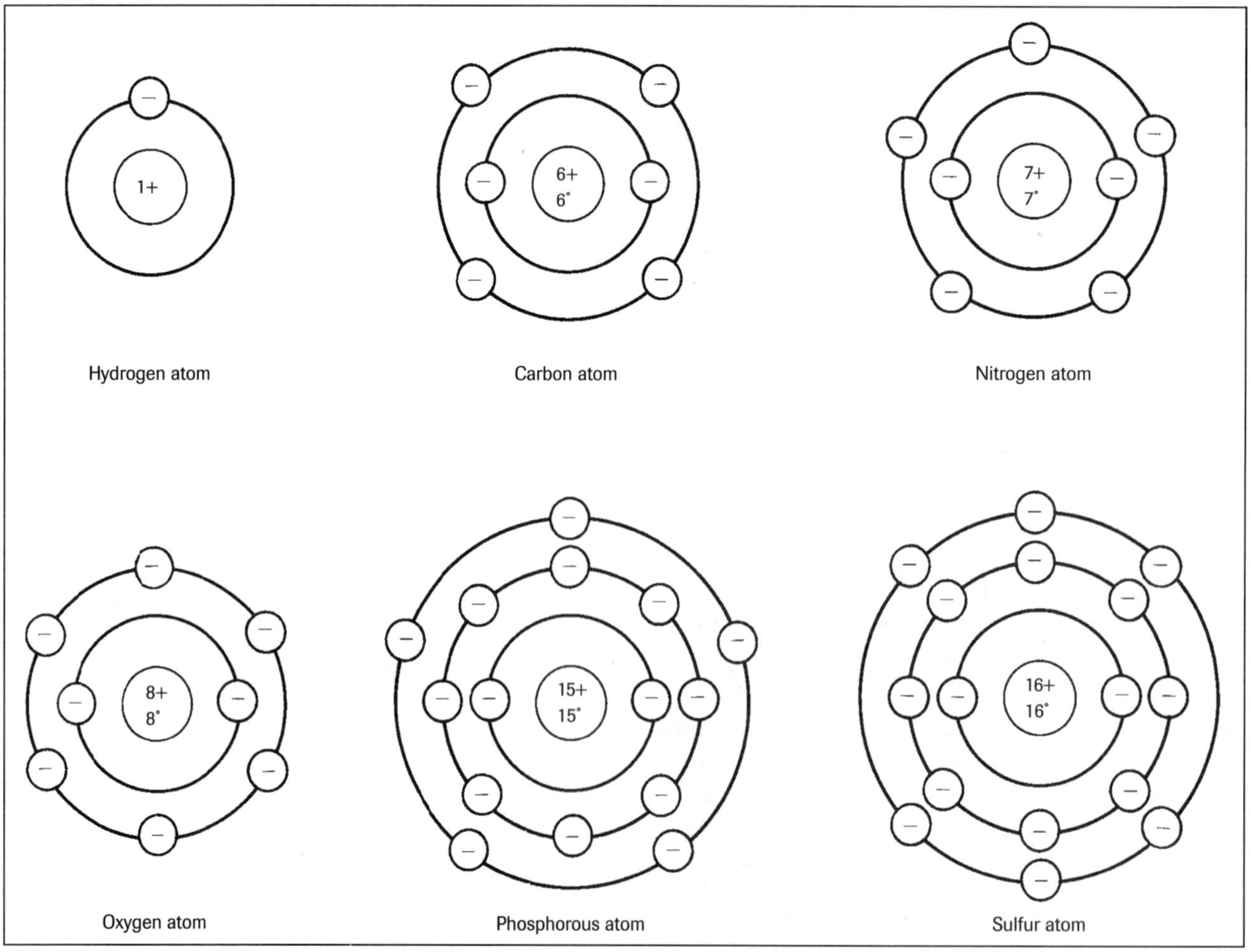

Figure 4.7 Diagrams of the atoms of the six major elements of life. Symbols: + = proton; O = neutrons; ϴ = electron.

bonding, that occurs between atoms during a chemical reaction. More specifically, the electrons in the *outermost* energy level, the level farthest from the nucleus, are involved in such interactions. These electrons, known as *valence* electrons, determine the chemical properties of an atom and are the electrons involved in chemical bonding (Fig 4.7). **Valence** is the bonding or combining capacity of an atom. Atoms with different numbers of valence electrons have different chemical properties, and atoms with the same number of valence electrons have similar chemical properties.

We have said that the first energy level surrounding any atomic nucleus holds a maximum of two electrons. The second energy level can hold eight electrons, and the third energy level holds eight as well. (For elements with an atomic number greater than 20, the third energy level can hold another ten electrons.) Atoms with a completely filled outermost energy level are more stable, or less reactive, than atoms with this energy level only partially filled. Helium, for instance, with only two electrons that completely fill its first—and, in this case, outermost—energy level does not tend to react with other elements. It has a very stable electron configuration. Helium is one of five elements known as rare gases, which are found in very small amounts in Earth's atmosphere and all exhibit this stable property. These five gaseous elements—helium, neon, argon, krypton, and xenon—tend to be less chemically reactive than all the other elements (Fig. 4.8). The outermost energy level—the highest occupied energy level—of each of these five gases, as you might expect from their unreactivity, is filled. Neon has ten electrons: two filling the first energy level and eight filling the second—its outermost—energy level. Argon, with eighteen electrons, has two electrons filling the first level, eight filling the second, and a full third energy level holding eight electrons. In contrast, the outermost energy levels of all other known atoms have fewer than eight electrons. Hydrogen has only one electron, located in its outermost (the first and only) energy level, which has the capacity to hold two electrons. Carbon has two electrons filling its first level and only four partially filling its outermost, second energy level. Atoms with partially filled outer energy levels tend to react with other atoms, trying to fill these incomplete outermost levels. They accomplish this either by transferring valence electrons between atoms or by sharing them. Herein lies the nature of chemical bonding. These interactions between atoms create an attractive force, or **chemical bond,** between the atoms. Any substance in which two or more different atoms are bonded together in a fixed proportion is called a **compound.** Let's look at some simple examples to see how this bonding occurs.

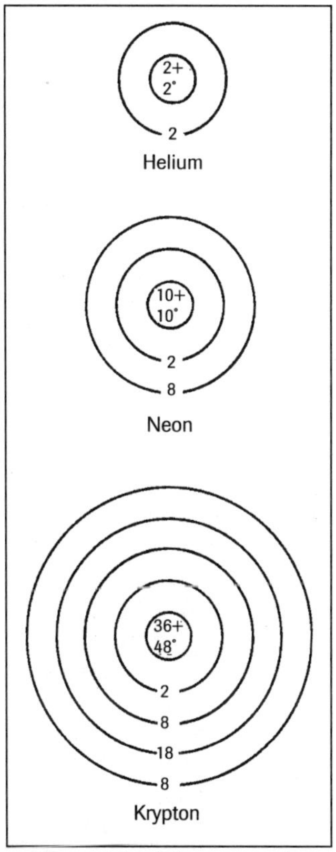

Figure 4.8 Diagrams of atoms of three rare gases. These atoms do not tend to react with other atoms because their outermost electron shells are filled to capacity.

Ionic Bonds

Sodium chloride, ordinary table salt, is designated by the symbol NaCl, which means it consists of one atom of the element sodium and one atom of the element chlorine held together by an attractive force. This force results from the transfer of electrons from one atom to the other. Any compound has properties different from those of the elements that comprise it. If NaCl dissociated into its constituent elements, its properties, such as its distinctive taste and grainy texture, would be lost.

The element sodium (Na) has an atomic number of 11, meaning one atom has 11 protons and therefore 11 electrons. The first energy level holds two electrons and the second energy level holds eight. In the case of sodium, then, there must be a single electron in the third, or outermost, energy level. An atom of chlorine (C1), with an atomic number of 17, would have two electrons filling the first energy level, eight filling the second, and seven in the outermost energy level. Sodium, therefore, has one valence electron and chlorine has seven. Chlorine, with seven valence electrons, has a greater relative power to attract electrons than does sodium. The relative ability, or power, of an atom to attract electrons to itself is called its **electronegativity.**

Certain atoms are more electronegative than others. In some reactions, the difference in electronegativity between the reacting atoms is so unequal that the more electronegative atom actually can attract and accept valence electrons from the outermost energy level of the other atom (Fig. 4.9). Such is the case in the reaction between the sodium and chlorine atoms. The more electronegative chlorine atom attracts the sodium valence electron, which is transferred to the outermost energy level of the chlorine atom. With this transfer, the outermost energy level of the chlorine atom now complete with eight electrons, and the outermost energy level (now *the second* level) of the sodium atom also is complete with eight electrons. In addition, all the electrons in both atoms are at the lowest possible energy level and have a very stable arrangement. A general rule of electron transfer is that atoms with one, two, or three valence electrons tend to transfer them, while atoms with five, six, or seven valence electrons tend to gain valence electrons.

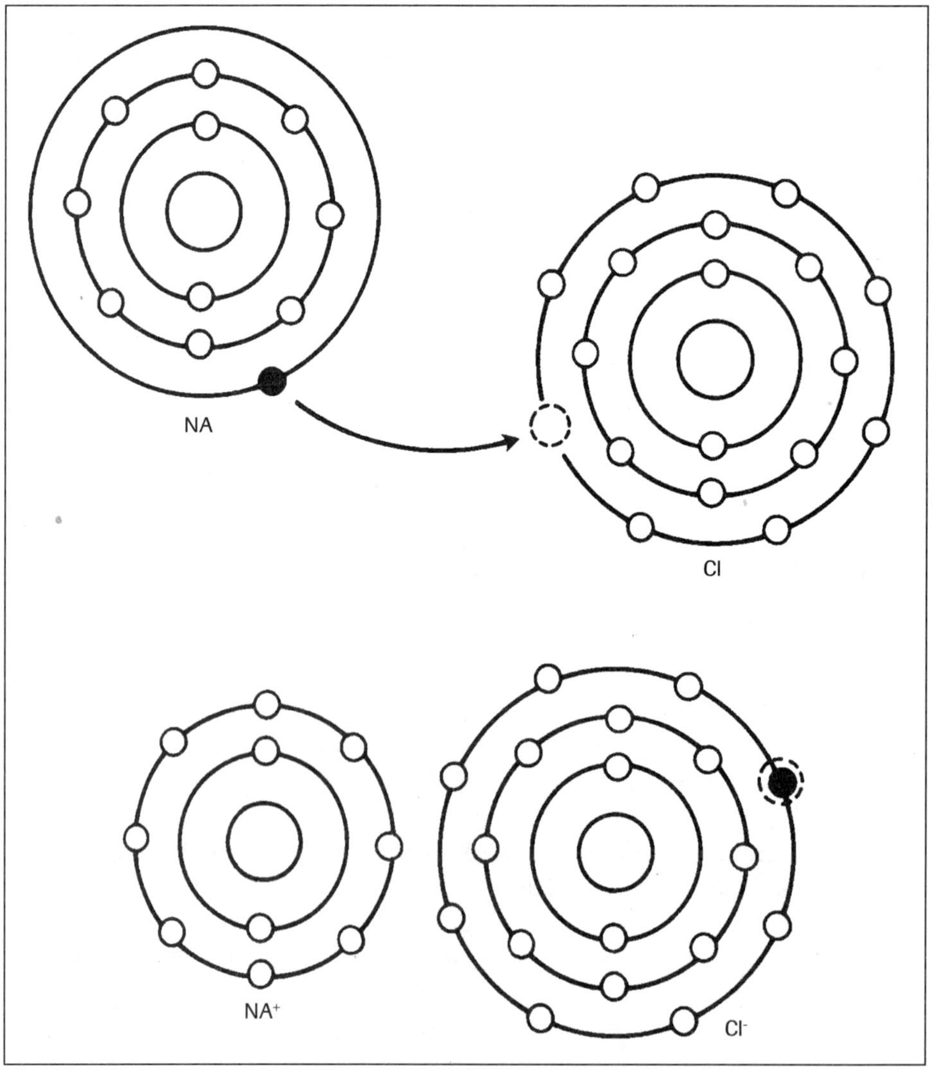

Figure 4.9 Reaction between sodium (Na) and chlorine (Cl) atoms to form the ionic compound sodium chloride (NaCl). In losing an electron (oxidation), the sodium atom becomes a positively charged ion; in gaining the electron (reduction), the chlorine atom becomes a negatively charged ion. Having opposite electrical charges, the two ions attract each other, forming an ionic bond.

As a result of this electron transfer, the sodium atom now has only 10 electrons, but still has 11 protons, which gives the atom a net electric charge of +1. By gaining an electron, the chlorine atom has 17 protons and 18 electrons, and acquires a net electric charge of −1. A charged atom or molecule is called an **ion.** The sodium ion is written Na+, indicating the +1 charge, and the chloride ion is written C1−, indicating the −1 charge. Sodium, then, has a valence, or combining power, of +1, and chlorine has a valence of −1. Electrically speaking, opposite charges tend to attract each other and like charges repel. Thus, the oppositely charged sodium and chloride ions attract each other, forming crystals of the salt sodium chloride (NaCl). The attractive force between ions of opposite charge is called an **ionic bond.**

Covalent Bonds

A second type of chemical bond can be formed from the *sharing* of valence electrons, rather than by electron transfer (Fig. 4.10). This is another way in which atoms fill their outermost energy level. An atom of hydrogen, for example, has one valence electron in its first and only shell. When two hydrogen atoms approach each other, each atom will share the valence electron of the other, forming hydrogen gas (H_2). Both electrons will move around the nuclei of both hydrogen atoms. Each atom now has, effectively, two electrons filling its outermost energy level—a highly stable configuration. The attractive force between these two hydrogen atoms resulting from the shared pair of electrons is called a **covalent bond** or, more specifically, a *single* covalent bond.

If two pairs of valence electrons are shared between atoms, a *double* covalent bond is formed. Oxygen, with an atomic number of eight, has two electrons filling its first energy level and six electrons in its second level. The second energy level holds a maximum of eight electrons; thus two electrons are needed to com-

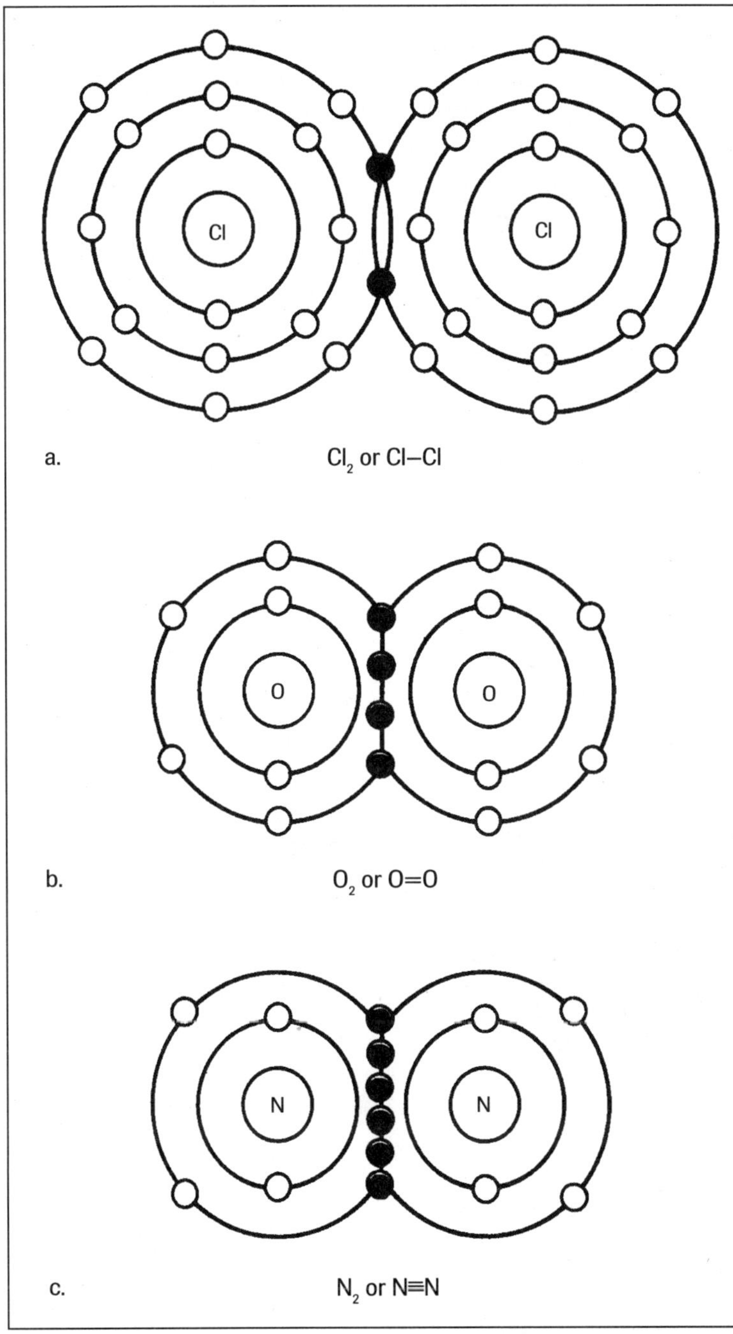

Figure 4.10 (a) By sharing one pair of electrons, the chlorine atoms are joined by a *single* covalent bond to form chlorine gas. (b) Oxygen gas is formed when the oxygen atoms share two pairs of electrons, forming a *double* covalent bond. (c) A *triple* covalent bond is formed when two nitrogen atoms share three pairs of electrons in forming nitrogen gas.

plete this level. Oxygen, then, is able to form two covalent bonds simultaneously. The covalent bonding of two oxygen atoms to form oxygen gas (O_2) results from each atom sharing two valence electrons with the other; that is, two pairs of electrons are shared.

Atoms also can share three pairs of valence electrons, as occurs in the formation of nitrogen gas (N_2). The atomic number of nitrogen is seven, indicating that it requires three electrons to complete its valence energy level and three covalent bonds can be formed between the two nitrogen atoms. The sharing of three pairs of valence electrons forms a *triple* covalent bond. Covalent bonds are relatively strong; that is, they are difficult to break, and are the major bonds that unite C, H, O, and N atoms in the compounds most commonly found in living cells.

The smallest unit of a compound that retains the properties of that compound is a **molecule** (Fig. 4.11). A molecule is composed of two or more atoms chemically bonded together and also is the smallest part of a compound that can have a stable existence. If we wish to designate one molecule of water, for example, we simply write H_2O; two molecules are designated as 2 H_2O; three molecules as 3 H_2O, and so on.

Hydrogen Bonding

While ionic bonds involve the transfer of electrons from one atom to another, and covalent bonds result from two atoms sharing one or more electrons, there is yet a third type of attractive force that occurs between atoms. This is called **hydrogen bonding** and does

Figure 4.11 One molecule of a chemical compound is formed by combining two or more *different* kinds of atoms.

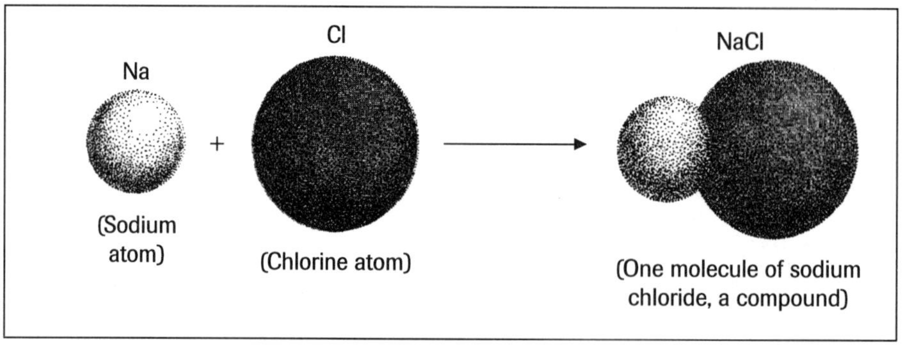

(a) and (b) Molecules also are formed by combining two or more atoms of the *same* element.

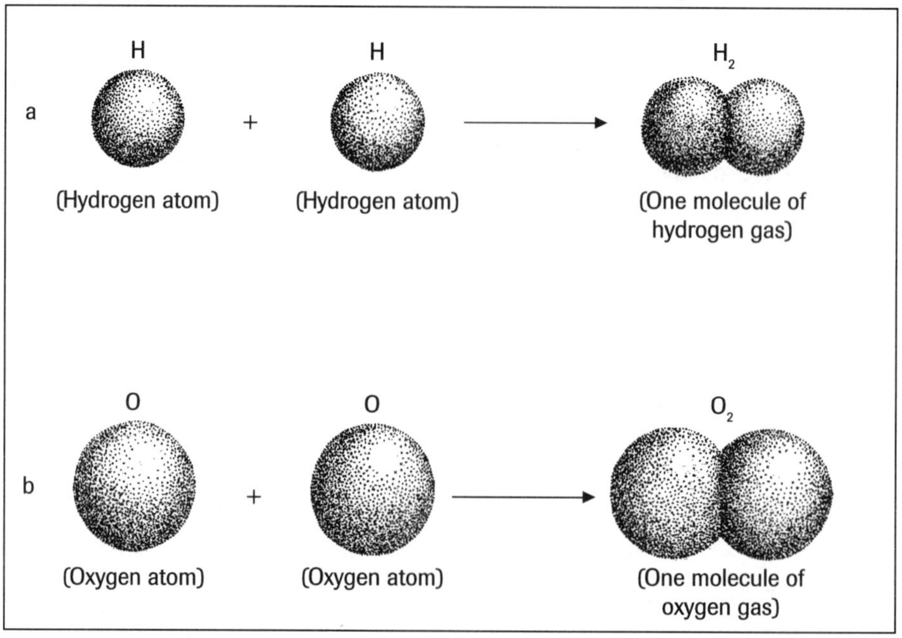

not actually join atoms together, but rather is a weak attraction between atoms (Fig. 4.12). A hydrogen bond can form between an atom of one molecule and a hydrogen atom that is already covalently linked to another atom. The hydrogen atom must be covalently bonded to an atom that is very electronegative; that is, the atom must have a strong attraction for electrons. In living matter, the hydrogen atom involved in hydrogen bonding is most often joined to either oxygen or nitrogen, both of which are electronegative. Water molecules, for example, exhibit hydrogen bonding in that the electronegative oxygen atom of one water molecule attracts the hydrogen atom of a second water molecule. Hydrogen bonding also occurs between atoms in the same molecule, such as a protein or DNA, and is responsible for maintaining the shape of the molecule.

Hydrogen, with just one electron and one proton, is the simplest element known and also is the most common element in the universe, accounting for some 90 percent of all matter. As you examine the chemical structures in this text, notice that hydrogen is covalently bonded to carbon, oxygen, and nitrogen in all the compounds of life. A variety of diverse compounds can serve as excellent energy fuels for organisms largely because they contain an abundance of high-energy hydrogen electrons, a concept we will pursue in Chapter 7.

Chemical Bonds and Energy

Now that you are familiar with chemical bonding, we can state, essentially, that a *chemical reaction* is the making and breaking of chemical bonds between atoms or mol-

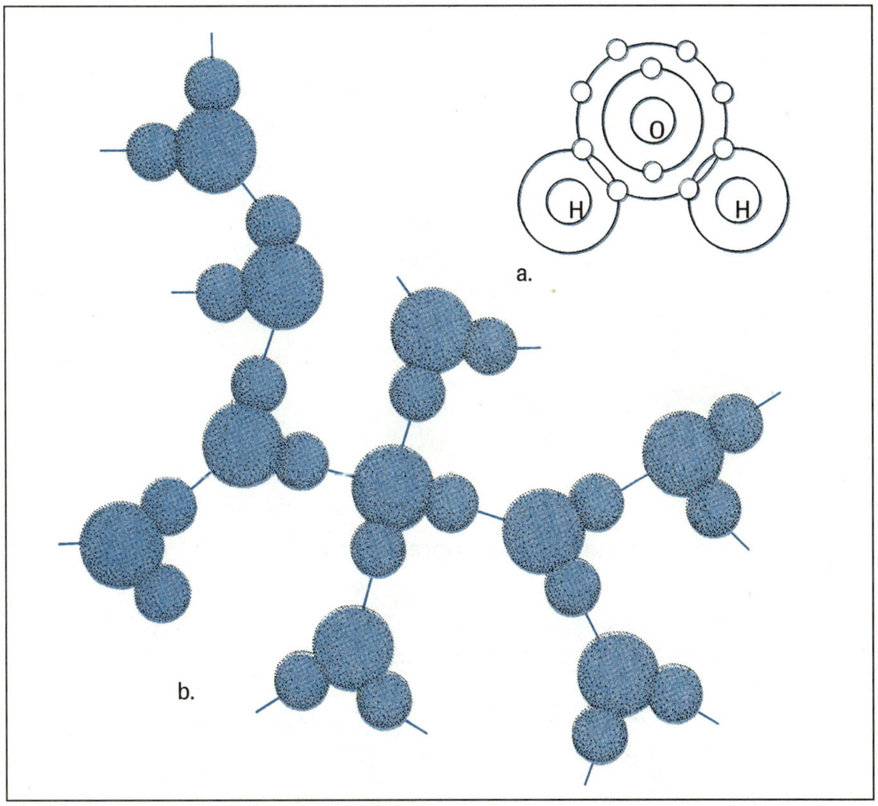

Figure 4.12 (a) Covalent bonding between one atom of oxygen and two atoms of hydrogen to form a molecule of water (H_2O). The large oxygen nucleus attracts the hydrogen electrons so that the oxygen atom acquires a weak negative charge. This leaves the hydrogen atoms with a weak positive charge. (b) Hydrogen bonds between water molecules in liquid water. Each positively charged hydrogen atom is attracted to a negatively charged oxygen atom in the area.

ecules. Chemical reactions involve electron exchange between atoms. One of the practical applications of your knowledge of chemical reactions involves their interrelationship with energy. We have seen that energy transactions are fundamental life processes that occur constantly. At this juncture you will begin to use this knowledge to build a clearer understanding of how life processes occur at the cellular level. The organic compounds—carbohydrates, lipids, and proteins—that you consume contain stored chemical energy by virtue of the chemical bonds that unite the atoms in each compound. When these bonds are broken, this stored, or potential, energy is released. You will recognize this as a *catabolic reaction*—the breaking of chemical bonds. The released energy, the energy that will actually do cellular work, is kinetic energy.

On the other hand, the making, or synthesis, of organic compounds requires an input of energy and involves formation of chemical bonds between atoms or molecules. This as an example of an *anabolic reaction*—the making of chemical bonds. The energy released from catabolic reactions can be harnessed by living cells and used to drive the anabolic reactions necessary for life. This is the essence of cellular metabolism.

We have discovered that energy can be stored and can be changed from one form into another. In every energy transaction, every chemical reaction, some portion of that energy is converted to heat. The measurement or unit of heat energy is the **calorie,** defined as the amount of heat required to raise the temperature of one gram of water 1° C, or from 14.5° C to 15.5° C. A calorie is a very small unit, usually too small for biological discussions. Instead, we will more often use the *kilocalorie,* which equals 1000 calories and is abbreviated *kcal.*

The anabolic and catabolic processes of metabolism require the participation of **enzymes.** Most enzymes are specialized protein molecules that act as catalysts in a given chemical reaction. A catalyst is a substance that speeds up the rate of a specific chemical reaction, but is not destroyed by that reaction. An enzyme does not cause a reaction to occur, it simply allows the reaction to proceed at an effective rate with a minimum expenditure of energy. Photosynthesis, cellular metabolism, digestion, muscular contraction, and virtually all other life processes depend on the regulating effects of enzymes.

Carbon: The Structural Element of Life

About three million compounds, which represent more than 90 percent of all known substances, involve the element carbon. Again, we have found unity in diversity, unity in the common participation of carbon in so many compounds, yet differences in its arrangement and combinations with other atoms that allow for the observed diversity of compounds. We indicated in Chapter Two that linked carbon atoms form the backbone of all organic compounds. There also are many inorganic compounds, such as carbon dioxide (CO_2), that contain carbon. Organisms are primarily composites of carbohydrates, lipids, proteins, nucleic acids, and water—all of which, excluding water, are constructed on a frame of carbon atoms. Carbon is a relatively light atom with six protons, six neutrons, and six electrons. With four valence electrons in its outermost energy level, carbon can form four covalent bonds simultaneously and does not tend to gain or lose electrons. Instead, the carbon atom completes its outermost energy level by sharing its valence electrons with a greater number of different atoms than any other element. In addition, a carbon atom can bond with another carbon atom to form the chain and ring structures around which organic molecules are often built (Fig. 4.13). No other earthly element has all these properties. Except for oxygen, silicon is the most abundant element by percentage (about 26 percent) in the crust and atmosphere of Earth. It is interesting that carbon, in contrast, constitutes less than 1 percent of the biosphere. Nonetheless, it is the uniqueness of carbon—with its four bonds, self- pairing ability, and low atomic number—that accounts for the basic structure of the four major organic compounds of life: *carbohydrates, lipids, proteins,* and *nucleic acids.*

The Major Compounds of Life

CARBOHYDRATES

Carbohydrates are used by cells as a source of energy, for food storage, and may even play a structural role. All carbohydrates contain the elements carbon, hydrogen, and oxygen in a certain ratio, generally some multiple of CH_2O. They are found in both animals and plants; in fact, these compounds account for about 95 percent of the dry weight (after removal of water) of most plants. Our interest in carbohydrates focuses on a few major types of sugars and related compounds. The simplest carbohydrate is a **monosaccharide,** or a single sugar molecule (Fig. 4.14).

Glucose is an important example of a monosaccharide. This six-carbon sugar, found especially in fruit, serves as the primary energy source for many organisms. The carbon, hydrogen, and oxygen atoms of glucose and other carbohydrates are united by covalent bonds. The enzymatic breakdown of these bonds during cellular metabolism releases energy that is ultimately stored in ATP. This release and storage of energy is one of the most important instances of the life feature of energy transactions.

The monosaccharide *fructose* is a six-carbon sugar found in honey and a variety of sweet fruits. Fructose, which is structurally quite similar to glucose, is also a ready source of energy for cellular metabolism. Another six-carbon monosaccharide is galactose, which is found in nervous tissue, sugar beets, seaweeds, and

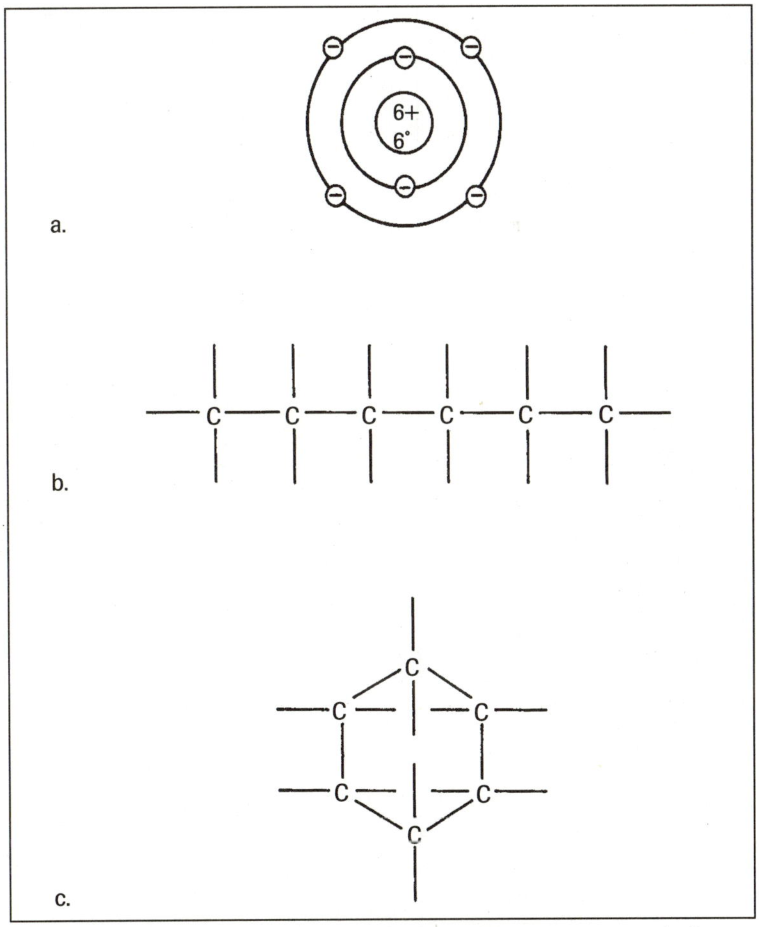

Figure 4.13 (a) Diagram of a carbon atom. With its four outer shell electrons, the carbon atom forms four covalent bonds simultaneously. (b) Carbon atoms are capable of bonding with each other to form chains or (c) rings—the structural "backbones" upon which biological compounds are built.

Figure 4.14A Three of the more common 6-carbon simple sugars (monosaccharides). Although each sugar is composed of the same number of carbon, hydrogen and oxygen atoms ($C_6H_{12}O_6$), their differences result from the manner in which these elements are organized. These sugars may exist in the form of straight chains or as ring compounds. Carbon atoms are located in each corner of the ring structures.

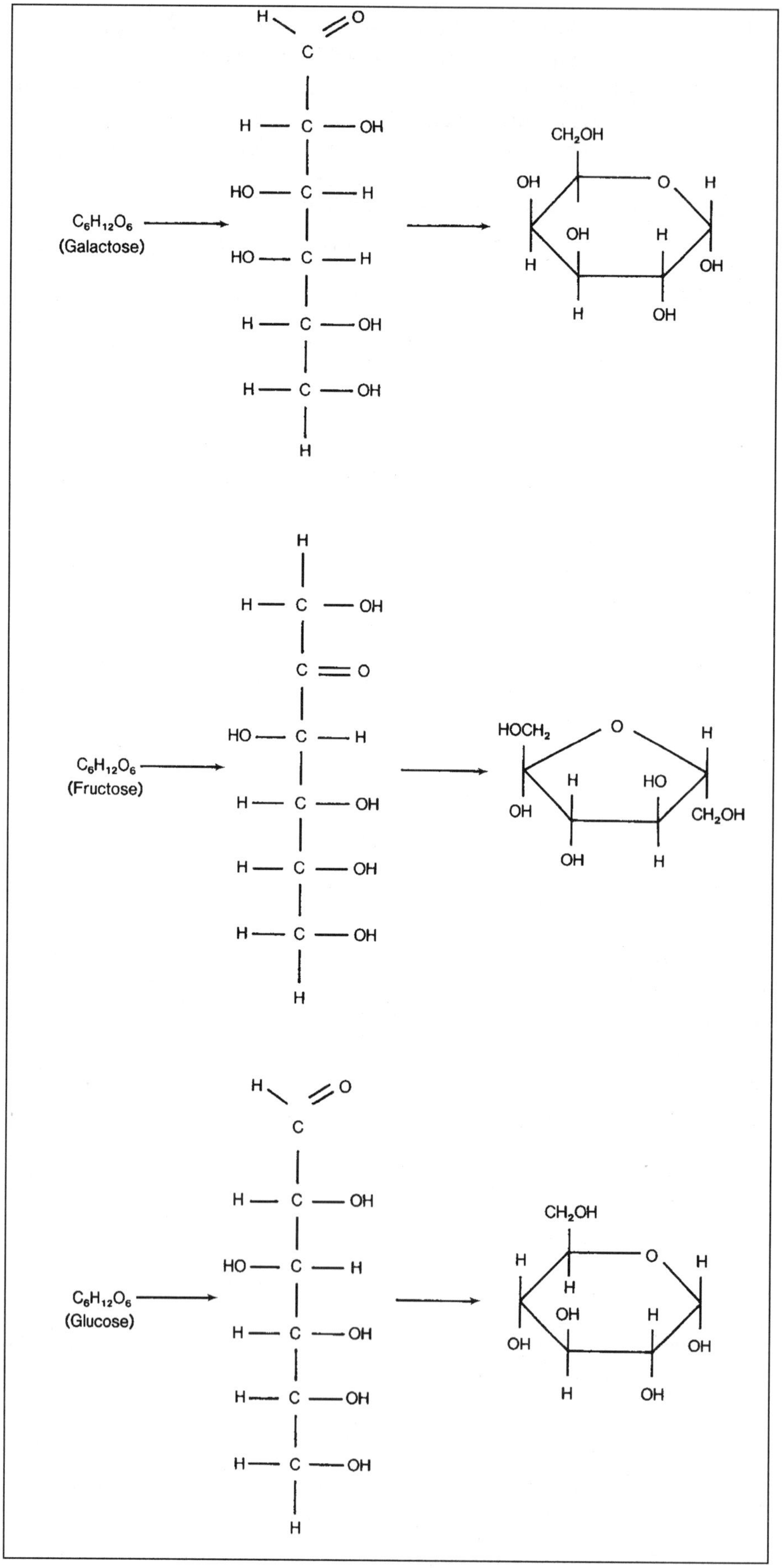

milk. All three of these simple sugars have the same molecular formula—$C_6H_{12}O_6$. But they are physically and chemically distinct because of the arrangement of the atoms in each molecule. Two five-carbon mono-saccharides of major significance are *ribose* and *deoxyribose*. These sugars are components of ribonucleic acid (RNA) and deoxyribonucleic acid (DNA), respectively; ribose also is a constituent of the high-energy molecule adenosine triphosphate (ATP).

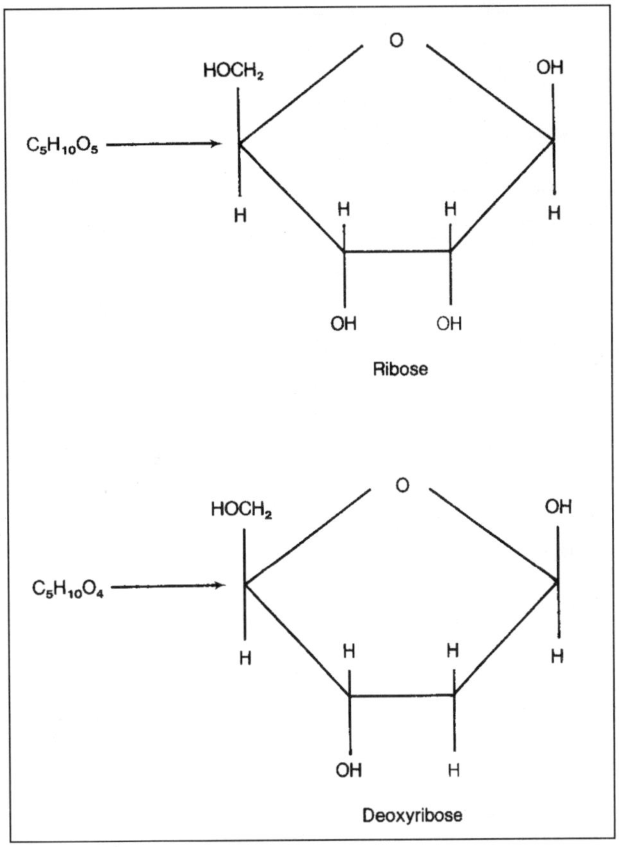

Figure 4.14B The two 5-carbon monosaccharides found in nucleic acids. Note that the two sugars are structurally identical except for one oxygen atom not present in deoxyribose.

The monosaccharides can bond together covalently to form a double sugar unit, or **disaccharide** (Fig. 4.15). A major disaccharide is *sucrose*—ordinary table sugar—which is formed from the covalent bonding of glucose and fructose. Sucrose is the form in which carbohydrates usually are transported in many plants. Another common disaccharide is *maltose,* or malt sugar, which is formed from the covalent bonding of two glucose molecules. In the digestive process of many animals, maltose is a product of the enzyme-catalyzed breakdown of starch—a food storage carbohydrate. Maltose can then be further degraded by another enzyme into its constituent glucose molecules, which is an energy-releas-ing process. A disaccharide derived from milk is *lactose,* which consists of glucose and galactose. Many people cannot consume milk or milk products because they lack the enzyme required to break down lactose into its component monosaccharides. Lactose thus tends to accumulate in the body.

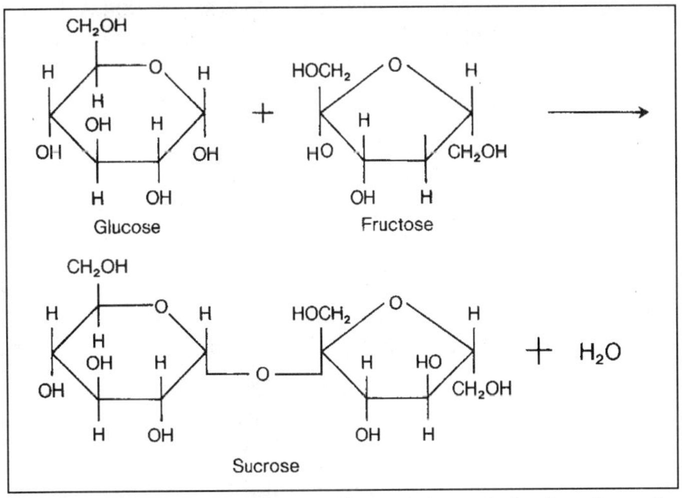

Figure 4.15 Disaccharides are formed through the bonding of two monosaccharides. In this illustration, glucose and fructose bond to form the disaccharide sucrose, ordinary table sugar. Note that a molecule of water is released in the reaction.

Long, linked chains of monosaccharides bonded covalently form **polysaccharides,** several of which are biologically important (Fig. 4.16). *Starch* is a common polysaccharide found in bread, crackers, cereals and potatoes, to list just a few examples. Photosynthesizing plants store their excess glucose in the form of starch. When needed, this starch can be broken down into glucose with a concurrent release of energy, which then can be used to fuel reactions that require an input of energy.

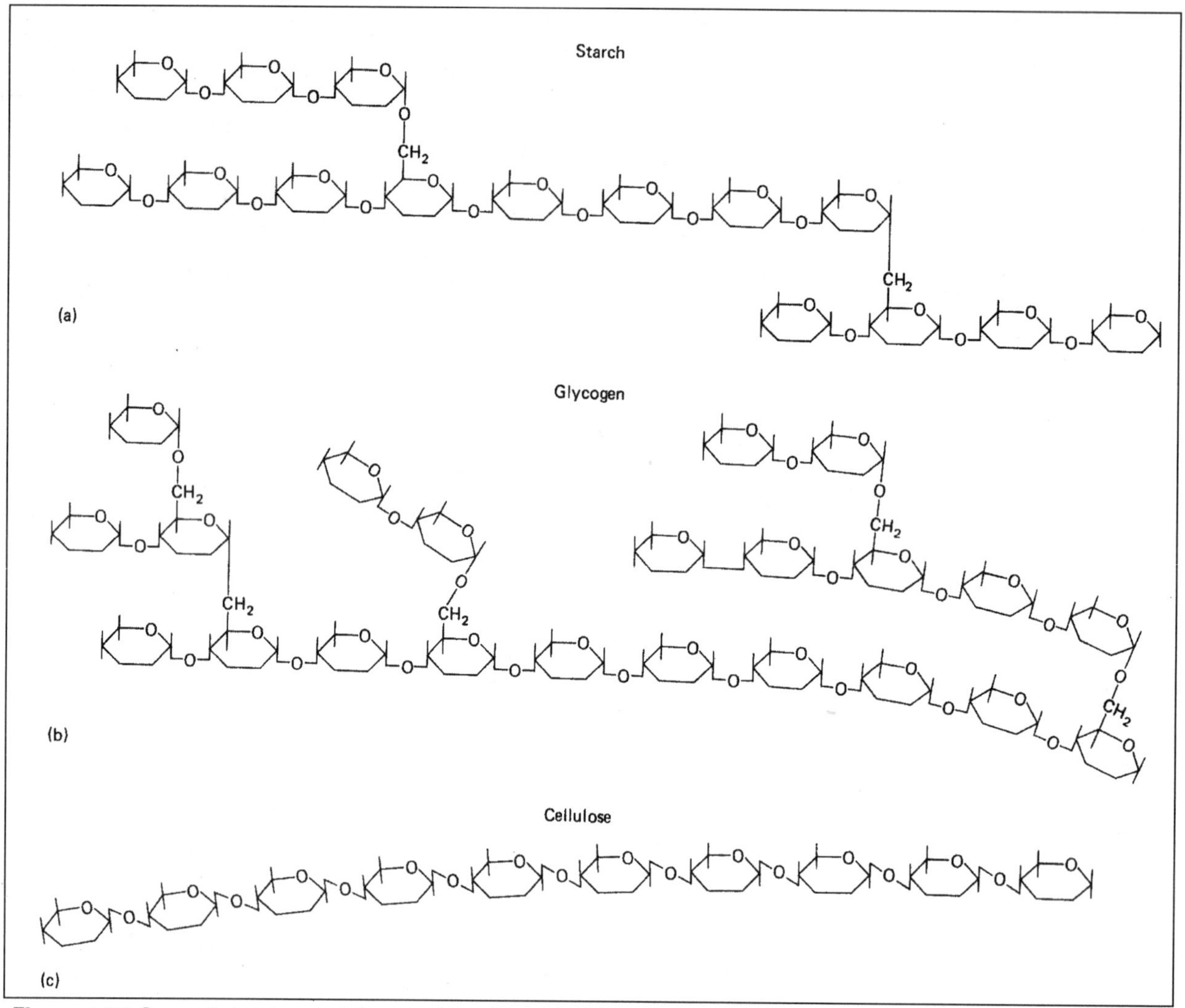

Figure 4.16 General structures of three major polysaccharides. All three are composed of glucose molecules, but the organization of these molecules in each polysaccharide is different. Cellulose is a straight-chain polysaccharide, but starch has branching side chains, and glycogen branches more frequenlty than starch.

Humans and various other animals store sugar that is not immediately needed for energy in the form of the polysaccharide **glycogen**. The primary storage sites for glycogen are the liver and skeletal muscles. If the level of glucose in the blood begins to fall, the body normally can replace the glucose from these storage sites. When the liver and muscles become saturated with glycogen, the excess glucose is converted into fat and stored in the body as adipose (fat) tissue. Plant cells and some protists contain the polysaccharide *cellulose* that imparts a degree of flexible toughness and support to the cell walls. The cell walls of fungi and the external body coverings of insects and crustaceans (crayfish, crabs, lobsters, and shrimp) are impregnated with the polysaccharide *chitin*, a nitrogen-containing carbohydrate (Fig. 4.17). Chitin is a remarkable substance, being light, but very tough, flexible, and waterproof.

The structures of these polysaccharides again remind us of the principle of organization in living systems. Although all four polysaccharides mentioned are essentially composed of chains of glucose molecules, each is functionally distinct because of the arrangement and bonding of the glucose molecules in each compound.

Lipids

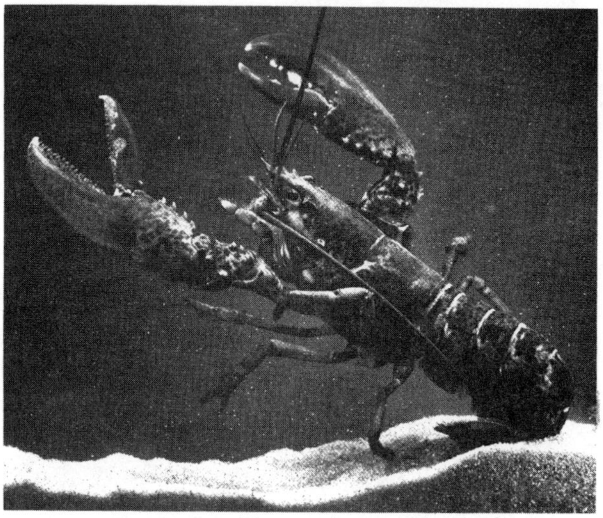

Figure 4.17 The tough exoskeleton of this spiny lobster is made of chitin.

Typically, lipids are fatty, oily, greasy, or waxy organic compounds, some of which function as reserve energy sources and some as structural components of living cells. The major constituents of lipids, as with the carbohydrates, are the elements carbon, hydrogen, and oxygen. Again, it is primarily the organization of these elements that provides the differences between these two groups of compounds. The general group known as lipids can be further divided into fats, oils, waxes, phospholipids, and steroids (Fig. 4.18).

Common *fats* include butter, lard, and cheese, which, by virtue of their carbon-hydrogen bonds, contain a large amount of potential energy. Fats are typically solid at room temperature. In addition to fats acquired in the diet, many animals produce fats from carbohydrates and a small amount from proteins. Fats may provide an adaptive advantage to some organisms by not only participating in energy storage, but also by forming many layers of adipose tissue that provide insulation against the cold temperatures of certain environments.

The *oils,* such as corn oil, peanut oil, and other vegetable oils, also store food energy and are typically liquid at room temperature. Fats and oils are constructed on a backbone of a three-carbon compound celled *glycerol,* bonded to three *fatty acid* molecules. **Fatty acids** are long chains of carbon atoms bonded at one end to a COOH, or acid group. Fatty acids with no double bonds in the carbon chain (C – C) are called *saturated* fatty acids. If there is one carbon-to-carbon double bond in the chain (C = C), the fatty acid is *unsaturated,* and if there are two or more of these double-bonded carbon atoms in its makeup, the fatty acid is said to be *polyunsaturated.* All of the common oils consist of unsaturated fatty acids, which contain one or more double bonds in the long fatty acid chains of each molecule.

These double bonds cause kinks in the long chains, which prevent close packing of the molecules and result in oils tending to be liquids at room temperature. Fats, though, usually contain saturated fatty acid chains that pack closely together and thus fats tend to be solid at room temperature.

Some plants store and secrete lipids in the form of oils, and many animals secrete oily substances that protect the skin and body covering. Certain skin cells of some waterfowl, for example, secrete an oily substance that coats the feathers, preventing them from sticking together. This allows the birds to take flight immediately, even after floating in water for hours.

Various plants form waxy secretions that waterproof and protect the surface of leaves and stems. These secretions, composed in part of long-chain fatty acids, are also important evolutionary adaptations. Surface waxes of leaves are quite important in reducing the evaporation of water from the epidermis. By preventing excess water loss, these waxes contribute to the growth and survival of many land plants.

The more complex lipids include the *phospholipids* and *steroids.* All the phospholipids, as the name indicates, contain the element phosphorus. Phospholipids also use glycerol as a backbone and contain two long, fatty acid chains. They are important structural components of cell membranes.

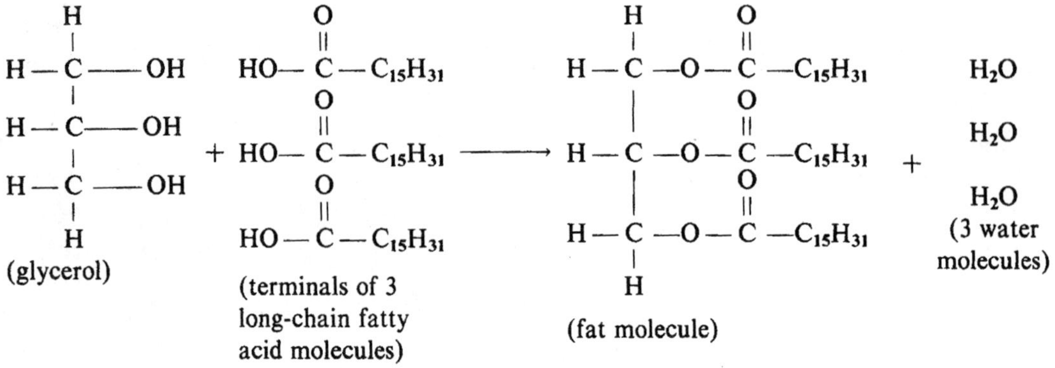

Figure 4.18A Fats are formed through the bonding of glycerol and three fatty acid molecules, with the release of water.

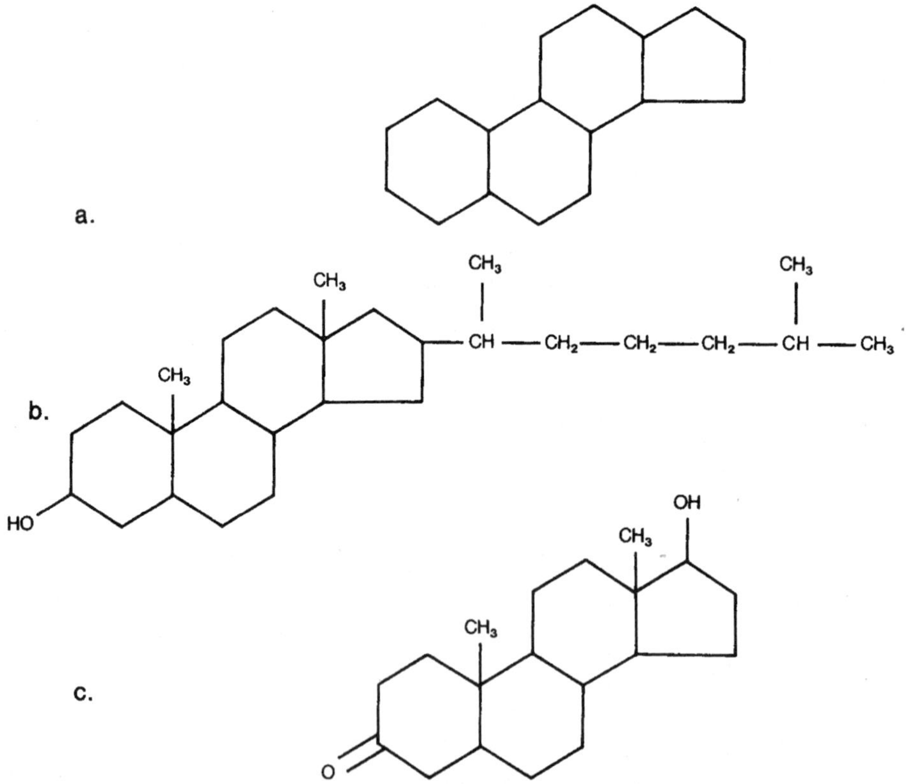

Figure 4.18B (a) The four interlocking rings of carbon atoms that form the basic structure of steroids. Steroids differ according to the particular side groups attached to the four carbon rings. (Carbon atoms are located in each corner of the hexagons and the pentagon.) (b) The steroid *cholesterol,* a component of cell membranes and nervous tissue. (c) The steroid *testosterone,* the sex hormone of human males.

Unlike any of the other lipids, the steroids are constructed of four interlocking rings of carbon atoms. One of the most important steroids is *cholesterol,* the parent compound from which most other steroids are synthesized. Cholesterol is found in cell membranes and nervous tissue, and is found in a converted form in bile secreted by the liver. Other steroids include vitamin D, sex hormones, and certain hormones secreted by the adrenal glands of many animals. The anabolic steroids, synthetic derivatives of the male sex hormone testosterone, are sometimes used to stimulate muscle building. Excessive use of the steroids, however, can cause health problems such as liver damage and heart disease.

PROTEINS

Much of the dry weight of animals is composed of the third group of organic compounds, the proteins (Fig. 4.19). Skin, muscle, hair, bone, teeth, cartilage, ligaments, and tendons of higher animals are all primarily proteins. All proteins contain the elements carbon, hydrogen, oxygen, and nitrogen. **Amino acids** are the building blocks of proteins, and bond together in a specific sequence to produce a specific protein. There are normally 20 amino acids found in living organisms. All of the 20 amino acids are composed of the same basic backbone structure: H_2N-CH-COOH. They differ, however, in the variety of chemical groups that are linked to the carbon atom at the center of this structure. The properties of these amino acids also differ based on the nature of this additional side group. For example, a side group might possess a negative or positive charge in certain

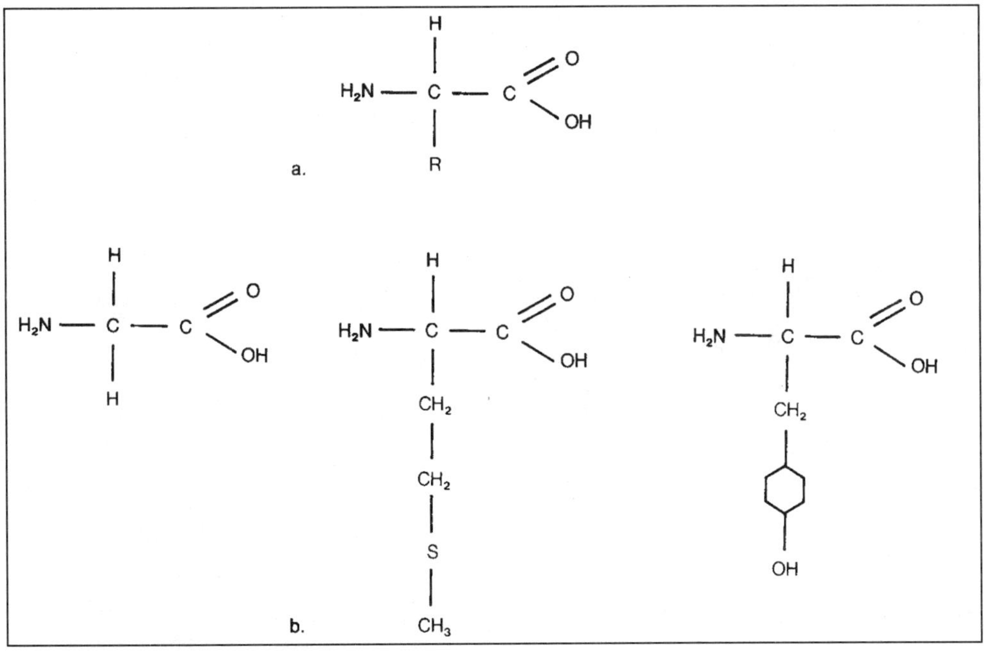

Figure 4.19 (a) the basic structure of an animal acid. Amino acids differ according to their side group, designated *R.*
The C (=O, —OH) group represents an acid and is called a *carboxyl group.*
The NH2 is called an *amino group.* (b) Examples of different amino acids, showing some of the variations in side groups.

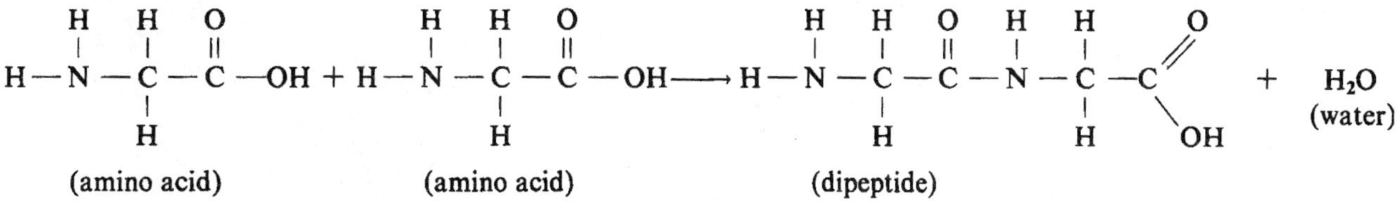

The covalent bonding of two amino acids forms a dipeptide, with the release of water.

solutions or be soluble or insoluble in water. These characteristics may be reflected in the different proteins that are constructed from these amino acids; for example, amino acids can differentially affect the shape of a protein or its solubility in water.

An average protein may consist of a sequence of fifty to fifty thousand linked amino acids. There are, however, several shorter and longer protein molecules. Animal proteins, such as those found in milk, eggs, fish, and meats, contain what are called *essential* amino acids; that is, the nine amino acids that must be obtained in the diet because the body cannot synthesize them. No single plant, however, contains all the essential amino acids in adequate amounts. The other eleven *nonessential* amino acids can be synthesized by the body in humans and many other animals.

The covalent bonding of two amino acids forms a *depiptide;* three bonded amino acids form a *tripeptide,* and a long chain of bonded amino acids forms a *polypeptide.* Critical to the proper functioning of a protein is its particular shape (Fig. 4.20). The specific sequence of the different amino acids in a protein is called its **primary structure.** In addition, the protein molecule usually is twisted into a spiral or pleated, which forms the **secondary structure,** and then folded over on itself to form the **tertiary structure.** A single protein that is comprised of two or more amino acid chains, such as the red blood pigment hemoglobin, for example, has a **quaternary structure.** The shape of a protein molecule is maintained by ionic, covalent, and hydrogen bonds that link the folds and preserve the turns and twists in the molecule. In fact, the shape of a protein molecule is essential for its proper functioning. A given kink or fold may actually render a given sequence of amino acids accessible to an essential interaction with another molecule.

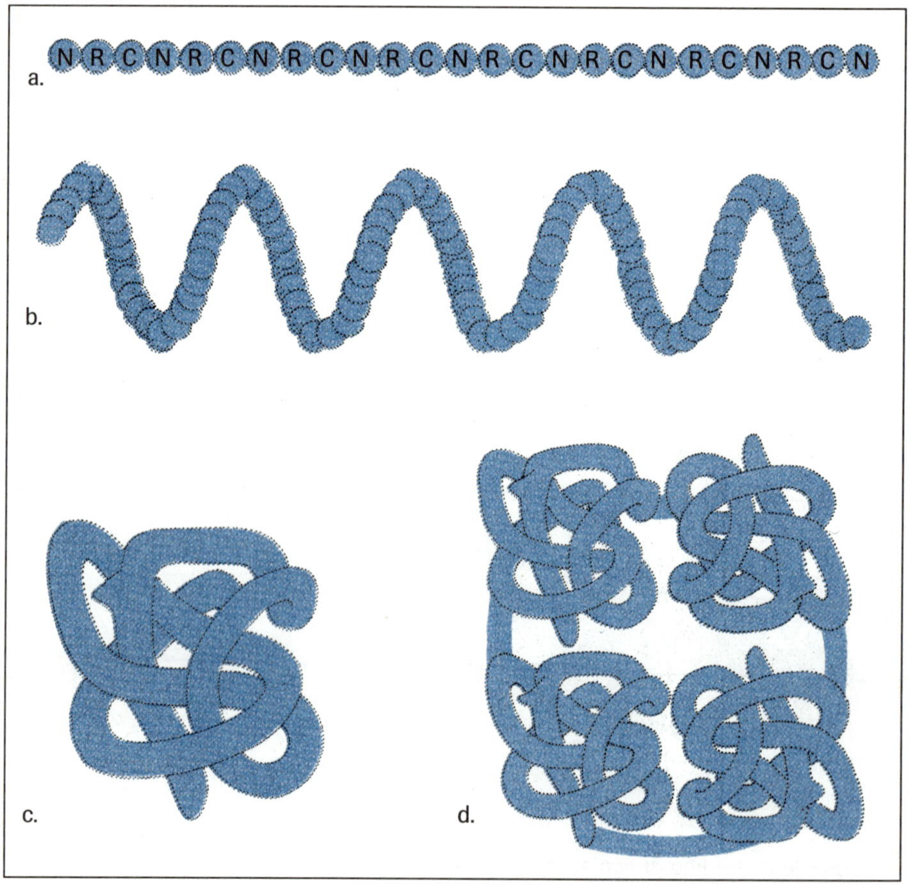

Figure 4.20 (a) The *primary* structure of a protein is determined by the sequence of amino acids in the chain. (c –n = peptide bond between amino acids.) (b) A common *secondary* structure of proteins is the formation of a helix. The turns in the helix are maintained by hydrogen bonds (not shown) between amino acids. (c) A folding of the helix results in the *tertiary* structure of a protein. The folds are united by chemical bonds between two sulfur atoms (not shown). (d) Some proteins, such as hemoglobin, mentioned in the text, are composed of several folded polypeptide chains; this association forms the *quaternary* structure of a protein.

If subjected to sufficiently high temperatures—usually from 50–55° C (122–131° F)—some of the chemical bonds separate and the protein molecule may unravel. The protein then is *denatured,* and once it loses its structure permanently, the protein can no longer function. This is one of the reasons very few organisms exist at temperatures above 55° C. No doubt you have seen the effects of high heat on protein structure in a fried egg. Frying the egg disrupts the chemical bonds of the egg protein, causing it to denature and coagulate. Other factors, such as chemicals, salt concentration, or a change in acidity or alkalinity can also denature a protein.

Earlier we commented on a special group of proteins—the *enzymes*—which are organic catalysts that function in regulating the rate of chemical reactions (Fig. 4.21). The substance on which an enzyme acts is called its **substrate,** and each enzyme has a specific substrate. This is because a given enzyme has a unique shape that limits its reaction to a specific substrate. Actually, an enzyme has a region called the *active site* which binds with the substrate. When the active site and the substrate bind, the enzyme is induced to alter its shape somewhat so that there is a close fit between the active site and the substrate. The substance formed from a reaction between an enzyme and its substrate is the **product(s).** For example, if an enzyme acts on starch to break it down into maltose molecules, starch is the substrate and maltose is the prod-

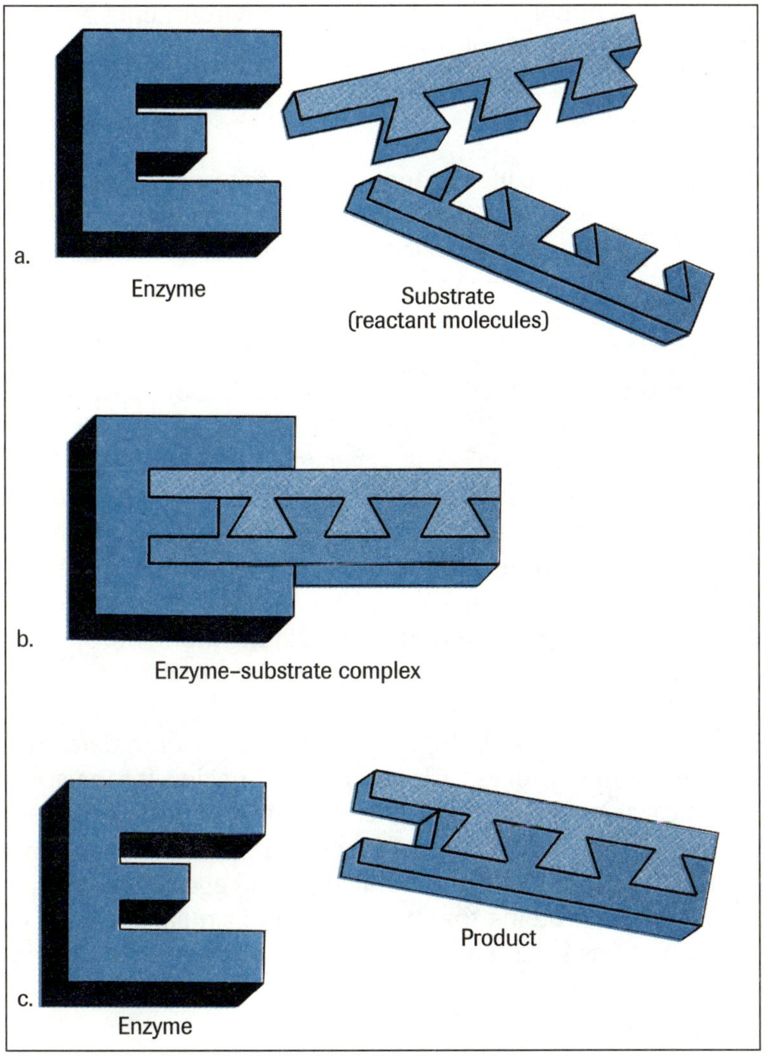

Figure 4.21 Generalized diagram of enzyme-substrate interaction. (a) The substrate, or reactant molecule, interacts with the enzyme molecule only at a specific active site on the enzyme to form an enzyme-substrate complex. (b) The result of the enzyme-substrate interaction is the formation of a product. (c) The enzyme is unaffected by the interaction and can be used again. In many reactions, the illustrated sequence is reversible, i.e, the enzyme catalyzes the *breakdown* of a substrate into separate molecules rather than *joining* them. In an actual biochemical reaction, the substrate induces changes in the shape of the enzyme molecule so that the substrate and enzyme fit together as efficiently as possible. When the product is released, the enzyme reverts to its original shape.

uct. There are also enzymes that catalyze the reverse reactions; that is, they catalyze the synthesis of starch from individual maltose molecules. Most metabolic reactions are reversible and a single enzyme can catalyze both the breakdown and the synthesis of the substrate. During a chemical reaction, an enzyme is not altered or destroyed and can be used again to catalyze another reaction.

Factors Affecting Enzymes. Three major factors influence the functioning of enzymes. First, there must be a sufficient amount, or *concentration,* of substrate molecules. If there are more enzyme molecules than substrate molecules, the reaction slows down. One way to speed up the reaction is to add more substrate. By so doing, more of the enzyme molecules become involved in the reaction. Once a sufficient substrate concentration is reached that involves all the enzyme molecules, the reaction will then proceed at its maximum rate. Beyond this point, however, adding more substrate will not further increase the rate of the reaction.

Second, the *temperature* of the cellular environment is crucial. At very low temperatures the rate of a chemical reaction is greatly slowed. If the temperature rises too high, the bonds holding the enzyme molecule intact are disrupted. As with other proteins, disruption of its bonds causes the enzyme molecule to unfold and lose its activity permanently. Enzymes affected in this manner are said to be *denatured.*

Third, every enzyme works most effectively at a certain pH. In your stomach, for example, some enzymes function at a pH well below 2; in your mouth, however, the enzyme in saliva is active at a pH of about 7. If either of these environments became too acidic or basic, the respective enzymes would cease to function.

Nucleic Acids

The fourth and final major group of organic life compounds are the nucleic acids (Fig. 4.22). They are constructed of chains of individual units called **nucleotides,** which consist of the elements carbon, hydrogen, oxygen, nitrogen, and phosphorus. Nucleic acids are constructed of chains of nucleotides, which are small individual units composed of a five-carbon ring structure called a sugar, a nitrogen-containing ring structure called a base, and a phosphate group (PO_4). Nucleotides are present in a number of important compounds, two of the most important of which are **deoxyribonucleic acid** (DNA) and **ribonucleic acid** (RNA). DNA is the major constituent of chromosomes, which carry the genetic information of a cell in functional units called genes. Encoded in the DNA of a cell are the instructions needed to synthesize all necessary proteins. DNA acquired part of its name from the constituent sugar of its nucleotides—the five-carbon monosaccharide *deoxyribose*. RNA has a structure somewhat similar to DNA, but differs in that it contains another five-carbon monosaccharide, *ribose*. RNA has the essential function of decoding the instructions contained in DNA and aiding in translating them into specific proteins. Thus, DNA and RNA function in information storage, information retrieval, and subsequent translation into cellular proteins.

Knowledge of the structure and function of DNA has proven vital in our understanding of life as revealed in the evolutionary process. Closely related species, for instance, have a greater proportion of similar nucleotide sequences in their DNA than do less closely related species. Evidence suggests that humans and chimpanzees may have diverged from a common ancestor as recently as 6–8 million years ago. Research has indicated that humans possess some proteins that are extremely similar to those of chimpanzees and gorillas. The proteins in humans and those of birds or amphibians, however, are markedly different.

Figure 4.22A Constituent nitrogen bases, sugars, and phosphate group of nucleic acids.

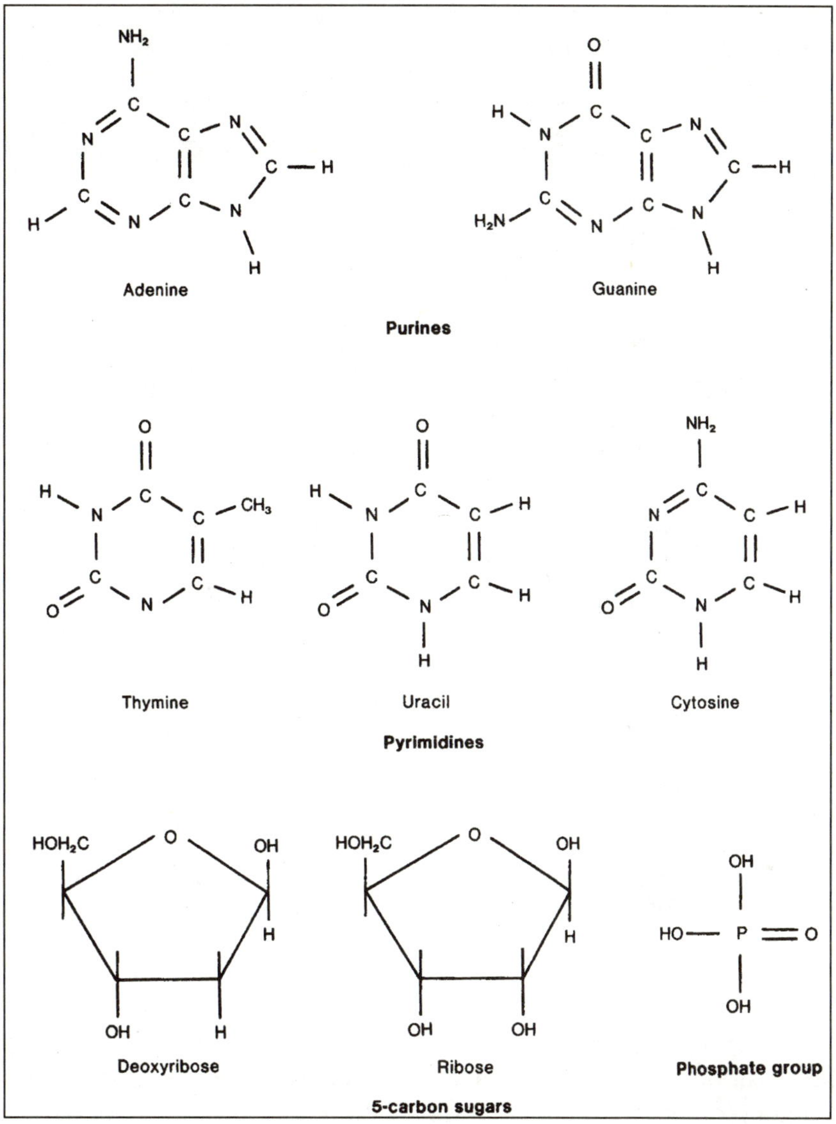

The Major Inorganic Compound of Life: Water

The major inorganic compound vital to life is water. On Earth, water and life are inseparable. Living cells are composed of 70–95 percent water and must be bathed continually in a watery solution to prevent dehydration. Water is a unique liquid, having life-supporting properties not possessed by any other liquid.

A water molecule (H_2O) consists of two hydrogen atoms covalently bonded to one atom of oxygen (Fig. 4.23). The oxygen atom has eight protons in its atomic nucleus, and each hydrogen atom has one proton in its atomic nucleus. Oxygen, therefore, has a stronger positive charge that tends to attract the electrons of hydrogen. These electrons spend more time around the oxygen nucleus than they do around the hydrogen nuclei. As a result of this electric tug-of- war, the oxygen atom acquires a rather weak negative charge, leaving the hydrogen atom with a weak positive charge. Due to this formation of two opposite electrical poles, water is referred to as a **polar molecule.**

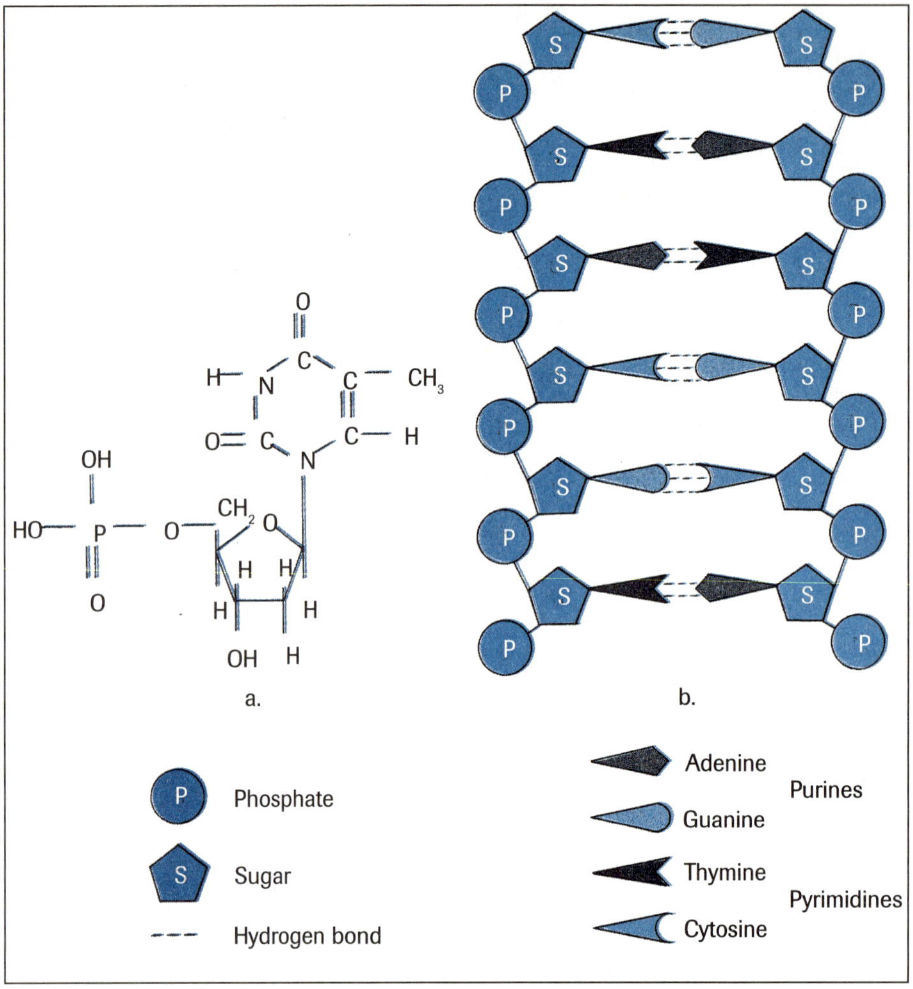

Figure 4.22B (a) Detailed structure of a thymine nucleotide taken from (b), the molecular structure of a section of DNA, showing base pairing and orientation of sugar-phosphate "backbone."

Many of the organic and inorganic substances in living cells dissolve, or go into solution, because the charged water molecules tend to separate them into their constituent ions. Sodium chloride (NaCl) readily dissolves because water separates the compound into $Na+$ and $Cl-$ ions, which then are attracted to the respective poles of the water molecule. Many other cellular constituents, also in ionic form, have a similar attraction for the charged water molecules. In fact, more different substances will dissolve in water than in any other liquid known. In the makeup of water, hydrogen bonds between oxygen and hydrogen atoms of different molecules serve as bridges, linking one water molecule with another water molecule forming a unique lattice. Many of the unique properties of water are attributable to the making and breaking of these hydrogen bonds.

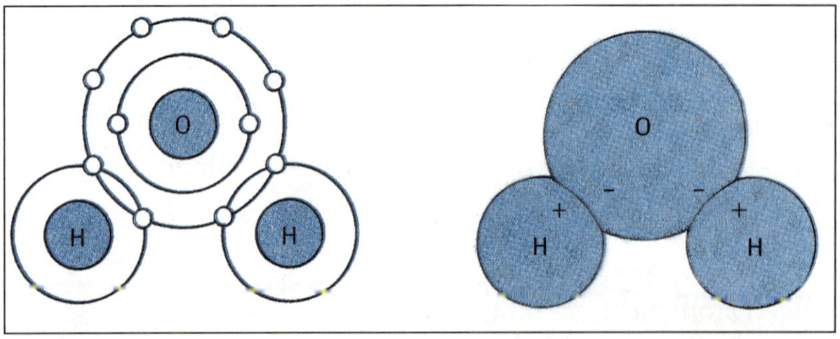

Figure 4.23 Water is a polar molecule because it forms oppositely charged electrical poles. The hydrogen electrons are pulled closer to the oxygen atom, which acquires a weak negative charge. This leaves the hydrogen atoms with a weak positive charge. Ionic compounds readily dissolve in water because their constituents are attracted to the electrically charged poles of the water molecule.

The movement of water within the roots and stems of plants is facilitated by the unique cohesive and adhesive properties of water. **Cohesion** refers to the tendency of molecules of the *same* substance to cling, or hold, together and **adhesion** is the ability of *different* substances to cling together. It is cohesion that produces surface tension, a property of water that makes it possible for small insects to run across the surface of a pond. **Surface tension,** which is greater for water than for most other liquids, results from the surface water molecules clinging to the molecules below them. The surface molecules thus form a tight "skin" over the surface of the liquid. Within a plant, water molecules tend to cohere to each other and to adhere to the inner walls of the water-conducting tubes of roots and stems. In this manner, a continuous unbroken water column is formed that helps conduct the water from the roots to other parts of the plant. Water molecules cohere because of **hydrogen bonding** and adhere because, being electrically charged, they are attracted to charged molecules of other substances.

Another property of water that contributes to its importance to life lies in its response to changing environments. The temperature of water does not tend to rise or fall drastically, providing a relatively stable environment for aquatic organisms and for the individual cells of all organisms. Water, then, has a high **specific heat,** meaning that it must either absorb a large amount of heat from warm air or release a large amount of heat to cold air to display even a slight change in its temperature. The specific heat of a substance refers to the amount of heat that must be absorbed or lost to raise or lower the temperature of 1 gram (gm) of that substance by 1 ° C. This heat energy is measured in calories. It requires quite a large amount of heat absorption, in comparison with that required by other liquids, to raise the temperature of water. The essential reason for this is that most of the heat energy applied to raise the temperature of water is used to break the hydrogen bonds between the water molecules.

Continuing our discussion of the properties of water in relation to temperature, water also has a high **heat of vaporization.** This means that water does not tend to evaporate readily; it requires a great deal of heat to change liquid water into a gas. This property, which allows water to absorb heat from the surface of an organism, is one of the cooling mechanisms of terrestrial animals and plants. For water to evaporate, that is, to change from a liquid to a gas, hydrogen bonds between the water molecules must be broken and the amount of heat necessary is much greater for water than for almost any other liquid. In contrast, rubbing alcohol evaporates almost instantly when applied to your body. Consequently, it absorbs very little heat and must be reapplied to keep you cool. Bodies of warm water absorb heat from the sun, break hydrogen bonds, and evaporate. Heat is released when gaseous water vapor condenses and falls as rain. This heating and cooling process helps in regulating the climate of Earth.

As the temperature drops, liquids tend to become denser as the molecules move more slowly and pack more tightly together (Fig. 4.24). Water, however, on reaching a temperature of 4° C (39.2° F), begins to expand to allow maximum hydrogen bonding between molecules. Water, in fact, becomes less dense and as the temperature decreases, it continues to expand until it freezes at 0° C. Ice, then, is less dense than liquid water, and this is the reason ice floats in water. In cold weather, ice always forms on the surface of ponds and lakes. In the arctic regions, ice is a constant presence in the surface waters of the oceans. If ice were denser than water, it would sink and bodies of water would freeze from the bottom up. Instead, ice on the surface provides a layer of insulation that prevents water below from freezing at all. If this were not the case, survival of aquatic life in the cold regions of Earth would be impossible. Life arose in the waters of Earth and it is water that has sustained life's evolutionary development.

Figure 4.24 The structure of ice. Water molecules are held in a hexagonal shape by hydrogen bonds. This arrangement prevents the molecules from becoming packed together, as they are in liquid water. Consequently, ice is less dense and lighter than water in the liquid state.

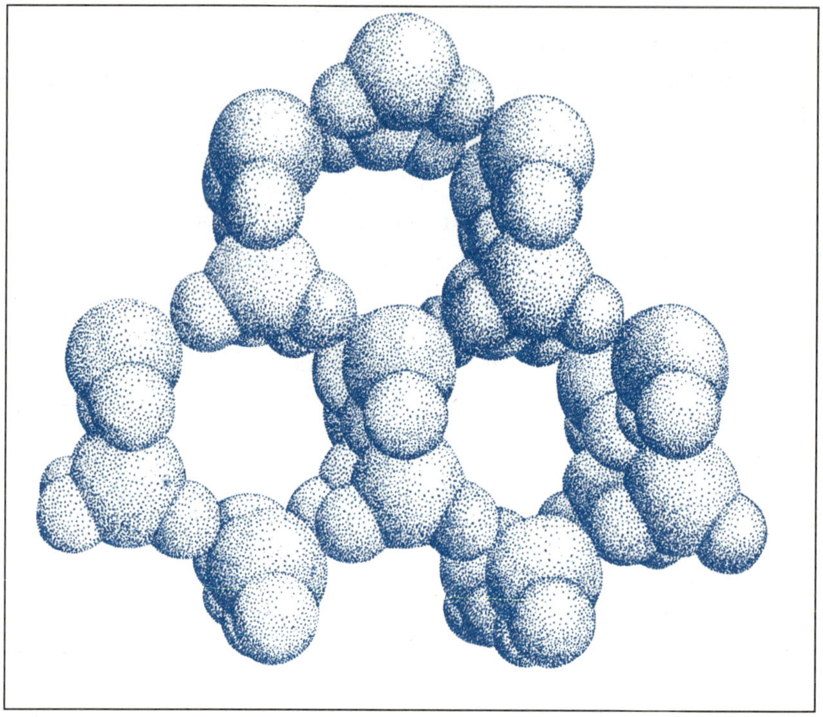

Solutions and pH

The fact that water is the liquid of life brings us to a consideration of the relationship between water and substances contained in it. Water mixed with another substance that is dissolved in it make up a **solution.** Water is the *solvent* and the dissolved substance is the *solute.* Solutions can be a uniform mixture of any two substances; in living organisms, however, the liquid solvent is always water. Perspiration and tears, for example, are mostly water containing dissolved ions. The chemical reactions we have mentioned take place in the water within and around living cells.

Pure water contains a fraction of ionized water. This means that a majority of the water molecules are in the form of H_2O; however, there is a small number of **hydrogen ions** (H +) and an equal number of **hydroxide ions** (OH–) present in the water. A small fraction of the water has dissociated into its two constituent ions. If a substance is dissolved in water, it can affect these ions. We previously stated that some substances can dissociate into their constituent ions when placed in water. For example, some substances dissolve in water and release hydrogen ions into the solution (Fig. 4.25). This increases the number of hydrogen ions in solution. A substance that releases hydrogen ions in solution is called an **acid.** When strong acids dissociate or break apart, they release more hydrogen ions into solution than do weaker acids when they dissociate. This is why a strong acid is more acidic than a weak acid. In contrast, if a solute dissolves in water and decreases the hydrogen ion concentration of a solution, it is called a **base.** Many bases, such as sodium hydroxide (NaOH), release hydroxide ions (OH–) in solution. These hydroxide ions take up, or combine with, the free hydrogen ions to form water. This lowers the hydrogen ion concentration of the water, making the solution more alkaline. Strong bases release more hydroxide ions in solution than do weaker bases.

The relative acidity or alkalinity of a solution is measured by its **pH value.** The pH value of a solution may range from 0 to 14, with pH 7 being neutral. Pure water, with

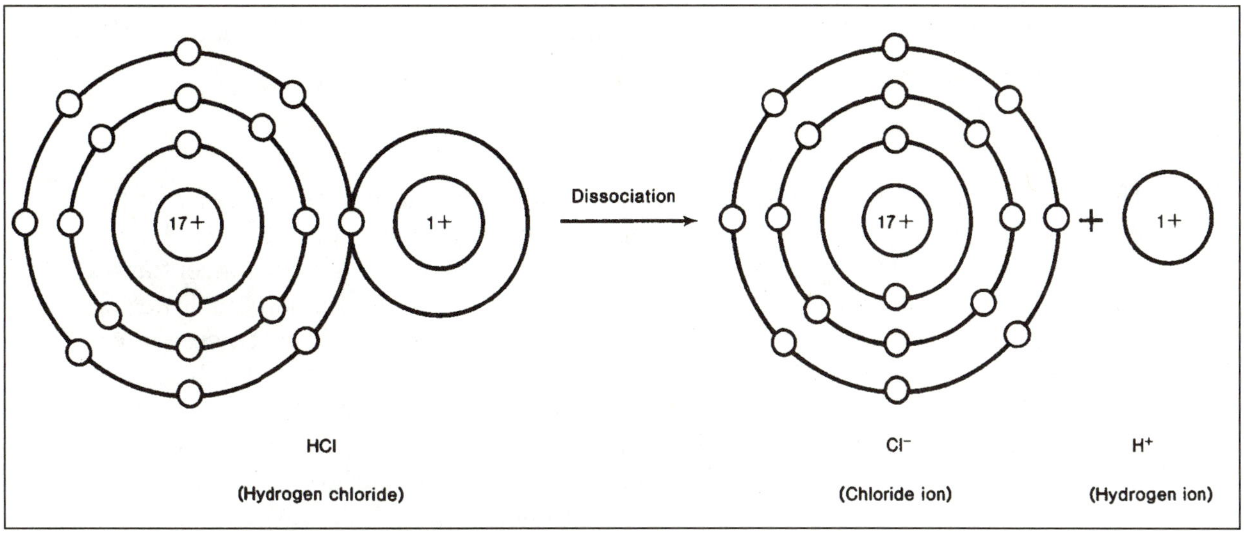

Figure 4.25 Dissociation of hydrogen chloride (HCl), or hydrochloric acid. The electron of the hydrogen atom remains with the chlorine atom, while the hydrogen ion (proton) is released. All acids release hydrogen ions when they dissociate with water.

equal concentrations of hydrogen and hydroxide ions, has a pH value of 7. Any solution with a pH value below 7 is an acid and any solution with a pH value above 7 is a base. Acidity increases in going from pH 7 down to pH 0 and alkalinity increases in going from pH 7 up to pH 14 (Fig. 4.26). The normal functioning of any organism is dependent on the pH value of its body solutions. All solutions in organisms are either acidic or alkaline and must be maintained within a certain pH range. Certain proteins, for example, will only function in an acidic environment.

The pH scale, by the way, is an exponential one. This means that a change in the pH of a solution by just one number to the next represents a ten-fold change in its acidity or alkalinity. For example, a solution with a pH value of 2 is ten times more acidic than a solution with a pH value of 3, and one hundred times more acidic than a solution with a pH value of 4. Similarly, a solution of pH 11 is ten times as alkaline as a solution of pH 10 and 100 times as alkaline as a solution of pH 9. The pH value of a solution reflects its relative concentration of hydrogen ions. The more

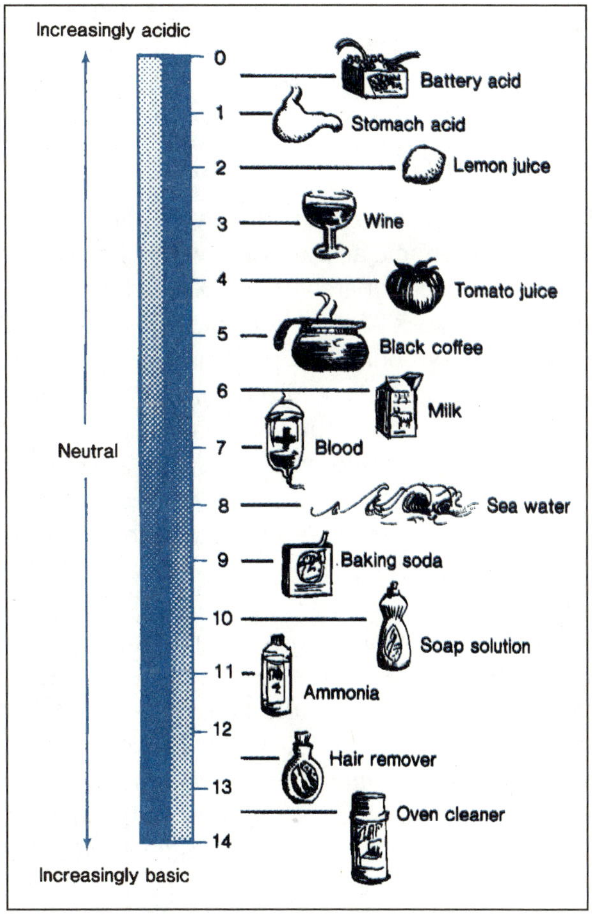

Figure 4.26 The pH scale. Acidic solutions (those with pH below 7) contain more free hydrogen ions (H^+) than hydroxide ions (OH^-). Basic solutions (those with a pH above 7) contain more hydroxide ions than hydrogen ions.

hydrogen ions that are present in solution, the more acidic the solution, and the lower its pH.

BUFFERS

The relative constancy of pH in living systems is essential to the maintenance of homeostasis. However, the metabolic reactions in living organisms continually add or remove hydrogen ions; if not countered, such fluctuations in hydrogen ion concentration tend to alter dramatically the pH of the body fluids. One of the fundamental mechanisms for maintaining a constant pH of the body involves chemical substances known as buffers. A buffer is capable of either bonding with hydrogen ions or bonding with hydroxide ions in solution. If a fluid becomes too acidic, a buffer "takes up," or removes, the excess hydrogen ions responsible for the increased acidity. Conversely, a buffer removes excess hydroxide ions if a fluid becomes too alkaline, or basic.

For example, one of the major buffer systems in human beings is the **bicarbonate buffer system.** All of the fluids of the body contain carbonic acid (H_2CO_3), a weak acid, and sodium bicarbonate ($NaHCO_3$), a weak base, in solution. If the acidity of the body fluids begins to increase, that is, if the hydrogen ion concentration in the fluids increases, the sodium bicarbonate takes up the excess hydrogen ions. Assume, for example, that excess hydrochloric acid (HCl) enters the body fluids. You will recall that hydrochloric acid dissociates into H^+ ions and Cl^- ions in solution. The reaction between HCl and $NaHCO_3$, occurs as follows:

$$HCl + NaHCO_3 \longrightarrow H_2CO_3 + NaCl$$

The products formed are carbonic acid (H_2CO_3), a weak acid, and sodium chloride (NaCl), a salt. As you can see, the hydrogen ions released by hydrochloric acid are taken up by sodium bicarbonate. Consequently, the acidity of the body fluids does not increase significantly.

If a base such as sodium hydroxide (NaOH) is added to the body fluids, it reacts with the carbonic acid. Recall that sodium hydroxide dissociates into Na^+ ions and OH^- ions in solution.

$$NaOH + H_2CO_3 \longrightarrow NaHCO_3 + H_2O$$

In this instance, the hydroxide ion (OH^-) of sodium hydroxide combines with a hydrogen ion from carbonic acid to form water. Removal of this hydrogen ion from carbonic acid forms the bicarbonate ion (HCO_3^-). When the sodium ion combines with the bicarbonate ion, the weak base sodium bicarbonate ($NaHCO_3$) is formed. By taking up excess OH^- ions, the bicarbonate ion prevents the body fluids from becoming too alkaline.

If carbonic acid and sodium bicarbonate are always present in the body fluids, why do they not greatly affect the acidity or alkalinity of the fluids? The key to the answer is that H_2CO_3 is a *weak* acid and $NaHCO_3$ is a *weak* base. Acids are designated as either weak or strong, depending on the degree to which they dissociate in water. Strong acids, such as HCl, dissociate more completely in water, whereas weak acids dissociate to a much lesser degree. Therefore, weak acids release fewer hydrogen ions in solution than do strong acids.

Similarly, strong bases form solutions that have a very high concentration of OH^- ions. These bases are corrosive and extremely irritating to living tissues. Weak basic solutions have low concentrations of OH– ions. Solutions of sodium hydroxide (NaOH) and potassium hydroxide (KOH) are examples of strong bases; magnesium hydroxide ($Mg[OH]_2$), known as *milk of magnesia,* and aluminum oxide (Al_2O_3) are weak bases.

Both of these weak bases are found in the antacids and buffered aspirin sold in drug stores. By taking up hydrogen ions, $Mg(OH)_2$ and Al_2O_3 help neutralize excess stomach acid. For milk of magnesia this neutralization reaction can be shown as follows:

$$Mg(OH)_2 + 2HCl \longrightarrow MgCl_2 + 2H_2O$$

(stomach acid) (magnesium chloride, a salt)

The hydrogen ions released by the dissociation of stomach acid (HCl) are taken up by the OH^- ions to form water, and magnesium and chloride react to form the neutral salt magnesium chloride ($MgCl_2$).

It is important to recognize the intimate relationship between living organisms and nonliving matter. We have seen that both are comprised of elements. A given element is the same element regardless of whether it is in a living organism or in nonliving matter. This fundamental unity provides the key to the evolutionary relationship between the living and the nonliving. Life arose from nonliving chemicals; chemical evolution preceded biological evolution. The physical link between nonliving chemicals and a metabolizing, reproducing life form is organization. The simplest level of chemical organization that manifests life is a single cell. No molecule, compound, or group of compounds below the level of a single cell can be considered a living organism. It is to the evolution, structure, and function of living cells that we next turn our attention.

Chapter Summary

1. Matter is anything that has mass and takes up space. The basic kinds of matter—the elements—are the same in both living organisms and nonliving material. The six most abundant elements in organisms are carbon, hydrogen, nitrogen, oxygen, phosphorus, and sulfur. According to the law of conservation of mass, matter is neither created nor destroyed. Thus, the amount of matter in the universe is constant.

   Key Terms: *matter, p. 48*
   element, p. 48
   property, p. 50

   Concept Question: What is meant by the statement, "Life exhibits a unity below the level of a single cell?"

2. Elements are comprised of atoms. Further, each type of atom has a specific number of protons, electrons, and neutrons. Electrons farther away from the atomic nucleus possess more potential energy than electrons found closer to the nucleus.

   Key Terms: *atom, p. 50*
   atomic number, p. 50
   energy level, p. 52
   valence, p. 55
   compound, p. 55

   Concept Question: Diagram the atomic nuclei and electron energy levels of the following atoms: (a) aluminum (atomic number = 13); (b) helium (atomic number = 2); (c) magnesium (atomic number = 12); (d) sulfur (atomic number = 16)

3. Atoms may react with other atoms, forming chemical bonds through participa tion of their outermost electrons. Reacting atoms either lose, gain, or share valence electrons in forming chemical bonds. Bonding results in the formation of molecules or compounds. Ionic bonds are formed through the loss and gain of valence electrons by participating atoms. Covalent bonds are formed by sharing electron pairs between atoms. This can result in formation of single, double, or triple covalent bonds. Hydrogen bonds are actually a weak attraction between an atom of one molecule and a hydrogen atom that is already covalently linked to another atom. Water molecules, for example, participate in hydrogen bonding

and these bonds influence the unique properties of water. Chemical bonds store energy, which is released when the bonds are broken. The making and breaking of chemical bonds in a reaction usually require the participation of enzymes.

Key Terms: *chemical bond, p. 55*; *electronegativity, p. 56*; *ion, p. 57*; *molecule, p. 58*; *calorie, p. 60*; *enzyme, p. 60*

Concept Questions: What is the relationship between potential energy, kinetic energy, and chemical bonds? In the chemical reaction between magnesium (atomic number = 12) and oxygen (atomic number = 8) to form magnesium oxide (MgO), which atom loses electrons? What is the electrical charge on each of the formed ions? Describe the formation of the covalent bond in chlorine gas ($C1_2$) (atomic number of chlorine is 17). How would the formation of this bond compare with that of fluorine gas (F_2)? (atomic number of fluorine is 9).

4. Living matter is constructed around the element carbon. Carbon is a light atom of low atomic number, is capable of bonding with other carbon atoms, and can form four covalent bonds simultaneously. Carbohydrates are an abundant energy source in living organisms and may also play a structural role in cells. Carbohydrates are composed of one or more sugar molecules and include monosaccharides, disaccharides, and polysaccharides. Glucose, a monosaccharide, represents the primary energy source for organisms. Lipids are organic compounds that function as reserve energy sources and as structural components of cells. Lipids include fats, oils, waxes, phospholipids, and steroids. Proteins are specific sequences of amino acids and are abundant constituents of living organisms, functioning in a wide diversity of roles including structural components and enzyme-catalyzed reactions. Nucleic acids are constructed of chains of nucleotides, which are composed of a sugar, a nitrogen-containing base, and a phosphate group. The two basic types of nucleic acids are DNA and RNA, which store genetic information and decode and translate this information into proteins, respectively.

Key Terms: *monosaccharide, p. 61*; *disaccharide, p. 63*; *polysaccharide, p. 63*; *glycerol, p. 65*; *fatty acid, p. 65*; *protein structure, p. 68*; *substrate, p. 69*; *nucleotide, p. 70*

Concept Questions: How do the structures of starch and glycogen illustrate the significance of the life feature of organization? Describe the general mechanism of enzyme function. How could this mechanism be disrupted?

5. Water is unique in its ability to support life. Water has a high specific heat, high heat of vaporization, strong cohesive and adhesive properties, and expands at 4° C.

Key Terms: *polar molecule, p. 72*; *surface tension, p. 72*; *hydrogen bonding, p. 73*

Concept Question: Explain how the freezing of water and the melting of ice illustrate the law of conservation of mass. Does the law apply to all ordinary chemical reactions? Explain.

6. A solution consists of a solvent and a solute. The acidity or alkalinity of a solution is expressed in terms of its pH. The pH scale ranges from O to 14, with pH 7 being neutral. Acids, which release hydrogen ions in solution, have a pH value of less than 7; bases, which accept hydrogen ions, have a pH value greater than 7.

Key Terms: *hydrogen ion, p. 74*
hydroxide ion, p. 74
acid, p. 74
base, p. 75

Concept Questions: If solution A has a pH value of 5 and solution B has a pH value of 2, which solution is more acidic and by what factor? Generally, why is the maintenance of a normal pH necessary for solutions in and around living cells?

Chapter Five

Cells: The First Life Forms

Reduced to its physical essence, life is a collection of chemicals. Indeed, as we have seen in Chapter 4, all forms of matter are constructed of chemicals. It is intriguing and, at the same time, perhaps disconcerting to view life in such terms. The intrigue stems from the inescapable fact that life, indeed, is constructed of nonliving chemicals, and yet collectively these chemicals are alive. While we cannot approach any philosophical misgivings scientifically, we can, however, objectively analyze the smallest collection of chemicals that exhibits life—a single cell.

The Cell Theory

Although he did not discover cells as such, the English scientist and architect Robert Hooke (1635–1703) is credited with the first use of the word *cell.* Using a crude microscope he had constructed, Hooke described the tiny empty compartments in a thin slice of cork as little boxes or *cells,* although dead cork actually has no living contents. The compartments reminded Hooke of the monks' cells in a monastery. The first person known to observe living cells was Anton van Leeuwenhoek (1632–1723), a Dutch lens grinder and Hooke's contemporary. Using an improved microscope of his own design, van Leeuwenhoek was able to observe many species of algae and protozoa, organisms later to be described as unicellular forms of life.

Discovery of the cell eventually brought about a revolution in biology, opening up a new field of inquiry into the structure and function of living systems. Aided by the studies and observations of many scientists over the years, the basis of the **cell theory** was formulated in 1839. The cell theory states that all living organisms are composed of one or more cells, and that the cell is the structural and functional unit of life. The theory was expanded in 1858 to include the observation that all cells arise from preexisting cells. In the years since, all subsequent interpretations involving the functioning of organisms have been based on this original theory.

Cells: Evolution and the Emergence of Prokaryotes and Eukaryotes

The scientific evidence available suggests that life on Earth arose from a particular combination of chemicals that, as a crude unit, acquired the ability to reproduce itself.

This crude unit was the apparent forerunner of all subsequent cell types. No one knows just when the first cells appeared on Earth; the oldest fossils of any kind of life, however, date back some 3.5 billion years. All of these are fossil *prokaryotes,* or **bacteria,** some very similar to modern bacteria. Although structurally less complex than all other known cell types, modern bacterial cells are most likely marvels of complexity compared to the earliest rudimentary cell-like structures. Scientists have discovered that analyses of modern-day bacteria provide much information and many intriguing suggestions about the structure and function of ancient bacteria, which may well have been the first living organisms on our planet.

In Chapter 2 we touched on some of the general theories of the origin of life. Regardless of its precise evolutionary origin, the first primitive cell-like structure on Earth evolved to become a more complex independent entity, eventually giving rise to all other life forms as we know them. Let's take a look at the emergence of prokaryotic and eukaryotic cells during evolutionary history and explore the exceedingly fine line that separates life from nonlife.

Almost 2300 years ago, the Greek philosopher Aristotle wrote: "Nature makes so gradual a transition from the inanimate to the animate kingdom that the boundary lines which separate them are indistinct and doubtful." Where, then, do we draw the biological line between life and nonlife? No single element, compound, or group of compounds below the organizational level of the cell can be considered a living system. It follows, then, that any organism must consist of at least a single cell. As a living entity, a cell is capable of metabolism, response, reproduction, and adaptation—the life features characteristic of any living system. Ultimately, the anatomy and physiology of an organism are attributable to the type of cell or cells it possesses and to what goes on within each cell.

The evolutionary history of cells is difficult to construct. We mentioned that the first known cell types were prokaryotes—bacteria that for 2 billion years were the only life forms on Earth. The first single-celled *eukaryote* made its appearance perhaps 1.5 billion years ago, apparently in the form of a protistlike cell. It seems certain that there were intermediate stages between prokaryotes and the first eukaryotes, although no evidence of these intervening stages has been discovered. Still, the evolution of eukaryotic cells may well have been an extended process. Interestingly, the origin of eukaryotic cells has been a rather controversial issue for some time. A fascinating theory proposes that eukaryotic cells initially arose from the integration of a group of ancient prokaryotes (Fig. 5.1). According to this theory, prokaryotic cells entered into a symbiotic relationship in which larger cells actually internalized smaller cells, coexist-

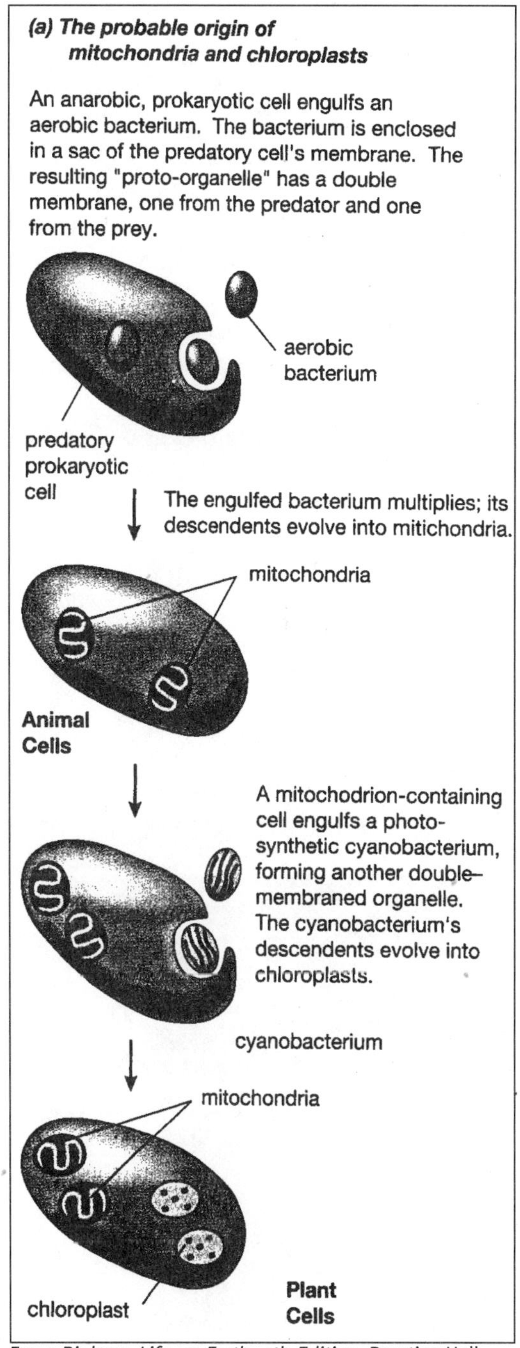

From *Biology: Life on Earth, 4th Edition,* Prentice-Hall.

Figure 5.1 Possible mechanisms for the appearance of organelles in eukaryotic cells.

ing together to their mutual benefit. Further, it is thought that certain organelles present in modern-day eukaryotic cells—specifically mitochondria and chloroplasts—are the descendants of those ancient internalized prokaryotic cells. Over the eons, it is possible that this close association of bacterial cells became so interdependent, they formed a new functional unit—a eukaryotic cell.

The evolutionary significance of eukaryotic cells is impossible to overstate and can be illustrated with a simple example. When a bacterial cell undergoes division—a process called *binary fission* ("splitting in two")—it produces two individual bacterial cells. This process of cell division serves only as a means of reproducing more bacteria. With the rise of eukaryotes, however, cell division began to serve a number of additional functions. While cell division in protozoa and algae, for example, still produces separate individuals, many dividing eukaryotic cells do not separate from each other or form two new independent cells. Instead, the cells remain together to form tissues of various types and functions. From tissues arise discrete structures, such as arms, teeth, stomachs, leaves, and eyes. Following the rise of eukaryotes, organisms could become truly **multicellular,** a condition leading to increased specialization and division of labor among cells. Eukaryotic cells provided the key to the subsequent evolution of all forms of life. With the rise of multicellular organisms about 1 billion years ago, biological evolution began to proceed at a rapid pace.

Prokaryotic Cells

All bacteria, including the ancient organisms represented in fossil remains, belong to the kingdom Monera (Gk. *moneres* single) and are referred to as **prokaryotes** (Fig. 5.2). The cells of these organisms consist of a semi-fluid cytoplasm bounded by a membrane. Most prokaryotes also have an external cell wall surrounding the membrane. Prokaryotic cells, unlike most other cell types, however, do not have a defined nucleus.

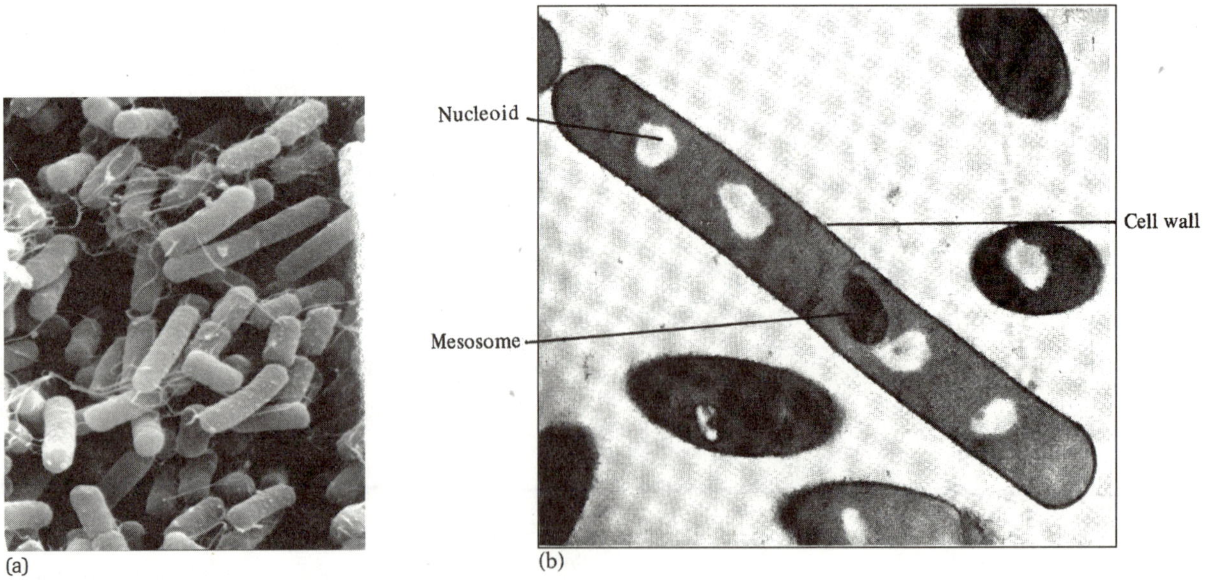

Figure 5.2 (a) A clump of bacterial cells (*Escherichia coli*) as seen in a scanning electron micrograph. (b) A bacterium, *Bacilus subtilis,* showing the prominent Cell wall surrounding the cell, mesosomes, and nucleoid. The mesosomes are infoldings of the plasma membrane.

A nucleus is a specific compartment within certain cells that is surrounded by a membrane and contains the genetic material, or DNA. The single circular strand of DNA found in prokaryotes, however, lies almost "naked," so to speak, within the cell. This simple organization is considered to be a primitive level of cellular design. Every other known cell type has a distinct membrane-bound nucleus at some stage in its existence. During their development, immature mammalian red blood cells, for example, possess nuclei, but these are shed when the cells mature.

In addition to lacking a true nucleus, bacterial cells also lack most of the organelles—membrane-bound compartments or "little organs" within a cell—that characterize the cells of all other organisms. Most bacteria are single-celled, although some species form pairs of cells or long filaments of cells, while still other species form colonies in which there is some division of labor among the cells. Figure 5.2 depicts the general structure and organization of bacterial cells. Prokaryotes are found virtually everywhere and have evolved diverse adaptations to survive in their particular environments. One group thrives in hot, acidic waters, while another group prefers extremely salty water. Various other bacterial species inhabit the intestines of animals or live in swamp water that contains no oxygen.

Organisms other than bacteria are called **eukaryotes,** a term meaning "true nucleus," and are constructed of eukaryotic cells. You are a eukaryote, as are amoebae, dung beetles, yaks, tumbleweeds, and prickly pears. The underlying architecture of all eukaryotic cells is basically the same, demonstrating another aspect of the fundamental unity of life. Eukaryotic cells are generally larger than prokaryotic cells—about ten to twenty times larger on the average—and possess a more complex internal organization, containing membrane-bound organelles and a nucleus surrounded by a nuclear membrane.

Eukaryotic Cells

The structural differences in prokaryotic and eukaryotic cells account for their particular functional characteristics. Our major focus in this chapter is eukaryotic cells; we will discuss the economic, medical, and ecological roles of bacteria in the following chapter. Structurally and functionally, a eukaryotic cell is quite similar to a miniature factory. Just as a factory consists of machines, assembly lines, and control centers all housed within concrete and steel, a cell contains metabolic "machinery," molecular "assembly lines," and a control center, housed within a surrounding cell membrane. Moreover, both factories and cells manufacture certain products and produce wastes, utilizing energy in the process. Often, both factories and cells are individual units of a larger structure. Many factories may compose an industry, and many cells make up the tissues of an organ. The human heart, for example, is an organ consisting of tissue that primarily is a collection of individual cardiac muscle cells.

Nucleus and Cytoplasm

Basically, a eukaryotic cell consists of a number of organelles suspended in a semifluid **cytoplasm** that is completely enclosed within a thin **plasma membrane** (Fig. 5.3). Plant cells have an additional structure that lies immediately outside the plasma membrane—a tough cell wall that imparts rigidity and provides support for the plant. The cell wall of plant cells consists mainly of cellulose, while most bacterial cell walls contain complexes of sugars and proteins. The cytoplasm of eukaryotic cells contains ions,

Figure 5.3 (a) Generalized diagram of an animal cell, showing the major cellular organelles. (b) Generalized diagram of a plant cell, showing the major cellular organelles.

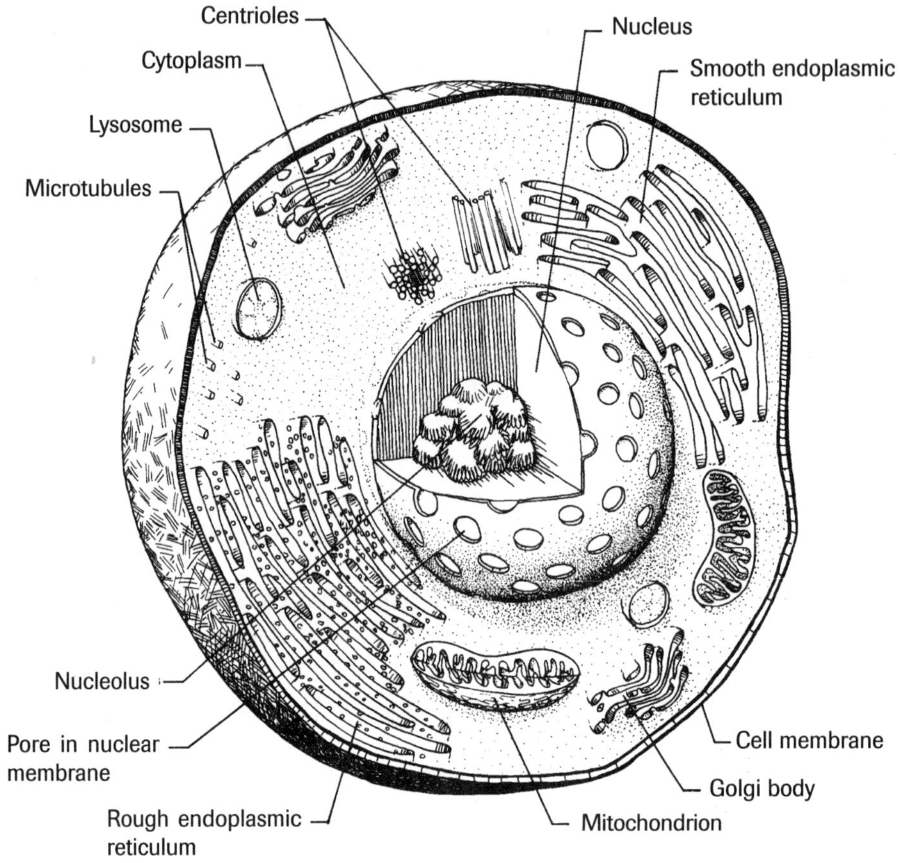

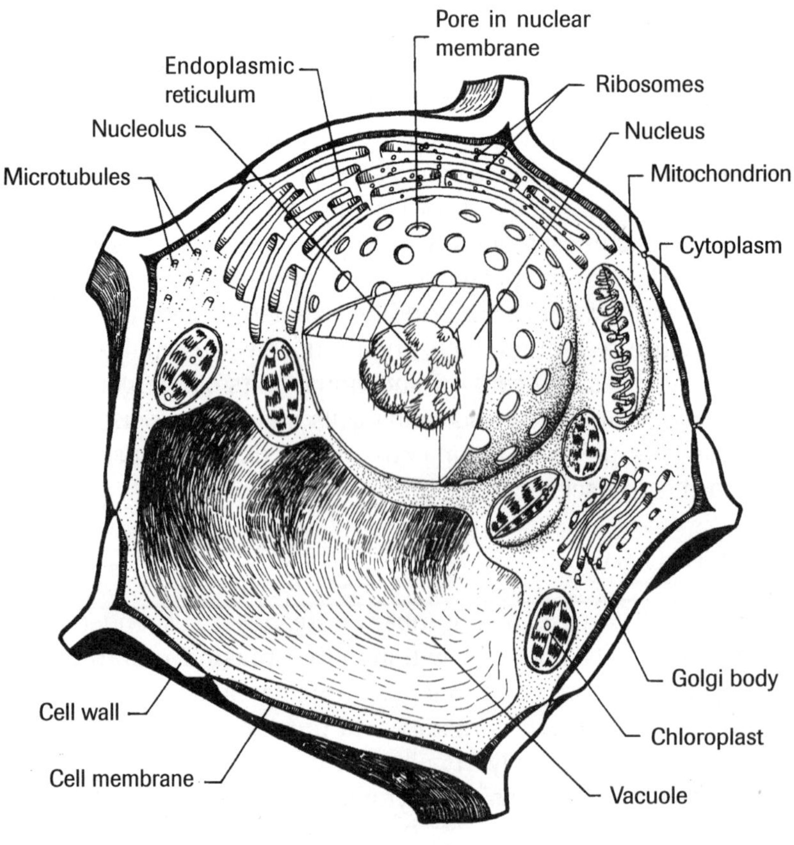

amino acids, various proteins, lipids, a small amount of carbohydrates, and waste products. Some of these substances are raw materials used by the cellular machinery during metabolism, and some are the products manufactured from these raw materials. The cytoplasm of a certain cell type contains specific structures or substances associated with the unique function of that cell. Your red blood cells, for example, contain the protein *hemoglobin,* which plays a vital role in the transport of oxygen throughout your body by the bloodstream.

A conspicuous organelle in the cytoplasm is the membrane-bound **nucleus,** the "control center" of the cell. The nucleus contains **chromosomes,** which consist primarily of *deoxyribonucleic acid* (DNA). The functional units of chromosomes are the genes that contain the information necessary for the synthesis of cellular proteins. In nondividing cells, the chromosomal material appears as a tangled mass of very fine threads known as **chromatin.** In cells that are undergoing division, the chromatin condenses to form distinct chromosomes that can be observed with a microscope. Some cells, such as those of skeletal muscle and some fungi, have several nuclei; that is, they are **multinucleate.**

Typically, the nucleus also contains two dense regions called **nucleoli.** Each nucleolus participates in the production of ribosomes, which are involved in protein synthesis. The nucleus is surrounded by a nuclear membrane that is composed of a double layer of phospholipid molecules, appropriately called a **lipid bilayer.** This membrane is penetrated by small pores, or passageways, that allow a regulated exchange of materials into and out of the nucleus. Further, the nuclear membrane separates the nuclear contents and the surrounding cytoplasm into two distinct environments. This separation, or compartmentalization, is characteristic of eukaryotic cell organization and is essential for normal cell function.

Endoplasmic Reticulum and Ribosomes

Throughout the cytoplasm within the cell is a network of membranous tubules and channels called the **endoplasmic reticulum** ("network within the cytoplasm"), abbreviated ER. The ER channels extend from the nuclear membrane to the plasma membrane, forming convoluted passageways through which materials can be distributed throughout the cell. Some portions of the ER membrane are studded with tiny spherical structures called **ribosomes,** which are also found "free" in the cytoplasm unattached to membranes. Ribosomes are constructed in the nucleoli, move from the nucleus to the cytoplasm, and are the sites at which proteins are built from individual amino acids. The regions of ER that are studded with ribosomes are called "rough" ER. Proteins assembled on these ribosomes are subsequently transferred across the ER membrane to the interior of this system of flattened membranous stacks and channels. During their passage through the ER, these proteins are packaged into tiny membrane-bound sacs that bud off the ER membrane and deliver the proteins to another stack of membranes called the *Golgi apparatus.* Portions of the endoplasmic reticulum that lack ribosomes are called "smooth" ER and participate in the synthesis of glycogen, fats, and other lipids. Smooth ER also aids the liver in detoxifying alcohol, drugs, and various poisons. Cells have relative amounts of rough and smooth ER, depending on the specialized function of a particular cell. For example, cells that are actively producing large amounts of protein possess large amounts of rough ER. Other cells, such as liver cells that inactivate poisons, contain large amounts of smooth ER.

Golgi Apparatus and Lysosomes

Products of the ER are processed, packaged, and redistributed by an organelle known as the **Golgi apparatus,** named for the Italian scientist Camillo Golgi who first described the organelle in 1898. We could refer to the Golgi as the "receiving, packaging, and shipping department" of eukaryotic cells. The Golgi apparatus consists of a stack of several flattened, membranous sacs. Materials arriving from the ER, along with products of the Golgi itself, such as certain polysaccharides, are processed and packaged in the stacked membranes. Tiny membrane-enclosed sacs, or *vesicles,* containing the packaged products pinch off the stacked membranes, and are routed to other areas of the cell. The Golgi apparatus is particularly prominent in secretory cells, which specialize in the synthesis of certain proteins and their subsequent export from the cell. Secretory products of plant and animal cells are transported within the Golgi sacs to the plasma membrane where the sac membrane fuses within the plasma membrane and the sac contents are released to the exterior of the cell. Other tiny sacs budding off the Golgi apparatus transport proteins that are components of the plasma membrane. These sacs move throughout the cytoplasm to the plasma membrane where they release their contents, which become incorporated into the structure of the plasma membrane.

Lysosomes, membranous vesicles derived from Golgi membranes, contain several **hydrolytic enzymes** that break down or degrade carbohydrates, lipids, proteins, and nucleic acids into their individual components. This process is important to the cell because these components can then be recycled to construct new carbohydrates, lipids, proteins, or strands of DNA. Lysosomes can also break down and destroy disease-causing organisms such as bacteria. A bacterial cell that enters an animal body is first engulfed by certain white blood cells and then enclosed within the cell. Once inside the white blood cell, the bacterium is exposed to the lysosomal enzymes and rapidly digested. The process by which a cell engulfs another cell is called **phagocytosis,** and is also utilized by many unicellular organisms as a method of feeding. Amoebae, for example, engulf food particles that their lysosomal enzymes subsequently digest into smaller component parts for use in other cellular processes. Phagocytosis is also a process by which large molecules, such as proteins and polysaccharides can be brought into the interior of a cell. Further, lysosomes are involved in the destruction of old worn out cells or cells no longer of use to an organism. Such is the case, for example, when a tadpole absorbs its tail as it develops into a frog. Lysosomes may even be involved in the process of animal aging and in degenerative diseases of the heart, brain, and skeletal muscles. Through release of digestive enzymes, lysosomes may contribute to the breakdown of body cells and the subsequent destruction of vital tissues.

Vacuoles

Other membrane-enclosed sacs found in the cytoplasm include **vacuoles,** which store a variety of substances, such as food, water, ions, pigments, and wastes. Typically, vacuoles are more developed in plant cells than in animal cells; plant cells often possess a large **central vacuole** that can fill more than 80 percent of the inside of the cell. This central vacuole stores various organic compounds, inorganic ions, metabolic wastes, and, in some plants, pigments that are responsible for the color of flowers and leaves. In addition, the central vacuole participates in increasing the size of a plant cell by absorbing water. Protozoa, algae, and some lower invertebrates possess small specialized vacuoles called *contractile vacuoles* that function as tiny hydraulic organs by repeatedly pumping excess water from the cytoplasm to the exterior of the cell. Found in a variety of cell types, *food vacuoles* store nutrients and also distribute the nutrients throughout the cytoplasm.

Mitochondria

The major energy-releasing "departments" of eukaryotic cells are the **mitochondria.** These organelles provide the energy for a multitude of chemical reactions that occur within a cell. Through a series of chemical steps, mitochondria are involved in the synthesis of the molecule adenosine triphosphate, or ATP, which stores potential energy in its phosphate-to-phosphate bonds. When ATP is involved in a reaction that requires an input of energy, one of its phosphate bonds is broken, releasing the stored energy. These mitochondrial reactions produce about 95 percent of the ATP in most eukaryotic cells. As you might expect, active cells with high energy demands, such as heart muscle cells, contain a large number of mitochondria. On the other hand, the number of mitochondria is much lower in relatively inactive cells, such as those of fat tissue. Mitochondria are surrounded by a smooth outer membrane and a very convoluted and folded inner membrane. Interestingly, mitochondria contain much of their own genetic machinery, including DNA, RNA, and ribosomes similar to that of bacteria, lending support to the theory that they evolved from ancient prokaryotic organisms once capable of an independent existence.

Microfilaments, Microtubules, and Intermediate Filaments

Three types of filamentous structures that play a role in intracellular support, movement, and cell division are **microfilaments, microtubules,** and **intermediate filaments.** Microfilaments are thin, elongated structures composed of protein and are thought to be present in all eukaryotic cells. Microtubules are hollow cylinders also comprised of protein and are somewhat thicker than microfilaments. A third type of protein filaments, intermediate in size, are called intermediate filaments. These three types of protein filaments are assembled within a cell to form a lattice throughout the cytoplasm. This network provides support, maintains cellular shape, and facilitates transport or movement of materials across the cell. This internal cell "skeleton" is appropriately referred to as the **cytoskeleton.**

Microfilaments and microtubules are also involved in other cellular processes. For example, in skeletal muscle cells, the sliding action of thick and thin microfilaments is responsible for contraction of the muscle. Microfilaments also participate in **cytoplasmic streaming,** evident in some plant and animal cells. Cytoplasmic streaming is the flowing motion of a cell's cytoplasm in a particular direction within the cell, possibly to facilitate the distribution of cellular materials throughout that cell. Microfilaments are also responsible for the ability of certain protozoans to extend "false feet" or pseudopodia—extensions from the cell body—for movement or prey capture.

Microtubules, in addition to providing support and strength to a cell, participate in the movement of chromosomes during cell division. When a cell divides, microtubules form a specialized structure called the *spindle,* which is associated with movement of the chromosomes. The spindle gradually shortens, effectively pulling the chromosomes to opposite ends of the cell. This insures that the two cells resulting from the division each have their proper complement of chromosomes. Another structure composed of microtubules is the *centriole,* located just outside the nucleus of a cell. Centrioles are usually found in pairs, lying at right angles to each other. Centrioles are thought to play a part in organizing the formation of the microtubular spindle. Some higher plant cells that lack centrioles, however, are able to form spindles during cell division in a manner essentially similar to that of other cells.

Finally, in eukaryotes, microtubules are the components of **cilia** and **flagella,** structures involved in cellular locomotion. These tiny hairlike projections are found on some of the cells of animals, as well as on the cells of some plants and many unicellular organisms, which use cilia and flagella for locomotion. In humans and many other animals, cilia line the respiratory and reproductive tracts. In your windpipe, for example, the beating or waving motion of many cilia aid in preventing foreign particles from entering the lungs. In the fallopian tubes of human females, cilia lining the walls help propel the egg from the ovary toward the uterus. The sperm cells of many organisms, including humans, have a single, long flagellum used for locomotion. Some plants have sperm cells with two or several flagella. Cilia and flagella in most eukaryotes are structurally identical; the only observable difference is that flagella are longer. It is interesting, and indicative of the theme of unity in diversity, that the basic microtubular structure of cilia and flagella is the same in almost all eukaryotes, whether protozoa, plants, or human beings.

Cell Walls and Plastids

In addition to the cellular components described, plant cells contain other specialized organelles and structures not characteristic of animal cells. We have mentioned that many plant cells contain large vacuoles. Plant cells are also surrounded by a tough **cell wall** lying outside the plasma membrane, which confers strength and rigidity to the plant. The cell wall is composed mainly of the polysaccharide *cellulose.*

Within the cytoplasm of plant cells are organelles called **plastids,** membranous sacs containing a variety of materials. There are three general types of plastids: **leucoplasts, chromoplasts,** and **chloroplasts.** Proteins, lipids, and starch are stored in leucoplasts, which are usually abundant in storage cells such as are found in plant roots. The red, yellow, and orange pigments displayed in autumn leaves, flowers, and fruits are stored in chromoplasts. The reactions of *photosynthesis* are carried out within the chloroplasts, which contain the green pigment *chlorophyll.* As you might expect, chloroplasts are abundant in cells such as leaf cells that carry out photosynthesis. Chloroplasts, leucoplasts, and chromoplasts are enclosed within a double-layered membrane. Chloroplasts, though, contain flattened, membranous sacs called **thylakoids,** typically arranged in pancakelike stacks. Each stack is called a **granum**. Chlorophyll and other photosynthetic pigments, along with enzymes and electron carrier molecules, are embedded in the thylakoid membranes. The area around the grana, the **stroma,** is partially filled with a protein-rich solution. Chloroplasts are the only other cellular organelle in addition to mitochondria to have their own DNA, RNA, and ribosomes, and can reproduce themselves by division. Many of the characteristics of chloroplasts are similar to those of bacteria—a similarity that is suggestive of the possible evolution of chloroplasts from an ancient photosynthetic prokaryote to the integrated eukaryotic cell organelle that it is today.

Cell Membranes

Like any building, a factory is partially separated from the outside environment. So it is with a living cell. Often there is quite a difference between the physical and chemical events occurring within the cell and those occurring in the surrounding environment. Similarly, activities within the organelles are usually quite different from those occurring outside of the organellar membrane in the cytoplasm. Like a walled room, the interior of each organelle, or the larger cytoplasmic interior of an entire cell, is a separate and distinct environment not separated by solid walls, but by surrounding membranes. These membranes serve to regulate and maintain these unique environments. The similar structure of cell membranes in virtually all organisms is a further indication of the unity of life.

Basically, cell membranes are constructed of a double layer of **phospholipids** in which a variety of protein molecules are embedded (Fig 5.4). These proteins carry out specific functions for the cell and, depending on its function, a protein may extend out from either the inner or outer layer of phospholipids or it may extend all the way through the double layer. Some of these proteins function in the transport of materials across the membrane; others recognize and bind to substances such as hormones, which initiate specific cellular events. The cell membranes of many animal cells have a sticky coat composed of short-chain carbohydrates. This coat helps cells stick tightly together and, similar to the cell wall of plant cells, adds strength to the cell. These carbohydrates on the cell's outer surface are also involved in cell-to-cell recognition; that is the ability of cells to distinguish between one type of cell and another. This is critical for example in the function of the immune system in which foreign cells entering the body must be identified so that they can be destroyed.

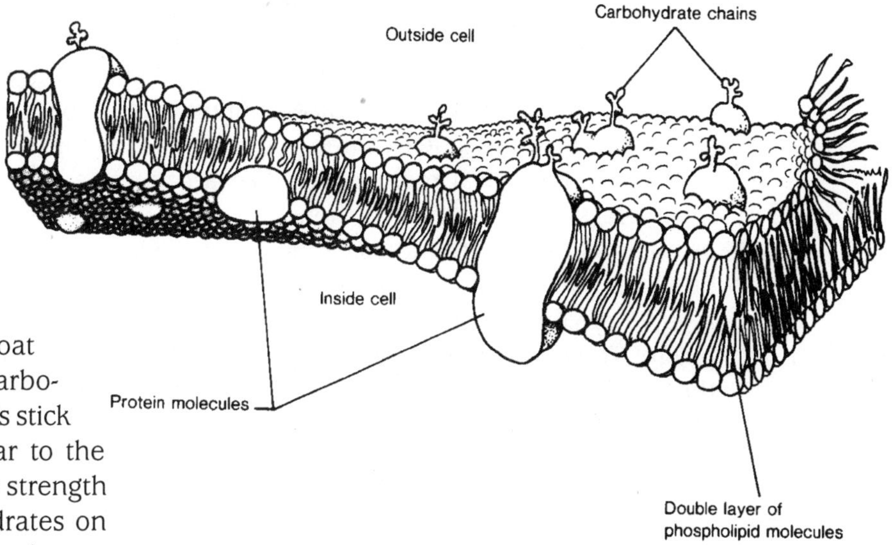

Figure 5.4 The "fluid mosaic" model of the cell membrane. The diagram illustrates the mosaic of protein molecules embedded in the double layer of phospholipid molecules (the two layers are separated at the right of the diagram). The term "fluid" refers to the apparent ability of some of the proteins to move laterally within the bilayer. The external carbohydrate chains are involved in various cellular interactions.

It might appear that membranes form an impenetrable barrier between the inside of a structure and its surroundings. This is not the case however in that the structure of membranes permits the selective passage of some substances while restricting passage of others. Thus membranes are said to be **selectively permeable.** By regulating what enters and exits a cell, the membrane maintains the cellular environment. We have already mentioned the relationship between structure and function in living systems. Cellular membranes exquisitely demonstrate the significance of this crucial relationship.

Cells and Water

Any discussion concerning living cells would be incomplete without referring to a most vital substance of cellular composition—water. Approximately 70 to 95 percent of a cell is water. The cells of many simpler organisms such as sponges and some cnidarians and protists are exposed to the fresh or salt water in which they live. In more complex organisms, many or most of their cells are not exposed to the outside environment but are surrounded by a watery medium. The unique properties of water make possible the vast array of chemical and physical activities of living cells. Part of this activity is possible because water unlike many other substances passes through biological membranes easily and rapidly.

Significance of Cell Size

As a general rule, cells are very small and do not vary significantly in size from one organism to another. There are, to be sure, larger and smaller cells; eukaryotic cells, however, are typically about 25 micrometers, or about one- thousandth of an inch, in diameter. A large cell, such as the human ovum, is one exception, having a diameter four and a half times larger than an average eukaryotic cell. Size plays an important role in the living world and this certainly is true of living cells. So why, then, are cells so small? There are at least three possible reasons for the observed size of eukaryotic cells. First, biological membranes, like a sack or a plastic bag, can tolerate just so much pressure within them without bursting. There is, then, a restriction on the quantity of materials the membranes can support. Second, there is an apparent relationship between the nucleus and the volume of cytoplasm it can control. The nucleus is the regulatory center of the cell and exerts influence on many activities out in the cytoplasm. The nuclear membrane is porous, allowing some materials to enter and leave the nucleus. The chromosomes are housed in the nucleus and thus it is the repository for most of the genetic information in the cell. From the nucleus, genetic messages are relayed out to the cytoplasm where they are translated into the specific proteins that make up an organism. If the volume of cytoplasm is too great, the nucleus may very well be unable to exert adequate control over protein synthesis and other cytoplasmic processes. Third, all cells must regulate their metabolism, which requires the entrance and delivery of nutrients, water, and oxygen, and the exit of carbon dioxide and other waste products. If the volume of a cell becomes too great, exchange of materials across the surface membrane may not occur rapidly enough to carry out metabolism and the cell will die. To meet these three requirements, cellular function is restricted within certain size limits by what is called the *surface area-to-volume ratio* of the cell.

As shown in Figure 5.5, we can illustrate this ratio using a hypothetical cell that is 1 centimeter (cm) in diameter and another equally hypothetical cell 2 cm in diameter. Cells are three-dimensional, so let us imagine they are both cubes. The surface area of one side of the first cell is 1 cm x 1 cm, or 1 square cm. This cell, hypothetically shaped like a cube, has six sides, so its *total* surface area is 6 square cm. The volume of the cell is 1 cm x 1 cm x l cm, or l cubic cm. The surface area-to-volume ratio for the first cell, then, is 6 to 1. For the larger cell, the surface area of one side is 2 cm x 2 cm, or 4 square cm. For all six sides, the total surface area is 4 square cm x 6, or 24 square cm. The volume of this second cell is 2 cm x 2 cm x 2 cm, or 8 cubic cm. Thus, the surface area-to-volume ratio for the second cell is 24 to 8, or 3 to l. Evidently, an increase in the diameter from 1 cm to 2 cm results in an eight-fold increase in volume from 1 cubic cm to 8 cubic cm, while the surface area only increases four-fold from 6 square cm to 24 square cm. The **volume** is eight times greater in the larger cell, but its **surface area** is only four times greater than that of the smaller cell. From this example we can see that as the diameter of a cell increases, its volume increases faster than its surface area. What does this mean? It means that larger cells have less surface area in relation to their volume and smaller cells have more. The volume of a cell

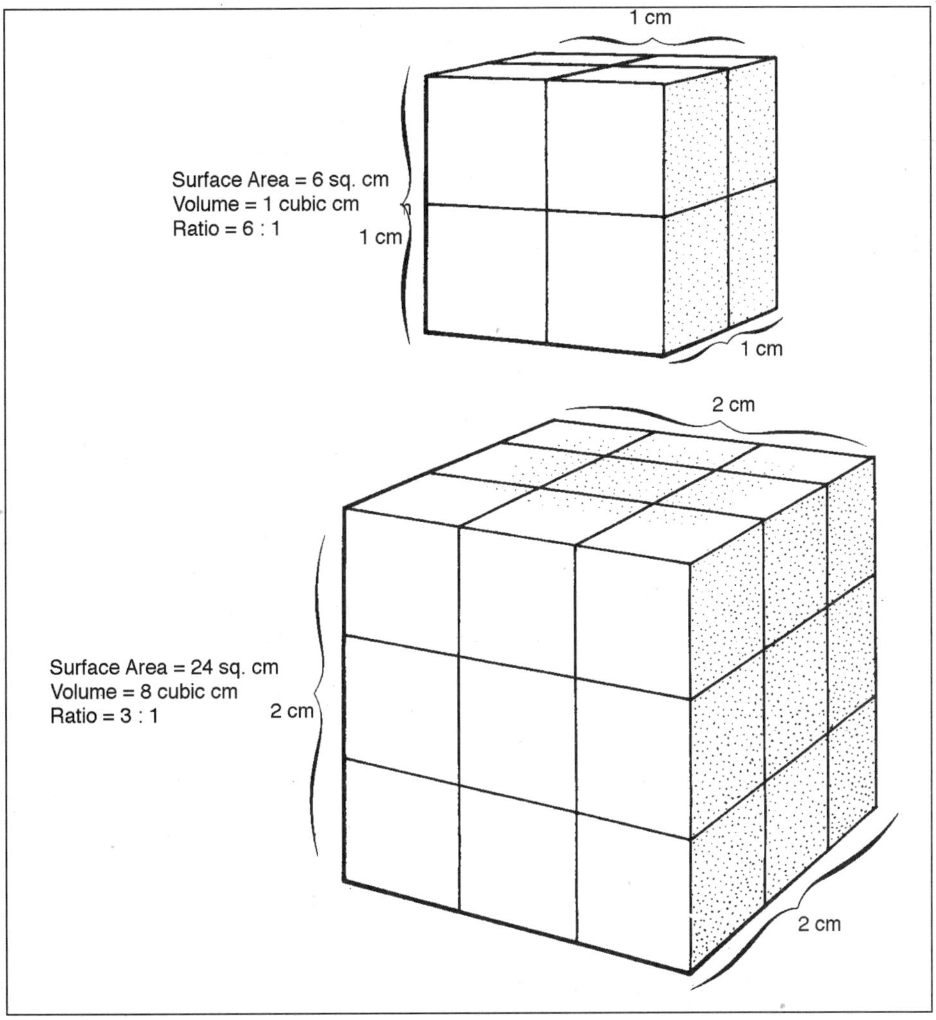

Figure 5.5 Relationship between surface area and volume in a 1-cm cube and a 2-cm cube. Like the cubes, living cells are three-dimensional stuctures. A smaller cell (represented by the 1-cm cube) has more surface area exposed in proportion to its volume than does a larger cell (the 2-cm cube).

determines the quantity of raw materials it can take in for use in metabolism and also dictates the quantity of wastes it can eliminate. Moreover, volume in part determines the rate at which distribution of materials within the cell occurs. The efficient entry and exit of cellular materials depend on the available surface area of the plasma membrane surrounding the cell. If the volume of a cell becomes too great, the exchange of materials across the plasma membrane cannot occur rapidly enough to carry out all necessary cell functions and the cell will die. In other words, as the size of a cell increases, more and more of its volume is farther away from the cell surface. And that, in essence, is why living cells are so small. Actually, surface area-to-volume ratios affect many biological processes. In your own body, for example, the proper functioning of blood vessels, intestines, and lungs depends on their large surface area-to-volume ratios. In the case of blood vessels, many tiny vessels can carry the same volume of blood as one large vessel, but the smaller blood vessels have many times the surface area of the large one. As with individual cells, this facilitates more rapid and efficient exchange of materials.

Cell Membranes and Transport

Through our exploration of the basics of cell function, we are actually considering how life functions in general. In unicellular organisms, all life processes are carried out by a single cell; in multicellular organisms, life results from the cooperative interplay of all the cells. To delve a little deeper into the phenomenon of life, we need to consider a few processes that determine how cells function and thereby sustain their existence. For now, we will focus on one of these critical roles—the movement of substances through cells and the role played by cell membranes.

We mentioned earlier that the nuclear membrane separates the environment within the nucleus from the surrounding cytoplasmic environment. Similarly, the plasma membrane separates the intracellular environment from the extracellular environment. These membranes help to control the environment inside the nucleus or inside the cell by regulating passage of substances across these membranes. The membranes enclosing some organelles such as mitochondria and chloroplasts function in much the same way. Organelle membranes separate the internal environment of the organelles from the cytoplasm. Membranes can accomplish this discriminatory activity partly because they are selectively permeable; that is, they allow the passage of some substances but not others. This function is facilitated by the structure of membranes—the phospholipid arrangement and the particular proteins embedded in the bilayer.

Diffusion

At any temperature above absolute zero (–273. 160° C or –459.7° F), the molecules of a substance are in constant random motion. This motion results in molecular **diffusion,** one of the most common and vital processes affecting cell function (Fig. 5.6). *Diffusion* is *the process whereby the particles of a substance will spontaneously move from an area where that substance is in higher concentration to an area of lower concentration, until there is a uniform distribution of particles throughout both areas.* In other words, a diffusible substance moves from where there is more of it to where there is less of it, resulting in an equal dispersion of this substance throughout the available space. Once the substance is equally distributed throughout an area, the molecules continue their random motion; however, the substance will remain uniformly distributed. Diffusion can occur through the air, through liquids, and through solids.

A familiar example of diffusion involves perfume. When applied to the body or as it exits a nozzle, the perfume molecules are in high **concentration;** that is, they are clustered together. The molecules then begin to diffuse away from each other, randomly moving outward into the surroundings where they are less clustered or concentrated. This movement results in the perfume molecules becoming equally distributed throughout the available space. Once this equal distribution, or equilibrium, is reached, **net** diffusion stops. The molecules will continue to move about randomly, but there is no net diffusion in any particular direction. You may also have observed diffusion through liquids when coloring eggs. A tablet of blue dye dropped into a glass of water soon colors the entire glass of water blue. Here again, net diffusion stops when the dye is evenly distributed throughout the available space. Diffusion through liquids is slower than through gases such as air, and even slower through solids. Diffusion is a **passive** process, meaning cellular diffusion does not require an input of energy by the cell; molecules simply move down a concentration gradient from a higher concentration to a lower one.

Figure 5.6 (a) Diffusion of a dissolving solute in water. Particles of the solute diffuse until they are equally distributed throughout the beaker. (b) Movement of molecules through the air (perfume) or (c) across cell membranes (as in *Paramecium)* occurs from areas where the molecules are in higher concentration to areas where they are in lesser concentration, until a state of molecular equilibrium is reached.

Diffusion is of great importance in cell function because it is one of the processes by which various substances move across cell membranes and move through the cytoplasm. For instance, the respiratory gases—oxygen and carbon dioxide—enter and leave a cell and its various organelles by diffusion across their respective membranes. Diffusion normally occurs quite rapidly over short distances. This fact can be correlated with the importance of cell size that we discussed earlier. Oxygen, for example, is required by almost all metabolizing cells and is used almost immediately once it diffuses to the interior of a cell. If the cell is too large, oxygen might diffuse through the interior of the cell for a relatively long time before it reaches the organelles where it is used. As an exaggerated hypothetical example, imagine your cells suddenly increasing in size as you approach the 13-mile mark of the Boston Marathon. You are in full stride, laboriously sucking in oxygen when whamo!—your muscles will not respond because oxygen cannot be delivered to your energy-producing mitochondria fast enough. At the same time, carbon dioxide, a waste product of cellular metabolism, needs to be removed rapidly from the cells. As its concentration increases in the cytoplasm, carbon dioxide begins to diffuse out of the cells, but this movement, too, is slowed due to the increased cell size. Sadly, you poop out and finish a distant and humiliating last. This hypothetical situation demonstrates the vital relationship between diffusion and cell size.

Osmosis

Cell membranes are involved in a specialized form of diffusion that affects cell function—**osmosis** (Fig. 5.7). By definition, osmosis is *the spontaneous movement of a pure solvent such as water across a selectively permeable membrane from an area where there is proportionately more water to one where there is less water, until there is an equal*

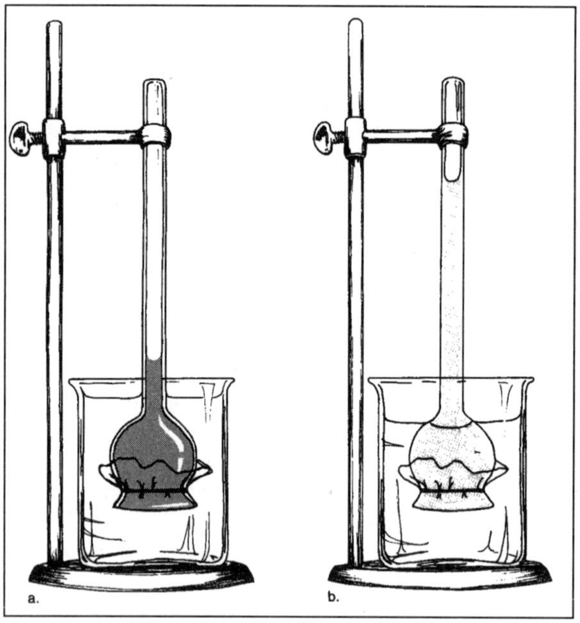

Figure 5.7 Demonstration of osmosis. (a) A thistle tube containing a concentrated sugar solution is immersed in a beaker of water. The bottom of the tube is covered with a semipermeable membrane. (b) Water diffuses through the membrane into the tube, lowering the concentration of the sugar solution and increasing the volume of liquid in the tube. The membrane prevents the passage of sugar molecules out into the beaker.

concentration of water on both sides of the membrane. To put it simply, osmosis is the diffusion of water across a selectively permeable membrane. Osmosis is a passive process that does not require the expenditure of cellular energy in moving water across a membrane. We have discussed the necessity of water for life and that all living cells are bathed in a watery **extracellular fluid.** In addition, the cytoplasm within a cell contains the watery **intracellular fluid.** There are, then, two solutions—one outside the cell and one inside—separated by the selectively permeable plasma membrane. If the two solutions are unequal in the concentration of a given solute, water will tend to move across the membrane to equalize the solute concentrations. If too much water enters a cell, it may swell and even burst; if too much water is lost, a cell may shrivel and dehydrate. Solutions, you will recall, consist of a solvent—water in living cells—and a solute, the substance dissolved in the solvent. Every solution has a certain *concentration* of solute; that is, there is a given amount of solute dissolved in a given volume of solution. If two solutions have the same concentration of solute—the same amount of solute per unit volume—they are **isotonic** solutions. For example, assume we put 1 gram (g) of salt into each of two glasses and fill them each with 100 milliliters (ml) of water. These two salt solutions, each with 1 g of salt per 100 ml, have the same 1 percent concentration of solute and thus are isotonic to each other. Now suppose we prepare another solution in the same way except we use 5 g of salt per 100 ml of water instead of 1 g (Fig. 5.8) This solution is more concentrated; that is, it is a 5 percent solution and thus contains more solute than the 1 percent solution that has only 1 g of salt. We can compare the 1 percent and 5 percent solutions by stating that the 1 percent solution is **hypotonic** compared to the 5 percent solution. When comparing two solutions, a hypotonic solution contains less solute per unit volume than does the other solution. Conversely, the 5 percent solution is **hypertonic** compared to the 1 percent solution; it contains more solute per unit volume. Isotonic,

Figure 5.8 Two beakers containing solutions of different concentrations. The more highly concentrated solution in beaker (a) is said to be *hypertonic* to the less concentrated solution in beaker (b). It also can be said that (b) is *hypotonic* to (a).

hypotonic, and hypertonic are *comparative* terms; these terms can be applied only when comparing one solution with another.

Now let us return to our two solutions of unequal concentrations separated by a selectively permeable membrane. One solution is the intracellular fluid and the other is the fluid bathing the outside of the cell. Let's say hypothetically that the intracellular fluid contains a 1 percent concentration of salt, and the cell is exposed to a salt concentration of 5 percent in the extracellular fluid. In other words, the extracellular fluid is hypertonic compared to the intracellular fluid. We will assume the plasma membrane separating these two solutions restricts passage of the salt crystals. So what, if anything, will happen to the cell? (Fig. 5.9) Returning to the definition of osmosis, water will tend to move across a membrane from a region of proportionately more water to a region where there is less water. In this case, water will move across the membrane *out* of the cell because there are proportionately more water molecules per unit volume in the 1 percent solution within the cell than there are in the surrounding 5 percent solution. With the loss of water, the cell will tend to shrink. The net movement of water out of the cell will continue until both solutions have equal solute concentrations. In general terms, this occurs when the water pressure on either side of the membrane is the same and water molecules, although continuing to move in and out of the cell, are not moving in any *net* direction. If the situation were reversed and the cell contained a 5 percent salt solution

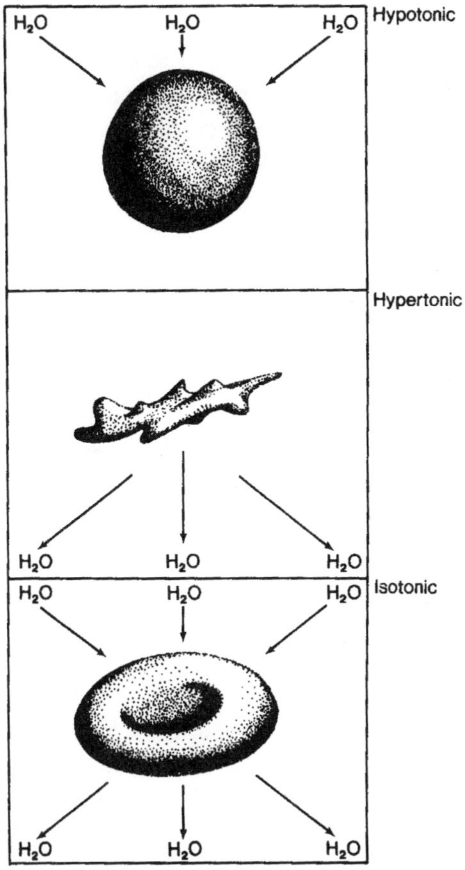

Figure 5.9 (Upper) A red blood cell exposed to a hypotonic solution takes up water and swells. (Middle) When exposed to a hypertonic solution, the cells shrink as water moves out. (Lower) In an isotonic solution the net inflow of water is equal to the net outflow, and the cell content is at equilibrium with its surroundings.

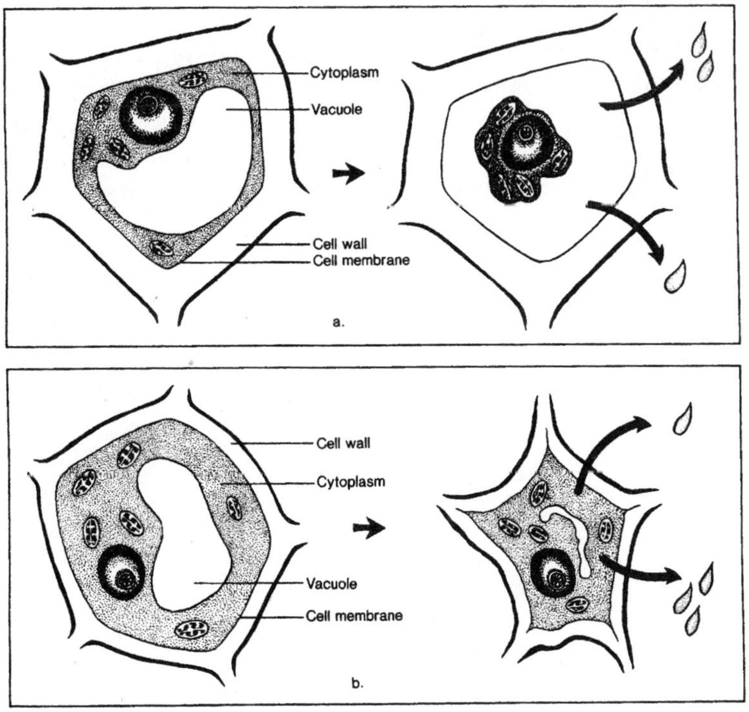

Figure 5.10 (a) Plasmolysis. In plants with a thick cell wall (trees and shrubs), loss of water by osmosis causes a shrinkage of the cytoplasm and vacuole. The pulling away of cytoplasm from the cell wall is called plasmolysis. (b) Wilting. In plants within a thin cell wall (lettuce and celery), loss of water also causes a shrinkage of the cell and vacuole. In addition, the decrease in turgor pressure causes collapse of the cell wall, or wilting.

with the extracellular fluid being a 1 percent salt solution, water would *enter* the cell, causing it to swell. If the concentrations of the two solutions had not equalized before the cell reached its maximum size, it would burst.

In contrast to the reaction of animal cells to hypotonic and hypertonic extracellular solutions, the exposure of plant cells to solutions of different concentrations produces somewhat different physical results (Fig. 5.10). Compare these results in animal and plant cells by referring to Figures 5.9 and 5.10. Remember, however, that the *principles* of osmosis remain the same for both cell types. Plant cells, though, are encased within a fairly rigid, strong cell wall that surrounds the plasma membrane. Freshwater and land plants are exposed continually to hypotonic solutions—pond and lake water and rain. Accordingly, plant cells tend to take up water by osmosis and swell. The presence of the cell wall, however, prevents the pressure of incoming water from breaking the cell. Eventually, the internal water pressure becomes great enough to prevent any further entry of water into the cell. At this point the cell is **turgid,** or firm, which is a structurally stable condition for most plant cells because it lends mechanical support to leaves and other soft structures. On the other hand, when exposed to hypertonic solutions, plant cells lose water by osmosis as do animal cells. But in plant cells, the plasma membrane collapses and pulls away from the cell wall, leaving the wall intact. This phenomenon in plant cells is called **plasmolysis,** and the cells are said to be *plasmolyzed.* Plant cells possessing more flexible, less rigid cell walls undergo drooping, or *wilting,* when they lose water. You have probably seen celery or lettuce that has wilted after being left in the refrigerator too long. Wilting in nature, however, is often a consequence of severe drought.

Phagocytosis and Pinocytosis

The ability of cell membranes to regulate the passage of select substances is attributable to their complex structure. Selective permeability is just one mechanism by which membranes control the movement of substances into and out of cells. Not only does water enter and exit the cell, but a variety of solute molecules must also cross the cell membrane (Fig. 5.11). Earlier we mentioned **phagocytosis,** the process by which large particles are surrounded and engulfed by extensions of the cell. The cell membrane that has extended around the particle now pushes inward, or invaginates, folding into the cell and forming a small sac or **vesicle** surrounding the particle. This vesicle, with the particle safely inside, pinches off from the plasma membrane and moves to the interior of the cell where it fuses with a lysosome. Basically, the particle-containing vesicle and the lysosome become one membrane-bound vesicle with the particle now

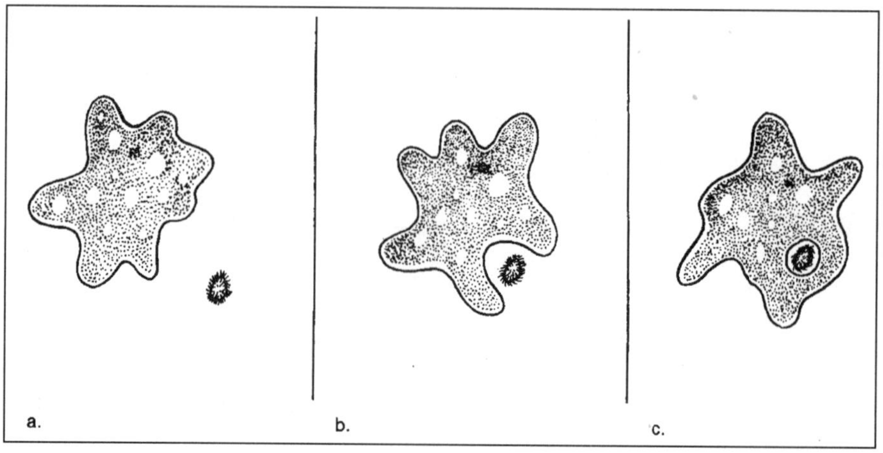

Figure 5.11A Phagocytosis involves the engulfment of a large molecule by cytoplasmic extensions of the cell. Following engulfment, the molecule is contained within a cytoplasmic vacuole.

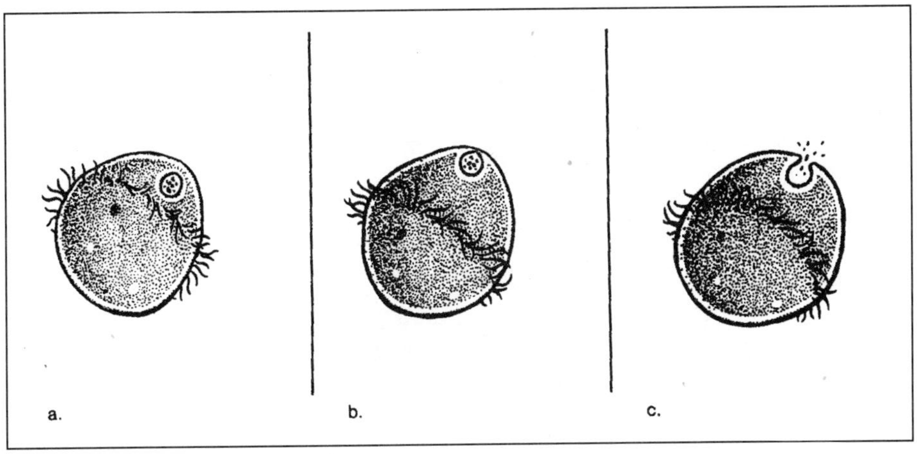

Figure 5.11B Exocytosis, the reverse of phagocytosis. When the vacuole and its contents move from the interior of the cell to the surface, the vacuolar membrane fuses with the cell membrane, and the contents of the vacuole are expelled to the outside.

exposed to the enzymes contained within the lysosome. Lysosomal digestive enzymes digest or hydrolyze the particles, the fragments of which are then released into the cytoplasm. The cell can use these components for its own metabolic processes. For example, the amino acids produced from the degradation of a phagocytized protein can be used to synthesize a new and different protein needed by the cell.

A somewhat similar process called **pinocytosis** occurs when fluids are taken up by a cell. Water containing dissolved solutes from the extracellular fluid can be taken into the cell via tiny vesicles on the surface of the plasma membrane. Pinocytosis occurs in many cell types, including white blood cells, intestinal cells, and plant root cells. Whereas phagocytosis involves engulfment of solid particles, pinocytosis involves the cellular uptake of materials dissolved in water.

Phagocytosis and pinocytosis may occur in reverse; that is, vacuoles with their contents move from the interior of the cell to the surface, where the vacuolar membrane fuses with the cell membrane, expelling the contents to the outside. This process is referred to as **exocytosis** and is employed by many cells in releasing their secretions and waste products to the exterior (Fig. 5.11B). Some of the vacuoles that transport substances from the interior of the cell to the cell surface are formed from membranes of the Golgi bodies.

Facilitated Diffusion

Many small solutes cannot penetrate cell membranes by ordinary diffusion. Most cell membranes contain membrane proteins that function as **carrier** or **transport molecules** to facilitate the transport of a specific solute across the membrane. One form of transport across membranes is called **facilitated diffusion** because carrier molecules assist in the movement of the solute particles from a region of higher concentration to a region of lower solute concentration, the same direction of movement we found with simple diffusion (Fig. 5.12). For example, if the solute concentration is higher outside the cell than inside, the solute would diffuse with the aid of the carrier molecules across the membrane toward the interior of the cell. Facilitated diffusion is a passive process, transporting various ions, amino acids, and sugars across membranes without an input of energy by the cell.

Active Transport

There are times when an organism requires the movement of a specific solute from a region of low concentration to a region of higher concentration across a cell membrane (Fig. 5.13). This is movement in a direction opposite to that occurring in diffusion. In effect, the movement of a substance from low to higher regions of concentration is analagous to water running uphill. How is this possible? Just as it requires an input of energy to get water to run uphill (using a water pump), the movement of solutes up a concentration gradient also requires an input of energy. Such movement occurs through the process of **active transport,** in which a membrane protein will transfer a specific solute across the membrane toward a higher solute concentration. Note that the energy for this active transport is derived from the breakdown of ATP and the release of energy stored in this molecule. Substances that require active transport into living cells include sodium ions, potassium ions, hydrogen ions, and some sugars.

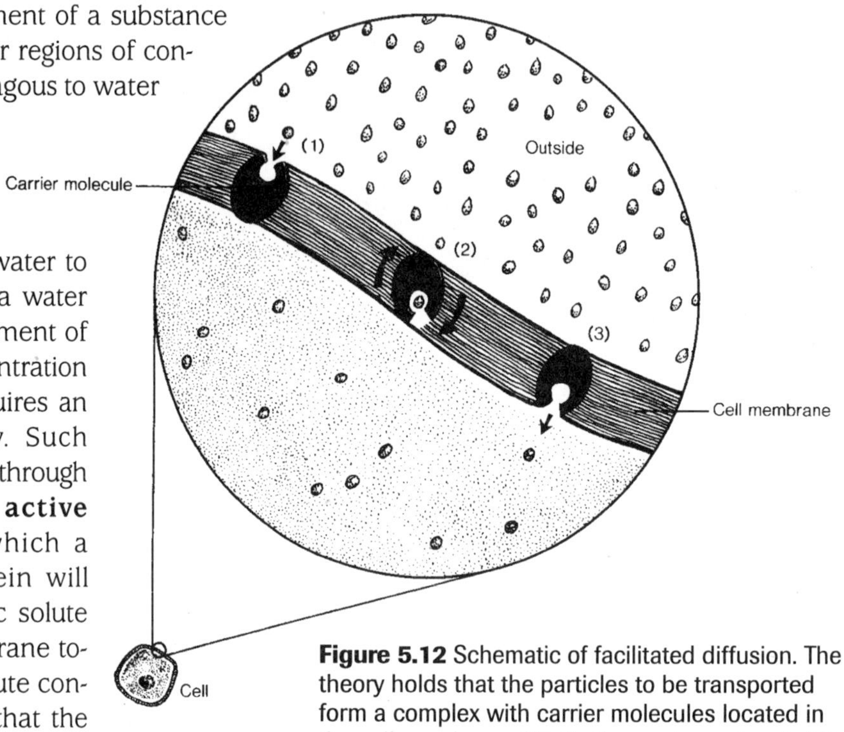

Figure 5.12 Schematic of facilitated diffusion. The theory holds that the particles to be transported form a complex with carrier molecules located in the cell membrane (1). By the carrier molecule (2), the particles are transported across the cell membrane and released (3). Facilitated diffusion transports substances from areas where they are in higher concentration to areas where they are in lower concentration.

The exchange of materials between cells and their environment is but one of the life-sustaining activities of these tiny units of life. As we continue our story of the evolution, processes, and interactions of living systems, we will return again and again to the cell theory as it applies to other life activities. We will build upon and expand this theory, integrating it into our explorations of virtually every field of biology.

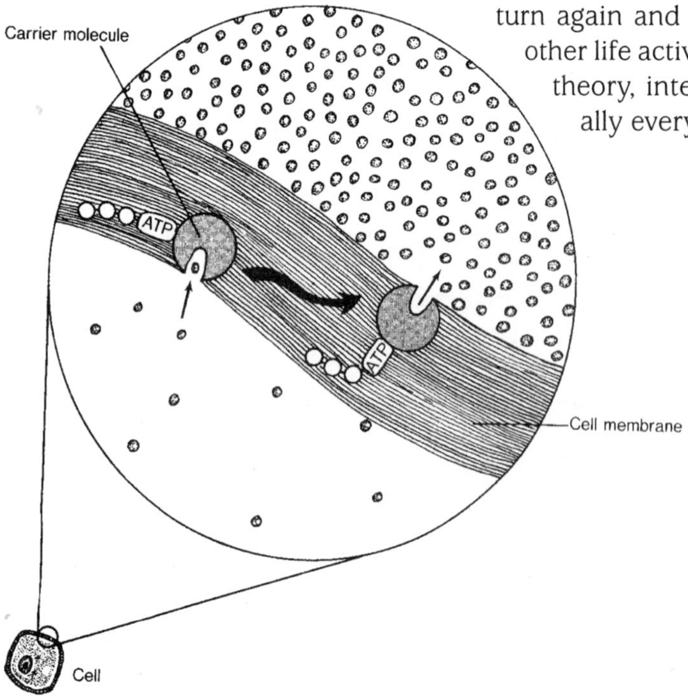

Figure 5.13 Schematic of active transport. The substances to be transported form a complex with carrier molecules located at the surfaces of the cell membrane. Apparently, the carrier molecule changes shape as it moves across the membrane, releases the transported substance, and then returns to its original position in the membrane. Active transport moves substances from regions of lower concentration to regions of higher concentration, i.e., *against* a gradient. This requires an input of cellular energy, which is supplied by ATP.

Chapter Summary

1. The cell theory states that the cell is the structural and functional unit of life and that all cells arise from preexisting cells. The earliest known cells are fossilized bacteria which, like modern bacteria, are prokaryotes. All other known forms of life are eukaryotes, which are proposed to have evolved from prokaryotes. One theory holds that eukaryotic cells arose from a symbiotic integration of prokaryotic cells. With the emergence of eukaryotes, multicellular organisms later appeared, and biological evolution began to proceed rapidly.

   Key Terms: *cell theory, p. 78*
   bacteria, p. 79
   multicellular, p. 80
   prokaryote, p. 80
   eukaryote, p. 81

   Concept Question: Do you think the cell theory applies equally to both prokaryotic and eukaryotic cells? Explain your answer.

2. Prokaryotic cells have a relatively simple organization, lack a distinct membrane-bound nucleus, and possess a single strand of DNA. Eukaryotic cells have a more complex internal organization including DNA in the form of chromosomes. The cytoplasm of both prokaryotic and eukaryotic cells is surrounded by a selectively permeable plasma membrane.

   Within eukaryotic cells, the cytoplasm is compartmentalized in part by the complex network of membranes called the endoplasmic reticulum. Rough ER contains ribosomes; smooth ER does not. Ribosomes are the sites of protein synthesis in cells and thus rough ER participates in protein synthesis and also protein secretion from the cell. Smooth ER contains enzymes involved in lipid synthesis, carbohydrate breakdown, and detoxification of harmful substances.

   The Golgi apparatus in eukaryotic cells packages and distributes products formed by the ER and by the Golgi itself. The Golgi apparatus is especially prominent in secretory cells. Lysosomes are thought to originate from the Golgi apparatus. Lysosomes contain digestive enzymes that break down various cellular contents.

   Key Terms: *cytoplasm, p. 81*
   chromatin, p. 83
   multinucleate, p. 83
   lipid bilayer, p. 83
   hydrolytic enzyme, p. 84
   phagocytosis, p. 84

   Concept Questions: About 60 different digestive enzymes have been isolated from the lysosomes of human cells. A variety of conditions, known as lysosomal storage diseases, result from a deficiency or malfunction of one or more of these enzymes. In a general sense, what effect might defective lysosomal enzymes have on cell function?

   What is the general significance of cell compartmentalization?

3. Eukaryotic cell vacuoles store a variety of cell substances and are more prominent in plant cells than in animal cells. Contractile vacuoles pump water from the interior of some cells and food vacuoles store and distribute nutrients within the cytoplasm.

Mitochondria are the energy-releasing organelles of eukaryotic cells. They are the major sites of ATP synthesis and contain their own genetic machinery. Interestingly, it is thought that these organelles may be derived from an ancient prokaryote that entered into an intimate and interdependent relationship with another, perhaps larger prokaryote.

Microfilaments and microtubules function in cellular support and movement. Microtubules also comprise the spindle and centrioles. Cilia and flagella are hairlike organelles used to filter and sweep particles and for locomotion. Plant cells are surrounded by a cell wall and contain a variety of plastids. Chloroplasts, which contain chlorophyll, grana, and the stroma, are the sites of photosynthesis. Like mitochondria, chloroplasts contain their own genetic machinery and are thought to have evolved from ancient prokaryotic cells. Other plastids store pigments and starch.

Key Terms: *central vacuole, p. 84*
cytoskeleton, p. 85
cytoplasmic streaming, p. 85
plastid, p. 86

Concept Questions: In what type of cells would you expect to find a large number of mitochondria? Of the Golgi apparatus? Explain your answers.

What are the basic structural features that differentiate plant and animal cells?

4. The internal environment of a cell is maintained by a selectively permeable membrane. Membranes are composed of a double phospholipid layer in which proteins are embedded. All living cells are bathed in a watery fluid.

   The size of living cells is limited by their surface area-to-volume ratios. Small cells have more surface area in proportion to their volume than do larger cells. Small cell size facilitates rapid and efficient exchange of materials across the plasma membrane. Surface area-to-volume ratios are an important consideration for many biological processes.

   Key Terms: *phospholipid, p. 87*
   selectively permeable, p. 87
   surface area, p. 88
   volume, p. 88

   Concept Questions: How is the selective permeability of membranes related to overall cell function? Why is it necessary that a plasma membrane surround and effectively compartmentalize a cell?

   In your own words, explain why small cells are more efficient biological units than large cells.

5. Diffusion is a passive process by which substances move from regions of high concentration to regions of lower concentration. Diffusion can occur through air, liquids, and solids.

   Osmosis is the passive diffusion of water across a selectively permeable membrane. Water movement results from the difference in concentration of solutions on either side of the membrane. Solutions are compared in terms of their relative solute concentrations as isotonic, hypotonic, and hypertonic.

   In animal cells, osmosis results in swelling or shrinking of the cells; osmosis in plants results in their cells becoming turgid, plasmolyzed, or wilted.

   Key Terms: *concentration, p. 90*
   passive, p. 90
   intracellular fluid, p. 92
   extracellular fluid, p. 92
   turgid, p. 94

Concept Questions: You notice one morning that your prize potted plant is drooping. You water it and return in the evening to find it upright and restored. Explain the biological phenomenon responsible for the plant's recovery. And by the way, why did it droop in the first place?

Seawater contains about 3 percent salt, while your body cells contain a 0.9 percent salt solution. Explain why sea water will not quench your thirst and what effect sea water would have on your cells.

6. Solute transport across cell membranes may occur by phagocytosis, facilitated diffusion, or active transport. Water droplets containing dissolved materials may be transported by pinocytosis.

Key Terms: *vesicle, p. 94*
carrier or transport molecule, p. 95
active transport, p. 96

Concept Question: Why is it necessary that a living cell employ at least three different mechanisms—phagocytosis, facilitated diffusion, and active transport—to transport solutes across cell membranes?

Chapter Summary Question: Christian de Duve, a Nobel Laureate in medicine stated: "We are sick because our cells are sick. We cannot make ourselves well unless we know what is happening inside our cells." What is the basis for de Duve's statement?

Chapter Six

From Cells to Organisms

Our final consideration under the common life feature of organization takes us beyond the level of an individual cell to broader, more inclusive levels of the natural world. You should remember, however, that these levels ultimately reflect the structural and functional characteristics of single cells. When a brick wall is constructed, the finished structure is an expression of the function of each individual brick. If the bricks fail to function as bricks, there is no wall. And so it is in the living world. We realize from the previous chapter that life is an expression of individual cells; if the cells fail, there is no life. As is true at the cellular level, organization at more complex levels is crucial to the proper functioning and maintenance of life itself. Following a brief discussion of tissues, organs, and systems, this chapter introduces the major groups of organisms in the five kingdoms.

Life: Levels of Organization

Plant Tissues

Many multicellular organisms are composed of groups of cells that function together to carry out a specific, specialized activity. These organized groups of cells are known as tissues. Plant tissues perform a number of functions including protection, support, conduction, storage, and metabolism. Quite often, a tissue may carry out more than one function; certain vascular plant tissues transport nutrients through the plant body, as well as provide mechanical support for the plant. Plant cells are organized into three general types of tissues—**dermal tissue, vascular tissue,** and **ground tissue** (Fig. 6.1).

Dermal Tissue

Dermal tissue includes the *epidermis* and, in some woody roots and stems, the *cork*. The epidermis consists mainly of an outer layer of tightly packed nonphotosynthetic cells that covers and protects the leaves, stems, and other plant structures. The epidermis also may contain *guard cells,* which, like many other plant cells, contain chloroplasts; guard cells function in regulating the size of tiny openings called *stomata*. These openings permit carbon dioxide to enter the leaf and oxygen to exit. The opening and closing of stomata also regulate the evaporation and retention of water by the leaf. Epidermal cells in leaves and stems secrete a waxy covering called the *cuticle* that aids in minimizing water loss from the plant. Root hairs, which are critical in the absorption of water and minerals from the soil, are specialized extensions of the epidermal cells of most roots.

In most woody plants, the epidermis of the roots and stems is eventually replaced by cork tissue. Cork is composed of cells that are important in wound healing, regeneration of lost cells, and in providing strength and support for some parts of the plant.

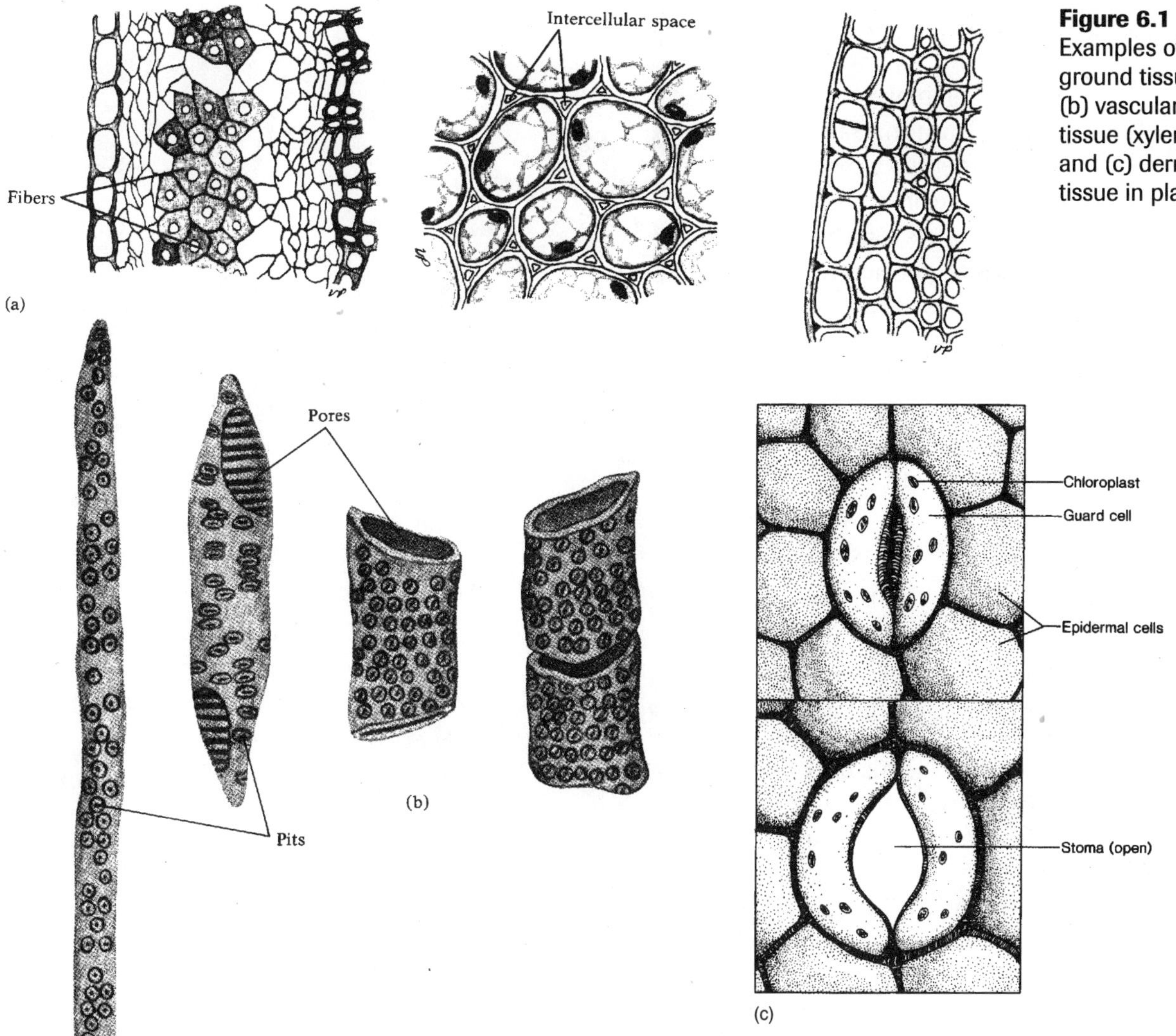

Figure 6.1 Examples of (a) ground tissue, (b) vascular tissue (xylem), and (c) dermal tissue in plants.

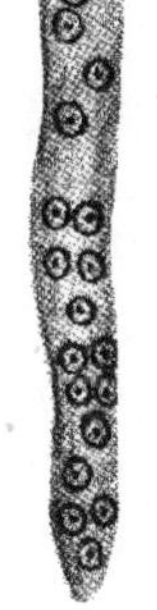

Vascular Tissue

The vascular tissues include **xylem,** which mainly functions in the transport of absorbed water and minerals from the roots upward to the stems and leaves, and **phloem,** which transports various organic nutrients, such as sucrose, throughout the plant. Both xylem and phloem provide mechanical support for the plant as well, and are visible as *vascular bundles* in the "veins" of leaves. Xylem and phloem are composed of numerous small, tubular cells that form a continuous vascular network throughout the plant. Some of these cells also function in storage of materials.

Ground Tissue

The bulk of a plant is composed of ground tissue that functions mainly in support, secretion, storage of materials, and in photosynthesis. Ground tissue cells may be long and slender or oval-shaped with thick or thin walls. There are different types of ground tissue as is evident if we compare the flesh of a fruit with a stringy stalk of celery.

Animal Tissues

In most animals, especially humans and other vertebrates, there are four main categories of tissues: **epithelial tissue, connective tissue, muscle tissue,** and **nervous tissue** (Fig. 6.2). These tissues are present in varying degrees in some invertebrates as well.

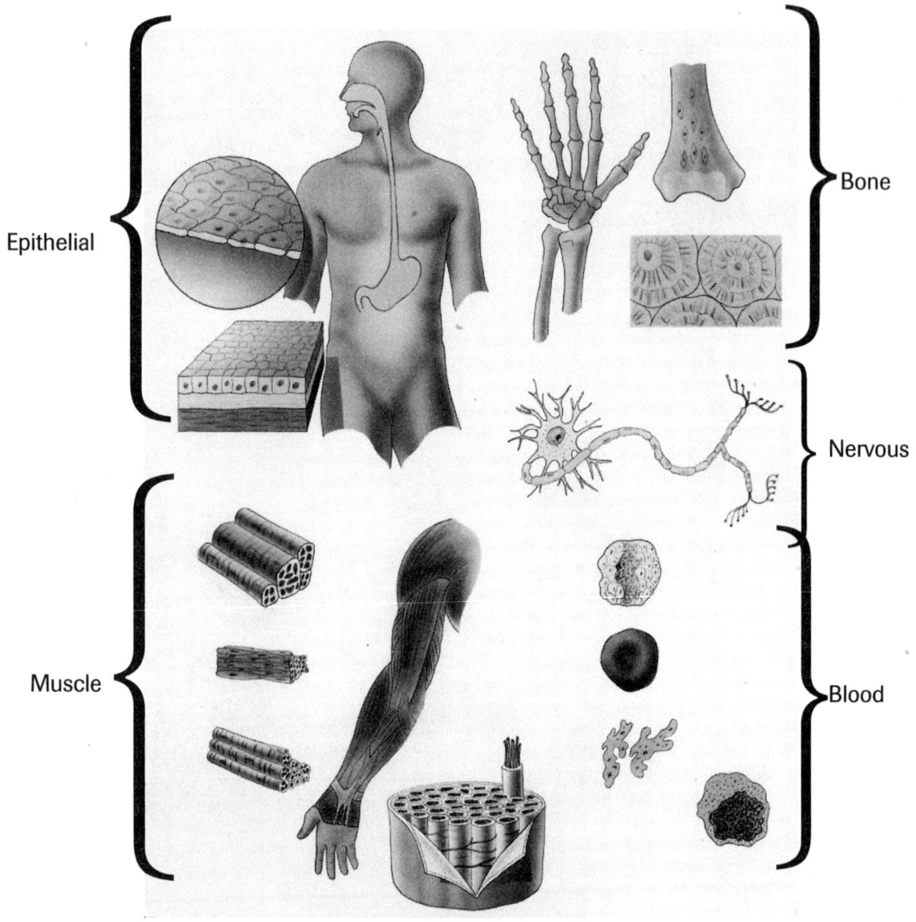

Figure 6.2 Types of body tissue.

Epithelial Tissue

Epithelial tissue forms the outer covering of an animal body and lines the inner walls of body organs and other internal structures. Layers of tightly meshed epithelial tissue cover the body, providing a protective barrier against harmful chemicals and microorganisms. Epithelial cells lining various internal structures, such as the stomach and the intestines, secrete a watery, lubricating mucus and a variety of digestive enzymes. These cells also absorb nutrients that were broken down in the digestive process. Some types of epithelia contain gland cells that secrete an oily substance onto the surface of the skin. The oil tends to soften and moisturize the skin, preventing it from drying and cracking. In some animals, cells lining the respiratory passageways possess cilia that help keep the lungs clear by trapping inhaled dust and other particles. In some vertebrates, cells in the deep layers of the epidermis produce *melanin,* a yellow-brown, brown, or black pigment. Melanin is primarily responsible for the color of the skin in humans; the epidermal cells of darker skinned individuals produce more melanin than do the cells of individuals with lighter skin. By absorbing ultraviolet light from the sun, melanin also helps protect the skin from burning and prevents damage to nuclear DNA.

Connective Tissue

Connective tissue is a general term for tissues that function to support body parts, protect underlying tissues, store fats, bind body parts together, and transport materials throughout the body. Connective tissue can be separated into three general categories: *connective tissue proper, fluid connective tissue,* and *supporting connective tissue.*

Connective Tissue Proper

Connective tissue proper refers to a diverse collection of cells and fibers that functions in a variety of ways. Fatty tissues are part of this group; they store lipids as energy reserves and cushion vital organs. Dense fibrous connective tissue also falls in this category. This tissue is tough and flexible and provides strength, support, and connections between body organs and structures. Examples of fibrous tissues include *tendons,* which attach skeletal muscles to bones, and *ligaments,* which attach one bone to another bone. Loose connective tissue consists of widely separated cells, which serve as the packing material of the body by filling spaces between organs. This tissue also consists of fibers that exhibit a certain amount of elasticity; that is, they are capable of stretching and returning to their original shape. This type of connective tissue lines the walls of blood vessels and provides support for many organs.

Fluid Connective Tissue

The two fluid connective tissues of many animals are *blood* and *lymph.* The liquid component of blood is *plasma,* while the cellular component includes *red blood cells, white blood cells,* and *platelets.* Red blood cells transport oxygen to cells and a small amount of carbon dioxide away from them; white blood cells are part of the immune system that protects the body against infection and disease. Platelets are tiny cell fragments that play an essential role in the blood clotting process.

Lymph is formed from tissue fluid, along with plasma water and small proteins that have leaked from blood capillaries. Lymph returns this excess tissue fluid and protein substances to the circulating blood through a system of lymphatic vessels.

Supporting Connective Tissue

The supporting connective tissues are *cartilage* and *bone.* Cartilage is found throughout most of the vertebrate body, and the various types are both tough and flexible to varying degrees. One type of cartilage anchors the respiratory passageways and also cushions the ends of bones where they meet at a joint. A more flexible cartilage is found in the flap of the outer ear and in the vocal cords of some vertebrates. A very tough form of cartilage forms the disks between vertebrae of the spinal column. The disks act like shock absorbers in preventing the vertebrae from jamming together.

Bone is much harder and stronger than cartilage because of its composition of combined fibers and calcium salts. Bone, of course, makes up the supportive skeleton of many animals and also protects internal organs, such as the heart and brain. In addition, some bones possess hollow cavities that contain bone marrow, which produces blood cells.

Muscle Tissue

Muscle tissue is made up of cells that are capable of contraction and thus is the tissue responsible for body movement, either **voluntary** or **involuntary**. The major muscle tissues in vertebrates are *skeletal, smooth,* and *cardiac.*

Skeletal muscle is attached to bones directly or by tendons. This type of muscle functions in voluntary motion and can also be found at the openings of some tubular structures such as the digestive and urinary tracts. Skeletal muscle is made up of long, multinucleate cells, each cell containing hundreds of nuclei. The cells are not capable of dividing to produce new muscle cells; rather, new cells can be produced by embryonic cells in surrounding tissues. Skeletal muscle cells are typically bundled together and enclosed in a sheath of connective tissue to form a "muscle." Skeletal muscle contraction normally requires stimulation by the nervous system.

Smooth muscle is largely involuntary, which means that an animal cannot control its contractions. Smooth muscle cells are long cells with tapered ends. Each cell contains a single nucleus and is capable of reproducing by division. Smooth muscle is found in the walls of internal organs, such as the stomach, in blood vessels, and in the intestines. Rhythmic contraction and relaxation of smooth muscle propels fluids and other materials through an organ; this action is controlled by nerves originating in the brain and spinal cord.

Cardiac muscle also is involuntary and is found only in the wall of the heart, which spontaneously contracts to pump blood through the body. Cardiac muscle cells are branched cells, each possessing a single nucleus. These cells cannot replace themselves if damaged. Specialized muscle fibers in the heart provide an internal rhythmic contraction that is not dependent on input from the nervous system. Certain nerves from the brain, however, control the rate and force of the heartbeat.

Nervous Tissue

Nervous tissue is comprised of nerve cells, or **neurons,** which are responsible for the sensation-response mechanisms of the body. In the brains of some vertebrates, nervous tissue also stores information. There are several different types of neurons found in nervous tissue. These cells are distinguished by their function; one type of neuron receives sensory information, other types relay or transmit nerve impulses. In addition, nervous tissue contains several different types of *glial* (*glia,* glue) cells that support and protect the neurons. For example, some cells provide a supporting framework for nervous tissue and other cells function as phagocytes that protect the tissue from invading microorganisms. Neurons cannot undergo reproductive division, and therefore most neurons cannot be replaced or repaired if they are damaged.

Plant Organs

Two or more different tissues functioning together form an **organ.** The major organs of higher plants include *roots, stems,* and *leaves,* along with the *reproductive structures.* Each of these organs performs a specialized function in a plant, as we will discover in the following sections.

Roots

Roots anchor the plant in the soil, store food, and absorb water and minerals, which move into xylem tissue for transport to the stems and leaves (Fig. 6.3). The root of a plant originates from the lower end of the plant embryo when the seed germinates. Growth of the root occurs at its tip, where cells in a region called the **apical meristem** undergo division to produce new root cells. The tip is covered by a tough, protective **root cap,** which is produced from cells in the apical meristem. Above the apical meristem is a zone of cells that elongates without dividing; elongation of the cells increases the length of the root and pushes the tip through the soil. The root cap protects the root as it pushes through the soil; in fact, cap cells damaged during this process coat the cap with a viscous film that facilitates root movement.

Epidermal cells above the zone of elongation form outgrowths called **root hairs** that increase the effectiveness of water and mineral uptake from the soil. A root hair is extremely tiny and is an efficient absorber because of its large surface area-to-volume ratio. In addition to producing epidermal tissue, the apical meristem also produces cells that give rise to the ground tissue. Within the root, other specialized cells derived from the apical meristem give rise to the vascular tissues xylem and phloem.

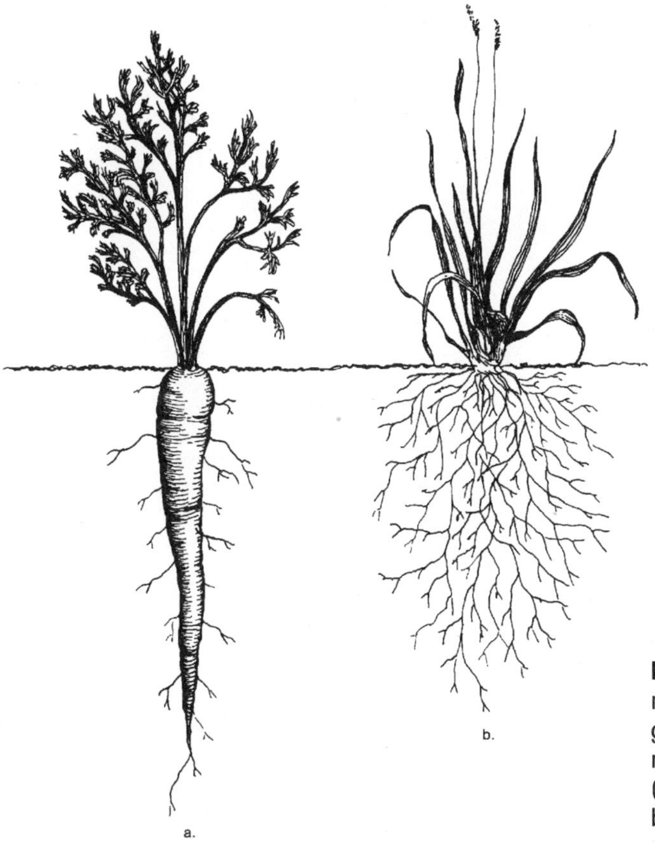

Figure 6.3A The two major types of root systems. (a) Taproot system. (b) Fibrous root system.

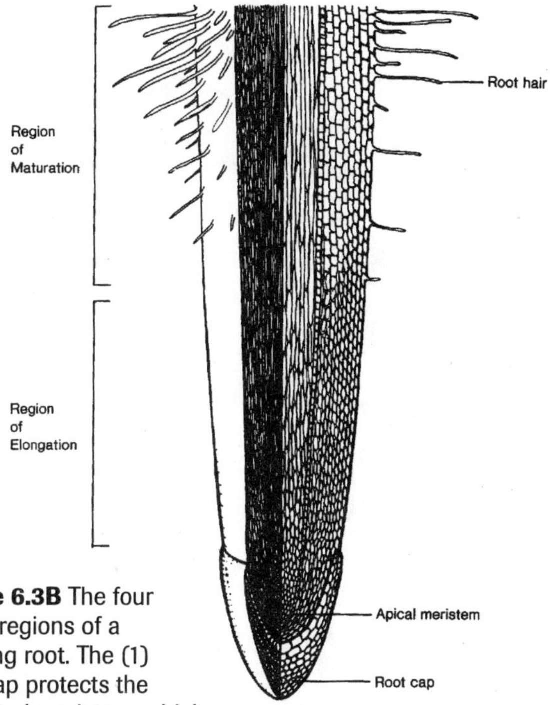

Figure 6.3B The four major regions of a growing root. The (1) root cap protects the (2) apical meristem, which, by cell division, produces the tissues of the root. The (3) region of elongation is responsible for the increase in length of the stem, and the (4) region of maturation is the area where the primary tissues arise through cellular differentiation. Epidermal extensions called *root hairs* greatly increase the absorptive surface area of the root.

Stems

Stems form the basic framework of higher plants, supporting the leaves and conducting food, water, and minerals. When a plant starts to grow, the tip of the stem has an apical meristem in which cells actively divide (Fig. 6.4). These cells produce the tissues of the stem and give rise to embryonic leaves. Below the apical meristem is a zone of elongating cells, which is involved in growth of the stem. As the stem grows, the apical meristem leaves behind pockets of meristematic cells that develop into lateral buds along the length of the stem. These buds develop into lateral stems. As in roots, the stem apical meristem gives rise to epidermal, vascular, and ground tissues.

Leaves

The primary function of leaves is to synthesize nutrients through the process of photosynthesis. A flat leaf consists of a small amount of cellular material relative to its total surface area. This large surface area-to-volume ratio makes a leaf a highly adapted structure for the capture of solar energy. This ratio, however, increases the likelihood of rapid water loss from the leaf surface. The leaf is covered by an upper and lower **epidermis** consisting of irregularly shaped cells that are tightly interconnected (Fig. 6 5). The epidermis protects the leaf from injury and invading organisms and also secretes a waxy cuticle that restricts water loss. The epidermis is perforated by tiny pores called

Figure 6.4A Major growth areas of stem. Vertical growth is a function of the apical meristem. Lateral buds are responsible for growth of branches, and nodes are the sites of leaf attachment. The spaces between nodes are called internodes.

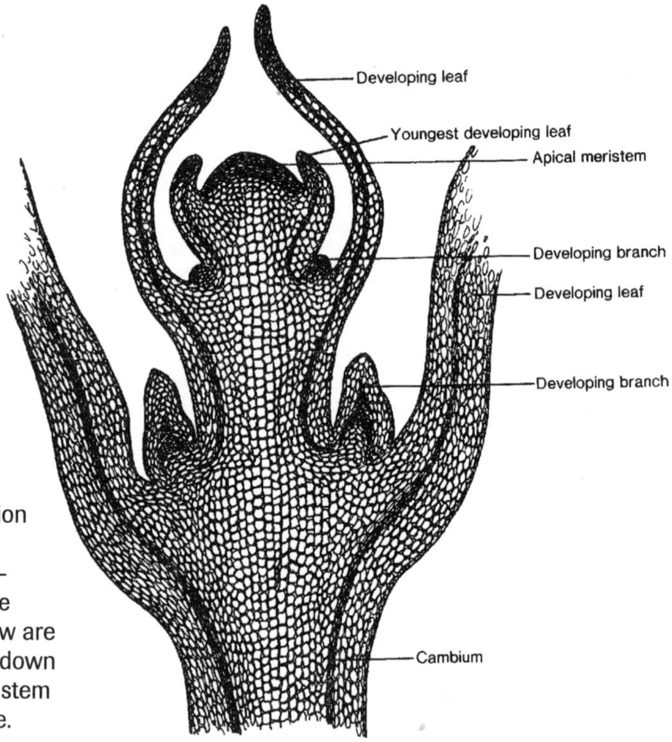

Figure 6.4B Longitudinal section through a growing stem. As dividing cells in the apical meristem increase the length of the stem, the older structures below are left behind. The cambium lays down xylem toward the inside of the stem and phloem toward the outside.

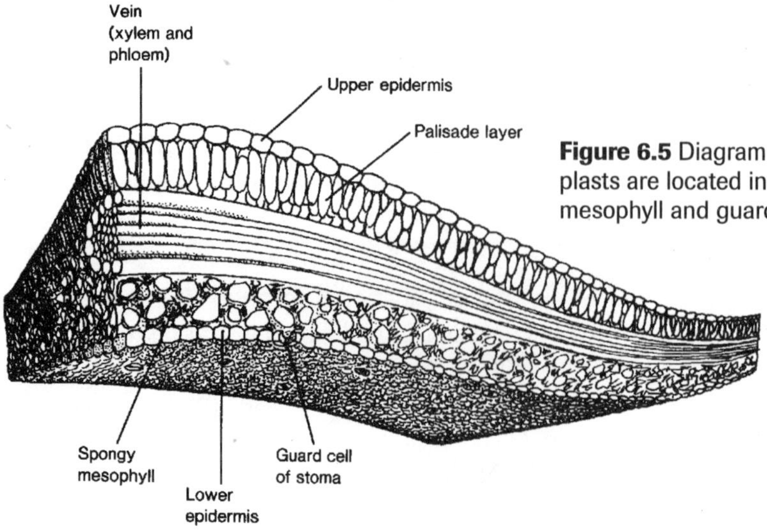

Figure 6.5 Diagram of leaf cross-section. Chloroplasts are located in the palisade layer, spongy mesophyll and guard cells.

stomata (singular, *stoma*), which allow gas exchange between the surrounding air and cells within the leaf. Water is also released through stomata as it evaporates from the plant. Stomata open to allow gas exchange and close to prevent excessive water loss. This process is regulated by a pair of **guard cells** located on either side of each stoma.

The vascular tissues—xylem and phloem—comprise the veins of a leaf and are continuous with the vascular tissues of the stem. The ground tissue consists of photosynthetic, storage, and supporting cells located between the two epidermal layers.

Reproductive Structures

Reproductive structures vary in the different groups of higher plants. Some angiosperms, or flowering plants, possess both male and female structures, while others have only a male structure or only a female structure. The male reproductive structure is called the **stamen** and the female structure is called the **carpel** (Fig. 6.6). The carpel consists of a slender stalk, the *style,* that widens at its base to form the *ovary*. Located within the ovary are one or more *ovules,* each of which contains an egg.

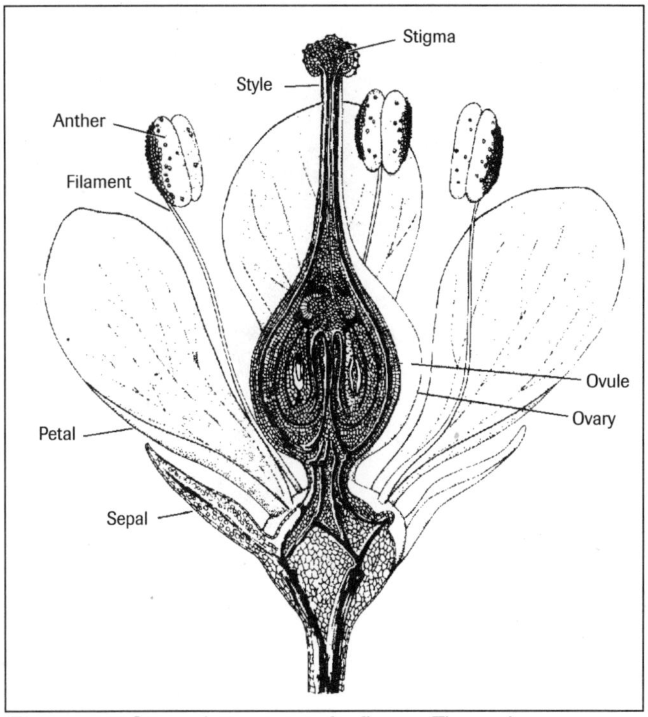

Figure 6.6 General structure of a flower. The male structure—the stamen—consists of the filament and anther; the female structure—the carpel—consists of the style and the ovary.

The stamen consists of a stalk, the *filament,* at the tip of which is the *anther.* Inside the anther, pollen grains develop that give rise to sperm cells. By a variety of mechanisms, including wind, bees, and birds, the pollen grains are carried from the anther to the stigma, the sticky tip of the female style. After the stigma collects the pollen grains, a pollen tube subsequently develops, carrying the sperm cells down through the style to the ovary where fertilization and seed development occur. Different flowering plants produce an amazing variety of structures to house seeds—fruits such as apples and oranges to peanuts to maple "helicopters"—hence, the name "angiosperm," which is Greek for "vessel seed" or a seed contained within a "vessel."

Angiosperms may have evolved in Gondwanaland, a giant land mass that separated from the supercontinent Pangaea some 190 million years ago. The evolution of flowering plants is thought to be correlated with that of their animal pollinators and reflects a fascinating relationship. This is evident in the structural and behavioral features of both groups of organisms. Some animals became beneficial to the plants by transferring pollen, and the animals benefited by feeding on sugary nectar or pollen, a rich source of protein. Over time, different flowers have evolved that have efficiently exploited their pollinators, and the pollinators have become increasingly adapted to exploit the flowers in obtaining food. The nectar of some flowers, for example, is found deep in the floral tube at the base of the petals where only birds with long tongues or beaks can reach. The head of the bird brushes against the reproductive structures of the flower, collecting pollen that is transferred to another plant as the bird searches for more nectar. This mutual adaptive change in two or more species over time is known as *coevolution.*

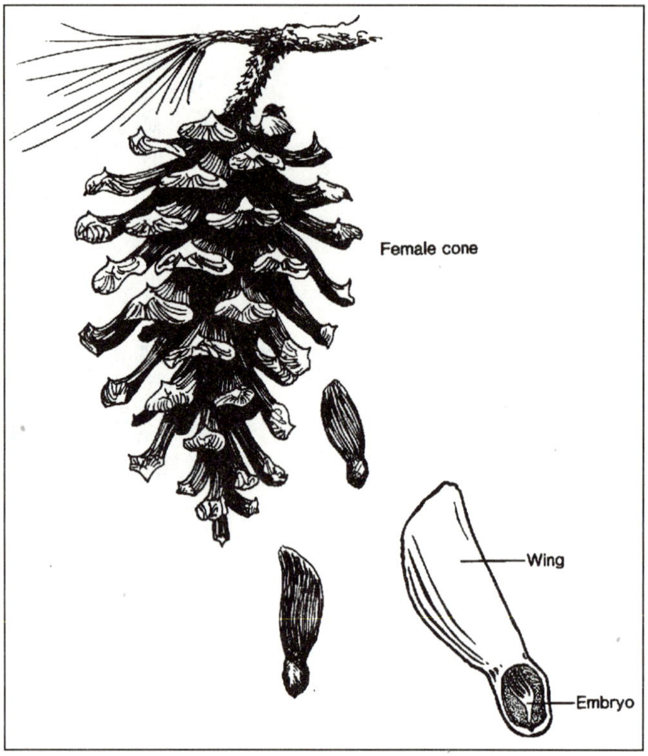

Figure 6.7 Female cone of pine, showing the winged seeds and a cutaway view of the pine embryo.

In contrast to the flowering plants, the other major group of vascular plants are the *gymnosperms,* which is a Greek word meaning "naked seeds." This group of plants includes pine trees, firs, and spruces. Gymnosperms possess separate male pollen cones and female ovulate cones in which tiny spores and eggs are produced, respectively (Fig. 6.7). The spores, carried by the wind, give rise to the sperm cells, which fertilize the egg cells within the female cone ovules. The developing plant embryo lies "naked" or exposed on the outer surface of the female cone until it eventually becomes enclosed by a tough seed coat.

Animal Organs and Systems

Among the invertebrate animals, only the sponges and cnidarians lack different tissues that interact to form specialized organs. Further, the majority of invertebrates also possess different organs interacting to form **body systems,** such as digestive, reproductive, and circulatory systems. All vertebrates possess both organs and systems. Each system normally functions through a coordinated interplay with other systems; that is, the systems are interdependent and rarely can one system function without input from one or more other systems. The muscular system, for example, usually cannot bring about coordinated or purposeful movement without input from the nervous system.

Humans and many other vertebrates possess a dozen different systems involving hundreds of different structures and organs, all composed of four basic tissue types. We should keep in mind, however, that these multiple and complex organizational levels ultimately reflect the structural and functional characteristics of individual cells. The progression of organisms from the cellular level up through the systems level underscores the common life feature of organization and illustrates the relationship between unity and diversity in the living world. We will examine some of the major vertebrate organs and systems in Chapters 10 and 11.

Survey of the Kingdoms of Life

We indicated in Chapter 3 that biologists have separated all organisms into five kingdoms—Monera, Protista, Plantae, Fungi, and Animalia. Monerans, as pointed out earlier, are the oldest known organisms, dating back about 3.5 billion years. The first eukaryotic fossils are about 1.5 billion years old. There is little fossil evidence available from the rise of eukaryotes to the earliest known fossils of marine plants and animals nearly 600 million years ago. Unfortunately, this 900 million-year gap in the fossil record prevents us from following the early evolution and expansion of the major plant and animal groups. It is thought that many of the organisms in existence more than 600 million years ago were soft-bodied and simply did not fossilize readily. It is also reasonable to assume that most of the fossils that formed were destroyed over many millions of years.

Kingdom Monera

The simplest organisms are the prokaryotic bacteria, organisms restricted, for the most part, to the cellular level of development. In some species of bacteria, two or more cells are grouped together, but these species are not considered to be truly multicellular because there is little significant division of labor among the cells. Bacterial cells are found in an assortment of shapes, including spherical, rod-shaped, or spiral-shaped (Fig. 6.8). The average size of prokaryotic cells, 2.5 micrometers (um), is only about one-tenth that of most eukaryotic cells (25 um). Most people probably associate bacteria only with infection and disease. Bacteria are responsible for a number of human diseases, including tuberculosis, strep throat, tetanus, diphtheria, scarlet fever, syphilis, and gonorrhea. The **pathogenic** (disease-producing) capability of bacteria was first suggested by the great French scientist Louis Pasteur (1822–1895) in the late nineteenth century. As the cell theory was being developed in the 1830s, scientists demonstrated that beer and wine fermentation was caused by the growth of yeast cells. Subsequently, Pasteur confirmed this by showing that fermentative activity is retained if yeasts and bacteria are transferred from one wine or milk culture to another. He also demonstrated that wine will spoil if contaminated by certain bacteria. Applying this reasoning to animals, Pasteur hypothesized that microorganisms also might be the causative agents of animal diseases. This was confirmed in the 1860s when he showed that microorganisms were indeed responsible for silkworm diseases. Pasteur developed the process of *pasteurization,* in which milk or wine is heated sufficiently to kill microorganisms but not enough to spoil the beverage.

But there is another side to the story of bacteria. The great majority of bacteria make numerous beneficial contributions to the biosphere; in fact, life on Earth would be impossible without some of them. For example, bacteria play a critical role in the processes of decay and nutrient recycling. You may have assumed that a dead plant or animal simply decays or rots as part of the death process. But the actual **decomposition** of organic matter is caused in large part by bacteria that break down complex molecules to simpler ones. As a result, various elements, such as carbon, nitrogen, and phosphorus, are released from the dead organic matter and **recycled** to be used again as nutrients. Bacteria, therefore, are instrumental in the principle of conservation of matter. In the recycling of carbon, for example, soil or water bacteria release carbon from dead organic matter in the form of the gas carbon dioxide (CO_2). This gaseous carbon may be utilized by photosynthesizing plants and incorporated into carbohydrates, such as glucose, and then stored by the plants as starch. When the plants die, the cycle is repeated. Were it not for the ability of bacteria and a few other organisms

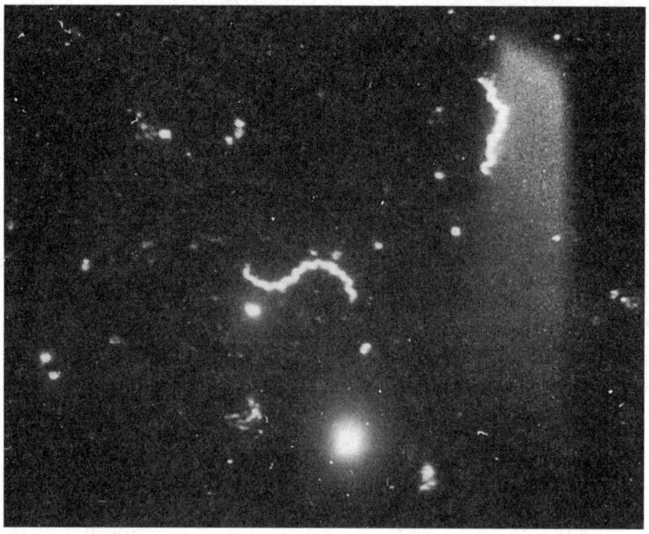

Figure 6.8(a) The basic shapes of bacterial cells: (a) spiral-shaped.

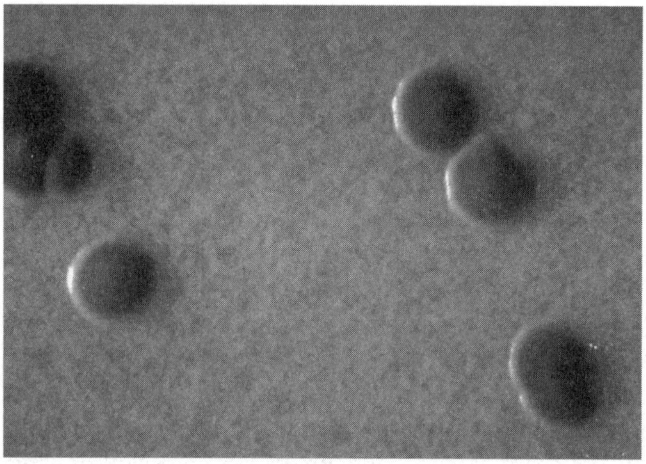

(b) Spherical-shaped bacteria.

(c) Rod-shaped bacteria.

such as fungi to decompose dead organic matter, the bodies of every dead plant or animal that ever existed would be piled on one another over the surface of Earth. We will have more to say about bacteria and the cycling of matter in Chapter 8.

Bacteria also have proved to be invaluable in many commercial industries. Vinegar, vitamins, dairy products, cattle feed, and various textiles are manufactured employing the metabolic activities of bacteria. Through genetic engineering, bacteria have been used to produce hormones such as insulin and human growth hormone, along with other organic compounds. Several species of bacteria produce some widely used antibiotics, such as streptomycin and tetracycline. The human intestine harbors a population of bacteria that synthesizes vitamins, such as vitamin K, and also metabolizes undigested food.

Kingdom Protista

The protists, which today include various unicellular, colonial, and multicellular algae, protozoa, and funguslike protists, first appeared perhaps 1 to 1.5 billion years ago. This group of diverse organisms has been assigned to the kingdom Protista, but their characteristics cannot be defined simply. Protists are all eukaryotic organisms; beyond this characteristic, they defy clear classification. They exhibit an "intermediate" nature; that is, like plants, some can make their own food through photosynthesis, but they move around like animals. Protists appear to have evolved from prokaryotes and are ancestral to the members of the other eukaryotic kingdoms, the plants, animals, and fungi. It was the rise of the first eukaryotic cells—unicellular protists—that led to the evolution of multicellular organisms and the subsequent flourishing of life on Earth.

Algae

We can divide the **algae,** or plantlike protists, into seven different phyla, which include the diatoms, dinoflagellates, euglenoids, and the green, gold, red, and brown algae (Fig. 6.9). These organisms are referred to as "plantlike" because most of them produce their own food through the process of photosynthesis.

There are three groups of protists that are called **phytoplankton** ("drifting plants"), the tiny algae floating on the surface of lakes and oceans. The phytoplankton group of organisms is a major food source for many freshwater and marine animals, and the photosynthetic forms produce much of Earth's oxygen. Diatoms are constructed of two silicon shells that fit together like the top and bottom of a shoe box. Over millions of years, diatom shells have accumulated on the ocean floor, forming deposits of "diatomaceous earth." This material is mined and then used in fertilizers, detergents, polishes, toothpaste, scouring powders, and a variety of other products.

Dinoflagellates, a term meaning "spinning whips," are mostly marine organisms and are another major group of phytoplankton. Some species are photosynthetic, and others are nonphotosynthetic parasites of animals. Most species of dinoflagellates have two flagella that, when beating, cause the organisms to spin like tops. On occasion, some species of dinoflagellates multiply to produce such vast numbers of organisms that they color the water red. These "red tides," usually appearing off the Pacific Coast and in the Gulf of Mexico, are responsible for the death of countless fish and oysters, the consequence of an extremely potent nerve toxin produced by these dinoflagellates.

Euglenoids are microscopic unicellular protists found mostly in fresh water; some species are photosynthetic and some are not. They move by means of two anterior flagella and possess a tiny "eyespot" that is light- sensitive, allowing the organism to respond to light in accordance with its photosynthetic activity.

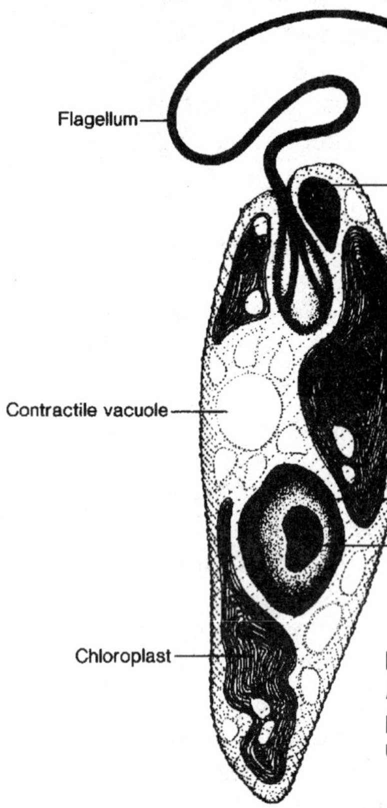

Figure 6.9A *Euglena,* a photosynthetic unicellular alga.

Figure 6.9B Electron micrograph of diatoms.

Other algae are separated into green, gold, red, and brown algae by virtue of the particular pigment found in each type. Algae are largely distinguished from higher plants in that they lack true roots, stems, and leaves. Green algae range in size from microscopic, single-celled forms to multicellular species over 7 meters (25 feet) in length. Multicellular species include the filamentous alga *Spirogyra*—the dreaded "pond scum," and the marine "sea lettuce" *Ulva.*

Green algae are of great evolutionary significance because they are thought to be the organisms that gave rise to all land plants. Moreover, one group of green algae provides some insight into the evolution of multicellular or colonial organisms from unicellular forms. The unicellular member of the group is *Chlamydomonas,* a microscopic, photosynthetic alga having two anterior flagella for locomotion (Fig. 6.10). This alga stands at the beginning of the evolutionary line that illustrates the possible progression toward more complex **colonial** or multicellular forms. Following *Chlamydomonas* in complexity is a form that exists in colonies of 4, 8, 16, or 32 *Chlamydomonas*-like cells. This type is followed by a sequence of increasingly complex forms, each displaying an increase in cell number. The final organism in this evolutionary sequence is the alga *Volvox,* a large, spherical colony consisting of 500 to 60,000 *Chlamydomonas*-like cells. This progression from *Chlamydomonas* to *Volvox* represents the stages by which a unicellular ancestor may have given rise to multicellular plants (Fig. 6.11). We cannot necessarily conclude that each alga in the sequence evolved from the preceding one, although that may be the case. Of additional evolutionary significance is the increasing interdependence and division of labor among the cells of the alga colonies; that is, the cells become increasingly dependent on one another and cannot exist separately from the colony. Some cells, for instance, become specialized for reproduction, whereas others become specialized for locomotion.

Red and brown algae are generally known as seaweeds and most species inhabit the oceans. Some red algae, however, are found in fresh water and in the soil. Some

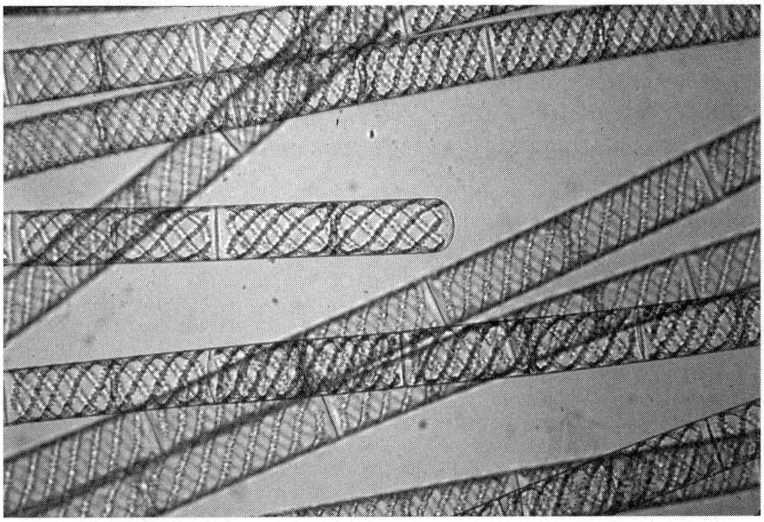

Figure 6.10B Spirogyra, a filamentous green alga.

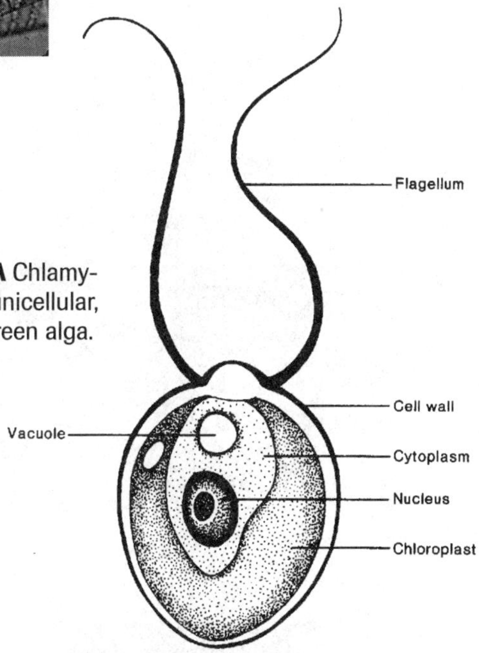

Figure 6.10A Chlamydomonas, a unicellular, freshwater green alga.

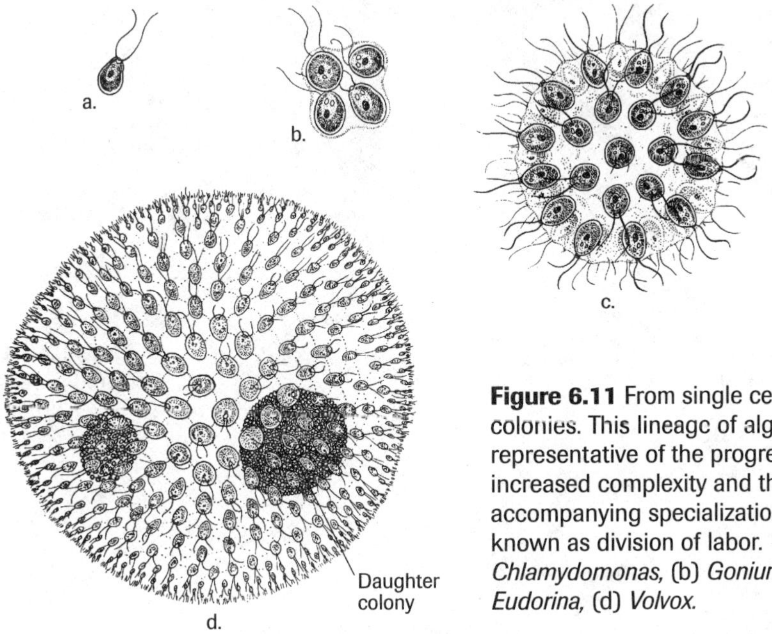

Figure 6.11 From single cell to colonies. This lineage of algae is representative of the progression of increased complexity and the accompanying specialization of cells known as division of labor. (a) *Chlamydomonas,* (b) *Gonium,* (c) *Eudorina,* (d) *Volvox.*

brown algae are extremely large sea-weeds, with a few species attaining a length of over 90 meters (300 feet).

PROTOZOA

The animal-like protists are the unicellular or colonial *protozoans,* represented by six different phyla (Fig. 6.12). Protozoans lack chloroplasts and thus cannot produce food by photosynthesis. They move by means of **flagella, cilia,** or **cytoplasmic streaming** (sending out cytoplasmic extensions called **pseudopodia,** or "false feet"). One phylum of protozoans is entirely parasitic, living within a host organism, and the adults are not capable of movement.

Among the flagellated protozoans is the dangerous parasite *Trypanosoma,* the agent of African sleeping sickness. This parasite enters the bloodstream via the bite of tsetse

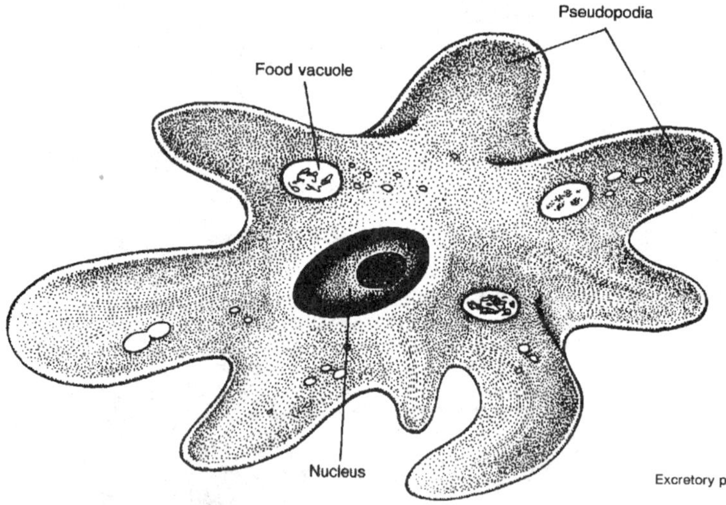

Figure 6.12A *Amoeba,* a protozoan. Pseudopodia function in movement and in phagocytosis, the engulfment of food particles.

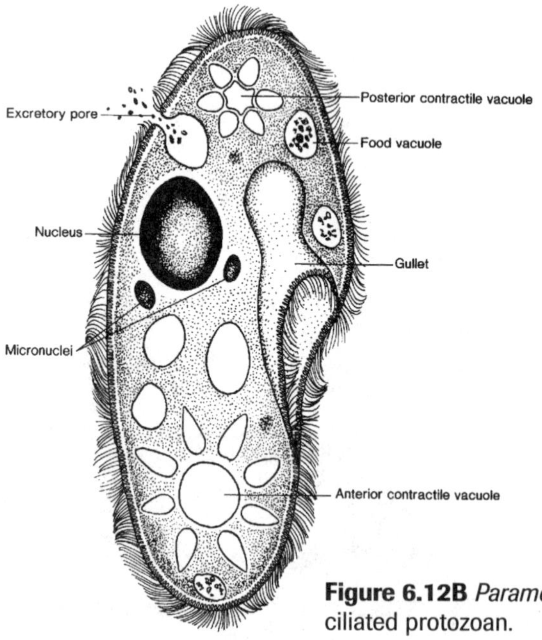

Figure 6.12B *Paramecium,* a ciliated protozoan.

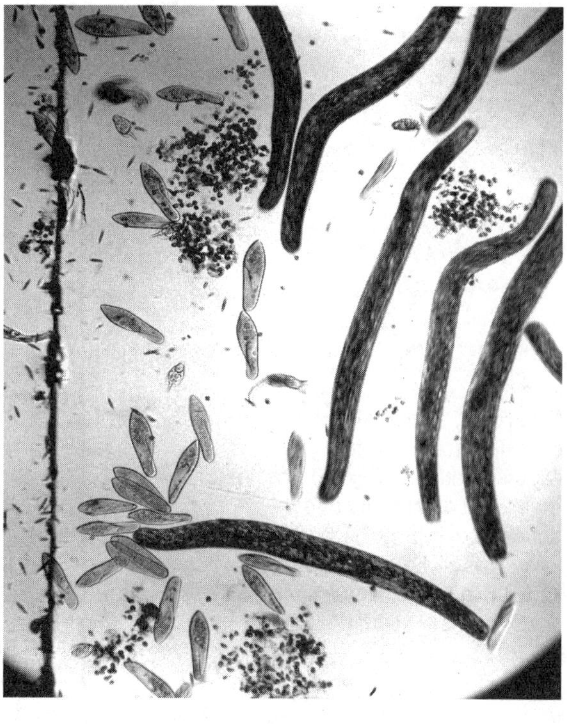

Figure 6.12C Species of single-celled protozoans.

flies. From there, these protozoans attack the brain. Some flagellated protozoans live symbiotically in the intestine of termites. These flagellates release enzymes that aid the termite in digesting the cellulose of wood.

The ciliated protozoans are named for the hundreds or thousands of tiny hair-like cilia extending from the cells of these organisms. Cilia function in both locomotion and obtaining food. Unlike other protozoans, ciliates possess two different kinds of nuclei—a large *macronucleus* and one or more smaller *micronuclei.* The macronucleus functions in growth and asexual reproduction, and the micronuclei are involved in a type of sexual reproduction known as conjugation. Conjugation involves the exchange of micronuclei between the two mating organisms, which produces individuals that are genetically different from the parent organisms.

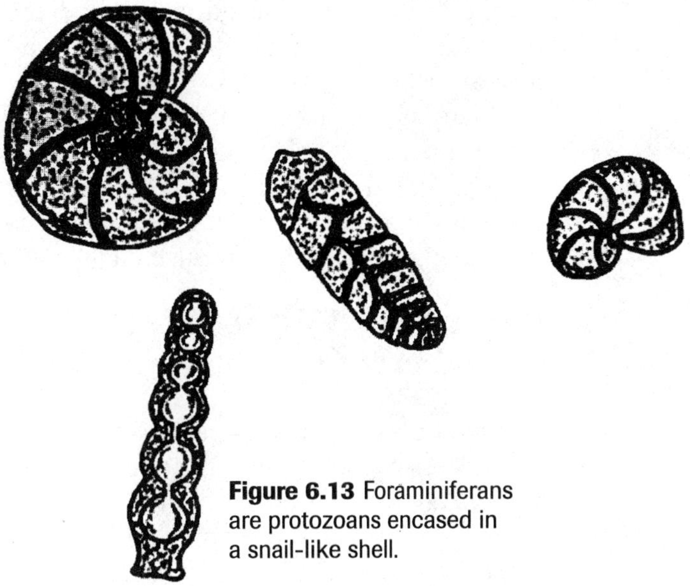

Figure 6.13 Foraminiferans are protozoans encased in a snail-like shell.

Perhaps the most familiar representative of the ciliates is *Paramecium,* the slipper-shaped protozoan commonly found in water containing decaying vegetation. An unusual ciliate is *Vorticella,* a freshwater species composed of a bell-shaped body attached to an elongated stalk. The organism anchors itself by the stalk, which can coil into a tight spiral. Should a tasty morsel happen by, the stalk uncoils in a "popping" movement, thrusting the body outward. The morsel, with no time to react, is then drawn into a ring of cilia encircling the open end of the bell.

Protozoans that employ pseudopodia as a means of locomotion and feeding include the amoebas, most of which are free-living, although there are parasitic forms. These forms include a species that causes an inflammation of the intestine called amoebic dysentery. This infection is characterized by vomiting, nausea, diarrhea, and abdominal pain. Other amoebic forms are encased within solid, often brightly colored shells (Fig. 6.13). There are a variety of types of these protozoans, including the foraminiferans, heliozoans, and radiolarians. These protozoans move using pseudopodia and their shells are the main ingredient of chalk.

Another phylum of protozoans are animal parasites, some of which infect not only humans, but reptiles, birds, and other mammals as well. These organisms cannot live separately from their host organism(s). The most infamous member of this group is *Plasmodium,* a species of which causes human malaria. The parasite is transmitted from one human being to another by the female *Anopheles* mosquito. The life cycles of *Plasmodium* and other members of this group are quite complex, usually requiring two hosts to complete one cycle.

Slime Molds and Water Molds

Organisms in this group include the slime molds and water molds. Slime molds are single-celled organisms that obtain nutrients by secreting digestive enzymes onto decaying organic matter and subsequently absorbing the digested matter into the cell. Members of one phylum of slime molds form an amoeboid mass that contains thousands of free nuclei. The mass feeds by phagocytosis as it creeps across rotting logs or dead leaves. If feeding or environmental conditions become unfavorable, the mass dries up and develops specialized stalked bodies containing spores at their tips. When conditions are again favorable, the spores are released, and two spores unite to produce a new amoeboid mass.

Water molds somewhat resemble fungi and exist on dead or decaying organic matter or as parasites on animals. They probably are most closely related to the amoebalike protists than to true fungi. The cell walls of water molds contain cellulose like that of most green plants, rather than chitin as found in the cell walls of true fungi. Also unlike true fungi, water molds produce flagellated reproductive cells.

Kingdom Plantae

All members of the plant kingdom are multicellular and produce their own food through the process of photosynthesis. They are classified into at least twelve different divisions. In the evolutionary history of both the plant and animal kingdoms, the movement of life from water onto land presented an assortment of adaptational problems, most of which centered on the need for water. The point was made in Chapter 4 that the temperature of water does not tend to rise or fall drastically; aquatic organisms, then, exist in a relatively temperature-constant environment. With terrestrial species, however, temperature fluctuations and other climatic variables require special adaptations. Further, many terrestrial organisms must have a system for bathing their cells in water and for transporting water from one part of the body to another. In some aquatic organisms, reproductive cells are released and fertilized in water. Subsequent development of the young also occurs in open water. Thus, in the evolution of most land plants and animals, new methods were necessary to allow reproduction and development to occur out of water. In addition, large terrestrial organisms require physical support and cannot depend on the buoyancy of water.

Land plants (although a few species do grow in water) are separated into the general categories of either **nonvascular** or **vascular** plants, depending on the presence or absence of the internal conducting tissues xylem and phloem. Nonvascular plants, such as mosses and liverworts, are not true land plants in the same sense as the higher vascular plants. In lacking true xylem and phloem, nonvascular plants do not transport fluids efficiently and must rely on the water immediately present in their moist environment. They possess a waterproof coat to prevent water loss and drying out. Also, water in the form of rain or dew is required for the transport of sperm from the male structures to the female. In the absence of supporting vascular tissue, nonvascular plants cannot grow very tall, and usually appear as green carpets spread over a surface. Mosses, for example, may coat the surface of wet rocks or cover the side of a tree trunk.

Vascular plants have been around for over 400 million years and include some small, primitive mosslike and fernlike groups and the more familiar true ferns and seed plants. In addition to the vascular tissues xylem and phloem, all vascular plants have true roots, stems, and leaves. The primitive vascular plants and the true ferns still require a film of moisture for transferring sperm and thus have not adapted fully to a terrestrial existence. The seed plants—**gymnosperms** and **angiosperms**—however, have completely adapted to land. Gymnosperms include the extinct seed ferns and the present-day palmlike cycads, the ginkgo tree, and the cone-bearing trees, such as pines, cedars, firs, and redwoods. Some redwoods of Northern California grow to a height of more than 100 meters (325 feet), and some bristlecone pines of the Sierra Nevada range are over 4000 years old.

Angiosperms are the flowering plants, the dominant and most successful form of plant life on Earth today. Almost any type of vegetable, flower, fruit, nut, cereal grain, grass, or hardwood tree you can think of is an angiosperm. They are characterized predominantly by the presence of flowers—the reproductive structures of angiosperms—and the **fruit,** which contains the seeds. The term "angiosperm" is a reference to the female tissues, the ovary, that cover, or enclose, the seed. The fruit actually consists of the ovary with the seeds inside (Fig. 6.14). In fruits such as apples, the core is the wall of the ovary, which surrounds the apple seeds. The fleshy part of the apple consists mainly of stem. In pea plants, the pod is the wall of the ovary and is, therefore, a fruit; the peas within the pods are the seeds.

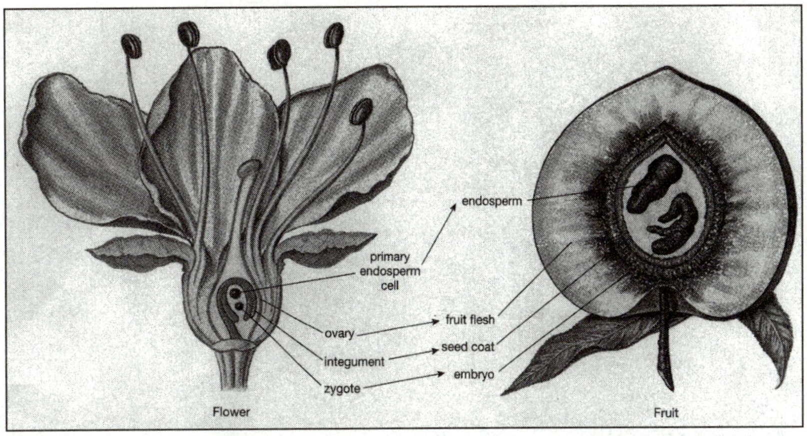

Figure 6.14 Development of fruit and seeds. The fruit and the seed within it develop from various parts of the flower. Starting at the outside, the ovary wall ripens into the fruit flesh, which may be soft and tasty like a peach, or variously hard, hooked, or tufted in other fruits. The integuments of the individual ovule, which surround the embryo sac, harden and become waterproof, forming the seed coat. These two fruit parts are derived from tissues of the parent sporophyte plant. Within the seed, two structures develop from cells of the fertilized female gametophyte. The triploid endosperm, a food-storage structure within the seed, and the zygote, which develops into the embryo. Reprinted from *Biology: Life on Earth,* Fourth Edition, by Gerald and Teresa Audesirk, 1996, Prentice-Hall, Inc.

In both angiosperms and gymnosperms, reproduction involves pollen grains that house the male reproductive cells. Pollen does not require water for transport and thus is a significant adaptation of plants to life on land. Seeds nourish and protect the plant embryo within and prevent it from drying out. Seeds can be blown by the wind, carried by flowing water, and transported in the fur and stomachs of animals. By many diverse methods, angiosperms and gymnosperms have been widely dispersed over the surface of Earth.

Kingdom Fungi

The evolutionary origin of fungi is uncertain but they are thought to have evolved from some type of protistan ancestor at least 400 million years ago. Being plantlike in appearance, fungi originally were placed in the plant kingdom, but their many unique features warrant classifying them in a kingdom of their own. The majority of fungi consist of a network of fine, slender **filaments** (except for the unicellular yeasts) and are not multicellular in the true sense of the word (Fig. 6.15). The cells along a filament are not partitioned completely into individual units, which allows the cytoplasm and its constituents to flow freely throughout an entire filament. The major component of fungal cell walls is not cellulose as found in plant cell walls, but rather **chitin,** the tough polysaccharide that forms the outer skeleton of crustaceans, insects, and other arthropods. Fungi lack chlorophyll and therefore are not capable of photosynthesis. Rather, they feed by secreting digestive enzymes onto an external food source. Fungi decompose living or nonliving organic matter; when broken down, the digested food is absorbed into the filaments through tiny rootlike structures. Thus fungi, along with bacteria, are involved in the decomposition of dead organic matter and the subsequent recycling of various nutrients. This cycling is essential for the continuation of life on Earth.

Fungi, like bacteria, display both harmful and beneficial effects. Some fungi are parasitic on plants and animals, including humans, and some are major agents of food

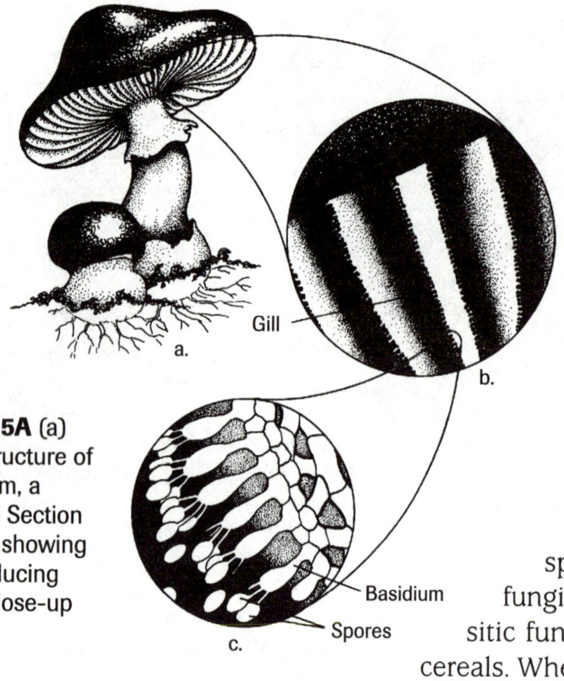

Figure 6.15A (a) General structure of a mushroom, a fungus. (b) Section of the gills showing spore-producing cells. (c) Close-up of spores.

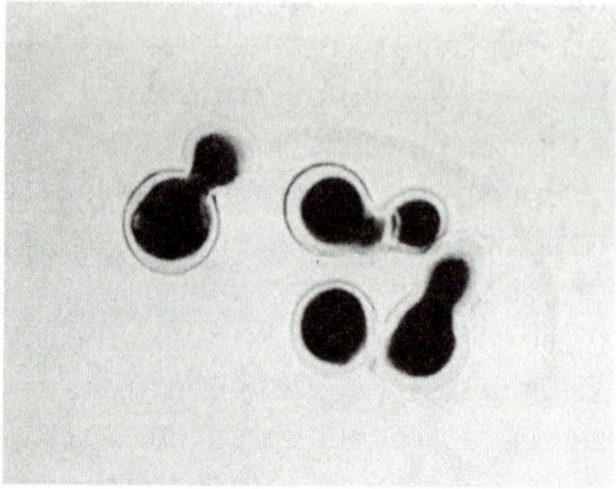

Figure 6.15B Yeast cells, a fungus, shown here budding.

spoilage. Several plant diseases caused by fungi also affect humans. Some species of parasitic fungi cause *ergot,* a disease of rye and other cereals. When eaten by humans, this contaminated rye may produce serious mental and physical complications. For example, one of the constituents of ergot is lysergic acid, a substance found in the psychedelic drug LSD. Various other plant diseases, such as Dutch elm disease and chestnut blight, also are caused by fungi.

Human diseases caused by fungi include ringworm, lung infections, and athlete's foot. The grayish mass growing on bread kept for too long is a fungus as well. Speaking of bread, the yeasts, the only unicellular fungi, are used to make bread dough rise. Yeasts also are instrumental in the production of beer and other alcoholic beverages. Other fungi are used in the manufacture of cheese and in the making of soy sauce. Mushrooms, which frequently appear on our dinner tables, are fungi. The antibiotic penicillin is obtained from a fungus, species of which also are used to enhance the flavor of Roquefort cheese. In addition to antibiotics, several other chemicals of great medical importance have been isolated from fungi. Drugs that prevent rejection of organ transplants, for example, are derived from chemicals extracted from some species of fungi.

Fungi are one of the organisms that comprise **lichens,** mosslike patches that grow on rocks and tree trunks. Lichens are organisms that result from the symbiotic association between a fungus and either a green alga or a blue-green bacterium. In the arctic tundra, the most prevalent form of food for grazing reindeer and caribou is "reindeer moss," which is not a moss at all but a type of lichen.

Kingdom Animalia

Invertebrates

Animals have been separated by biologists into two broad categories: **invertebrates** and **vertebrates.** Invertebrates lack a backbone; vertebrates have a backbone consisting of either cartilage or bony vertebrae. There are at least eleven different phyla of invertebrates (Fig. 6.16). The vertebrates belong to the phylum Chordata and thus constitute a subphylum, under which there are seven living classes: agnathans, two classes

Figure 6.16 A few of the representatives of the major phyla of invertebrates.

A glass sponge.

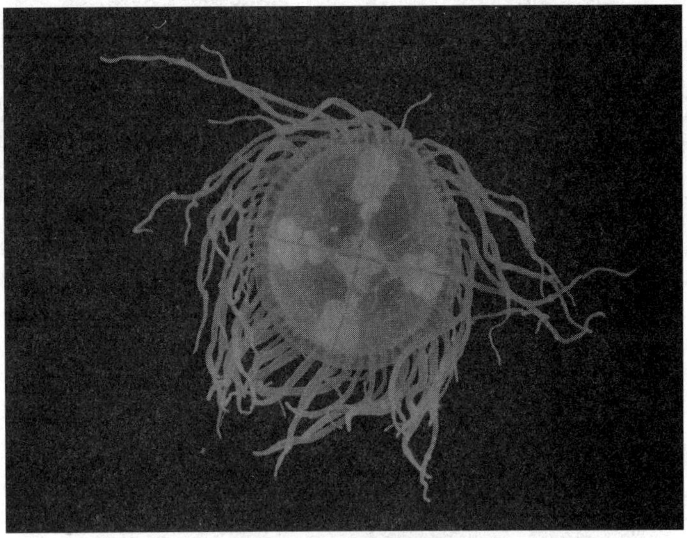

A jellyfish, a cnidarian.

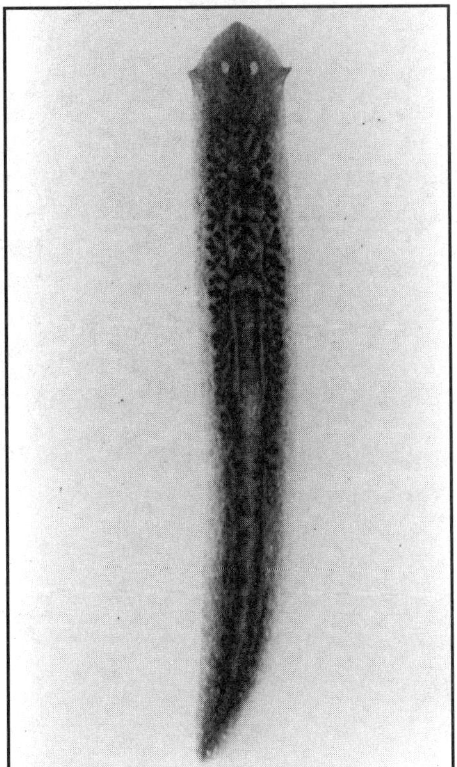

A freshwater planarian,
a common flatworm.

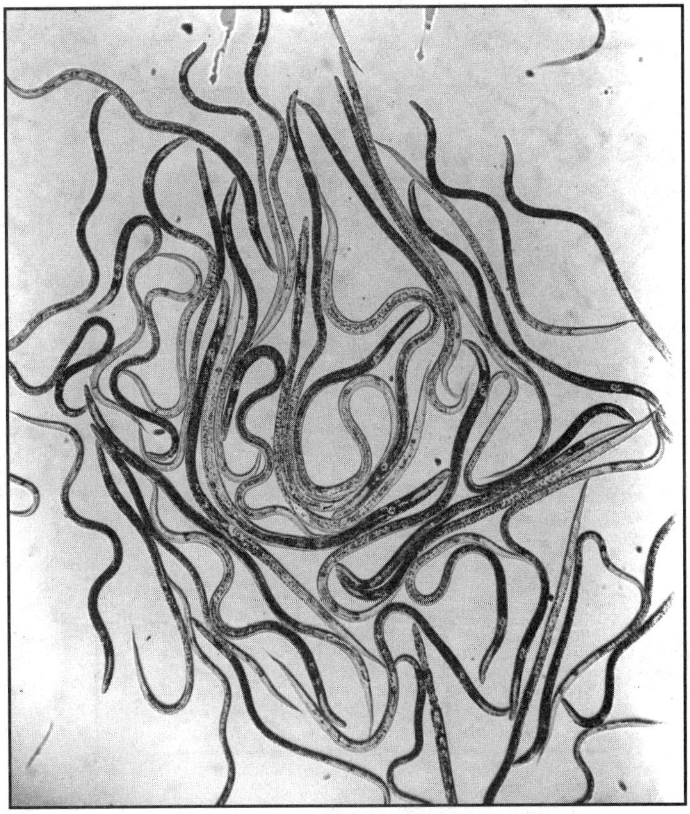

Vinegar eels, a species of roundworms.

An octopus, a mollusk.

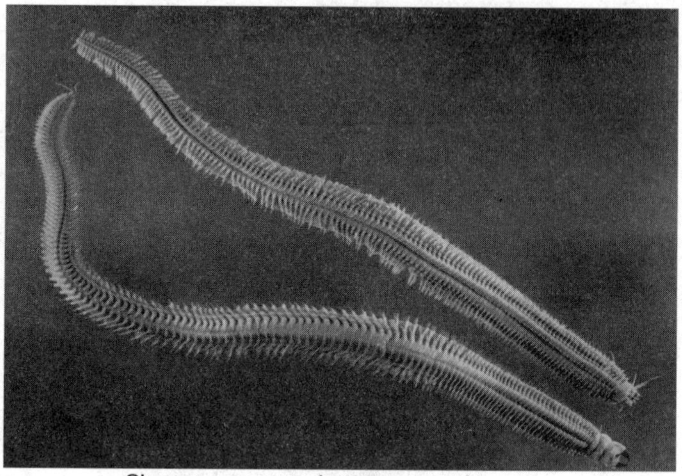
Clamworms, a marine segmented worm.

A praying mantis, an arthropod.

A group of representative echinoderms.

of fishes, amphibians, reptiles, birds, and mammals. There is some controversy concerning the evolution of many of the animal groups, but all animals are multicellular and have attained at least a tissue level of development, with the exception of sponges. Further, animals cannot produce their own food; they must rely on taking in nutrients produced by other organisms. Animal development always includes a time period for development of the embryo, and most animals are capable of some form of locomotion. Let's now take a look at the different levels of organization in this kingdom, starting with simple sponges and continuing through to the more complex mammals, examining evolutionary trends, characteristics, and representative organisms along the way.

Sponges

Although sponges represent a multicellular level of development, they lack organs and their cells are loosely organized and relatively unspecialized. They are the only animals without a nervous system; sponges, however, can sense and respond to environmental stimuli. Most sponges are marine, although some species are found in fresh waters. Sponges obtain food by moving water through the pores on their bodies and trapping particles suspended in the water.

Sponges have been used rather extensively in studies on **regeneration,** the replacement of lost body parts. If a section of the sponge is cut away or broken off, the organism is able to replace, or regenerate, the missing part. One of the classic experiments with these animals involves passing the sponge through a silk mesh, separating it into tiny pieces, perhaps even individual cells. Eventually, the cells and pieces will rejoin to reform, the sponge. In fact, if two different species of sponges are treated in this manner and their pieces mixed together in the same container, the fragments of one species will aggregate only with other fragments of that species. By some unknown recognition mechanism, the cells of one species reorganize only with each other and not with the cells of other species.

Sponges commonly reproduce asexually; that is, fragments broken from a sponge will develop into a separate animal. Sexual reproduction also occurs, with sperm cells transferred from one sponge to another by water currents. The young, developing sponge is motile, possessing flagella, but will become nonmotile as an adult.

Cnidarians

Cnidarians, which include hydras, jellyfish, sea anemones, and corals, are mostly marine animals that possess tissues, but lack organs. In other words, the cells of cnidarians are organized into tissues but these tissues, in turn, are not organized to function together to carry out a specific task for the organism. As a group, cnidarians display two different body forms; one type is a hollow, cylinderlike form called the **polyp** and the other form is somewhat umbrella-shaped and is known as the **medusa** (Fig. 6.17). All cnidarians have elongated tentacles used for defense, capturing prey, and eating their capture. Tentacles are equipped with tiny

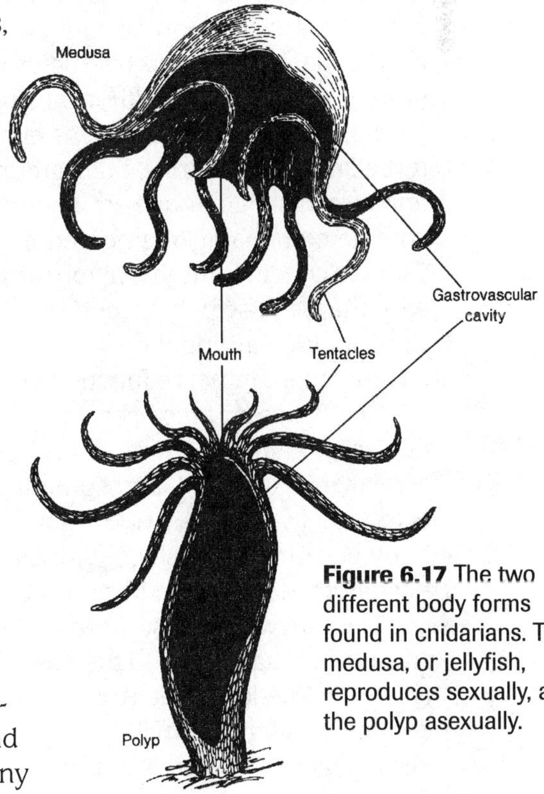

Figure 6.17 The two different body forms found in cnidarians. The medusa, or jellyfish, reproduces sexually, and the polyp asexually.

specialized cells containing a fine thread. When ejected, the thread releases a sticky substance or a paralyzing toxin that incapacitates prey organisms. After capture, the prey is pushed past the mouth opening by the tentacles and into the hollow gut. This process allows them to eat food particles considerably larger than those eaten by sponges. Jellyfish, such as the Portuguese man-of-war, capture and paralyze fish using a nerve toxin secreted by cells in the tentacles.

Corals, which are tiny colonial organisms, secrete a protective limestone skeleton around themselves. Massive accumulations of these skeletons create the many coral reefs found in the South Pacific and other parts of the world. Coral reefs provide habitats for a great number of fish and marine invertebrates.

Sea anemones, or "sea flowers," are beautiful animals with brightly colored tentacles that, when expanding and contracting, give the appearance of the opening and closing of flower petals.

Like the sponges, cnidarians exhibit tremendous powers of regeneration. The freshwater hydras, for example, have been used extensively in laboratory regeneration studies for 250 years. A hydra can be cut into several pieces and each piece will regenerate into a completely formed organism. Hydras, incidentally, possess a primitive nerve network, with cells that can both receive and respond to external stimuli.

Flatworms

The first major phylum of invertebrates to attain an organ level of organization and development are the flatworms. Included in this group are the mostly free-living *turbellarians* and the parasitic *flukes* and *tapeworms.* Many of the flatworms possess muscular, digestive, excretory, and reproductive systems.

Flatworms are *bilaterally symmetrical;* that is, the right half of the organism is a mirror image of the left half. Further, flatworms are the first group of animals we have looked at thus far that exhibit three distinct tissue layers during embryonic development. This is a very important evolutionary step, allowing development of a variety of differentiated, specialized structures. For example, flatworms possess a crude excretory system not present in either the sponges or the cnidarians.

It is thought that flatworms evolved from a form of ancient cnidarians. Most of the turbellarians are marine, but a group known as *planarians* are found in freshwater ponds and streams. Planarians are another group of animals that is able to regenerate into a complete animal if fragmented into several pieces. In addition, planarians have been used in conditioning behavior (trial-and-error learning) studies, being trained to navigate mazes in search of food. Flatworms may be the lowest animal group having the capacity for learning through simple conditioning. Such activity indicates considerable advancement in nerve function compared with that in any lower animal group.

The flukes and tapeworms are most notable as the agents responsible for a variety of diseases affecting other animals, including humans. Blood flukes, for example, cause a weakening, painful, and sometimes fatal disease called *schistosomiasis.* Flukes characteristically possess two or more suckers for attachment to the **host** and have a thick body covering that is highly resistant to the digestive enzymes in the intestines of the host. Both of these features are significant adaptations to a parasitic mode of existence.

The tapeworms also have a resistant body covering and possess several anterior suckers for attachment to the host. Some tapeworms may grow to a length of 23 meters (75 feet). The long, flat body of a tapeworm is composed of individual segments that contain male and female reproductive structures and tens of thousands of fertilized eggs. If tapeworms are eaten by a cow, hog, fish, or some other vertebrate, eggs within the body segments are carried by the blood to the muscles of the host where they encyst. If humans eat the raw or poorly cooked meat of these infected animals, the tapeworm eggs pass to the intestine, where they mature into adult tapeworms. The tapeworm attaches to the intestinal wall with suckers or hooks; growing tapeworms can cause considerable pain and also nutritional problems by consuming some of the nutrients eaten by their human host.

Roundworms

The roundworms exhibit some advancement over the flatworms in having a complete digestive system. By far the most numerous of the roundworms, indeed among the most numerous of all animals are the nematodes. These worms have thin, elongate bodies, usually tapered at both ends. They are found in abundance in soil and water; some are parasites of plants and animals. Most nematodes, however, are harmless. One of many species parasitic in humans is *Trichinella,* the causative agent of the disease trichinosis. The intestines of hogs or other mammals may harbor adult *Trichinella* worms, the females of which release **larva** (sexually immature form) worms into the blood and lymph of the host. The young worms circulate to skeletal muscles where they grow and become enclosed in a resistant cyst, or sac. Humans may acquire the encysted worms by eating poorly cooked pork. The worms mature and reproduce in the human intestine, after which the newly produced larval worms circulate in the blood to the skeletal muscles where they encyst. The resulting disease—*trichinosis*—is characterized by high fever, general weakness, and extreme muscular pain. Trichinosis can be prevented simply by cooking pork thoroughly to kill the encysted worms.

Other common parasitic nematodes include *Ascaris,* species of which are found in the intestines of humans, horses, and pigs; and pinworms and hookworms, which can be acquired by larvae contacting the skin or by hand-mouth contact.

Mollusks

The mollusks (L. *mollis,* "soft," a reference to their soft bodies) constitute one of the largest of all animal phyla and include such organisms as clams, oysters, scallops, snails, slugs, squids, and octopuses. Many mollusks have a hard, outer shell surrounding and protecting their soft bodies, although the squid shell is small and internal, and the octopus is missing it altogether. A feature of most mollusks is a *muscular foot* used in locomotion. In addition, mollusks are characterized by a soft *visceral mass* that contains most of the internal organs, and by the *mantle,* a sheath of tissue that enfolds the visceral mass. Also present in most mollusks are external respiratory structures called *gills.*

The shelled mollusks include bivalves ("two-valved shell"), such as clams, oysters, mussels, and scallops; and the gastropods ("stomach foot") which include snails, slugs, nudibranchs, whelks, abalones, and conchs. In the course of their evolution, slugs and nudibranchs have lost their shells. In the bivalves, the internal surface of each valve is composed of *nacre,* or mother-of-pearl. At times, a grain of sand or some other irritating substance becomes lodged in a valve, causing secretion of several layers of nacre around the irritating substance, and a pearl is produced. Although an object of pleasure in the human realm, a pearl to a bivalve is but a product of annoyance.

Most gastropods are marine, but some are found in fresh water and others are terrestrial. Gastropods are the largest molluscan group, and many of these animals have eyes located at the tips of tentacles that extend from the head. Land snails are among the few invertebrate groups that have adapted fully to a terrestrial existence. These animals, unlike most aquatic gastropods, lack gills and utilize the cavity within their shells as a lung for the exchange of respiratory gases.

Cephalopods ("head foot") are named because the foot, in the form of arms and tentacles, is located in the head region. All the cephalopods—squids, octopuses, cuttlefishes, and the chambered nautilus—are marine organisms and are among the most advanced invertebrates. It is difficult to believe that cephalopods belong to the same large group of animals as clams, snails, and most other mollusks. The squid, for example, is freed of a cumbersome shell and moves in any direction with speed and dexterity. With its ten "arms," the squid can grasp and manipulate its prey. It has a very highly developed brain and a pair of eyes that nearly rival our own. Squids also possess a tough beak for tearing their prey and employ an ink sac that expels a black fluid for concealment. They can even change body color as a means of camouflage. Although most squids are of moderate size—about 75 centimeters (30 inches) long, some are less than 5 centimeters (2 inches) long. The North Atlantic giant squid—the largest of all living invertebrates—may approach 15 meters (50 feet) in length.

Contrary to popular legend, the eight-armed octopuses are neither gigantic nor ferocious. Their bodies usually measure less than 0.3 meters (1 foot) in diameter and they are shy animals, preferring a solitary existence along the ocean shore. For many millions of years, the earliest cephalopod mollusks were covered by heavy shells; today, the only shelled cephalopod is the chambered nautilus.

Segmented Worms

The segmented worms include, among others, the common earthworm, the clam worm, and leeches. These worms are found in a variety of habitats, including soil and marine and fresh waters. They are the first animals we have encountered that possess a discrete circulatory system that transports oxygen and nutrients throughout the body. They also possess a centralized nervous system and an excretory system.

Earthworms not only serve as fish bait, but also are otherwise economically and ecologically beneficial. With their underground twisting and churning, earthworms turn over millions of tons of earth every year, and thereby aerate the soil and help maintain its fertility.

Clam worms and most others of its kind are marine organisms. The clam worm resembles the earthworm somewhat except for the presence of fleshy, lateral appendages along the length of the body and a well-developed head. Relatives of the clam worm—the brightly colored fanworms—live in tubes constructed of mucus and sand.

Leeches are found in marine and fresh waters and on land. Some leeches are parasites on vertebrates and usually possess anterior and posterior suckers used for locomotion and for attachment to the host. Leeches possess sharp teeth composed of the polysaccharide chitin to penetrate the host skin. Their salivary glands then produce an anticoagulant called *hirudin* that prevents clotting of blood. In bygone days, leeches were used by physicians for bloodletting in the treatment of disease. Today, leeches have made a comeback, being used in surgical procedures to stimulate blood circulation. Some leeches are capable of consuming such enormous quantities of blood that they may survive without feeding again for several months or even a year.

Arthropods

It has been estimated that two out of three known animal species on Earth are arthropods. The word *arthropod,* meaning "joint foot," refers to the jointed legs that characterize all members of this phylum. These animals have developed adaptive features that have placed them at the peak of invertebrate evolution. Arthropods probably evolved from a type of segmented worm or an early ancestor that gave rise to the segmented worms.

The diversity of arthropods is enormous, including such representatives as crabs, crayfish, barnacles, lobsters, mites, ticks, spiders, scorpions, centipedes, millipedes, and the largest group of all the animals, the insects. Most arthropods have a protective, waterproof exoskeleton partially composed of chitin. Periodically, however, in order to grow, this **exoskeleton** must be shed, a process called *molting*. Arthropods also have an elaborate arrangement of muscles that permits a variety of body movements, a highly developed nervous system, an extensive array of sense organs, and, of course, some of them can fly. Arthropods occupy virtually every known habitat on Earth, some as free-living forms and some as parasites. Although it would be impossible to comment on all the structural and functional variations displayed by even a few of these organisms, we can highlight some of the more distinctive and unusual features that make the arthropods such a biologically successful and fascinating group of animals.

One of the most highly developed sense organs found in many arthropods is the *compound eye* (Fig. 6.18). This organ may be composed of as few as twenty or as many as several thousand individual "little eyes." Although not as acute as the human eye, the compound eye is better for detecting rapid motion. Arthropods with compound eyes actually see more "slowly" than we do.

As evidence of their complex nervous systems, arthropods on occasion demonstrate some unusual and even bizarre behavioral patterns. Imagine, for instance, a rendezvous between a pair of amorous praying mantises. As the male makes his advances, the female enfolds him in her forearms and proceeds straightway to chew off his head. His lechery unthwarted, the now headless male breaks free of the female and proceeds to mate with her. Apparently, the loss of certain areas in the anterior portion of the nervous system triggers the mating response in the male. After copulation, the female may devour the entire body of the male.

The vast array of physical adaptations developed by the arthropods includes structures such as claws, fangs, poison glands, silk-secreting glands, sensory receptors, swimming appendages, and mouth parts modified for biting, piercing, sucking, and grinding. Many arthropods are capable of communicating with each other by means of sight and sound and through release of chemical substances called **pheromones.** Pheromones are secreted not only by arthropods but by a variety of other animal groups as well. They function as a signal to other members of the same species, influencing behavior and development. By means of their odors, pheromones act as sex attractants, trail markers, and warning signals.

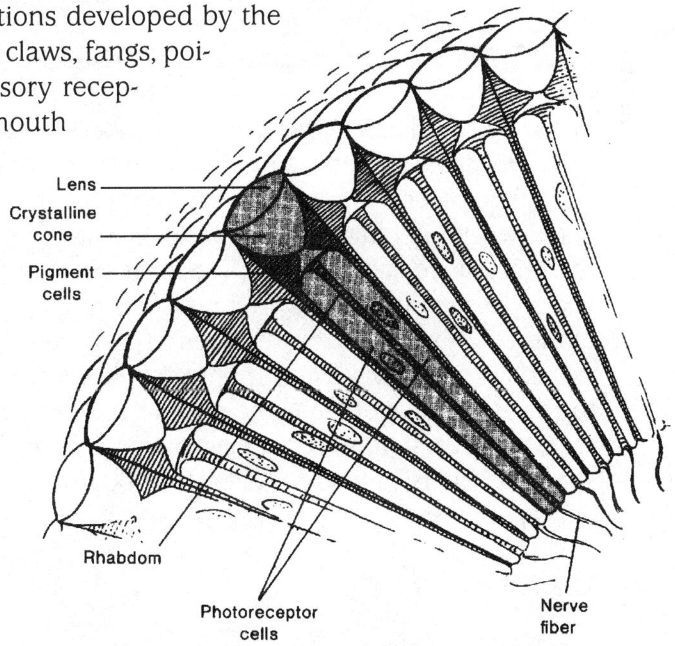

Figure 6.18 Section of a compound eye of an insect. A compound eye is composed of individual units called *ommatidia,* the number of which may vary from 20 to many thousands, depending upon the species. Each ommatidium is covered with a lens, and is connected to a nerve fiber leading to the brain by a light-sensitive central rod, or *rhabdom.*

Echinoderms

The echinoderms ("spiny skin") are the final phylum of invertebrates to be considered. They are all marine organisms. Their name is derived

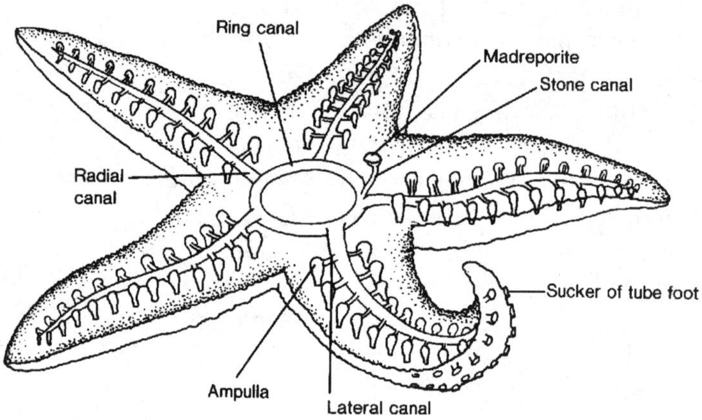

Figure 6.19 Internal water vascular system of a starfish. Water enters through the madreporite, eventually filling all the canals. The function of the system is locomotion, respiration, and capturing food.

from possession of an internal skeleton bearing small spines that project through the surface skin. Included in the echinoderm phylum are starfish, brittle stars, sea urchins, sand dollars, sea cucumbers, feather stars, and sea lilies. The body of an echinoderm generally is constructed on a five-part plan, with each part or arm radiating outward from a central point.

A feature unique to echinoderms is the **water vascular system,** an arrangement of internal canals that radiates outward from the center of the body and extends into each arm (Fig. 6.19). Each canal has a row of tiny, hollow tube feet that extends along its length. Each tube foot has a sucker that is used for attachment and locomotion. Water enters the canals through a small porous button on the body surface. When water is forced into the tube feet, they extend, allowing the suckers to attach to some object. In starfish, the water vascular system is used for a unique method of feeding. By straddling a clam or oyster, the starfish attaches its tube feet to both shells and begins to pull them apart. Once the shells are slightly separated, the starfish everts part of its stomach through its own mouth and into the slight opening in the clam or oyster shell. Enzymes secreted by the stomach of the starfish begin to digest the soft body within the shells. When the meal is finished, the starfish withdraws its stomach and replaces it to the interior of its body.

Starfishes and brittle stars characteristically have five arms, although some starfish may have twenty or more. Sea urchins, sand dollars, and sea cucumbers do not have arms, but they do have five rows of tube feet. The bodies of sea urchins are covered with moveable spines that, in some species, contain irritants or poisons.

Sea lilies, some of which look more like plants than animals, have stalked bodies with long, branching, feathery arms. Sea lilies and feather stars are considered to be the most primitive echinoderms; fossils of these animals date back well over 500 million years. Like some other invertebrates, echinoderms exhibit remarkable powers of regeneration. A starfish, for example, can regenerate an entire new arm. Equally amazing is the ability of many sea cucumbers, after expelling some of their internal organs through the anus, to regenerate all the missing structures.

This expulsion, which apparently is a normal seasonal activity in some species of sea cucumbers, allows regeneration of internal organs and structures that may be damaged.

Vertebrates

Vertebrates—animals with a backbone—constitute a subphylum under the phylum called *Chordata,* which also includes two small subphyla of the so-called invertebrate chordates (Fig. 6.20A and B). Vertebrates and the invertebrate **chordates** are categorized together because at some stage in their development—during embryonic life or during the adult stage or both—all chordates share several distinguishing characteristics. These include: (1) a thin, flexible rodlike structure called the **notochord,** which extends most of the length of the animal along its dorsal, or upper, surface; (2) a dorsal, hollow **nerve**

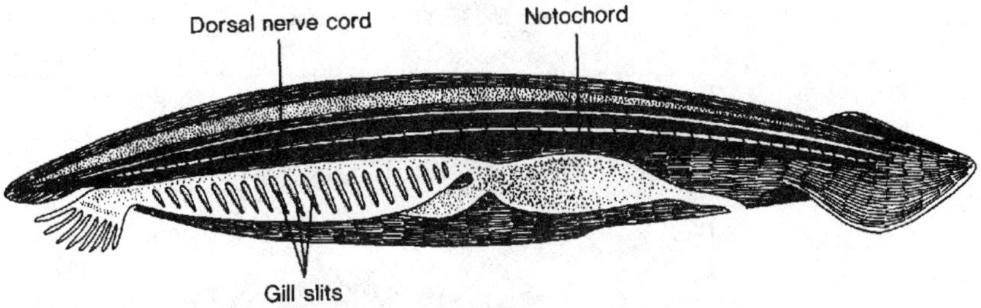

Figure 6.20B An adult amphioxus or lancelet, another invertebrate chordate, showing the three distinguishing chordate characteristics. In amphioxus all three characteristics are retained throughout life.

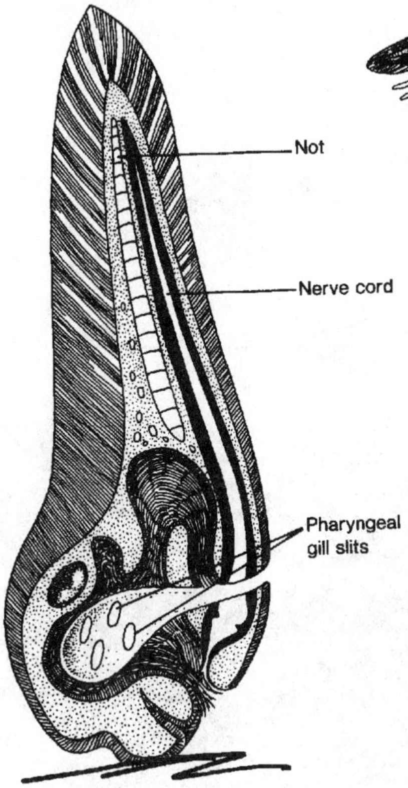

Figure 6.20A A tunicate larva, showing the three distinguishing chordate characteristics. In the adult tunicate, only the gill slits remain.

cord located above the notochord; (3) **pharyngeal slits,** which are tiny openings from the pharynx, or throat, region to the outside of the body; and (4) a **postanal tail,** which extends past the anus.

The invertebrate chordates, found in the oceans, are the tunicates, or "sea squirts," and the lancelet. As a group, the chordates are considered to be closest to the echinoderms in terms of evolutionary kinship. That is not to say that chordates evolved from echinoderms, but that both groups probably arose from a common ancestor. Our major focus is on the vertebrates, which include agnathans, fishes, amphibians, reptiles, birds, and mammals (Fig. 6.21).

Agnathans

Agnathans (meaning "without jaws") constitute the most primitive class of living vertebrates and include the scavenging hagfishes and the only living parasitic vertebrates—the lamprey eel. These animals do not have scales or paired fins, and their internal skeletons are composed of cartilage. Lampreys are **ectoparasites** on fishes, attaching to the outside of their host by a round, suckerlike mouth. Horny teeth lining the mouth are used to scrape away the scales of the fish, after which the blood and body fluids are sucked out. Hagfishes scavenge the oceans, feeding on dead fish or various invertebrates. The oldest vertebrate fossils on record are a group of jawless, bony-scaled, fishlike animals—the **ostracoderms**—that existed 350 to 500 million years ago (Fig. 6.22).

Figure 6.21 A representative of each of the classes of vertebrates.

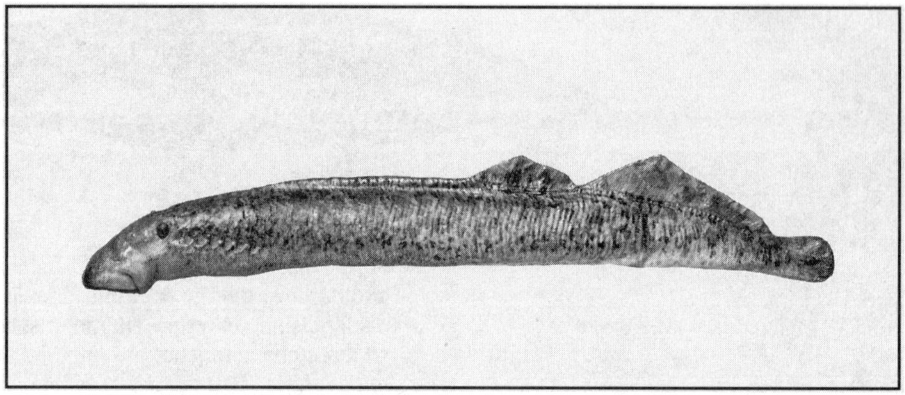

An agnathan—the lamprey eel—a jawless vertebrate.

A sunfish.

African clawed frog, an amphibian.

African crocodile, a reptile.

Figure 6.21 (continued)

Broad-tailed hummingbird.

An armadillo, a mammal.

Fishes

Fishes are separated into two major classes based on their internal skeletons. One class—the cartilaginous fishes—includes sharks, skates, and rays, all of which have an **endoskeleton** of cartilage. Skates and rays are characterized by flattened bodies and enormous lateral fins. Some rays have an organ that produces an electric shock, and others possess poisonous spines as defensive protection. The largest known species of fishes, the whale shark, can exceed 13 meters (41 feet) in length. In fact, sharks are the largest living vertebrates, except for the whales. The whale shark also lays the largest egg of any living animal.

Figure 6.22 An ostracoderm. Fossils of these jawless, armored fish are the first vertebrate fossils known, dating back 500 million years.

The second class—the bony fishes—is extremely diverse and includes salmon, bass, swordfishes, electric eels, seahorses, lungfishes, minnows, and piranhas. Almost all of these fishes have an endoskeleton of bone and are found in marine and fresh waters. Bony fishes display a wide range of sizes and physical adaptations. The Russian sturgeon, for example, may exceed 8 meters (25 feet) in length, whereas certain freshwater fishes found in the Philippines are less than 0.8 centimeters (0.3 inches) long.

A few of the evolutionary adaptations found among bony fishes include the ability of some to change color, which enables them to blend with their background as a means of camouflage; development of electric organs that emit shocks strong enough to incapacitate prey; and the unusual arrangement in seahorses and pipefishes in which the males possess a pouch for carrying eggs until they hatch. In addition, flying fishes can glide through the air; a catfish from Southeast Asia can "walk" on land using its gills

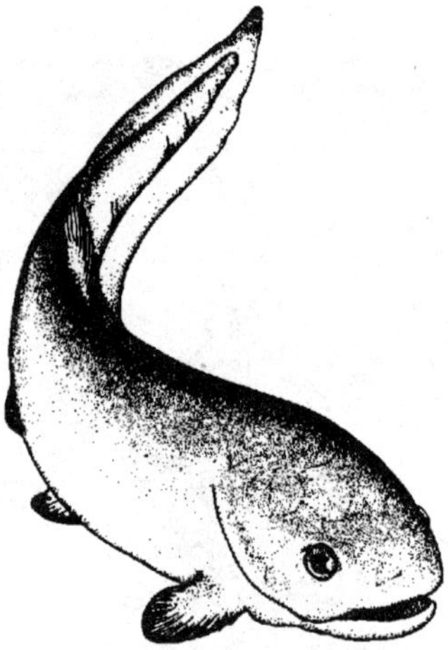

Figure 6.23 A placoderm. Members of this extinct group were the first known vertebrates with jaws.

and spines; a species of perch found in India climbs trees; and the male angler fish is but a tiny parasite attached to the body of the much larger female. Both the cartilaginous and bony fishes may have arisen from a group of jawed, bony-plated fishes—the **placoderms**—an extinct group that first appeared about 360 million years ago (Fig. 6.23). The early placoderms had a hinged jaw that radically altered feeding from the filtering of small food particles to ingesting much larger pieces of food. Not surprisingly, perhaps, some of these early fishes subsequently grew to tremendous sizes.

Amphibians

The first fossil record of amphibians dates back about 350 to 400 million years. They apparently evolved from a group of ancient freshwater fishes. The name of this class of vertebrates is derived from the Greek words *amphi,* "of two kinds," and *bios* "life." Many of these animals, then, lead "two kinds of lives;" that is, part of their lives is spent in water and part is spent on land. Amphibians are not considered true land vertebrates since most of them never have adapted fully to a terrestrial existence. There are species, however, that are totally aquatic or totally terrestrial. The ability to live at least part of the time on land represents a major evolutionary step. The invasion of land by totally aquatic organisms resulted in the gradual appearance of a variety of physical traits, such as legs and lungs, that were necessary for a terrestrial existence. The lungs of amphibians that possess them are small and relatively inefficient, but most species compensate for this inefficiency by exchanging respiratory gases across their moist skin.

The most familiar amphibians are the frogs and toads, along with salamanders. In addition there is one group of rare tropical amphibians called caecilians that, having no legs, somewhat resemble oversized worms. As adults, many amphibians live on land, but many species must return to water to reproduce. The fertilized eggs of some frogs develop into gill-breathing tadpolelike forms that undergo a metamorphosis to produce the adults. Some species, however, have developed methods for laying their eggs that do not require a return to water. Some rather unusual examples include a species of frogs in which the male carries the developing young in his mouth; the female of another species of frogs carries the young in her stomach.

As a protective measure, some frogs have the ability to change body color to adapt to their background. As a means of defense against predators, many amphibians secrete a toxic chemical produced in specialized skin glands. In fact, one of the most deadly venoms known is derived from skin secretions of the so-called arrow-poison frog of western Colombia, South America.

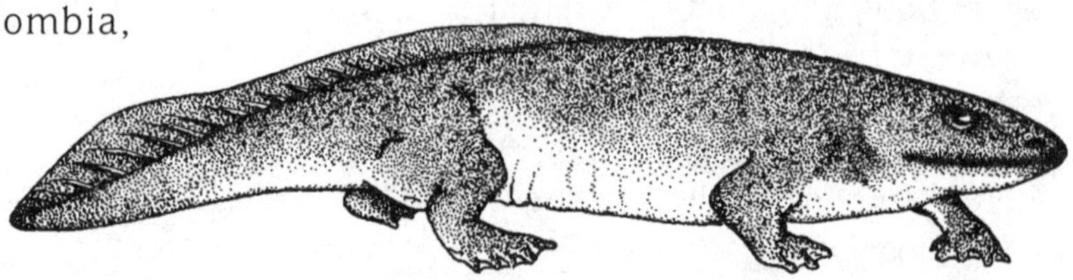

Figure 6.24 A stegocephalian. Considered to have evolved from ancient lobe-fin fish, stegocephalians were an early amphibian group that gave rise to the reptiles.

Reptiles

Reptiles arose from amphibian ancestors about 300 to 350 million years ago (Fig. 6.24). For over 200 million years, reptiles were the reigning terrestrial vertebrates on Earth. Modern reptiles include snakes, lizards, turtles, tortoises, alligators, and crocodiles. These animals have a body covering of waterproof scales and all four limbs are equipped with claws. Unlike amphibians, reptiles cannot exchange respiratory gases across their surface skin. Consequently, reptiles have evolved lungs that have a great surface area for the efficient exchange of respiratory gases. Reptiles are true land vertebrates, having adapted fully to a terrestrial existence. One of the primary reptilian adaptations to land is the **amniotic egg.** Covered by a tough, leathery shell that prevents it from drying out, the egg contains a watery fluid that bathes and supports the developing embryo. In a reproductive sense, the reptiles, "carry their water with them," so to speak The amniotic egg can be laid on land, often far from a source of water.

Figure 6.25 The tuatara of New Zealand.

The largest living reptiles are certain species of crocodiles that average about 4.5 meters (15 feet) in length and weigh over 450 kilograms (1000 pounds). Turtles and tortoises are shelled reptiles that lack teeth, although many have sharp, horny jaws. Tortoises probably live the longest of all vertebrates. Perhaps the most famous, or infamous, reptiles are the snakes. Only a few species, however, are dangerous to human beings. Snake venom is produced in salivary glands and released through specialized teeth, or fangs. Most venoms paralyze the nervous system or destroy the red blood cells of the victim. The flicking tongue of snakes is an organ of smell and is not poisonous. Snakes have good vision and can detect sound, and many species can move quite rapidly. The largest of all snakes, the anaconda of South America, averages over 6 meters (20 feet) in length.

Figure 6.26 *Brontosaurus.* This great plant-eating "thunder lizard" weighed 30 tons and measured over 65 feet from its nose to the tip of its tail. *Brontosaurus* inhabited the earth some 126 million years ago.

One group of reptiles includes only one species—the *tuatara* (meaning "back spine")—of New Zealand (Fig. 6.25). Referred to as a "living fossil," this lizardlike reptile is the only surviving species of an ancient group of reptiles older than the dinosaurs. A true lizard, the Gila monster of the Southwest, is the only poisonous lizard in the United States.

The most "glamorous" of the reptiles, at least in fable, were the now- extinct dinosaurs, the "terrible lizards" that first stalked Earth more than 200 million years ago (Fig. 6.26). Although dominating Earth along with other reptiles, the dinosaurs mysteriously became extinct over a relatively short period of time about 65 million years ago. The disappearance of the dinosaurs, along with untold numbers of plant and other animal species, is generally attributed to a cooling period on Earth at that time. It also has been proposed that Earth was struck by an immense meteorite during this period, and a huge cloud of dust resulting from the collision blocked out the sun. This could have

Figure 6.27 Fossilized remains of the ancient birdlike *Archaeopteryx*.

altered the climate even more severely for an extended period of time, perhaps long enough to hasten the extinction of so many living species.

Birds

Birds evolved from a group of dinosaurs or from some reptilian ancestor over 175 million years ago. In 1991, however, two partial bird skeletons were discovered in rocks dated to be about 220 million years old. The bird was named *Protoavis* ("first bird"), and was about the size of a pheasant.

The amniotic egg of birds is similar to that of the reptiles and was a critical adaptation to life on land. An ancient birdlike fossil from Germany—the so-called lizard bird *Archaeopteryx*—had features characteristic of both reptiles and modern birds (Fig. 6.27). However, it is not considered the ancestor of modern birds. *Archaeopteryx* had reptilian teeth, a lizardlike tail, and clawed digits, but it also had wings and feathers. Feathers are thought to have evolved from reptilian scales. Modern birds reveal their reptilian heritage in having scales on their legs similar to the scales of reptiles.

The structural adaptations of birds are related primarily to their capacity for flight. Their bones are hollow, and there are numerous air sacs throughout the body that make the animals buoyant and light. All birds have a covering of feathers that aids in decreasing the loss of water and body heat. Birds have massive breastbones to which the powerful wing muscles are attached. Styles of flight among the birds vary from the 90 wing beats per second of some hummingbirds to the soaring flight of vultures that may continue for hours without a single wing beat. Some birds, such as the penguin, kiwi, and ostrich, are incapable of flight.

Figure 6.28 A therapsid, a mammal-like reptile from which modern mammals probably arose.

Birds vary in size from the African ostrich, which stands 2.75 meters (9 feet) tall, to the bee hummingbird that measures a little over 5 centimeters (2 inches) in length. Most birds have extremely acute vision, many times more so than that of humans. The peregrine falcon and the spinetailed swift are the fastest animals known, having been clocked at speeds of over 322 kilometers (200 miles) per hour. Some penguins can swim at a speed of over 32 kilometers (20 miles) per hour under water.

Mammals

The dominant animals on Earth today are the mammals, a group that arose from a stock of carnivorous, mammal-like reptiles over 190 million years ago (Fig. 6.28). As mammals evolved, they departed from the crawling gait of their reptilian ancestors, acquiring limbs that lifted the body high off the ground—a distinct advantage in chasing and being chased. Mammals have hair or fur and mammary glands that, in females, secrete milk. The mammalian brain is extremely large in proportion to the rest of the body and the entire nervous system is highly developed. Mammals have keen senses and demonstrate degrees of intelligence unequaled in the animal kingdom.

Although existing along with the dominant reptiles for millions of years, mammalian ancestors managed to survive, giving rise to the true mammals that began to spread across Earth about 50 million years ago. One surviving group of mammals is the *monotremes,* the only group of egg-laying mammals. Although prominent millions of years ago, the only present-day monotremes are the duckbill platypus and the spiny anteater, both found only in Australia and New Guinea. After hatching from the egg, the young monotreme is nourished with milk secreted by the mammary glands of the female.

A second group, the pouched mammals, or *marsupials,* includes kangaroos, koala bears and wombats found only in Australia; and the opossum, which is restricted to North and South America. When born, the young marsupial is developmentally immature and must crawl up the abdomen and into the mother's pouch where development is completed.

By far the largest group of mammals is the *placental* mammals, in which the young develop within the female for a relatively long period of time and often are well-developed at birth. The **placenta,** an organ unique to mammals, is formed from the lining of the uterus and the outer layer of the embryonic sac surrounding the fetus. The placenta serves to exchange nutrients, oxygen, and wastes between the blood of the mother and that of the fetus. Some placental mammals, such as rabbits and cattle, are able to move around on their own within minutes after birth; others, including squirrels and humans, are virtually helpless at birth. Placental mammals are quite diverse and are found on land, underground, in trees, in marine and fresh waters, and in the case of bats, flying through the air. The largest animal on Earth and the largest ever known—the blue whale—is a placental mammal. These mammals average over 30 meters (100 feet) in length and weigh over 175 tons.

One group of placental mammals, the primates, is of particular interest because, along with monkeys, chimpanzees, gorillas, baboons, orangutans, and several others, it includes a species known as *Homo sapiens,* human beings. All primates have highly developed brains, eyes in the front of the head, opposable thumbs for grasping, and stereoscopic, or three-dimensional, vision.

As we have seen in this brief survey, organization is a unifying principle behind the theme of diversity. Whether the organism is single-celled or multicellular, the principle of organization is evident. To sustain this organization, the fundamental requirement of any organism is a continual input of energy. Using the sun as its source, life has evolved basic mechanisms for the production and use of energy by all organisms. In the following chapter, we will look at these mechanisms and discover how they, like organization, attest to the unity of the living world.

Chapter Summary

1. The structural and functional characteristics of the tissue, organ, and system levels of organization are reflected in the structural and functional makeup of individual cells.

   The basic types of plant tissues are epidermal, vascular, and ground tissues. Epidermal tissue covers and protects plant structures. Vascular tissue, which includes xylem and phloem, is involved in transport and support. Ground tissue functions in support, storage, and photosynthesis.

The major animal tissues are epithelial tissue, connective tissue, muscle tissue, and nervous tissue. Epithelium covers, lines, and protects body structures. Connective tissue protects against disease, and plays a role in storage, support, and anchoring of body structures. Blood and lymph are fluid connective tissues. Muscle tissue is concerned with body movement and heart function, and nervous tissue controls the sensation-response mechanisms of the body.

Key Terms: *dermal tissue, p. 100*; *vascular tissue, p. 100*; *ground tissue, p. 100*; *epithelium, p. 102*; *connective tissue, p. 102*; *voluntary muscle, p. 103*; *involuntary muscle, p. 103*; *neuron, p. 104*

Concept Question: Generally, what is the significance of the fact that some forms of life have a tissue level of organization? An organ level of organization? Apply your answer by comparing the structure and function of any plant and animal tissue of your choice.

2. The major organs of higher plants are roots, stems, leaves, and reproductive structures. Roots anchor the plant, store food, and absorb water and minerals. Stems are the supporting structures of plants, and leaves are the major photosynthetic organs. The reproductive structures of angiosperms are the male stamen and the female carpel. Gymnosperms possess male and female cones.

   Organs are present in most invertebrates and all vertebrates. Different organs function together to comprise the body systems of animals.

   Key Terms: *apical meristem, p. 104*; *root cap, p. 104*; *root hairs, p. 104*; *epidermis, p. 105*; *stamen, p. 107*; *carpel, p. 107*; *body system, p. 108*

   Concept Question: In what ways are roots, stems, and leaves functionally advantageous to a plant?

3. The kingdom Monera includes all bacteria. Bacteria are characterized as single-celled prokaryotes.

   Key Terms: *pathogenic, p. 109*; *decomposition, p. 109*; *recycling, p. 109*

   Concept Question: What are the characteristics that distinguish the kingdom Monera from the other four kingdoms of life?

4. The kingdom Protista includes algae, protozoa, and slime molds. Unicellular protists, the first known eukaryotic cells, gave rise to multicellular organisms.

   Key Terms: *algae, p. 111*; *phytoplankton, p. 111*; *colonial, p. 112*; *flagella, p. 114*; *cilia, p. 114*; *cytoplasmic streaming, p. 114*; *pseudopodium, p. 114*

   Concept Question: Compare the basic characteristics of the Protista with those of the Monera.

5. The kingdom Plantae includes multicellular photosynthetic organisms that are found predominantly on land. Nonvascular plants lack the conducting

tissues xylem and phloem that are characteristic of vascular plants. Vascular plants also possess true roots, stems, and leaves.

Key Terms: *vascular, p. 116*
nonvascular, p. 116
gymnosperm, p. 116
angiosperm, p. 116
fruit, p. 117

Concept Question: Why are angiosperms considered to be at the peak of plant evolution?

6. The kingdom Fungi includes nonphotosynthetic organisms, most of which exist either as parasites or by obtaining nutrients from dead or decaying organic matter.

Key Terms: *filament, p. 117*
chitin, p. 117
lichen, p. 118

Concept Question: At one time, the fungi were included in the plant kingdom but are now classified in a separate kingdom of their own. Explain the reasons for this latest classification of fungi.

7. The kingdom Animalia includes all invertebrates and vertebrates. All animals are multicellular and, except for the sponges, have attained at least a tissue level of organization. All invertebrates lack a backbone.

Key Terms: *regeneration, p. 121*
polyp, p. 121
medusa, p. 121
host, p. 122
larva, p. 123
exoskeleton, p. 125
water vascular system, p. 126

Concept Question: Why are the arthropods considered to be at the peak of invertebrate evolution?

8. Vertebrates are animals with backbones and are classified as a subphylum under the phylum Chordata. Vertebrates and the invertebrate chordates share the common characteristics of a notochord, dorsal nerve cord, pharyngeal slits, and a postanal tail at some stage in their development.

Key Terms: *chordate, p. 126*
ectoparasite, p. 127
endoskeleton, p. 129
amniotic egg, p. 131
placenta, p. 133

Concept Question: Compare the evolutionary adaptations to land of the vascular plants and the terrestrial vertebrates.

PART FOUR

Life Feature 2: Energy Transactions

Chapter Seven

Energy and Basic Life Processes

Energy: The First and Second Laws of Thermodynamics

An old proverb asserts that you cannot get something for nothing. This is applicable to the processes of life in that every activity of life exacts a price and that price is **energy.** Digestion, excretion, circulation, movement, secretion, growth, and all other life functions require an input of energy from an outside source. Energy, by definition, is the capacity to do work, to bring about some action or effect. Stored energy is called *potential* energy; the energy that actually does work, the energy of motion, is called *kinetic* energy.

Energy exists in several forms, including mechanical, chemical, electrical, magnetic, atomic, light, and heat energy. In most cases, one form of energy can be converted into another form. Rushing water (mechanical energy) turns a generator to produce electrical energy, which, in turn, is converted to light energy in a light bulb, which gives off heat energy.

As we saw in Chapter 4, the law of conservation of mass states that matter cannot be created or destroyed. Energy, too, is subject to a law of conservation, which states that energy can be converted from one form to another, but cannot be created or destroyed. This is known as the **first law of thermodynamics.** Moreover, the conversion of one type of energy into another is not a 100 percent efficient process; the amount of *useful* energy decreases during a given conversion. This concept is reflected in the **second law of thermodynamics,** which states that, in any energy conversion process, some of the energy is released as a less useful form of energy, usually as heat. It follows, then, that the amount of high quality, useful energy in the universe continues to decline, while the amount of less useful energy increases. This is why physicists tell us that the entire universe is "running down"; eventually, there will be no more energy available to do work and there will be no way to convert the heat energy in the universe into a more useful form of energy. The sun, however—our primary source of energy—is not scheduled to burn out for at least another 5 billion years.

You have probably noticed the heat released when a car engine runs, the heat felt when you rub your hands together, or the heat that warms a saw from cutting wood. Although heat escapes from any system—living or nonliving—it is critical to remember that, according to the first law of thermodynamics, the total amount of energy (including heat energy) entering and leaving the system remains the same (Fig. 7.1). Natural laws cannot be violated. The summary point here is that only a certain percentage of the energy available to a nonliving system or to an organism performs useful work; the remaining percentage is released as heat. Some of the heat released from a system may be useful—the heat that helps maintain your body temperature for example—but there is always a release of heat that ultimately cannot be used. In any system, the ratio of useful energy to the total energy entering the system is called **energy efficiency.** Due to the energy loss through heat, it is apparent that the energy efficiency of any system will always be less than 100 percent.

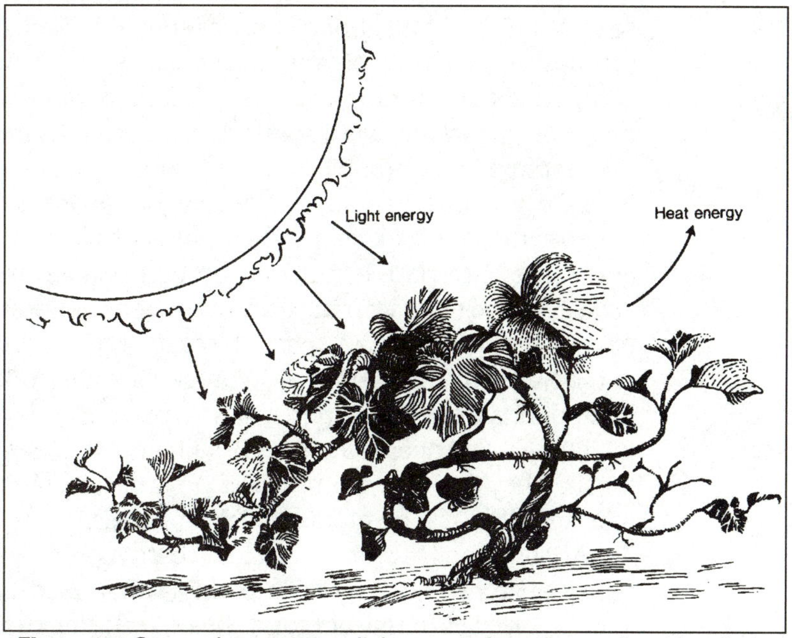

Figure 7.1 Green plants convert light energy from the sun into chemical energy. In the conversion process, some of the energy is dissipated to the environment as heat energy, which is unavailable to do work (second law of thermodynamics). However the *total* amount of energy entering and leaving the system (green plant) is the same (first law of thermodynamics).

Our interest in energy, of course, centers on its use and effects in organisms. The ultimate source of energy for life on Earth is light from the sun, our nearest star. Organisms, however, cannot use light energy directly as an energy source. To maintain the basic life processes, light energy must be converted to a usable form. In fact, on early Earth, the ability of some organisms to convert light energy into energy forms capable of doing work was a great impetus to the progress of biological evolution.

Heterotrophs and Autotrophs

First appearing about 3.5 billion years ago, bacteria apparently were the only life forms on Earth for at least 2 billion years. During their evolution, bacteria developed several methods for obtaining energy from various sources. Bacteria and all other organisms can be separated into two general categories depending on their method of obtaining food. An **autotroph** (Gk. *autos* meaning "self" and *trophos,* meaning "feeder") can synthesize its own food using energy from the sun or from some inorganic substance. Green plants, for example, are autotrophs that convert the light energy of the sun into energy used to synthesize the necessary organic (carbon-containing) nutrients. A **heterotroph** (Gk. *hetero,* meaning "different") must acquire its food by feeding on other organisms or on another food source. Humans and a host of other organisms are heterotrophs; they cannot use sunlight as a direct energy source and, for the most part, must feed directly on autotrophs or on other heterotrophs that feed on autotrophs. In feeding on other organisms, heterotrophs acquire not only a source of energy but also molecules and atoms that serve as building blocks for growth.

Most bacteria are heterotrophs that exist on dead or decaying organic matter. A few are parasites that obtain energy and nutrients from living organisms. The earliest bacteria on Earth probably were heterotrophs, absorbing organic nutrients from the ancient seas. With the gradual depletion of these nutrients, bacteria evolved mechanisms to synthesize their own organic nutrients by oxidizing inorganic matter, such as sulfur compounds and ammonia. These bacteria were the first autotrophs, although their methods apparently were somewhat different from those of present-day green plants. The earliest heterotrophic and autotrophic bacteria were probably **anaerobic** organisms—organisms that did not use oxygen—since the atmosphere of ancient Earth contained virtually no oxygen. The process of photosynthesis in modern plants, protists, and some bacteria releases oxygen (O_2) derived from the dissociation of water molecules. In ancient bacterial autotrophs, however, water probably was not used and thus oxygen was not released. Instead, some of the earliest bacterial autotrophs are thought to have used the inorganic gas hydrogen sulfide (H_2S) as an energy source; hence, sulfur compounds were released instead of oxygen. Today, bacteria inhabiting sewers, volcanic flumes, and deep ocean vents still employ this method of autotrophic nutrient synthesis. They produce the required organic substances from inorganic substances and release sulfur as a by-product of this process.

At some point over 2.5 billion years ago, however, bacterial evolution led to the appearance of photosynthetic bacteria that used light as an energy source and split water molecules during photosynthesis, releasing gaseous oxygen as a by-product. These organisms were the first blue-green bacteria—the apparent ancestors of modern blue-green bacteria. The rise of oxygen-releasing bacteria was a milestone in the history of biological evolution. Free oxygen gas first saturated the oceans and then escaped into the atmosphere. For the first time, oxygen permeated Earth's atmosphere. Over time, a protective ozone (O_3) shield developed that filtered out most of the harmful gamma, X-ray, and ultraviolet radiation from the sun. In addition, the accumulation of oxygen made possible the later evolution of aquatic and terrestrial **aerobic** organisms, organisms that are able to use oxygen to generate energy from organic compounds. No longer bombarded by lethal radiation, the shallow waters and surface of Earth became habitable. The stage was set for the explosive evolution of life that was to come.

Photosynthesis: Starlight to Sugar

An Overview of Photosynthesis

Photosynthesis—"putting together with light"—is one of the most important natural processes on Earth. Without it, aerobic life is not possible, only the anaerobic bacteria and yeasts would survive. The renowned biochemist and Nobel Prize winner Albert Szent-Gyorgyi stated: "What drives life is . . . a little electric current, kept up by the sunshine." That "little electric current" is part of a never-ending drama between the energy of the sun and almost all photosynthesizing organisms on our planet. Through photosynthesis, Earth's green plants annually produce billions of tons of sugar, or simple carbohydrate, which is the original source of nutrition for the living world. Photosynthesis occurs in green plants, algae, various protists, and some bacteria. Plant photosynthesis is an extremely complex process, but the basic principles underlying it can be presented in a fundamental manner. That is our intent in this discussion.

In Chapter 6, we were introduced to the major organ of photosynthesis in plants: the leaf. A single leaf can contain hundreds of millions of **chloroplasts,** the organelles in which the photosynthetic reactions take place (Fig. 7. 2). Chloroplasts contain the green pigment **chlorophyll,** which traps light energy from the sun and converts it to electrical energy (the movement of electrons). The capture and conversion of light energy are the initial steps in the photosynthetic process. In addition to chlorophyll, chloroplasts also contain membranous, disc-shaped **thylakoids** arranged in stacks called **grana** (singular, granum). Grana have been compared in appearance to stacks of pancakes. The fluid-filled space around the grana is the **stroma.** The entire organelle is surrounded by a lipid bilayer membrane.

Photosynthesis proceeds by two sets of separate reactions. The first set requires light as an energy source and constitutes the **light-dependent reactions;** the second set uses energy generated during the light reactions and does not require light to occur,

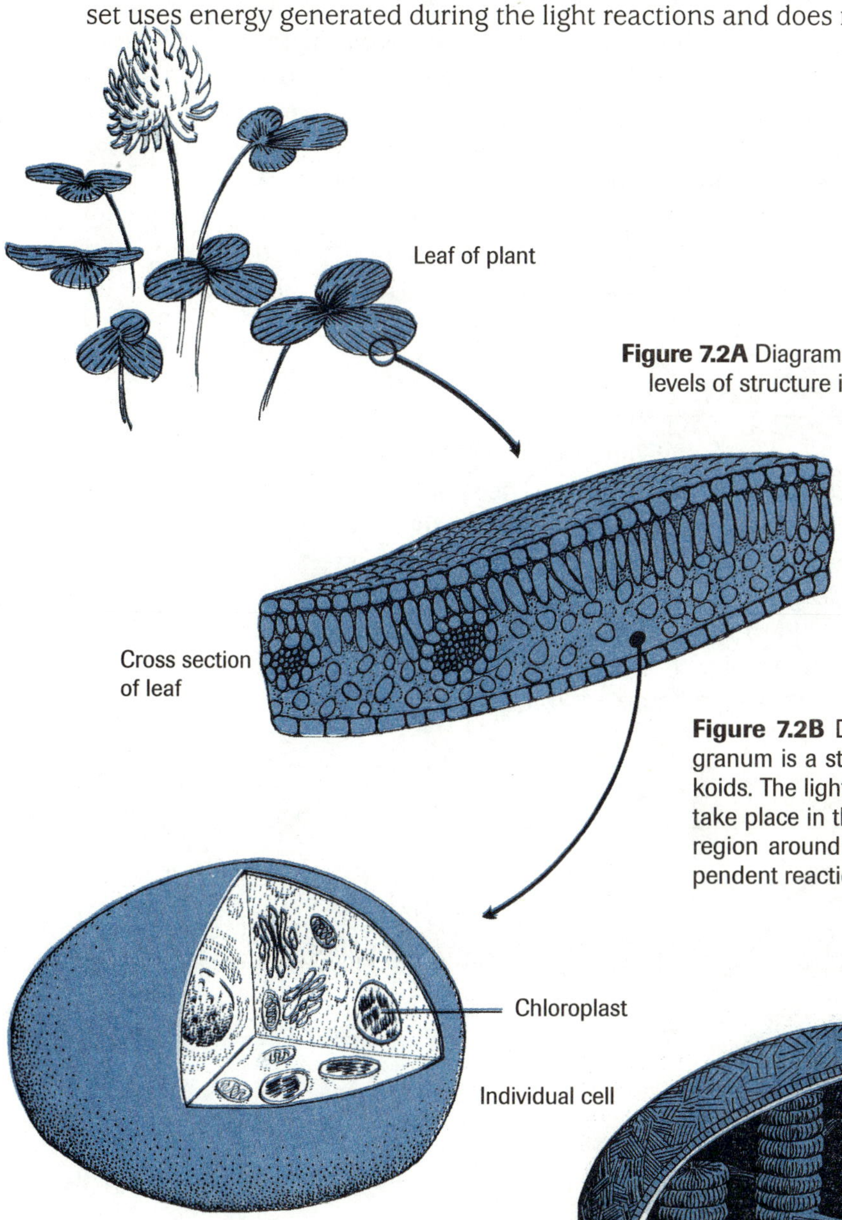

Figure 7.2A Diagram illustrating levels of structure in a leaf.

Figure 7.2B Diagram of chloroplast structure. Each granum is a stack of membranous, disc-shaped thylakoids. The light-dependent reactions of photosynthesis take place in the thylakoids. The stroma is a fluid-filled region around the thylakoids in which the light-independent reactions occur.

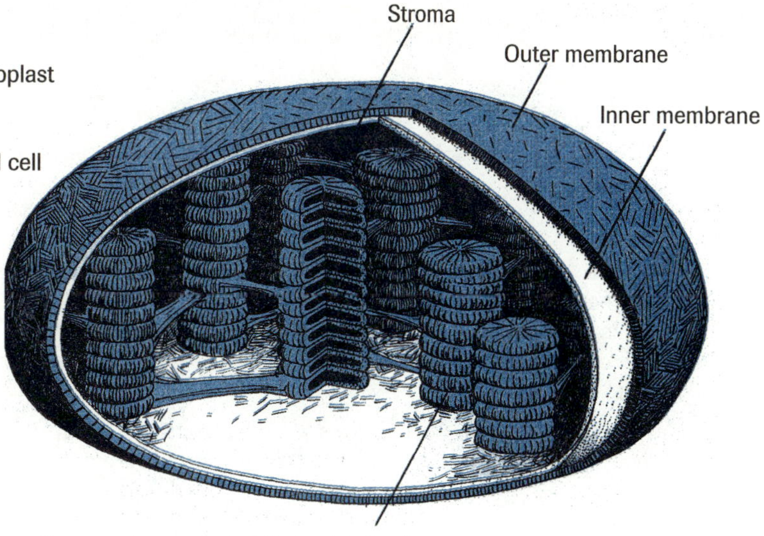

so these reactions are called the **light-independent reactions.** The latter reactions are also referred to as the Calvin cycle—named after the chemist Melvin Calvin who discovered most of the reactions and shared a Nobel prize for this work. The light-dependent reactions are carried out in the thylakoid membranes, which also contain the chlorophyll; the light-independent reactions take place in the stroma.

In its simplest form, the overall reaction for photosynthesis can be written as follows:

$$6CO_2 + 12H_2O + \text{light energy} \rightarrow C_6H_{12}O_6 + 6O_2 + 6H_2O$$

$$\text{Carbon dioxide} + \text{water} + \text{light energy} \rightarrow \text{Glucose} + \text{oxygen} + \text{water.}$$

Although glucose is shown as the organic product of the photosynthetic reaction, other carbohydrates—some derived from the glucose sugar—are also produced.

The whole point of photosynthesis is to reduce carbon dioxide, that is, to add hydrogens to it to form the sugar glucose and other simple carbohydrates. At the same time, water is oxidized by removing hydrogen from it and releasing free oxygen. The entire photosynthetic process requires energy and a source of hydrogen atoms (along with their accompanying electrons). In the process of photosynthesis, the source of hydrogen atoms is water and the initial source of energy is light. These requirements are on the left-hand side of the photosynthetic reaction.

The Light Energy of Photosynthesis

Sunlight is radiant energy consisting of different wavelengths of light (Fig. 7.3). Each wavelength possesses a certain amount of energy. The shorter the wavelength, the more energy it has. The wavelengths of light we see—**visible light**—are interpreted by the brain as various colors. From the longest wavelengths to the shortest, visible light consists of the colors red, orange, yellow, green, blue, indigo, and violet. Thus, violet light possesses the most energy in the spectrum of visible light and red light possesses the least energy. When striking an object, wavelengths of light may be absorbed by, reflected from, or transmitted through an object (Fig. 7.4).

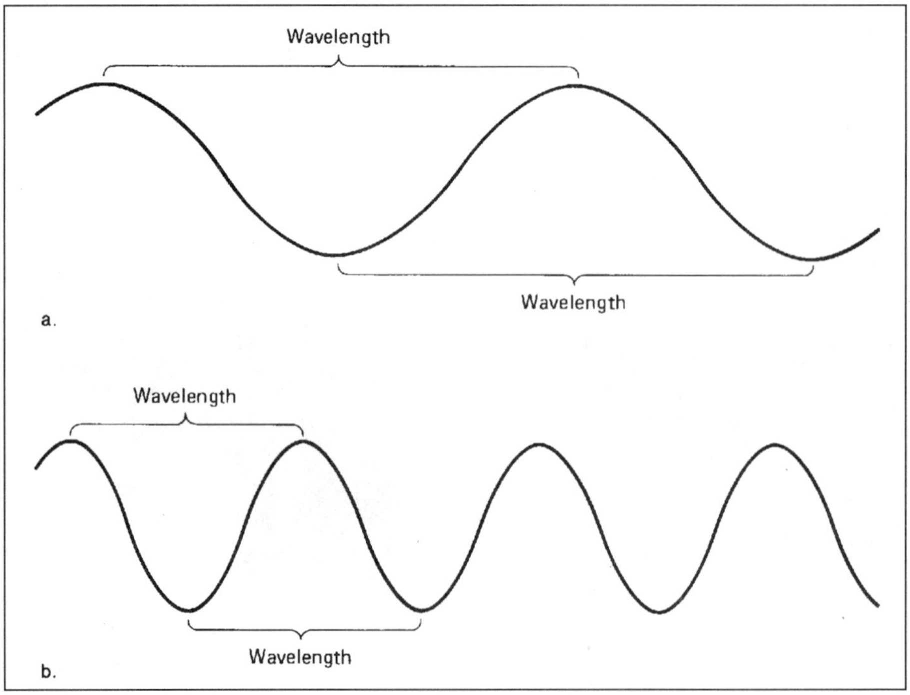

Figure 7.3 Light is the movement of photons in streams of varying wavelengths. Longer wavelengths of light (a) have less energy than do (b) shorter wavelengths.

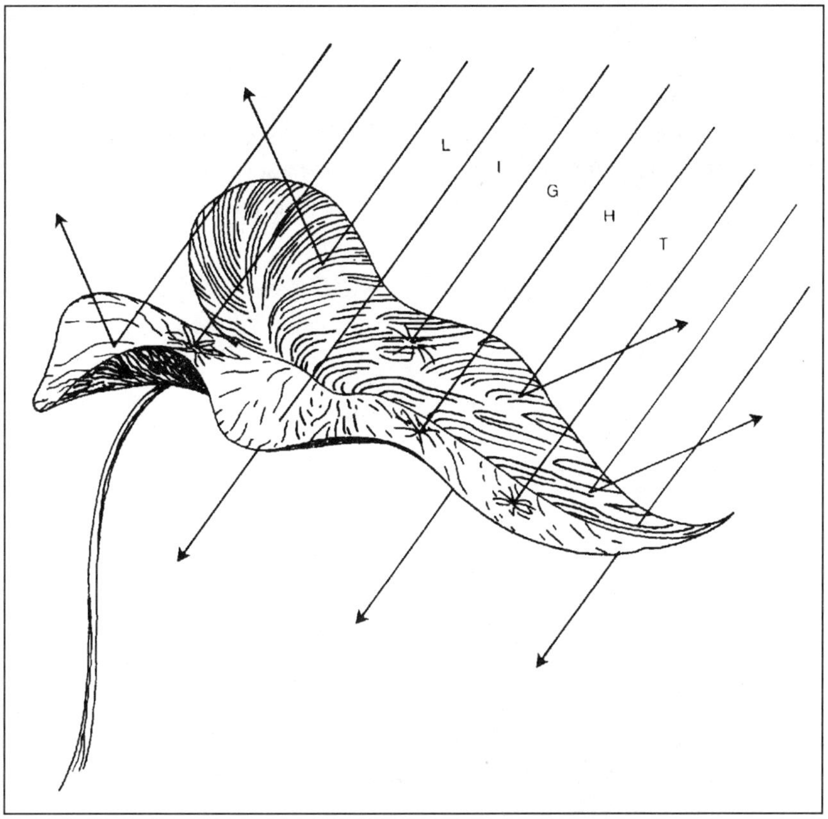

Figure 7.4 Some of the wavelengths of light striking a green leaf are reflected, some are transmitted, and some are absorbed. Chlorophyll primarily absorbs red, blue, and violet wavelengths, which provide the energy for photosynthesis. The leaf appears green because the green wavelengths are reflected.

The pigment chlorophyll primarily absorbs red and blue wavelengths of light, which are used as the energy sources for the light-dependent reactions of photosynthesis. Green light is reflected by chlorophyll—which explains why a green leaf appears green to us. Other wavelengths of light are transmitted through the chloroplast, although some of the wavelengths are absorbed by various yellow and orange pigments located within this organelle. These pigments, however, pass this trapped light energy on to chlorophyll. What happens, then, when effective wavelengths of light are absorbed by the chlorophyll molecule?

Initially, light energy "excites" electrons in the chlorophyll molecule, elevating them from their normal energy levels to a higher energy level. Boosting electrons to a higher level increases their potential energy. If they drop back to their original energy levels, this stored energy is released from the electrons. In photosynthesis, however, the elevated electrons do not return to their original levels in the chlorophyll molecule. Instead, they are trapped and held briefly at an elevated state by another molecule called the **primary electron acceptor,** located in the thylakoid membrane. In this first step of the light-dependent reactions, then, light energy is converted to electrical energy through the energizing of electrons.

The potential energy of the electrons that have been elevated to higher energy levels will be used to generate two high-energy compounds: **ATP** (adenosine triphosphate) and an electron and hydrogen ion carrier abbreviated **NADPH** (nicotinamide adenine dinucleotide phosphate). Formation of these two compounds is the essential function of the light-dependent reactions. As we will see, it is ATP and NADPH that provide the energy and electrons to power the light-independent reactions.

Let's now take a look at the high-energy molecule ATP and the biological use of hydrogen ions (H +) and electrons in the photosynthetic process.

ATP: Energy Currency

Known generally as the "energy currency" of living cells, ATP is often the immediate source of energy for cellular work (Fig. 7.5). In your body, much of the energy you derive from metabolizing pizza, milkshakes, or broccoli is ultimately stored in your cells

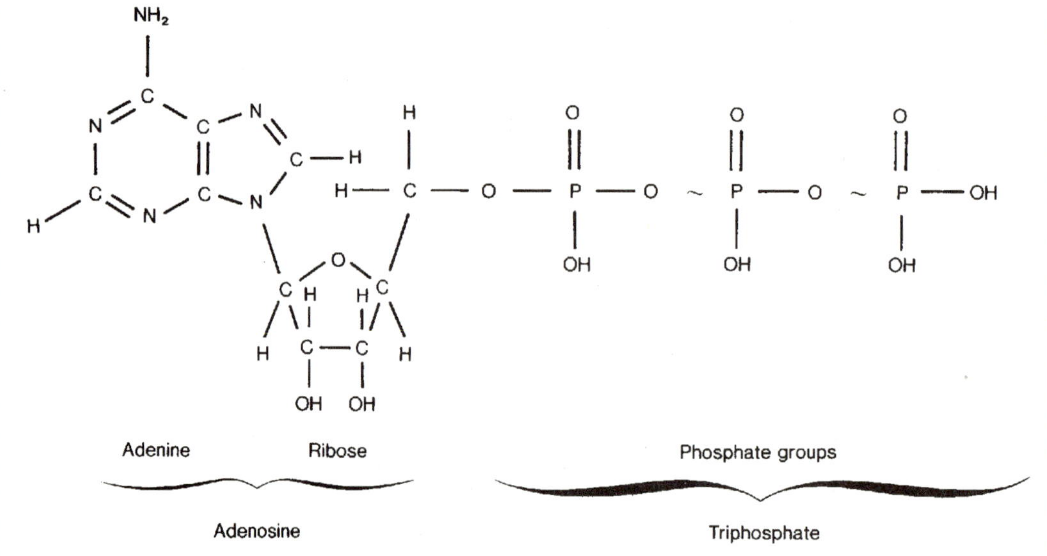

Figure 7.5 The complete structure of ATP, the "energy currency" of living cells. The high energy bonds are designated by the symbol ~. Removal of the terminal phosphate group forms adenosine diphosphate (ADP); removal of the middle and terminal phosphate groups from ATP forms adenosine monophosphate (AMP).

in the chemical bonds of ATP. Metabolizing plant cells also store energy in the bonds of ATP; indeed, virtually all known living things do so. More specifically, most of the energy is stored in the terminal two phosphate bonds of the three phosphates contained in an ATP molecule.

ATP consists of a nitrogen base called *adenine* attached to the five-carbon monosaccharide *ribose*. Bonded to ribose are three phosphate groups (PO_4) We can depict ATP in simplified form as adenine-ribose-P ~ P ~ P. The two terminal phosphate bonds—in contrast to the covalent bonds in the rest of the molecule—are so-called "high-energy" bonds and are used to store energy (~ represents a "high-energy" bond). If the terminal phosphate group of ATP is removed, ATP releases the energy in that bond and becomes ADP (adenosine *tri*phosphate becomes adenosine *di*phosphate). This released energy may be used by the cell to drive an energy-requiring reaction.

On the other hand, the synthesis of ATP in living cells entails the use of energy acquired from catabolized nutrients: carbohydrates, lipids, and proteins. Essentially, using the energy released from the breakdown of these nutrients, the terminal phosphate group is linked to the diphosphate ADP to form ATP. ADP is manufactured in the cell, where it is found along with free phosphate groups. The basic reaction, then, is: ADP + P + energy → ATP. This reaction is the ultimate reason why organisms must take in and metabolize nutrients: the energy released from the breakdown of these nutrients in the cells is used to carry out the ATP-synthesizing reaction.

Now, what happens when the energy stored in ATP is released? In most of the routine energy transformations in living cells, this involves breaking the terminal bond with the aid of an enzyme, which removes the terminal phosphate group. This bond, once broken, releases a relatively significant amount (about 7 kcal) of energy. This breakdown reaction in its simplest form is: ATP → ADP + P + energy.

The stored, or potential, energy released when the bond breaks becomes kinetic energy—in this case, the energy that does cellular work. The released phosphate group can be transferred and bonded to another molecule. This is the way ATP transfers some of its stored energy to another molecule, allowing that molecule, such as muscle protein, to do work.

As mentioned, the light-dependent reactions of photosynthesis also generate ATP—a process which involves the use of high-energy electrons. Thus, ATP is formed in two ways: (1) through the use of high-energy electrons in the light-dependent reactions of photosynthesis and (2) from the release and conversion of chemical bond energy stored in organic nutrients. Nonphotosynthetic organisms, such as humans and other animals, must feed either on plants or on another animal that fed on plants to acquire these nutrients to generate ATP. This is simply a restatement of our discussion earlier in this chapter. Heterotrophs cannot make non-protein organic compounds, and thus must obtain these compounds mainly by eating autotrophs or other heterotrophs that have eaten autotrophs. In your cells and in those of all other organisms, energy released from metabolizing nutrients is used to synthesize ATP.

Cellular synthesis of ATP from glucose and from other organic molecules falls under the heading of cellular respiration, which is our subject of discussion following our look at photosynthesis. As the energy currency of living cells, ATP can be drawn on like money in a bank as an immediate source of energy. Unlike bank accounts, however, an organism's cache of ATP cannot come up short. Without ATP, the cell cannot do most of its work, which translates into an organism being unable to function.

Coenzymes and Electron Acceptors

The Coenzyme NADP+

NADP+ (nicotinamide adenine dinucleotide phosphate) is an organic compound; further, it is called a **coenzyme**, which simply refers to an organic molecule that helps certain enzymes perform their specific activities. Coenzymes often accept or donate a small molecular group that is involved in a particular reaction. In some reactions, coenzymes assist enzymes in the transfer of electrons and hydrogen ions from one molecule to another. This is the manner in which NADP+ is used in photosynthesis. Two similar coenzymes—NAD+ (nicotinamide adenine dinucleotide) and FAD+ (flavin adenine dinucleotide)—serve the same basic function in cellular respiration.

To help us apply these coenzymes in practice, the transfer of electrons and hydrogen ions between molecules in a chemical reaction can be expressed using the terms *oxidation* and *reduction*. The loss, or removal, of electrons and hydrogen ions from a molecule is called **oxidation;** that is, the molecule has been oxidized. The gain, or acceptance, of electrons and hydrogen ions by a molecule is called **reduction,** and we say that the molecule has been reduced. Oxidation and reduction always occur together; if one molecule is oxidized, another molecule must be reduced. An electron lost by one molecule will be transferred to another molecule.

The NADP+ coenzyme accepts hydrogen ions and electrons removed from other molecules. These latter molecules, then, have lost hydrogen ions and electrons and become oxidized while NADP+ becomes reduced. The reduced form is written as NADPH to indicate the gain of hydrogen; notice that NADPH does not have a positive charge. The electrons removed from a molecule represent a source of potential energy. By accepting electrons and hydrogen ions, NADP+ becomes an energy storage molecule in the form of NADPH; that is, the accepted electrons and hydrogen ions represent sources of energy. As we will see, this stored electron energy is released and harnessed to do cellular work. The hydrogen ions are also released and can be used to reduce other molecules and to power the synthesis of ATP. After giving up its hydrogen atoms and electrons, the reoxidized NADP+ can be used again in a new round of reactions.

Electron Transport Chains

In addition to coenzymes, there are other molecules that accept electrons in cells. These electron acceptor molecules make up what is called an **electron transport chain.** Each molecule in the chain successively passes its electron "downhill" to the next molecule (from a higher energy level to a lower one) until the electrons are taken up by a final electron acceptor. The electron transport chain, then, functions through of a series of oxidations (loss of electrons) and reductions (gain of electrons).

An electron transport chain functions to release the potential energy of electrons gradually and in a controlled fashion. If the electrons "fell" from the top of the electron transport chain to the bottom, their energy would be released all at once. This would produce a burst of heat that could damage a cell and also make it impossible for the cell to control energy and to channel it to do useful work (Fig. 7.6). You might envision an electron being passed down the transport system as a ball bouncing step-by-step down the stairs, instead of being tossed from the top of the stairs to the bottom. At the bottom of the electron transport chain (at the final electron acceptor), most of the energy of the electrons has been released.

The Light-Dependent Reactions of Photosynthesis

The "little electric current" that "drives life" is generated in the light-dependent reactions of photosynthesis. In summary form, these reactions result in both the flow of energy from low energy electrons released from water molecules, and the activation of two chlorophyll systems from which chlorophyll electrons are released. Some of the chlorophyll electrons provide energy for ATP synthesis and some are accepted by NADP+. The accepted electrons possess high potential energy. In addition, hydrogen ions, also released from water molecules, are ultimately taken up by the same NADP+. By accepting the electrons and hydrogen ions, NADP+ is reduced to NADPH. Thus, as long as the sun is shining, the "little electric current" flows continuously from water molecules to NADP+. Energy flow in the light reactions results from the transport of electrons down the electron transport chain, a process that plays a critical role in photosynthesis. Let's take a look at the basic scheme by which light, water, and chloroplasts bring about the synthesis of NADPH and ATP.

Photosystem I and Photosystem II

Located in the thylakoid membranes of chloroplasts, along with other pigments, is a specific chlorophyll molecule that becomes reactive when stimulated by light energy. This chlorophyll molecule, along with its primary electron acceptor, is called a **photosystem**. Two different types of these photosystems are found in the thylakoid membranes, and are referred to simply as *photosystem I* and *photosystem II.* The essential difference between the two photosystems is that they absorb slightly different wave-

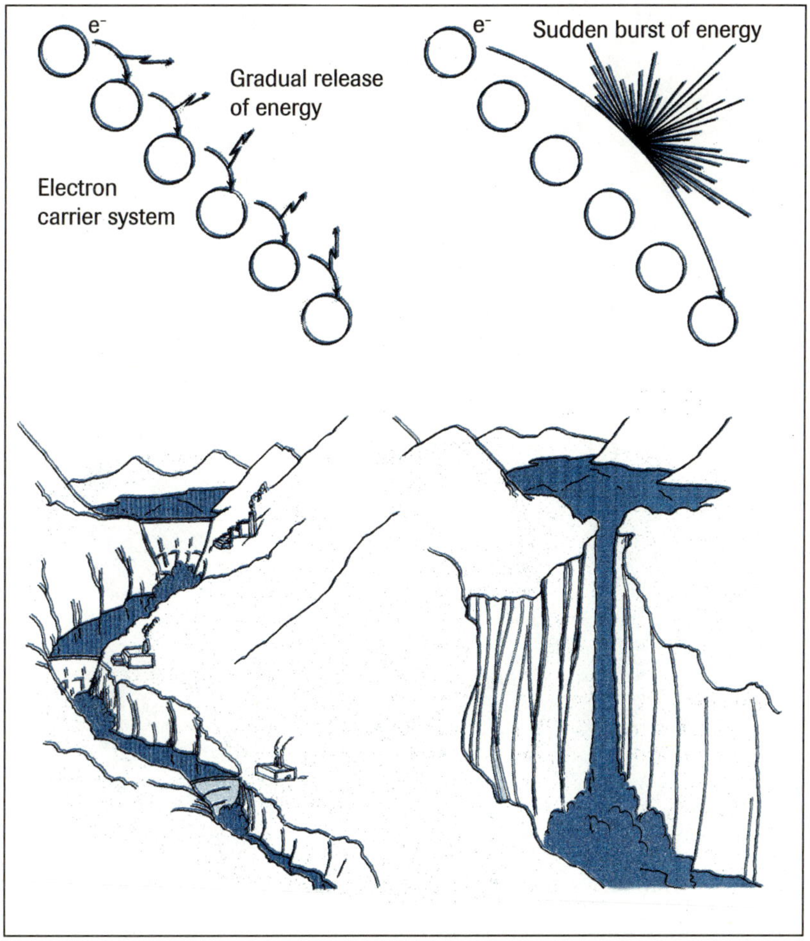

Figure 7.6 By descending stepwise down an electron carrier system, an electron e⁻) releases its potential energy gradually. This permits the cell to utilize the energy efficiently in the performance of useful work. An electron descending abruptly down the carrier system releases its stored energy all at once. This energy cannot be efficiently controlled and produces excessively high temperatures.

This scheme is analogous to downward rushing water, the energy of which can be controlled and utilized through the use of dams. If left to descend abruptly, the water dissipates its energy all at once, which is highly inefficient.

lengths of light. Both photosystems will be involved in our discussion of photosynthesis. There is a simpler form of photosynthesis that involves only photosystem I and does not produce oxygen, but we will not consider it here. The photosynthetic events in the light-dependent reactions are illustrated in Figure 7.7.

Figure 7.7 illustrates the initiation of the light-dependent reactions, in which light strikes the chlorophyll molecules in photosystems I and II; this "excites" chlorophyll electrons and boosts them to a higher energy level. The electrons are trapped and held by a primary electron acceptor, one in each photosystem. Chlorophyll, then, is oxidized and the acceptors are reduced. As stated earlier, these electrons now possess potential energy as a result of being raised to a higher energy level.

The oxidized chlorophyll molecule in photosystem II is a voracious reductant, strongly attracting electrons and acquiring them—with the aid of an enzyme—by splitting water molecules. These electrons from water actually replace the chlorophyll electrons that

Figure 7.7
Summary of the light-dependent reactions of photosynthesis.

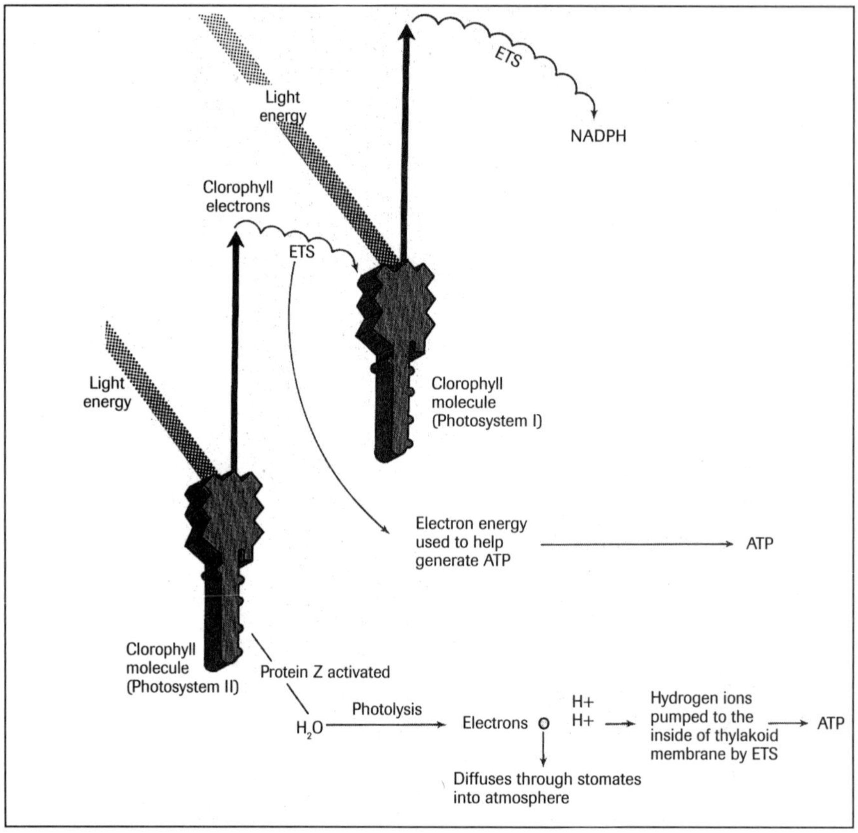

were boosted out of their energy levels in photosystem II and trapped by the primary electron acceptor.

Each water molecule is dissociated into two hydrogen ions (H +), an oxygen atom (O), and two electrons. The oxygen atom combines with another oxygen atom to form gaseous oxygen (O_2), which escapes from the plant into the atmosphere. This is the oxygen that we, as aerobic organisms, utilize every day. Oxygen is shown as one of the products of the light-dependent reactions in the overall photosynthetic reaction: $6CO_2 + 12H_2O$ + light energy $\rightarrow C_6H_{12}O_6 + 6O_2 + 6H_2O$. All that remains of each of the split water molecules are the two hydrogen ions, which are released into the compartment of the thylakoid.

The electrons held by the primary electron acceptor in photosystem II are transferred to an electron transport chain and release their potential energy as they descend the chain to lower and lower energy levels. Upon reaching the last electron carrier in the chain, the electrons are passed to photosystem I. These electrons replace chlorophyll electrons that were previously boosted to higher energy levels in photosystem I. The energy released by the photosystem II electrons as they move down the electron transport chain is used in association with hydrogen ions present in the chloroplasts to synthesize ATP.

We mentioned that the hydrogen ions split from water molecules accumulate in the thylakoid. These hydrogen ions, along with hydrogen ions present in the stroma, establish a *concentration gradient;* that is, a difference in hydrogen ion concentration on either side of the thylakoid membrane. The released energy of the descending photosystem II electrons is used to transport, or "pump," the hydrogen ions in the stroma across the thylakoid membrane to the interior of the thylakoid (Fig. 7.8). Once hydrogen ions are pumped inside the thylakoid, they accumulate in high concentration and tend to diffuse across the membrane back into the stroma through a facili-

Figure 7.8 Generation of ATP in the thylakoid. In the light-dependent reactions of photosynthesis, hydrogen ions (H^+), derived from the photolysis of water and from the stroma, accumulate in the interior of the thylakoid disc. The hydrogen ions can pass back out through the thylakoid membrane only at specific areas where there are enzyme-containing particles. The energy of the outward moving hydrogen ions is involved in the synthesis of ATP from ADP and inorganic phosphate (P_i). the locations of Photosystems I and II, and the electron transport system (ETS) in the membrane also are shown.

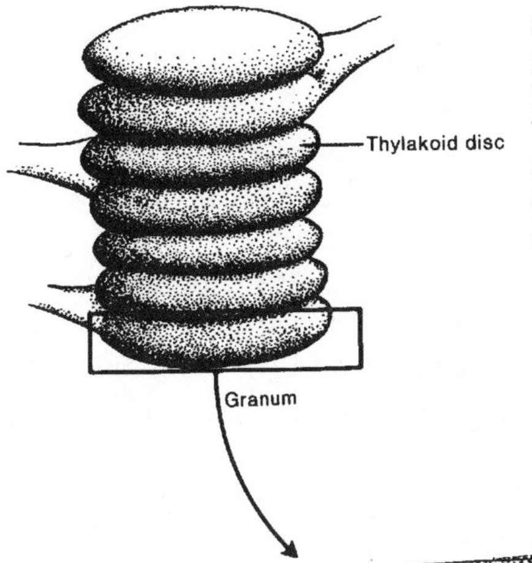

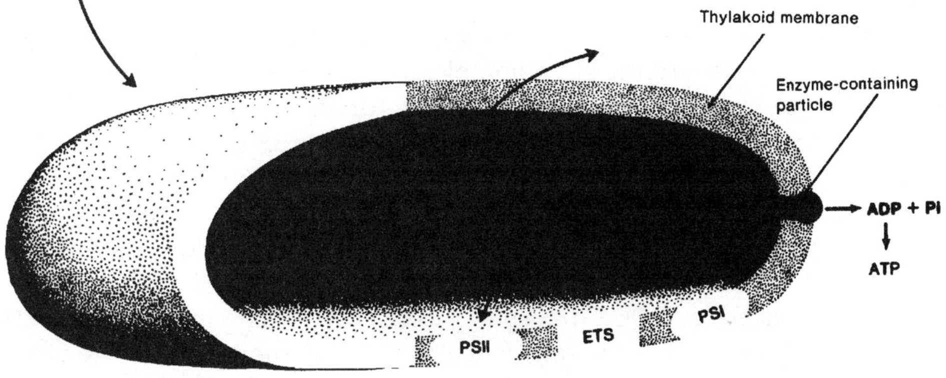

tated diffusion channel. These channels in the thylakoid membrane contain ATP-synthesizing complexes, or enzyme groups, through which the hydrogen ions must diffuse to leave the thylakoids and pass into the stroma. As they pass through these complexes, the hydrogen ions provide energy to bond a phosphate group to ADP, forming ATP in the stroma. The hydrogen ions, now in the stroma, are picked up by **NADP +** , forming NADPH.

The energy of the electrons held by the primary electron acceptor in photosystem I is not used to generate ATP; instead, these electrons are passed to NADP + , reducing it to NADPH. Thus, NADP + accepts the hydrogen ions present in the stroma and also accepts the photosystem I electrons. The result is that NADPH becomes a reservoir of high-energy electrons—a source of energy the plant cell can use. The ATP and NADPH molecules remain in the stroma of the chloroplast, where they will be used to power the light-independent reactions.

In summary, light energy stimulates photosystems I and II and boosts electrons to higher energy levels, where the electrons are held by primary electron acceptors. Electrons split from water molecules replace the electrons removed from photosystem II; the oxygen split from water diffuses out of the plant and the hydrogen ions from water remain in the thylakoid. After being passed down the electron transport chain, the electrons from photosystem II replace the missing electrons in photosystem I. As they pass down the electron transport chain, the photosystem II electrons release energy that pumps hydrogen ions in the stroma across the thylakoid membrane to the interior of the thylakoid. The facilitated diffusion of hydrogen ions back across the thylakoid membrane into the stroma powers the synthesis of ATP. NADP + is reduced to NADPH by accepting the photosystem I electrons and the hydrogen ions in the stroma.

The Light-Independent Reactions of Photosynthesis

The light-independent reactions of photosynthesis do not require light as an energy source, and therefore do not need to take place in the light. They are, however, powered by the ATP and NADPH produced in the light-dependent reactions. Recall that the light-independent reactions produce carbohydrates by reducing atmospheric carbon dioxide. NADPH supplies the reducing power by donating electrons and hydrogen ions, and ATP provides the energy to synthesize the carbohydrates and other organic molecules. The sequence of events in the light-independent reactions is shown in Figure 7.9.

The light-independent reactions begin when molecules of atmospheric carbon dioxide diffuse into the stroma of a chloroplast through the tiny openings, called **stomata,** located in the epidermis of a plant leaf (Fig. 7.10). In the stroma, carbon dioxide bonds with molecules of a five-carbon sugar abbreviated **RuBP** (ribulose bisphosphate). This sugar, which has to be regenerated repeatedly in the stroma, is formed from previous light-independent reactions.

The bonding of carbon dioxide with the five-carbon RuBP forms a very unstable six-carbon compound that immediately splits apart, forming two three-carbon molecules of an acid referred to as PGA (phosphoglyceric acid). Using the energy of ATP and the hydrogen ions and electrons from NADPH, the PGA molecules are converted to two three-carbon sugars abbreviated **PGAL** (phosphoglyceraldehyde).

These three-carbon sugars are the building blocks for the synthesis of larger organic molecules and thus are the most critically important molecules produced in photosynthesis. Molecules of PGAL, for example, can link together to form six-carbon sugars such as fructose, which can be enzymatically converted to glucose. Glucose molecules can be joined together to produce starch, which is the form in which plants store sugar. Alternatively, glucose molecules can bond to produce cellulose, the polysaccharide that composes part of the plant cell wall. PGAL molecules can also be acted upon by enzymes to produce the disaccharide sucrose, the form in which carbon is transported from the leaves to other parts of the plant. Not only does PGAL form carbohydrates, but it can enter other metabolic pathways in the cell to produce fatty acids, amino acids, and other compounds. The radiant energy of starlight now has been stored as chemical energy in the bonds of carbohydrates and other organic nutrients.

Finally, some PGAL molecules must be used to regenerate more of the five-carbon sugar RuBP. Since RuBP must be present to bond with carbon dioxide to initiate the

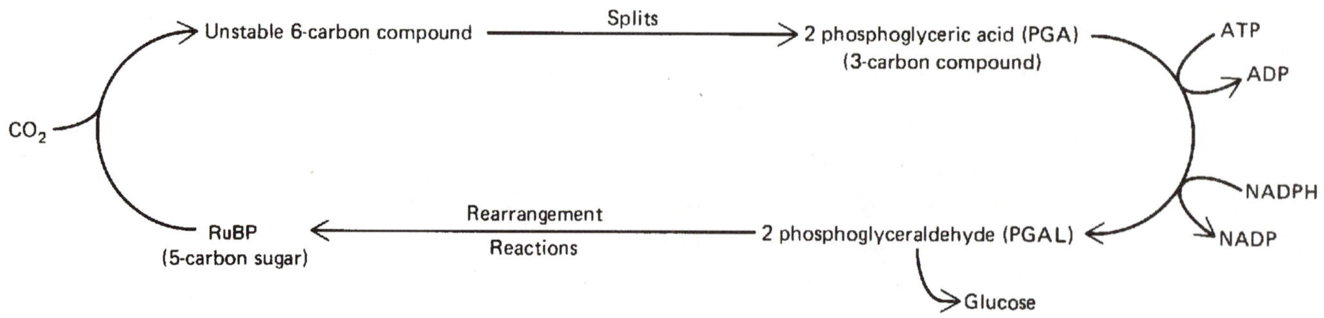

Figure 7.9 The sequence of events in the light-independent reactions of photosynthesis. The reactions begin with the diffusion of CO_2 into the plant, and are completed with regeneration of the 5-carbon sugar RuBP.

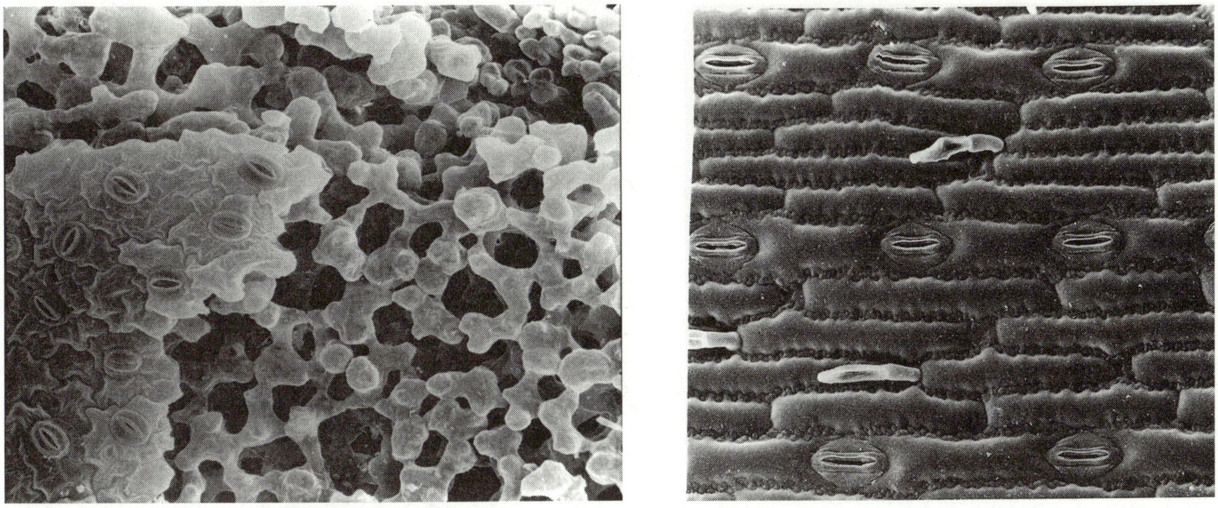

Figure 7.10 Stomata in the epidermis of leaves.

light-independent reactions, the cycle could not continue unless this sugar is regenerated repeatedly. With the regeneration of RuBP, the light-independent reactions are completed. Having released their stored energy, ATP and NADPH, now in the oxidized form of ADP and NADP + , respectively, return to the thylakoids to await another beam of sunlight.

In summary, carbon dioxide enters a leaf, bonds with RuBP, and forms an unstable six-carbon compound. This compound splits into two molecules of PGA. ATP and NADPH facilitate the conversion of PGA to PGAL molecules, which serve as the building blocks for synthesis of carbohydrates and other organic compounds. Some PGAL molecules are used to regenerate RuBP. ADP and NADP + return to the thylakoids (Fig. 7.11).

The scheme of photosynthesis reveals not only how green plants make food, but also provides insight into the fundamental phenomena operative in the living world. Life on Earth ultimately depends on the energy of sunlight. The process of photosynthesis confirms this by demonstrating how light energy is converted to other forms of energy and is eventually stored as chemical energy in the form of carbohydrates and other organic compounds. In the following section, we will see how this stored energy is put to work in cells. We will revisit the movement of electrons via an electron transport chain; ATP generation using hydrogen ions; oxidized and reduced coenzymes; conversion of molecules from one form to another; and the repeated regeneration of an original compound in a cycle. These are universal phenomena operating in all known forms of life.

Cellular Respiration

Living organisms exist because they are capable of converting energy from one form to another. We have just explored an example of an energy conversion in the cells of green plants. All cells require a reasonably steady supply of organic nutrients in the form of carbohydrates, lipids, and proteins. Autotrophs synthesize some of these nutrients for their own use; heterotrophs must acquire most of them directly or indirectly from autotrophs. These nutrients contain chemical energy—the energy source that must be tapped for life to continue.

The production and release of chemical energy in living cells occurs through the process of **cellular respiration,** an aspect of cellular metabolism. In this context, respi-

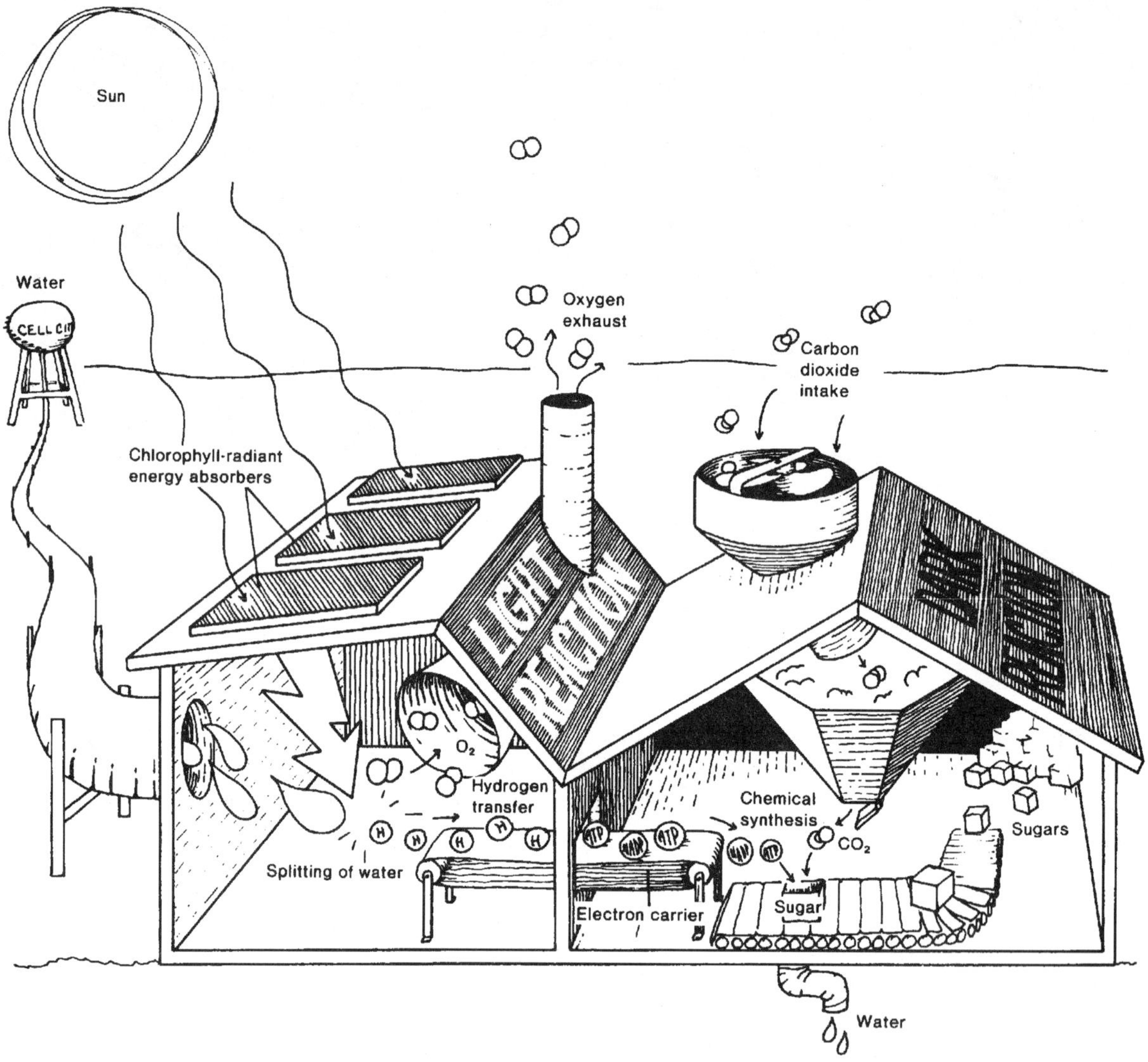

Figure 7.11 The manufacture of carbohydrates by green plants (photosynthesis) can be likened to a factory that requires an input of energy and certain raw materials to synthesize a product.

Summary of Basic Raw Materials and Products of Photosynthesis

	Raw Materials	Products
Light-Dependent Reactions	Light energy Water Chlorophyll NADP + H^+ ATP + P	Oxygen ATP NADPH
Light-Independent Reactions	Carbon dioxide Ribulose biphosphate ATP NADPH	PGAL Carbohydrates ADP + P NADP + H^+

ration refers to the oxidation of food molecules by cells. We were initially introduced to **metabolism** in Chapter 3; metabolism refers to all of the chemical processes that occur in a living organism, including **anabolism** and **catabolism.** Anabolic processes require energy to produce larger molecules from smaller ones; catabolic processes release energy as they break down more complex molecules into smaller, simpler molecules. The primary source of metabolic energy for most organisms is the simple sugar glucose. Our focus in the next section will be to observe the breakdown, or catabolism, of glucose by a living cell.

An Overview of Glucose Metabolism

The basic overall energy-generating metabolic process in cells involves a series of steps, or reactions, in which glucose is broken down to produce carbon dioxide and water. During this process, energy is released. Some of the energy extracted from glucose breakdown is used to synthesize ATP and some, in accordance with the second law of thermodynamics, is released as heat. Organisms, of course, may subsist on a variety of lipids, proteins, and carbohydrates other than glucose. We will discover that all of these foods can be used to make ATP by entering the various metabolic pathways we will be discussing.

First, however, we need to look at the major pathways involved in the breakdown of a single glucose molecule. The complete catabolism of glucose can be summarized by the following reaction:

$$C_6H_{12}O_6 + 6O_2 \rightarrow 6CO_2 + 6H_2O + \text{energy}$$
$$\text{glucose} + \text{oxygen} \rightarrow \text{carbon dioxide} + \text{water} + \text{energy}$$

Notice in this reaction that glucose is *oxidized* (hydrogen is removed) to form carbon dioxide and oxygen is *reduced* (hydrogen is gained) to form water. The entire catabolic sequence actually requires three distinct but continuous sets of reactions. The first set is called *glycolysis,* which refers to the "splitting of sugar," and these reactions break down the six- carbon glucose molecule into two molecules of a three-carbon compound called pyruvic acid. The second set of reactions is referred to as the **Krebs cycle,** or the **citric acid cycle,** in which pyruvic acid, in a converted form, is completely broken down into carbon dioxide molecules, hydrogen ions, and electrons. The third set of reactions involves an **electron transport chain,** which uses the energy of the electrons removed from glucose to power the synthesis of ATP. Collectively, these are the reactions known as cellular respiration.

Glycolysis

The reactions of glycolysis take place in the cytoplasm of cells, where specific enzymes catalyze all the reactions required to split the six-carbon glucose molecule into two molecules of *pyruvic acid,* a three-carbon compound (Fig. 7.12). Glycolysis can occur in the absence or presence of oxygen. The essential function of glycolysis is to strip the glucose molecule, step-by-step, of its hydrogen atoms and electrons, which are passed to electron acceptors. In each of the steps, the stored chemical energy in the glucose molecule is released.

To initiate glycolysis, two ATP molecules in the cytoplasm transfer a phosphate group to each end of the glucose molecule; this prevents glucose from passing out through the plasma membrane and also makes the molecule more reactive. Subsequent reactions rearrange the atoms of the glucose molecule and eventually split the six-carbon glucose molecules into two three-carbon molecules. These reactions involve both the use of ATP as an energy source and the later production of ATP. At the end of the glycolysis sequence, these energy transformations result in a net gain of two ATP molecules. Production of ATP in glycolysis does not involve the use of electron energy as we will see in the case in the final stage of cellular respiration. Instead, a phosphate group from molecules formed in the glycolysis pathway is transferred to ADP to form ATP with the help of enzymes.

As the breakdown proceeds, the hydrogen ions and electrons split from glucose are taken up by the coenzyme NAD+, reducing it to NADH. The NAD+ coenzyme has a function analogous to the NADP+ coenzyme involved in photosynthesis. In fact, the only difference between the two molecules is the presence of the phosphate group on NADP+. NAD+, then, accepts hydrogen ions and electrons transferred from other molecules. As a result, NAD+ becomes reduced and is referred to as NADH. With formation of pyruvic acid at the end of glycolysis, a total of two NADH molecules have been generated.

Overall, then, one molecule of glucose is split into two molecules of pyruvic acid. There is a net energy gain of two molecules of ATP and two molecules of NADH. In

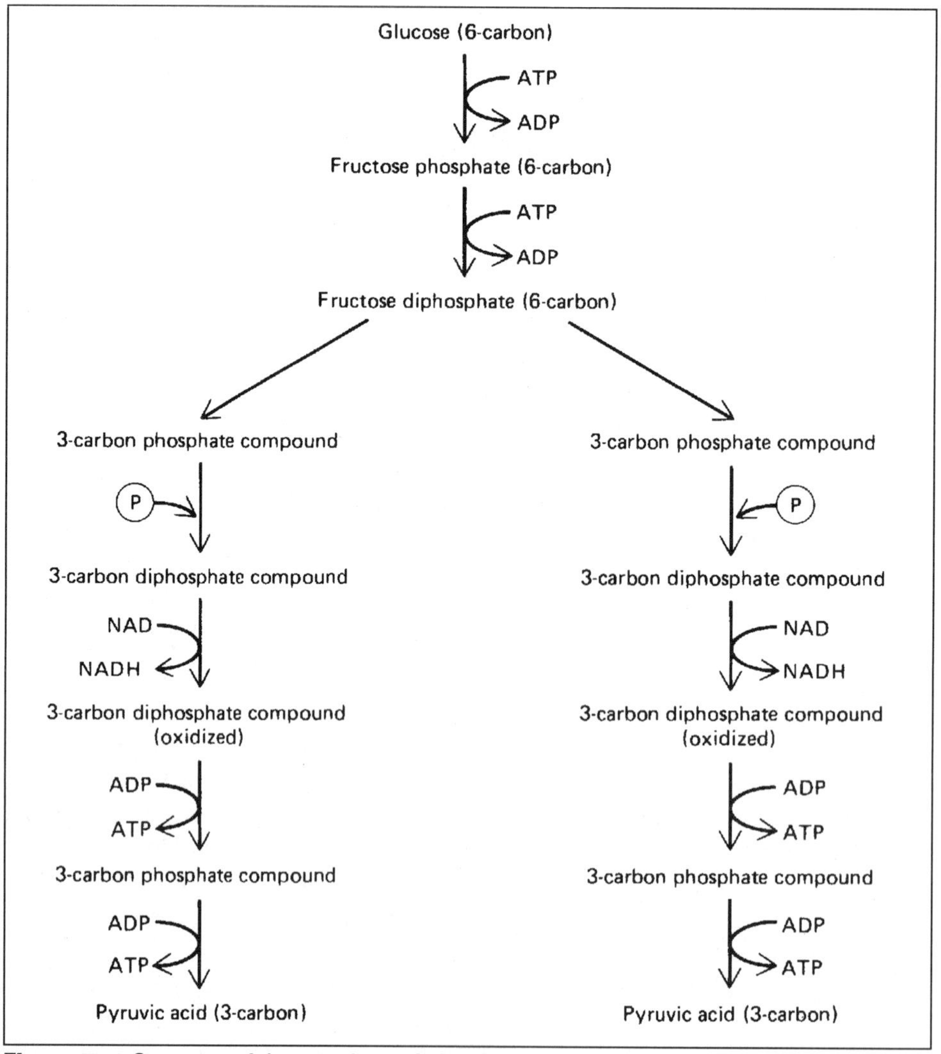

Figure 7.12 Summary of the reactions of glycolysis. One molecule of glucose is split into two molecules of pyruvic acid (pyruvate). The net energy gain in the sequence is two molecules of ATP and two molecules of NADH.

other words, some of the energy originally stored in the chemical bonds of glucose is now stored in the form of high-energy electrons in NADH and in the phosphate bonds of ATP. Most of the energy in the glucose molecule, however, has not been released, but remains in the two molecules of pyruvic acid.

Many carbohydrates, after being converted to glucose, enter the glycolysis pathway. Sugars, including the disaccharides sucrose and lactose, are converted into glucose and then broken down into pyruvic acid. Polysaccharides also enter the pathway after being broken down into simple sugars. If sufficient oxygen is present, the pyruvic acid molecules, after being converted in form, enter the Krebs, or citric acid, cycle. Here, they are broken down completely. We will see that this releases the remainder of the energy stored in the original glucose molecule and completes its breakdown. But what if sufficient oxygen is not available to complete the breakdown of the original glucose? What happens to the pyruvic acid molecules under anaerobic conditions?

Fermentation: Anaerobic Respiration

The breakdown of any organic molecule in the absence of oxygen is called **fermentation**, examples of which are shown in Figure 7.13. In a few organisms, principally some bacteria and yeasts, fermentation is the only stage of glucose metabolism. In yeasts, for example, pyruvic acid is converted to ethyl alcohol and carbon dioxide, which are the end-products of yeast anaerobic metabolism. We take advantage of this metabolic process in the wine, beer, and bread-making industries. In the brewing of beer, yeast is added to a watery mixture of cereal grains, usually barley and some other grain. The fermenting yeast cells metabolize the glucose in the grains, forming ethyl alcohol and carbon dioxide in the process. In the baker's yeast used in breadmaking, it is the carbon dioxide produced during fermentation that causes the dough to rise. You have probably seen the effects of carbon dioxide gas production in the pockets or holes in a slice of baked bread.

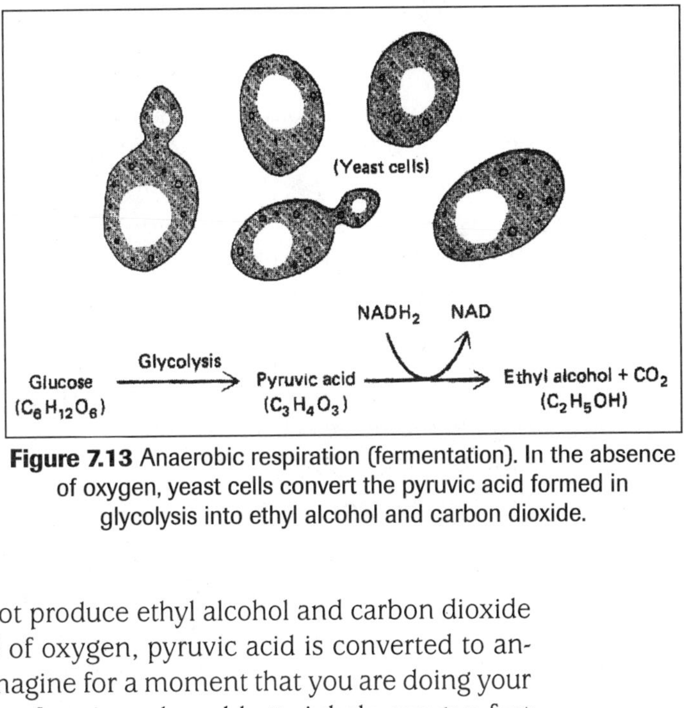

Figure 7.13 Anaerobic respiration (fermentation). In the absence of oxygen, yeast cells convert the pyruvic acid formed in glycolysis into ethyl alcohol and carbon dioxide.

In humans and many other organisms, however, anaerobic glycolysis does not produce ethyl alcohol and carbon dioxide from pyruvic acid. Instead, in the absence of oxygen, pyruvic acid is converted to another three-carbon acid called *lactic acid.* Imagine for a moment that you are doing your daily windsprints and find yourself gasping for air and unable to inhale oxygen fast enough to maintain sprint speed. Your blood cannot deliver oxygen to your muscle cells fast enough; hence, the cells begin to break down glucose anaerobically. As they do so, lactic acid is produced. Muscle cells can use this fermentation pathway for a short time but eventually you must stop sprinting (or collapse) and breathe deeply. The accumulation of lactic acid lowers the pH of the muscles, causes inefficient muscle contraction, and often produces stiff, sore muscles. You have incurred what is called an **"oxygen debt"** to the body and you must pay it back. You pay the debt by stopping and breathing deeply until adequate oxygen again reaches the cells. The lactic acid that was produced, however, accumulates in the circulatory system and is delivered to the liver.

When the oxygen debt is paid and sufficient oxygen is again available, the lactic acid is converted back to pyruvic acid by the liver cells and cellular respiration proceeds, as we will discuss shortly.

In terms of energy release, fermentation is an extremely inefficient process, extracting only about 7 percent of the available energy stored in the glucose molecule. It should be kept in mind, however, that anaerobic metabolism serves the yeasts and bacteria quite well. Some of the energy released from glucose, of course, is lost as heat and some is used to convert pyruvic acid to alcohol or lactic acid. Consequently, the net energy gain from anaerobic glycolysis, the breakdown of glucose to lactic acid, is very small: only two molecules of ATP and two molecules of NADH.

But this is not the end of life's metabolic journey; it is just the beginning. Some type of anaerobic metabolism was employed by primitive prokaryotes millions of years ago. Following the advent of oxygen-producing photosynthesis, however, there evolved a new mechanism of metabolism, the reactions of which follow directly after those of glycolysis. This aerobic process involved the Krebs cycle and the electron transport system of mitochondria.

Krebs Cycle and Electron Transport Chain: Aerobic Respiration

In the absence of oxygen, glucose is incompletely catabolized. At the completion of glycolysis, most of the energy in glucose is still stored in pyruvic acid molecules. When sufficient oxygen is present, catabolism continues and pyruvic acid is broken down completely in the next stage of cellular respiration. This stage is called the Krebs cycle in honor of the biochemist and Nobel laureate Hans Krebs, who elucidated and described most of its reactions. It is also referred to as the citric acid cycle because citric acid is the first product in this series of reactions. This reaction sequence is referred to as a "cycle" because it begins with a specific compound, which is broken down and is then ultimately reformed at the end of the cycle. This cycle and the final stage of cellular respiration—the electron transport chain reactions—occur in the mitochondria, the "powerhouse" organelles of eukaryotic cells. These oxygen-requiring steps of cellular respiration are outlined in Figure 7.15.

Krebs Cycle Preparation: Acetyl Coenzyme A

Before the molecules of pyruvic acid can enter the Krebs cycle, each must be converted to a compound called **acetyl coenzyme A,** or simply acetyl CoA. Pyruvic acid formed in glycolysis enters the mitochondria and is converted to acetyl CoA in the presence of oxygen. During this conversion, one carbon atom and two oxygen atoms—in the form of carbon dioxide—are split from each pyruvic acid molecule and diffuse out of the cell. The two-carbon fragment (the acetyl group) remaining is bonded to coenzyme A to form acetyl CoA. Coenzyme A makes the acetyl group more reactive and is essential for carbohydrate and fat metabolism. In addition, each pyruvic acid molecule loses electrons and hydrogens that are transferred to two separate $NAD+$ molecules, reducing each to NADH.

Keep in mind that the original six-carbon glucose molecule, split into two three-carbon molecules in glycolysis, is now in the form of two two-carbon molecules of acetyl CoA. Step-by-step, the energy of the glucose molecule is being extracted; some of this energy

Figure 7.14 Anaerobic respiration in skeletal muscles. A deficiency of oxygen in the muscles of a runner results in the conversion of pyruvic acid into lactic acid. This deficiency, known as *oxygen debt,* must be "repaid" for further metabolism to continue.

Repayment is made when the runner stops and breathes deeply.

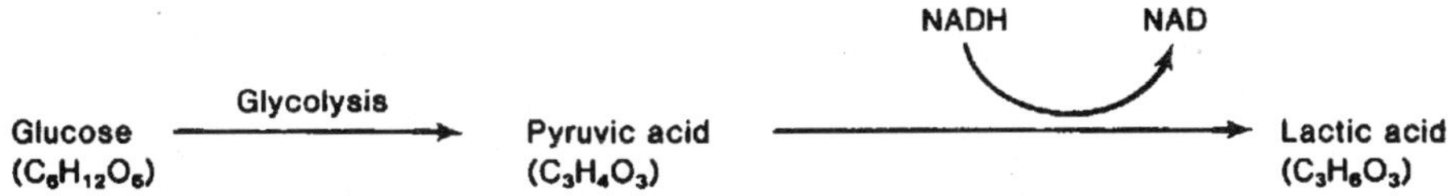

has been stored in the phosphate bonds of ATP and some of it as high-energy electrons in NADH. In the Krebs cycle, the two-carbon acetyl group, the last remnant of the original glucose molecule, is broken down.

While this conversion reaction may appear a bit complicated, it is important to understand its significance in the overall metabolic scheme. Acetyl CoA is the form in which the original glucose molecule enters the Krebs cycle; in fact, it is the form in which most molecules enter the Krebs cycle. Some fats and proteins, when needed for energy, can enter the Krebs cycle by different metabolic pathways in the cell. After these molecules are broken down, they, like glucose, can be converted to acetyl CoA and then further metabolized in the Krebs cycle. Acetyl CoA, then, is the "entry" compound into this cycle that allows the continued extraction of energy from a variety of partially metabolized nutrients.

The Krebs Cycle

The essence of the Krebs cycle is to continue the breakdown of glucose, now in the form of two molecules of acetyl CoA, and to simultaneously release carbon dioxide, hydrogen ions, and electrons from the two molecules. The cycle also generates a molecule of ATP for each pyruvic acid molecule that enters it and initially forms a four-carbon acid (oxaloacetic acid) that, like RuBP in photosynthesis, must be regenerated repeatedly. It requires two complete cycles—one for each pyruvic acid molecule—to complete the breakdown of one molecule of glucose.

NAD

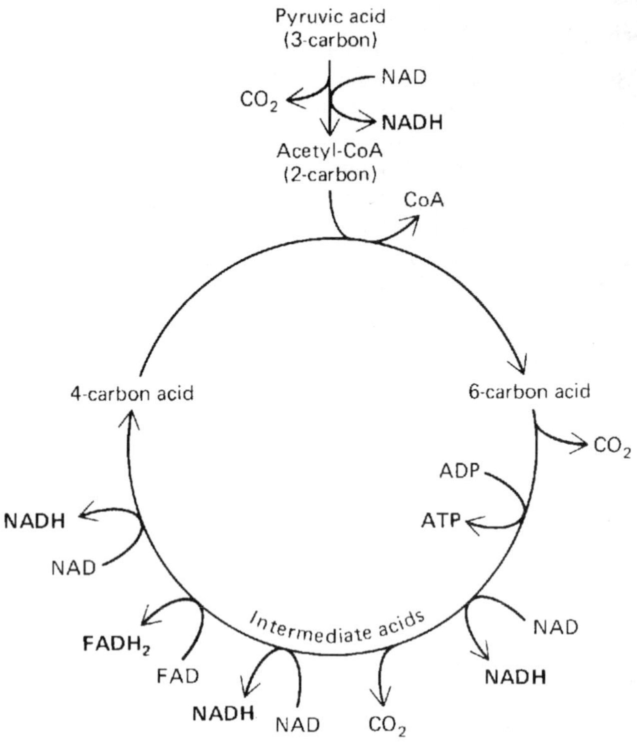

Figure 7.15A Generalized summary of the conversion reaction and Krebs cycle. Both processes release energy that is stored in coenzymes or in ATP. The Krebs cycle reactions continually regenerate the 4-carbon acid so that the cycle can be repeated.

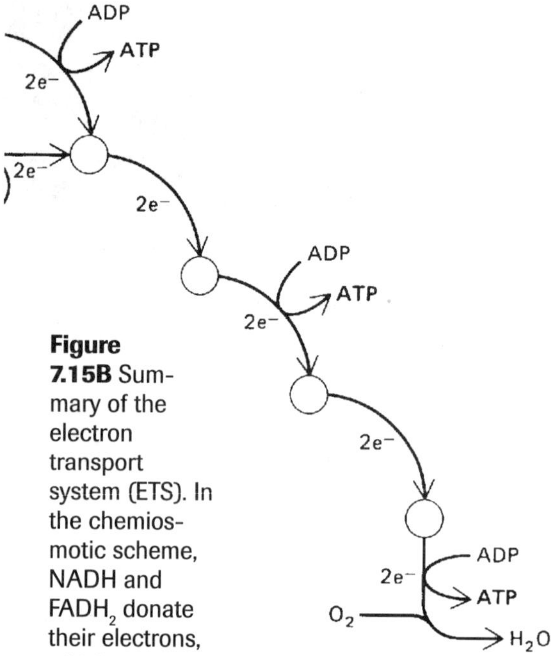

Figure 7.15B Summary of the electron transport system (ETS). In the chemiosmotic scheme, NADH and $FADH_2$ donate their electrons, which are passed "downhill" (i.e., to lower energy levels) along a system of electron carrier molecules. In moving from higher to lower energy levels, each pair of electrons (2^{e-}) releases potential energy that is used to generate ATP. At the "bottom" of the system, hydrogen ions from NADH and $FADH_2$ combine with oxygen to form water (H_2O).

On entering the Krebs cycle, coenzyme A is immediately split away from acetyl CoA and is now free to bond with another acetyl group derived from a new molecule of pyruvic acid. The remaining two-carbon acetyl group bonds with the four-carbon oxaloacetic acid to form citric acid, which undergoes a series of reactions until, at the end of the cycle, the reactions have regenerated the original four-carbon acid (oxaloacetic acid). In other words, before citric acid can be formed, oxaloacetic acid must be continually regenerated in the Krebs cycle. This reaction series releases carbon dioxide, hydrogen ions, and electrons. Carbon dioxide is the major waste product of cellular respiration. It is, of course, the same carbon dioxide you exhale some 20,000 times a day.

The hydrogen ions and electrons are passed to the coenzymes NAD+ and **FAD+** (flavin adenine dinucleotide) reducing them to NADH and $FADH_2$, respectively. At one stage in each of the two turns of the Krebs cycle, ADP receives a phosphate group from a cellular molecule, yielding a total of two molecules of ATP for the breakdown of one glucose molecule.

With the formation of the original four-carbon oxaloacetic acid at the end of the cycle, all the available energy in the glucose molecule has now been extracted. The molecule no longer exists, and all that remains of glucose in the cell is its released energy stored in ATP, NADH, and $FADH_2$. The regenerated four-carbon acid can now accept a new acetyl group from acetyl CoA to continue the cycle and complete the breakdown of another pyruvic acid molecule from glycolysis.

In summary, pyruvic acid is converted to acetyl CoA and this two-carbon acetyl group enters the Krebs cycle by bonding with a four-carbon acid. A series of reactions takes place, resulting in the regeneration of the same four-carbon acid at the end of the cycle. In the process, carbon dioxide is given off, and energy in the form of electrons and hydrogen ions released during the cycle is stored in ATP, NADH, and $FADH_2$. It is now time to put this energy to work.

The Electron Transport Chain

The final stage of cellular respiration is the payoff, where the energy harvested from glucose catabolism is finally used. The theme of this stage—the transport of electrons and pumping of hydrogen ions—is similar to the one we encountered in the light-dependent reactions of photosynthesis. The chain of electron carriers is located in the inner mitochondrial membrane (Fig. 7.16). Electrons held by the coenzymes NADH and $FADH_2$ are passed to the first carrier in the chain and then successively passed "downhill" to other carriers in a sequence of oxidations and reductions. As they descend from higher energy levels to lower ones, the electrons release their potential energy.

As the electrons are being passed down the chain, the hydrogen ions from NADH and $FADH_2$ are released to the interior of the mitochondrion and are pumped, using the energy released from these electrons, into the space between the inner and outer membranes. The inner mitochondrial membrane is relatively impermeable to hydrogen ions, which is why they must be pumped across it. The pumping of hydrogen creates a high concentration of hydrogen ions in the intermembrane space, with few hydrogen ions left in the interior of the mitochondrion. Accordingly, the ions tend to diffuse across the inner membrane and back into the interior to balance the ion concentration. In order to move back across the membrane, the hydrogen ions must pass through gatelike **ATP-synthesizing complexes,** which are found in parts of the inner membrane. As they pass through these membrane gates, the hydrogen ions are involved in the bonding of phosphate to ADP, forming ATP. Finally, the hydrogen ions, which are now back again in the interior of the mitochondrion (their short journey having produced ATP), combine with oxygen and the electrons used in the electron transport chain to form water.

At the end of the chain, the electrons are passed from the last carrier molecule to oxygen, the **final electron acceptor** in the sequence. This reduces the oxygen, which, along with the addition of hydrogen ions, forms water molecules. Without oxygen to ultimately accept these electrons, the electron transport chain cannot work. That is the reason why organisms undergoing anaerobic fermentation cannot utilize electron transport. In fact, that is why, in aerobic organisms, suffocation causes death; without oxygen, the electron transport chain cannot function and thus adequate ATP cannot be synthesized and the energy demands of the cell cannot be met.

We now have accounted for all the products in the overall summary reaction $C_6H_{12}O_6 + 6O_2 \rightarrow 6CO_2 + 6H_2O$ + energy. The carbon dioxide is produced mainly in the Krebs cycle, while water and most of the energy (ATP) are produced in the final stage of cellular respiration, the electron transport chain. This is the cache of ATP the cell can draw on, obtaining energy to carry out its life-sustaining functions.

The net energy gain from hydrogen ion flow across the inner mitochondrial membrane is estimated to be 32 ATPs for each molecule of glucose broken down. Adding this number to the 2 ATPs from glycolysis and the 2 ATPs from the Krebs cycle (one for each pyruvic acid molecule) yields a grand total of 36 ATPs for the complete catabolism of

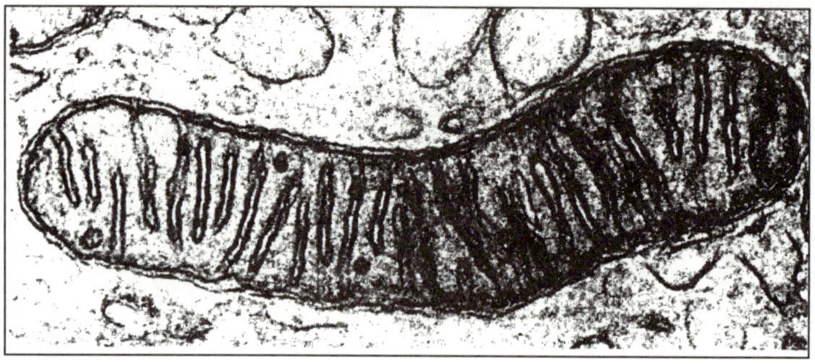

Figure 7.16 A mitochondrion, the site of the reactions of the Krebs cycle and electron transport system. Note the elongated folding of the inner membrane.

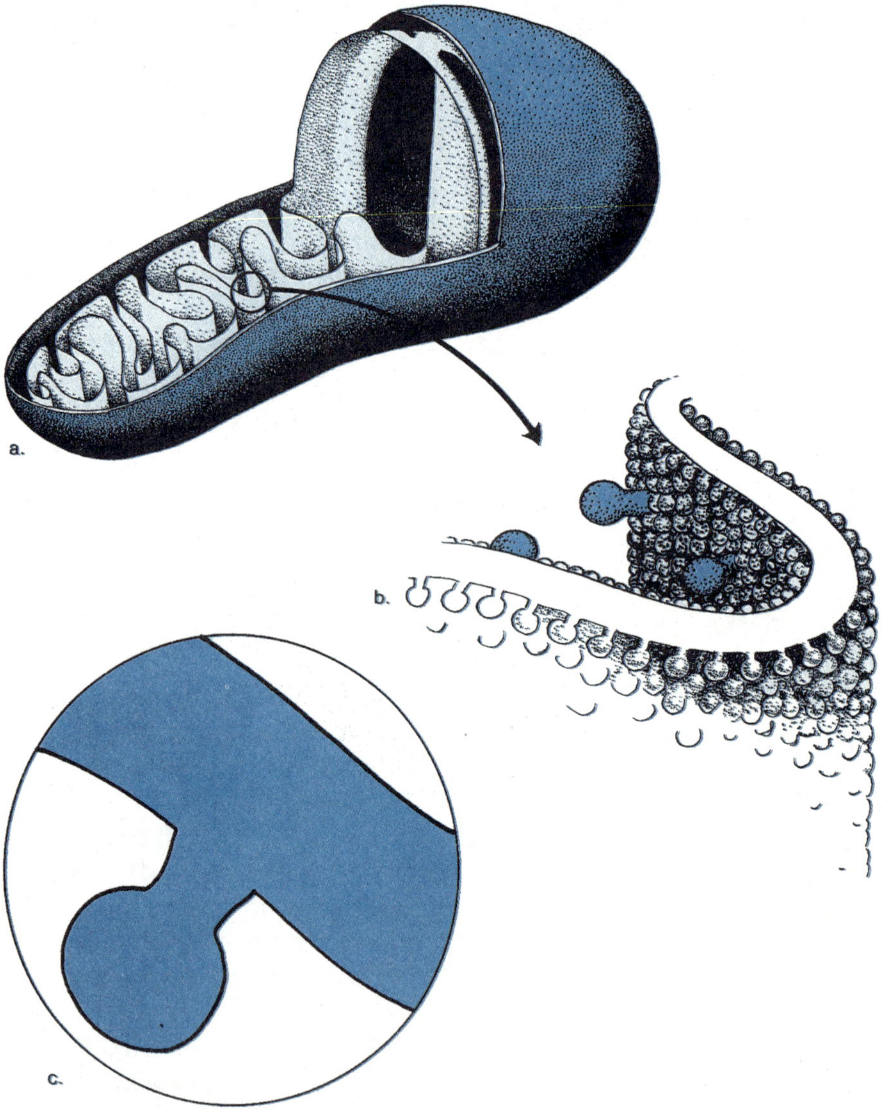

(a) Cutaway diagram of a mitochondrion, showing the outer membrane and the foldings, or *cristae,* of the inner membrane. (b) A section of the inner mitochondrial membrane showing the large, enzyme-containing stalked particles that project inward from the membrane. The numerous smaller projections embedded in the inner membrane represent the carrier molecules of the electron transport system. (c) Enlargement of enzyme-containing stalked particle. Hydrogen ions in the outer compartment of the mitochondria pass by diffusion to the inner compartment through these particles. Passage of the hydrogen ions generates ATP.

one molecule of glucose. Compared to glycolysis alone, which produces only 2 ATPs, it is evident that aerobic respiration is at least 18 times more efficient.

It was pointed out earlier in this chapter that the efficiency of a given system is the ratio of useful energy to the total energy of the system. It was also mentioned that no system can be 100 percent efficient because of the concomitant release of heat (second law of thermodynamics). How, then, does cellular respiration stack up in terms of energy efficiency? To find out, all we need to do is compare the useful energy derived from ATP to the total energy put into the cell. For one molecule of glucose catabolized, the total energy released is 686 kcal; that is, 686 kcal are available to do cellular work. The energy stored in the terminal phosphate bond of ATP is approximately 7.3 kcal. For 36 ATPs, the total useful energy is 36 ATP x 7.3 kcal = 262.8 kcal. Thus, the efficiency of this system is: 262.8 kcal (useful energy)/ 686 kcal (total energy) = 0.383 x 100 = 38.3 percent. By comparison, a simple steam engine may have an energy efficiency as low as 10 percent.

Energy efficiency is very poor in anaerobic organisms such as yeasts when compared to that of aerobic organisms. With only two ATPs produced in fermentation, the total useful energy is only 14.6 kcal. The efficiency of fermentation, therefore, is 14.6 kcal/686 kcal, or about 2 percent. The same is true for anaerobic exercise. The limited supply of ATP explains why strenuous bursts of energy, such as the 100-meter dash, cannot be sustained for a long time period. With over 60 percent of the energy from aerobic metabolism and 98 percent from anaerobic metabolism lost as heat, you can appreciate the tremendous energy investment required to maintain life on this planet.

When we consider the phenomenon of life, it comes down in a large part to understanding energy transactions: the relationships between energy input and energy release. Or, since the energy for cellular metabolism resides in chemical bonds, it comes down to the making (anabolism) and breaking (catabolism) of chemical bonds. Energy is the capacity to do work, but until it is in a usable form and then channeled to do useful work, it remains only a capacity. On Earth, photosynthesis and cellular respiration are the fundamental processes by which solar energy is converted in form and channeled to do the work of sustaining life.

Chapter Summary

1. Energy is the capacity to do work; stored energy is potential energy, and kinetic energy is the energy that does work. The first law of thermodynamics states that energy cannot be created or destroyed; the second law of thermodynamics states that energy conversions are not 100 percent efficient; they involve the release of heat energy. The ratio of useful energy to the total energy of a system is called energy efficiency.

   Organisms are either autotrophs or heterotrophs. Some of the earliest autotrophic bacteria on Earth used hydrogen sulfide gas as an energy source, liberating sulfur compounds as a by-product of metabolism. Subsequent evolution produced autotrophic blue-green bacteria that used water in place of hydrogen sulfide gas, releasing oxygen gas into the seas and atmosphere. Over time, oxygen saturated the atmosphere and eventually formed the protective ozone layer.

   Key Terms: *energy, p. 137* · *energy efficiency, p. 138* · *autotroph, p. 138* · *heterotroph, p. 138* · *anaerobic, p. 139* · *aerobic, p. 139*

Concept Questions: Compare the methods of obtaining energy in ancient heterotrophs and modern heterotrophs, and in ancient autotrophs and modern autotrophs. What is the general application of the first and second laws of thermodynamics to these methods?

2. ATP is the energy currency of living cells, with most of its energy stored in the two terminal phosphate bonds. When a terminal phosphate bond is broken, energy is released and the terminal phosphate group is transferred to another molecule. NADP+ is a coenzyme that is reduced to NADPH by accepting hydrogen ions and electrons. The hydrogen ions can be released to reduce other molecules, and the electrons are passed to an electron transport chain where the energy of the electrons is used to synthesize ATP.

   Metabolism refers to all the chemical reactions within an organism. Glucose is the primary energy molecule metabolized by living cells. The complete breakdown of glucose in cells is carried out by cellular respiration. Fats, amino acids, and various carbohydrates also can be metabolized after entering glycolysis and the Krebs cycle.

   Key Terms: *ATP, p. 142*
   NADPH, p. 142
   oxidation, p. 144
   reduction, p. 144
   electron transport chain, p. 145

   Concept Questions: How is life dependent on ATP and NADP+? What is the specific role of an electron transport chain in living cells?

3. Photosynthesis takes place primarily in green leaves, which contain millions of chloroplasts—the photosynthetic organelles. Chloroplasts are membranous sacs containing the green pigment chlorophyll, stacks of thylakoid discs, and the stroma.

   The initial energy source for photosynthesis is light. The entire photosynthetic sequence includes the light-dependent and the light-independent reactions. Oxygen, derived from the splitting of water, is a by-product of the light-dependent reactions of photosynthesis. Photosynthesis produces carbohydrates by reducing carbon dioxide, using NADPH and ATP produced in the light-dependent reactions as energy sources.

   Key Terms: *chloroplast, p. 139*
   thylakoid, p. 140
   visible light, p. 141
   primary electron acceptor, p. 142
   RuBP, p. 149
   PGAL, p. 149

   Concept Questions: Explain Szent-Gyorgi's statement: "What drives life is a little electric current, kept up by the sunshine." Why is water critical in the photosynthetic process? What is the reciprocal relationship between RuBP and PGAL?

4. The reactions of glycolysis take place in the cell cytoplasm and involve the splitting of a six-carbon glucose molecule into two three-carbon molecules of pyruvic acid. Under anaerobic conditions (fermentation), yeasts convert pyruvic acid to ethyl alcohol and carbon dioxide, while humans and many other animals convert pyruvic acid to lactic acid.

   Under aerobic conditions in the mitochondria, pyruvic acid is converted to acetyl coenzyme A, which is a pivotal compound by which fats and proteins, along with carbohydrates, enter the Krebs cycle to be broken down. The acetyl group, the remaining remnant of the original glucose molecule, is broken down completely in the Krebs cycle. These reactions release carbon dioxide, hydrogen ions, and electrons; the ions and electrons are accepted by NAD+ and FAD+, reducing them to NADH and $FADH_2$, respectively, carbon dioxide is a waste product of the Krebs cycle reactions.

Key Terms: *NADP+, p. 148*
catabolism, p. 152
fermentation, p. 154
oxygen debt, p. 154
FAD+, p. 157

Concept Questions: Why might glycolysis be considered a relatively inefficient process? What is the significance of this efficiency to organisms? Explain basically why the Krebs cycle reactions are essential in cellular respiration.

5. In the mitochondria, the electrons and hydrogen ions are passed from NADH and $FADH_2$ to the electron transport chain. The energy released from the electrons is used to pump hydrogen ions from the interior of a mitochondrion into the space between the inner and outer mitochondrial membranes. The last acceptor in the chain passes the electrons to oxygen. The hydrogen ions diffuse through ATP-synthesizing complexes in the inner membrane and back into the interior of the mitochondrion. Movement of the hydrogen ions through the complexes is involved in bonding a phosphate group to ADP, forming ATP. Finally, the hydrogen ions and electrons combine with oxygen to form water.

   Cellular respiration produces a total of 36 ATP molecules for each molecule of glucose broken down. The energy efficiency of glycolysis is about 2 percent, the entire respiratory scheme is about 38 percent energy efficient.

Key Terms: *final electron acceptor, p. 157*
ATP-synthesizing complex, p. 157

Concept Questions: What is the ultimate reason living cells must receive a steady supply of oxygen? Compare the roles of electron carrier molecules in photosynthesis and cellular respiration.

Concept Summary Questions

How can we apply the first and second laws of thermodynamics to the reactions of photosynthesis and cellular respiration?

Explain, using examples, why photosynthesis is an anabolic process and why cellular respiration is a catabolic process.

Chapter Eight

Energy Flow

Energy Conversions

It is remarkable that the conversion of one form of energy into different forms sustains life and that living organisms can convert the energy that gives them life. Equally remarkable, the original source of this life-giving energy is light from the sun. From its time of origin, the sun, which represents over 99 percent of the total mass of our solar system, has possessed enough energy to last 10 billion years and still possesses energy to last 5 billion years more. The capture of light energy by autotrophs and its eventual use by all organisms involve complex processes, as we have discovered with the processes of photosynthesis and cellular respiration. Energy, however, was present in our solar system and on Earth long before the appearance of life. We saw in Chapter 2 that energy in the form of thermonuclear explosions accompanied the formation of the solar system, and that light energy from the sun is transformed into heat energy when it is absorbed by an object.

With the introduction of organic nutrients into the natural world through photosynthesis, energy from the sun can be passed to all living cells and eventually stored in chemical compounds such as ATP. For autotrophs, solar energy is acquired directly and used to synthesize nutrients and ATP. Heterotrophs, however, cannot utilize solar energy directly and fulfill their energy requirements by feeding on autotrophs or other heterotrophs. Energy, then, enters the living world from the sun and is passed in converted form from one organism to another, a phenomenon known as **energy flow** (Fig. 8.1).

The availability of a usable form of energy is one of the factors that ultimately determines the reproductive capability and, hence, the very existence of a population. A *population* is a group of organisms of the same species that lives in a given area. An adequate supply of energy translates into food sources, which determine the health of a population and the subsequent probability of its individuals producing offspring. This in turn affects the reproductive capability of other populations that occupy the same geographic area—the community. There is, then, interdependency within a community, with each population contributing to community structure and function.

Ecosystems: Functional Units of the Biosphere

Ecology is the study of the interactions of organisms with each other and with their nonliving environment. These interactions take place within **ecosystems,** which consist of communities and their surrounding physical environment. An ecosys-

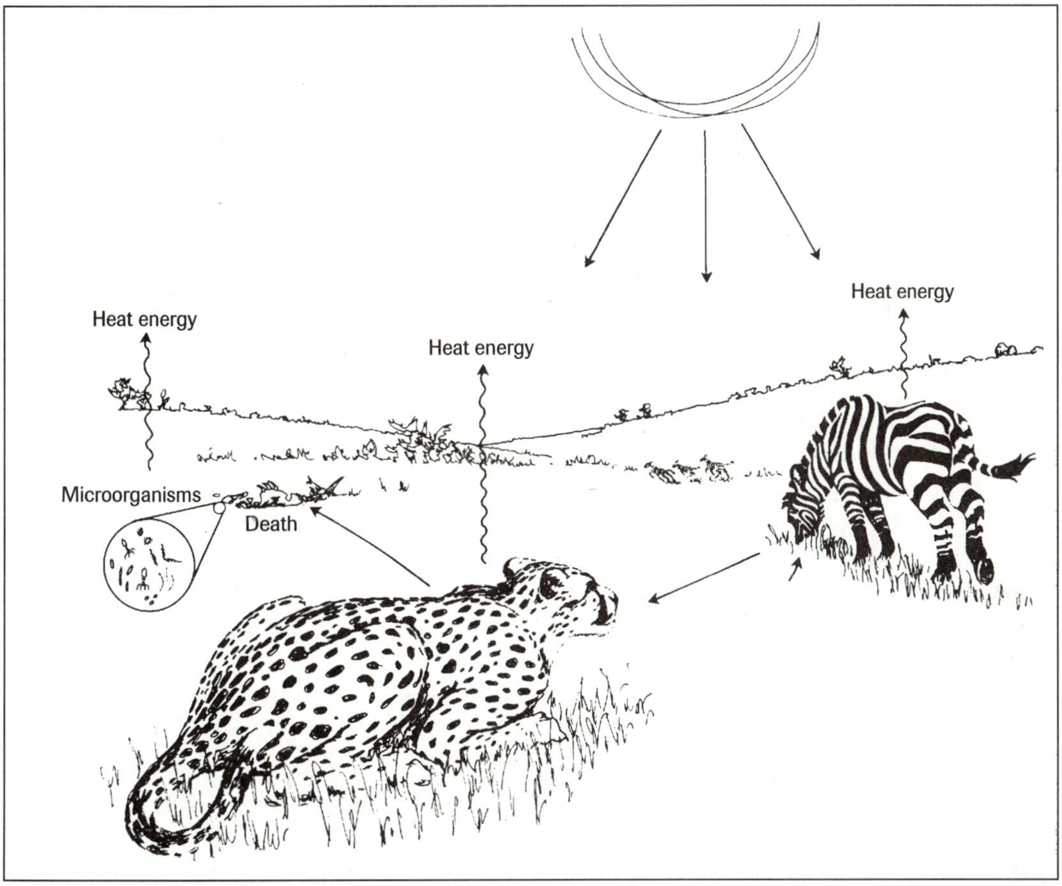

Figure 8.1 Energy flow on Earth. Utilizing light energy from the sun, green plants manufacture organic compounds eventually taken into the bodies of animals that feed on the plants. These animals in turn are fed on by other animals. Microorganisms obtain energy by feeding on the bodies of plants and animals that have died. In the energy conversions that occur within the plants, animals, and microorganisms, some of the energy is used for cellular metabolism and some escapes to the environment as heat.

tem is the highest level of structure and function in the biosphere. A lake, a thicket, or an open field, with their associated living and nonliving components, are familiar examples of natural ecosystems. Often it is difficult to determine where one ecosystem ends and another begins, since their boundaries usually overlap. Earth is actually a series of overlapping smaller ecosystems that form one huge global ecosystem (Fig. 8.2).

In general, there are two principal types of ecosystems: *terrestrial* and *aquatic*. Major terrestrial ecosystems found in a certain geographic area are called **biomes;** they are char-

Figure 8.2 The Earth's crust consists of the land (lithosphere), the water (hydrosphere), and the air (atmosphere). Together, these constitute the biosphere, the areas on Earth where life is found.

acterized by a dominant type of vegetation, which is affected significantly by climate and soil type. Representative terrestrial biomes include the frozen tundra in the northern regions of Earth, prairies, arid deserts, and tropical rain forests (Fig. 8.3). Biomes vary with their location on Earth and can even vary with altitude. Below the snow-capped peaks of tall mountains, for example, there may be a tundra biome and, at the base of the mountain, a tropical rain forest.

Aquatic ecosystems can occur in either marine or freshwater environments. The oceans, with their great variety of life, provide the most varied types of aquatic ecosystems. Physical factors, such as depth, pressure, salinity, light penetration, and temperature, generate a diverse variety of marine habitats; that is, the different types of areas within an ecosystem where a given organism will usually live. These different habitats, in turn, determine the kinds of organisms that can be supported. For example, surface waters of the oceans and lakes of the world are inhabited by billions of single-celled bacteria, microscopic algae, and protists that constitute *phytoplankton* ("drifting plants"), along with tiny crustaceans, fish eggs, and invertebrate larvae that make up *zooplankton* ("drifting animals") (Fig. 8.4). Both forms of plankton are the major sources of food for the swimming animals that inhabit the surface waters. In addition, the photosynthetic phytoplankton produce an estimated 40 percent of Earth's atmospheric oxygen.

The ocean bottom, another ocean habitat, is a cold region lying in total darkness and is inhabited by a few species of fish, various invertebrates, and bacteria. In this deep-sea habitat, organisms have been discovered existing in hot, oxygen-deficient vents of volcanic origin. With light unavailable as an energy source, bacteria in these vents oxidize inorganic compounds such as hydrogen sulfide (H_2S) or ammonia (NH_3) as energy sources. These bacteria, in turn, serve as food for the invertebrates and fishes.

Figure 8.3A A desert biome.

Figure 8.3B A deciduous forest.

Figure 8.4 A collection of unicellular diatoms, some of the organisms that constitute drifting phytoplankton in the seas and oceans.

Solar rays may penetrate to a depth of about 600 meters (2000 feet) in the ocean, but below this depth there is perpetual darkness. From the surface waters to a depth of about 200 meters (650 feet), the amount of light penetrating is sufficient for photosynthesis to occur. Below this zone, the light begins to fade gradually until there is only darkness. Since plant life in the oceans is limited to the area penetrated by light, deep-dwelling organisms depend to a great extent on nutrients that drift down from the surface waters. The intertidal zone, where land and sea merge, is another habitat providing an environment rich in plant and animal life.

Freshwater systems—ponds, lakes, rivers, and creeks—also have unique physical characteristics that determine the kinds of organisms that can exist in these environments. Phytoplankton and zooplankton ordinarily are more abundant in the quiet standing waters of ponds and lakes. The shallow waters of ponds permit the growth of bottom-dwelling plants, which are important food sources for aquatic animals. Compared with those in the oceans, freshwater ecosystems do not contain as diverse a variety of aquatic organisms, a result of the more limited diversity of freshwater habitats.

Maintenance of any terrestrial or aquatic ecosystem is dependent on the interactions and interrelationships among the various species of organisms. Such dependency produces both short-term and long-term effects that tend, overall, to stabilize the ecosystem. Procurement of food and water, availability of energy, control of population size, reproductive potential, and other factors are normally regulated by many *biotic* (living) and *abiotic* (nonliving) processes acting in concert. A simple example in a terrestrial ecosystem is the relationship between plant growth and adequate rainfall. Most of the water taken up by the roots of a plant eventually evaporates from the leaves through the stomata and a small amount of water is necessary for photosynthesis. We saw in Chapter 7 that plant leaves exchange carbon dioxide and oxygen across their surfaces. Carbon dioxide from the atmosphere cannot diffuse into the leaf cells without first being dissolved in water. Further, the turgor pressure of water within the plant cells maintains the plant in a firm, upright position. In the absence of adequate water, some plants wilt and, of course, eventually die. All of these factors affect normal plant growth and are related to the availability of water. These interrelationships and interactions constitute the functioning aspect of an ecosystem. Some of these processes will be considered in this chapter and others in Chapter 12.

The types of interactions possible in an ecosystem are based on how it is structured. We would expect specific interactions in an aquatic ecosystem, such as the surface waters of an ocean, to be different from interactions in a terrestrial ecosystem, such as a desert. With regard to energy, the different environmental factors in each ecosystem determine the relative availability and amount of energy. Many sea animals feed on plankton, which is abundant in the surface waters of the ocean; in contrast, some deserts do not have even year-long vegetation as a source of food because of low rainfall. This places greater limits on the availability of energy in deserts as compared to that in the oceans. The type of ecosystem, however, does not alter the principle of energy flow. The flow of energy in any ecosystem follows what is called a **food chain,** a sequence of organisms, each serving as energy sources for the next group of organisms in the chain. This is the essential and fundamental interaction in every ecosystem. As we examine the ecosystem in its role as the structural and functional unit of the biosphere, we also will consider some of the factors that may alter this structure and function.

Food Chains

Interactions among the different organisms in an ecosystem are based to a great extent on their feeding relationships. Every organism fits into the "who eats whom?" scheme, which passes energy from one level to the next up the food chain of organisms (Fig. 8.5). The chain of energy starts with the producers, who capture light energy to make their own nutrients. This energy is passed to the next level, the primary consumers, who eat the autotrophs to obtain nutrients. The energy is then passed to the next level; secondary consumers eat primary consumers and so it continues. The decomposers—bacteria and fungi—break down the remains of dead organisms as well as the waste products of living organisms.

Figure 8.5 An arctic food chain. The first link in the chain always is an autotroph, in this case, a photosynthetic cranberry bush. The herbivorous meadow vole, the primary consumer, is fed on by the ermine, a carnivore and secondary consumer. The final link in the chain, the snowy owl, also is a carnivore and a tertiary consumer. After death, each of these organisms is fed on by decomposers.

Producers

In addition to water, the basic commodity vital to every organism is food. In all ecosystems, food chains begin with autotrophs, also called **producers**. These are usually the photosynthetic organisms, such as algae and plants, which are capable of capturing solar energy and using it to synthesize organic nutrients. As mentioned, the photosynthetic phytoplankton are the major food producers in the surface waters of marine ecosystems. As we have mentioned, an alternative to the use of solar energy occurs in ecosystems found at great depths on the ocean floor. In this eternal darkness, certain bacteria utilize chemical reactions involving sulfur and nitrogen compounds to provide food for themselves and other bottom dwelling species. In addition, photosynthetic bacteria supply a limited amount of nutrients in various soil and water environments.

Consumers

Photosynthesis uses only about 0.1 of 1 percent of the light energy striking Earth. Autotrophs themselves use from 10 to 50 percent of this small fraction for their own metabolism. The useful energy remaining—the net productivity—is available to all other organisms in the food chain. The next link in the chain includes the *herbivores* (plant eaters), or **primary consumers,** which are heterotrophs that feed directly on the autotrophs. These may include such diverse organisms as insects that eat leaves, birds that feed on berries, or gorillas that feed on fruits and vegetables.

Animals that feed on herbivores are called *carnivores* (flesh eaters), or **secondary consumers.** Carnivores obtain their energy from plants through an indirect route. A fish or a frog, for instance, may devour a leaf-eating grasshopper, or a lion may feed on a grazing antelope. Grasshoppers and antelopes, being herbivorous, incorporate organic nutrients from the plants into their own bodies and, when eaten, pass these nutrients to the carnivores.

Still other carnivores feed on the secondary consumers and thus are **tertiary consumers.** An owl, for example, would be a tertiary consumer if it ate a snake (a secondary consumer) that had devoured a field mouse (a primary consumer), which feeds on grains.

Finally, various animals in a food chain can be **quaternary consumers.** If the owl in the example just cited were eaten by a hawk, the hawk would qualify as a quaternary consumer. These carnivores, then, feed on the tertiary consumers and are one step farther removed from the initial link in the food chain—the producers.

Decomposers

The ultimate benefactors at some point in every food chain are the **decomposers**—bacteria and fungi that break down the remains of all organisms that have died. Even the remains of the decomposers are broken down by other decomposers. As pointed out in Chapter 6, these microorganisms constitute an irreplaceable link within an ecosystem by returning organic matter to the environment to be taken up by other organisms and used again. In this way, the atoms making up the nutrients are recycled within the ecosystem. We will have more to say about decomposers and their roles in the ecosystem later in this chapter.

Figure 8.6 A land-based food web. The criss-crossing of food chains illustrates the sources of nourishment for each organism in a chain. The plants are photosynthetic and provide the first link in each food chain. Eventually, the remains of the plants and animals are fed on by decay microorganisms in the soil (not shown).

Food Webs

Food chains outline the basic pattern of energy flow through an ecosystem; it is usually more accurate, however, to relate energy flow and nutrient cycling to **food webs** (Fig. 8.6). Food webs are complex, interconnected food chains that often crisscross or branch from one food source to another. Organisms often feed at several different levels in a food chain; humans, for example, eat producers (plants), primary consumers (beef or poultry), secondary consumers (fish), and decomposers (mushrooms). In addition, a given species of organisms may consume a variety of different foods, all belonging to the same food chain energy level. A grazing herbivore, for example, may feed on several different species of grasses instead of only one or two. This is important because the availability of multiple food sources tends to stabilize and perpetuate an ecosystem. If a species has extremely limited food source, the loss of that source could be disastrous, not only for that species, but for a variety of dependent species as well.

The Ten Percent Rule

As energy flows through any food chain, from producers to consumers to decomposers, the amount of useful energy derived from each successive food source in the chain continually decreases (Fig. 8.7). This means, for example, that less energy is available to herbivores than to autotrophs, and still less energy is available to carnivores. Behind this phenomenon is the second law of thermodynamics. Metabolic activities of the organ-

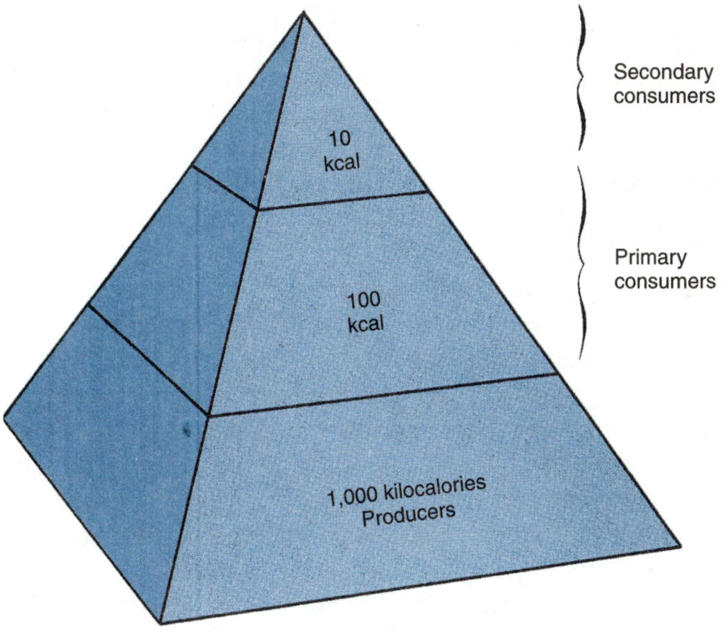

Figure 8.7 Illustration of the 10 Percent Rule. Of a net primary productivity of 1000 Kcal, only 100 Kcal are available to the primary consumers, and just 10 Kcal are available to the secondary consumers. At each trophic level, 90 percent of the energy is dissipated as heat.

isms at each link in the food chain release some of their energy as heat that cannot be used or recaptured. In fact, approximately 90 percent of the energy at each link in the chain escapes from the ecosystem as heat. Thus, only 10 percent of the total energy at any given level in a food chain is available to the next level. This is known generally as the **10 percent rule.**

As an illustration, assume a hypothetical ecosystem in which the energy available from the producers is 5000 kcal. The next link in the chain—the primary consumers—would obtain only 10 percent, or 500 kcal of the available energy; the secondary consumers, in turn, would capture only 50 kcal, and the tertiary consum-

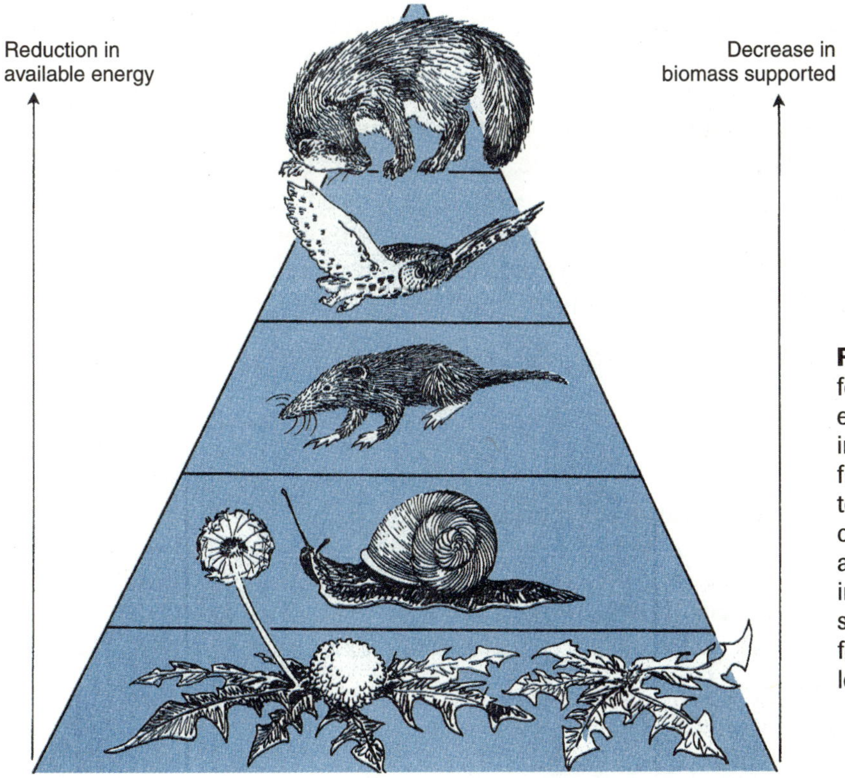

Figure 8.8 Illustration of feeding levels in an ecosystem. The pyramid indicates that, in moving from a lower feeding level to a higher one, there is a corresponding reduction in available energy. Accordingly, less biomass can be supported at a higher feeding level than at a lower level.

ers would derive just 5 kcal of the energy originally provided by the autotrophs. This example clearly illustrates why a food chain typically has no more than four levels. The amount of energy available for a fifth level is rarely sufficient to support an organism.

Any ecosystem would be short-lived without a continuous input of energy from the sun. Further, all the levels in a food chain are quite dependent on the amount of energy produced by the autotrophs (Fig. 8.8). What is the practical outcome of this loss of energy at each level in the food chain? Availability of energy determines the total mass of organisms that can be supported at each level. Ecologists refer to the total dry weight (organic matter with water removed) of all the organisms that feed at any level in the food chain as **biomass.** Biomass, then, represents the total amount of energy stored in all the organic matter at each food chain level. Because of the availability of greater amounts of energy, producers (at the first level) usually have the greatest biomass in an ecosystem, followed by the primary consumers, which have less biomass, followed by the secondary consumers, and so on. This is not always true, as some aquatic ecosystems have a large biomass of consumers feeding on a smaller biomass of producers. This is possible only because the producers—usually phytoplankton—reproduce so rapidly. Regardless of the ecosystem, however, the availability of energy is a limiting factor in the existence of all organisms.

We can apply the information above to an issue confronting human beings in many parts of the world—the expense and efficiency of producing enough food. Human beings, like various other animals, are both herbivores and carnivores—in a word, **omnivores**. As we have seen, carnivores have less total food energy available to them than do herbivores. By feeding on herbivores, the carnivores receive only about one-hundredth of the energy originally available from the producers, but the herbivores receive one-tenth of this energy. In terms of energy efficiency, therefore, feeding on plants conserves more food energy (calories) than does feeding on animals.

As an example, the food energy that humans obtain from beefsteak originally came from photosynthesizing autotrophs, mainly green grass. In consuming this grass, cattle use a portion of the energy stored in the plant for their own metabolism and a portion escapes as heat. According to the 10 percent rule, it would require about 10 pounds of plant food to produce one pound of human being if that person fed directly on plants. If that person is a carnivore feeding on cattle, however, the situation changes dramatically. It would take about 100 pounds of green plants to produce 10 pounds of cow to produce one pound of human being. Thus, it is ten times more energy efficient to subsist on plants rather than on meats. In areas of the world where there is little agricultural land available, feeding crops to animals and then consuming the animals represents a tremendous loss of food energy.

Cycles of Matter

According to the law of **conservation of matter,** the amount of matter in the universe is constant. This means, of course, that the matter constituting living organisms cannot be destroyed or created anew. It also means that, if life is to continue, the material for new organisms must come from preexisting material. The reuse of matter is made possible by various cycles, in which inorganic substances are converted in form and repeatedly cycled through the living and nonliving worlds. Among the most important cycles that underlie the continued existence of our ecosystems are those for carbon, oxygen, nitrogen, and water.

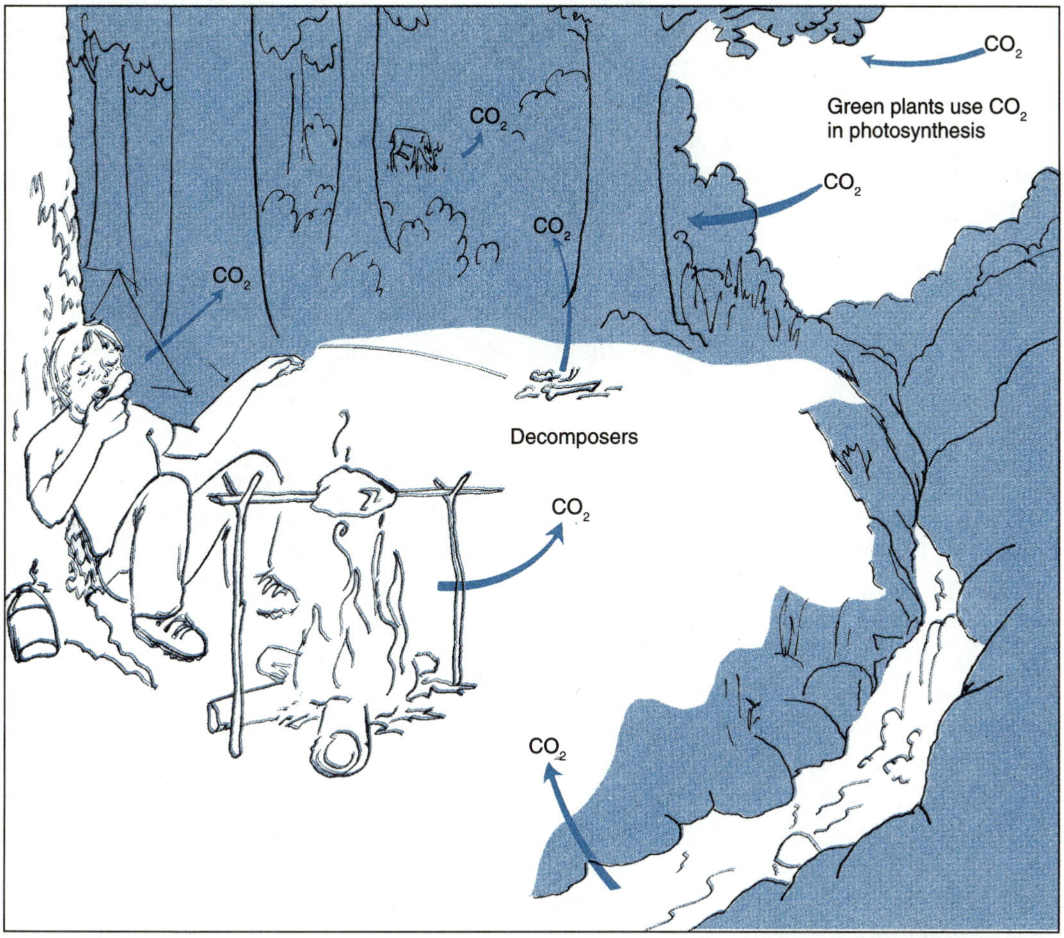

Figure 8.9 The carbon cycle. The drawing illustrates the various sources of carbon in the form of carbon dioxide (CO_2). The respiratory activities of humans, land and water animals, and bacterial decomposers release CO_2 into the atmosphere. Also, CO_2 is released by burning and by dissolving carbonates in water. The released CO_2 then is available to photosynthesizing plants.

The Carbon Cycle

In Chapter 4 we referred to carbon as the fundamental structural element in all organic compounds. Carbon atoms cycle from the nonliving environment through organisms and back to the environment again (Fig. 8.9). We can begin the carbon cycle with the bonding of carbon and oxygen atoms to form the inorganic atmospheric gas carbon dioxide (CO_2). Carbon is incorporated into living organisms initially through the chemical attachment of carbon dioxide to the sugar RuBP in the light-independent reactions of photosynthesis (Chapter 7). Sugars and other organic compounds produced by photosynthetic autotrophs are consumed by herbivores, which incorporate the carbon into their cells. Carbon is passed further along the food chain, as we have described, with every organism in the chain incorporating the carbon into its body and subsequently releasing carbon dioxide as a by-product of cellular metabolism. In addition, decomposers—bacteria and fungi—release carbon dioxide into the atmosphere during their metabolism of dead organic matter. Thus, carbon passes to the nonliving world in the form of carbon dioxide and through the living world as a constituent of organic compounds. This endless recycling conserves the available carbon for use by all living organisms.

Most of the carbon on Earth is in the form of inorganic calcium carbonate ($CaCO_3$), a salt found in substances such as limestone, marble, and chalk. In water, some of the calcium carbonate slowly dissolves, releasing carbon dioxide into the atmosphere. Dead marine plants and animals often sink to the bottom of oceans, carrying their fixed carbon with them. At great depths, where there is little oxygen and where the temperature and pressure are suitable, the accumulated carbon of these dead organisms forms deposits of oil, coal, and natural gas. Formation of these energy-rich fuels requires many millions of years. When these "fossil fuels" are burned, a tiny percentage of carbon is returned to the atmosphere in the form of carbon dioxide. In this manner, carbon that had long since been removed from the cycle is returned to the cycle once again.

The Greenhouse Effect

Although carbon dioxide constitutes only about 0.03 percent of atmospheric air at sea level, it is the single source of carbon for all organisms. But carbon dioxide plays an additional role in the biosphere. Solar energy heats the surface of Earth and some of this energy tends to radiate back out into space. Atmospheric carbon dioxide, however, along with water vapor, forms a gaseous blanket or shield around Earth that retains much of this heat (Fig. 8.10). This gaseous covering prevents excessive heat loss and thereby contributes significantly to warming Earth's surface. This phenomenon is called the **greenhouse effect** and without it, the average temperature of Earth, about 60° F (15.5° C),would drop about 100 degrees.

Over the past 150 years, however, there has been an estimated 20 percent increase in the concentration of carbon dioxide in the atmosphere. This increase, along with increases in the atmospheric concentrations of methane. chlorofluorocarbons, and other gases, has caused concern over its possible effect on significantly increasing the average temperature of Earth in a process called *global warming.* The issue of global warming is somewhat controversial and has been accompanied by fervid argument. One argument contends that increased quantities of the "greenhouse gases" in the atmosphere will raise the average temperature of Earth, resulting in climatic changes that will damage vegetation and certain crops, dry up rivers, melt polar regions, raise sea levels, and flood certain areas. Another argument maintains that increasing the temperature of Earth will actually benefit certain regions by making fallow land productive and increasing the amount of photosynthesis. Yet another argument contends that global warming will have little, if any, significant effect on Earth. All of these positions, of course, are speculative. The only conclusion we can draw at present is that we do not know which argument is valid.

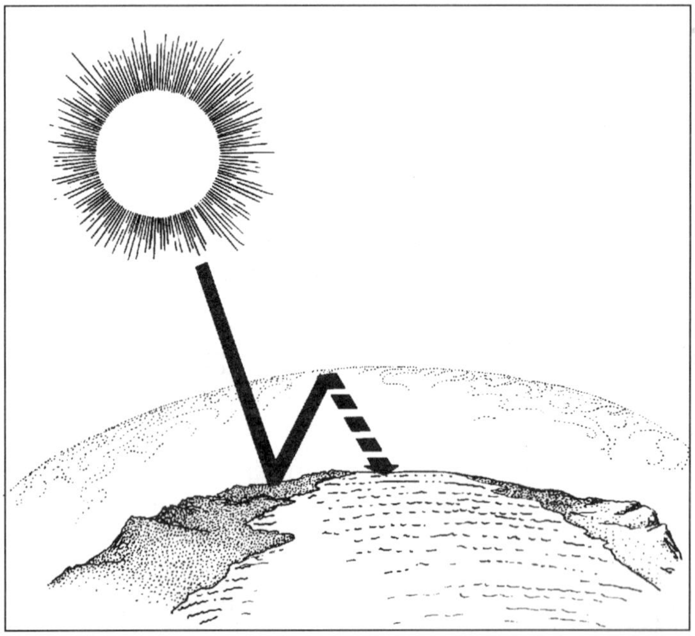

Figure 8.10 The greenhouse effect. Solar energy heats Earth, and some of the heat energy radiates back out into space. A layer of CO_2 and water vapor in the atmosphere absorbs the heat energy, preventing excessive loss of heat, and thereby contributes to warming Earth's surface.

It has been estimated that the concentration of atmospheric carbon dioxide will double by the middle of the next century. Scientists predict that this could increase the average temperature of Earth from about 2 to 9° F. There is limited time, therefore, to observe the effects of a warming trend, if indeed that is the case, and to plot a strategy to minimize these effects should they appear to be detrimental.

The Oxygen Cycle

Figure 8.11 The oxygen cycle. Land and water plants release O_2 as a by-product of photosynthesis. This O_2 is used in respiration by land and water animals.

The oxygen and carbon cycles are closely interrelated (Fig. 8.11). Atmospheric oxygen (O_2) has surrounded Earth for about 2 billion years and presently constitutes about 21 percent of atmospheric air at sea level. This renewable reservoir of oxygen, a by-product of photosynthesis by plants and phytoplankton, is available to all aerobic organisms for use in cellular respiration. Oxygen is also present in gaseous carbon dioxide, water, organic nutrients, dissolved ions, such as carbonate (CO_3-), and in ocean sediments and deposits. Oxygen enters photosynthesizing autotrophs in the form of carbon dioxide, which is subsequently incorporated into carbohydrates and other organic molecules. The oxygen released from these autotrophs is a product of the dissociation of water. In animals, oxygen may enter the body through lungs, gills, skin, or tiny openings in the body wall. Many aquatic organisms, such as protozoa and algae, take up oxygen dissolved in the surrounding water by diffusion. In all aerobic organisms, cellular respiration incorporates this oxygen as the final electron acceptor in the production of ATP. Other metabolic pathways use oxygen in the synthesis of organic nutrients. These nutrients, with their contained oxygen, are passed through the food chain from producers to consumers to decomposers. Ultimately, these nutrients are metabolized and converted to carbon dioxide and water, both of which contain oxygen that can be reused in the synthesis of new compounds.

The Nitrogen Cycle

As a constituent of proteins and nucleic acids, the element nitrogen is vital to all organisms. Atmospheric air consists of about 78 percent nitrogen in the form of nitrogen gas (N_2). Nitrogen gas does not tend to form compounds readily with other elements and so must be converted in form, or "fixed," before it can be used by most organisms (Fig. 8.12). The fixing of nitrogen gas is carried out mainly by various species of bacteria and to a lesser extent by some fungi. Species of **nitrogen-fixing** bacteria live in soil and water, while others inhabit the root nodules of legumes, such as peas, beans, clover, and alfalfa. These bacteria fix atmospheric nitrogen through chemical reactions that reduce it (add hydrogen atoms) to ammonia (NH_3). In turn, other aerobic bacteria are able to utilize this ammonia, oxidizing it to nitrite (NO_2) and then to nitrate (NO_3).

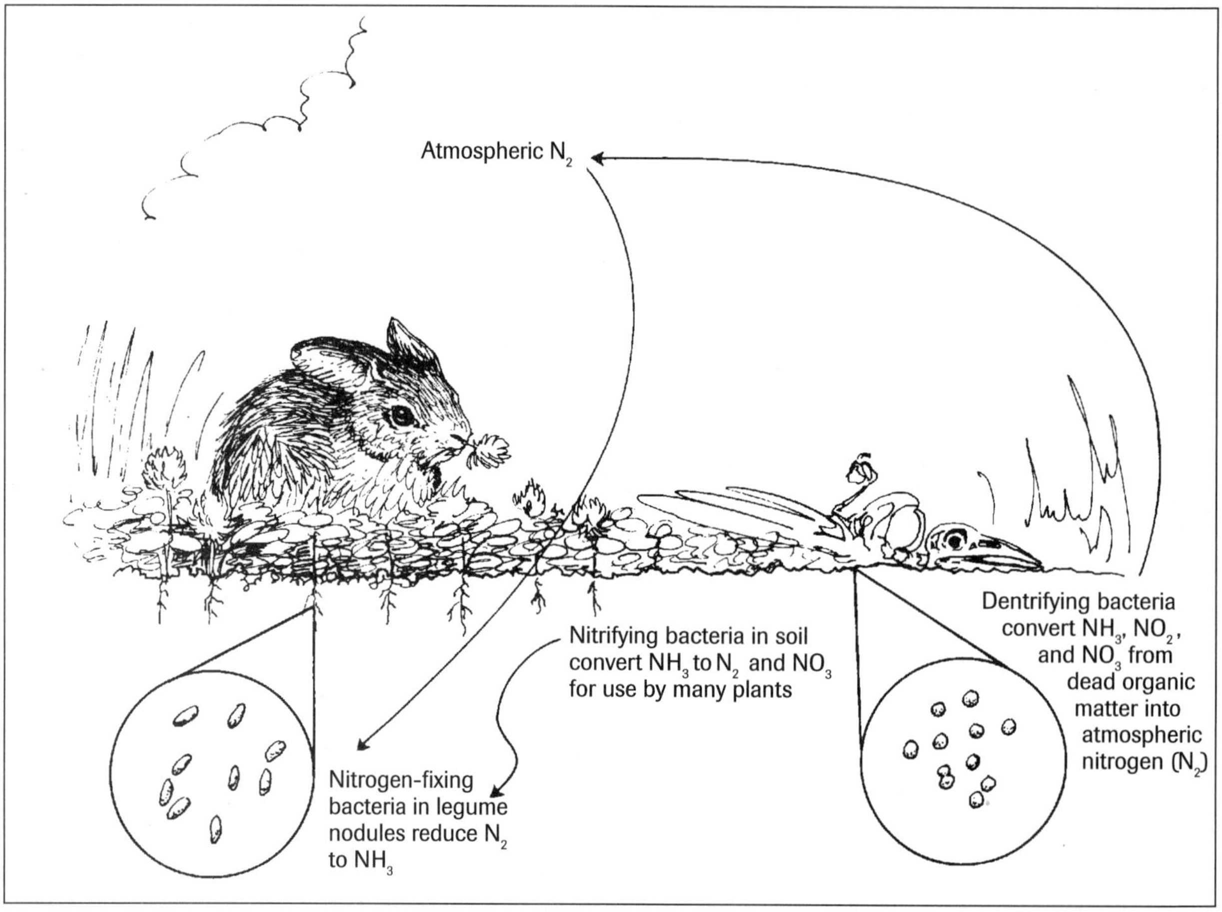

Figure 8.12 The nitrogen cycle. Before it can be used by living organisms, atmospheric nitrogen (N_2) must be converted into nitrogenous compounds. This is accomplished by various species of bacteria that, in their metabolic processes, convert N_2 to NH_3 (ammonia), NO_2 (nitrite), or NO_3 (nitrate). Plants take up these compounds, which become incorporated in the bodies of animals when the plants are eaten. When the plants and animals die, the nitrogen compounds in their bodies are converted to N_2 by certain species of soil and water bacteria. The N_2 is released from the bacteria into the atmosphere to be used again.

These nitrogen-containing compounds are then taken up by plants and incorporated into proteins, nucleic acids, and other organic compounds. These compounds are passed along the food chain to herbivores and carnivores. At death, plants and animals are decomposed by bacteria that also convert the nitrogen compounds back to ammonia, nitrite, and nitrate. Other bacteria metabolize nitrate back to nitrogen gas, which returns to the atmosphere, completing the cycle.

Only **legumes** and a few other plants serve as hosts to nitrogen-fixing bacteria; these plants possess nodules on their roots that house the bacteria. The plants, however, benefit from this association by obtaining a usable form of nitrogen from the bacteria. The bacteria, in turn, receive carbohydrates from the plant and thus profit from the association as well. Important agricultural crops, such as wheat, rice, and corn, do not possess root nodules or symbiotic bacteria. These plants are supplied with nitrogen fertilizer. In the past few years, however, research on wheat seedlings has resulted in the production of chemically induced nodules on their roots. Scientists subsequently infected the nodules with a species of bacteria that appeared to have the ability to fix atmospheric nitrogen. Studies are continuing to determine if the wheat will be able to incorporate the fixed nitrogen into organic compounds. If the bacteria can persist for an extended time in the nodules, and if the fixed nitrogen can be used by the wheat long enough to be of benefit, it could be possible in the future to develop a variety of food crops that harbor their own nitrogen-fixing bacteria.

Eutrophication

Ordinarily, the amount of nitrogen fixed by bacteria is balanced by the amount released back into the atmosphere. The addition of excessive amounts of fertilizer to the soil and the high concentrations of sewage in some areas, however, have caused nitrogen compounds to be washed into lakes and streams, resulting in what is called **eutrophication,** a term meaning "well nourished." In large, open fields, precipitation dissolves nutrients and fertilizers and often washes them away. This agricultural **runoff** can eventually be deposited in reservoirs, lakes and streams, greatly accelerating the rate of eutrophication. A consequence of eutrophication is the drastic overgrowth or "bloom," of algae in a lake or stream. As this tremendous excess of algae die, they are decomposed by bacteria present in the water. In the process, the respiring bacteria draw on oxygen dissolved in the water, greatly reducing the amount normally available to fish and other aquatic organisms. Eventually, these latter organisms may die from lack of oxygen, leaving nothing but bacteria and stagnant, deoxygenated water.

Although eutrophication may have its harmful effects on an ecosystem, it is, nonetheless, a natural process characteristic of many ecosystems. Like all living units, ecosystems age, often beginning as relatively unproductive systems and gradually becoming nutrient-rich productive systems. In other words, they become eutrophic. A deep, cold lake, for instance, may contain very little organic matter and thus can support only a small biomass. Over time, however, the plant population in the lake increases as a result of nutrients entering the lake through runoff and precipitation. With an increase in the number of plant species, the lake now can support more diverse animal populations. With the influx of silt and other particulate matter over time, the lake gradually becomes a more productive ecosystem.

A significant effect of natural eutrophication in most ecosystems is the change in the plant and animal life in the area. With the slow but constant influx of nutrients by the wind and by water seepage, an ecosystem is gradually altered in structure and function. Although it is possible for natural eutrophication to result in unfavorable changes in an ecosystem through an excessive influx of nutrients, this ordinarily does not occur. If it does, the process usually requires enormous spans of time. Pollution, however, greatly accelerates the process, often all but destroying an ecosystem in a few years.

The Water Cycle

As the most vital compound on Earth, water, too, must be recycled continually through the living and nonliving worlds (Fig. 8.13). Powered by the energy of the sun, water evaporates from soil, from the bodies of plants and animals, and from the oceans, rivers, and streams of the world. As water is heated, it forms water vapor, which rises and forms clouds through condensation. When sufficiently cooled high in the atmosphere, the clouds precipitate the water as rain or snow. Some of the water returning to Earth is taken in by plants and animals and some trickles down into the soil. Eventually, water in the soil finds its way into the oceans, lakes, and streams of the world. From these sources, water again evaporates into the atmosphere. Many animals release water in the form of urine, sweat, and exhaled water vapor. This water, too, eventually evapo-

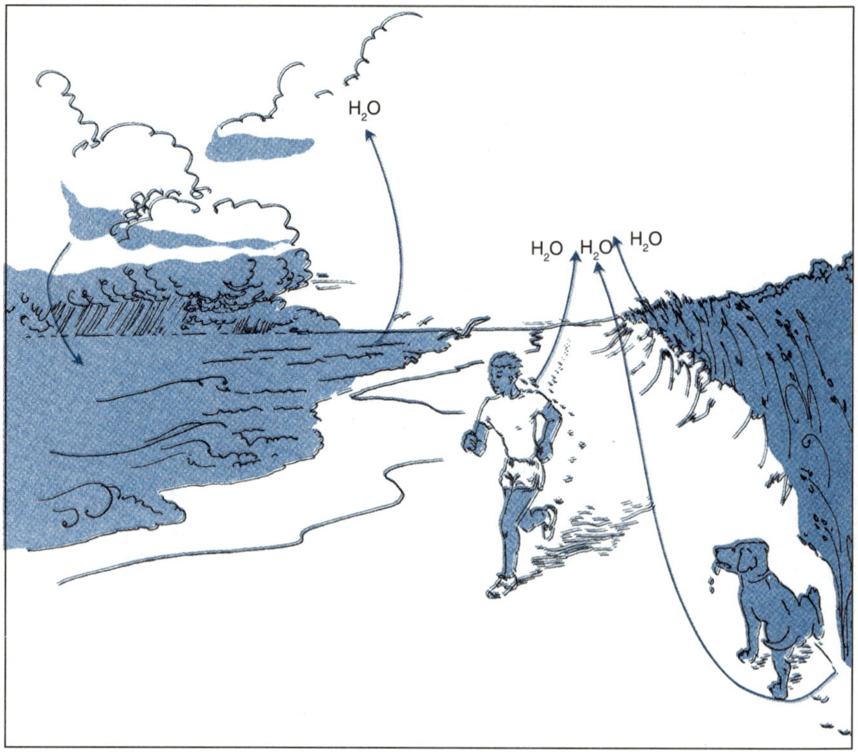

Figure 8.13 The water cycle. Water evaporates from the surface of lakes and oceans, from the soil, and also from the bodies of animals and plants. In the atmosphere, the evaporated water condenses to form clouds. Eventually, the water falls back to earth in some form of precipitation.

rates. Land plants lose water primarily by *transpiration,* the evaporation of water through the stomata.

In the endless evaporation-condensation process of the water cycle, it is possible for a variety of substances dissolved or suspended in the water to be redistributed over wide areas of Earth. This can result in both beneficial and harmful effects. On one hand, the water cycle distributes some of the nutrient minerals in the soil; on the other hand, some chemical contaminants carried by water can be distributed to areas where they present a hazard to physical structures as well as to living organisms. An example of such a contaminant is acid rain, which can erode metals and cement and, among other effects, reduces plant growth and lowers the pH of water to intolerable levels in aquatic ecosystems.

Biological Magnification

The flow of energy is integrated into cycles along the food chain. Carbon, present in foods, for example, is passed from autotroph to herbivore to carnivore and then to the decomposers. At each link in the chain, energy is used, heat is given off, and carbon, as we have noted, is released back into the atmosphere as carbon dioxide.

In the same manner, toxic substances may be passed along a food chain (Fig. 8.14). Consider, for instance, the release of a potentially toxic substance into a lake or stream. In time, the substance may be broken down and incorporated into the natural cycles of matter and thereby rendered harmless to organisms; that is, the substance may be **biodegradable.** In contrast, the substance may not be biodegradable or it may be altered by natural processes into a different, biologically harmful form. If taken into the bodies of one group of organisms, the substance may be passed to organisms at the

Figure 8.14 Biological magnification. A chemical may increase in concentration as it moves through the food chain. The concentration may reach a fatal level in the top predators of the food chain.

next level of the food chain. These organisms in turn may be fed on by organisms at the next level, and so on.

As it is passed along, the substance becomes more and more concentrated, or **magnified**, in the bodies of each successive group. If the substance is a biological hazard, such as mercury or some other toxic chemical, a high concentration of the chemical could prove life-threatening to some organisms in the food chain. This is particularly true of the terminal organisms in the chain because the chemical may be in low concentration at the primary level in the autotrophs, for instance, but in very high concentration by the time it reaches the carnivores. In some cases, of course, the top food chain level is humans, who may experience ill effects from ingesting the chemical.

Living organisms are composed of elements that must be continually recycled for life on Earth to continue. These cycles illustrate the interactions between organisms and their physical environment. Carbon, oxygen, and nitrogen are essential elements of both living and nonliving matter; water, which itself is recycled, is often a critical vehicle in the cycling of these elements. The replenishment of these elements depends on the activities of decomposers, which convert complex organic compounds to simpler forms and return them to the nonliving world. Thus, the land, water, and air contain reservoirs of inorganic nutrients that are accessible to the autotrophs and ultimately to every other form of life.

Energy Flow: Pollution and Intervention

The normal flow of energy through an ecosystem can be upset by many factors, including contamination of the environment known generally as *pollution*. For decades, many of the freshwater systems of the world have been used as dumping sites for a variety of chemicals, acids, pesticides, and other forms of potentially toxic materials. With the addition of these pollutants, natural ecologic interactions can be interrupted or eliminated entirely. It is known that various synthetic materials, such as PCBs (polychlorinated biphenyls) from plastics, as well as acetates and rayons, are not biodegradable. When released into lakes and streams, many of the synthetic compounds can accumulate in the bodies of organisms, sometimes leading to injury or death. Cardboard and paper, in contrast, are wood products that eventually are biodegraded and are not toxic.

In the oceans and seas of the world, the major pollutant is oil. Spills from wrecked oil tankers and drilling platforms have reduced oyster populations and soaked sea birds and other animals, causing death by suffocation. If extensive enough, oil spills or dumped chemicals can destroy massive numbers of fish eggs, plankton, and other forms of marine life. The oceans represent the final sink for all the wastes dumped into the rivers and streams of the world; there is no place else for the wastes to go.

Pollution is evident not only in the oceans, but is also a major problem in the atmosphere. The major sources of air pollution are the **fossil fuels**—coal and oil—that release several potentially harmful chemicals when burned. Two of the most toxic chemicals are sulfur dioxide (SO_2) and sulfur trioxide (SO_3). Sulfur dioxide is a colorless, corrosive gas that causes nose, throat, and lung irritation when inhaled. In polluted air, sulfur dioxide reacts with oxygen to form sulfur trioxide, which in turn reacts with water vapor to form sulfuric acid (H_2SO_4). Acid rain, mentioned earlier, is partly the result of sulfur dioxide and sulfur trioxide reacting with atmospheric water vapor. Sulfuric acid is a very strong acid and, in pure form, dissolves living tissues of both plants and animals. The corroding effects of sulfuric acid can be seen in the deterioration of buildings, statues, and other structures.

The burning of fossil fuels also releases nitric oxide (NO), a colorless gas, and nitrogen dioxide (NO_2), a brown corrosive gas. Both compounds are highly reactive and capable of doing extensive damage to living tissue. The thick brown smog hanging over many cities is nitrogen dioxide. In urban areas with heavy traffic, the high concentration of nitric oxide released from automobile exhaust combines with oxygen to produce large amounts of nitrogen dioxide. In warm regions, nitrogen dioxide absorbs ultraviolet light from the sun and breaks down into nitric oxide and oxygen radicals (O).

Ozone (O_3) is produced by the dissociation of oxygen gas and the recombination of three oxygen atoms. As pointed out in Chapter 2, the ozone layer filters out 95 to 99 percent of the potentially lethal UV radiation that could destroy life on Earth. Ozone is also generated by the electrical discharge of lightning. You may have detected the pungent odor of ozone immediately after a thunderstorm. Ozone is quite damaging to various materials and to organisms when it comes in direct contact with them. Therefore, ozone present in the lower levels of the atmosphere is a dangerous pollutant. The reaction of ozone with various organic compounds produces a variety of pollutants, including PANs (peroxyacyl nitrates) that are highly irritating to the eyes, nose, and throat.

Figure 8.15 A lush tropical rainforest.

Tropical Rain Forests

Energy flow through an ecosystem also can be upset by severely altering the physical environment. One of the more immediate ecologic concerns is the mounting destruction of tropical rain forests (Fig. 8.15). Over the past two or three decades, forest cover in the Philippines, Ghana, Brazil, and Indonesia has diminished enormously. Much of these areas has been cleared for cattle ranching and farming, which have not proved economically profitable because of the poor productivity of the land. Burning vast areas of timber and destroying more through careless and inefficient logging has resulted in economic disaster. The loss in commercial timber alone has amounted to billions of dollars.

Tropical forests serve as habitats for about half of the world's species of plants and animals. In some parts of the world, **deforestation** could result in the eventual extinction of a large number of plant species. In addition, a very high percentage of harvested wood in tropical countries is used for fuel. Forests also contain habitats for animals used for meat and are the sources of fruits, nuts, medicinal plants, and other commodities vital to local residents for their own use and for export to other countries. Excessive deforestation has been shown to promote soil erosion, leach minerals from the soil, and to interfere with natural water recycling. Sedimentation from soil erosion has damaged fisheries along some tropical rivers.

The loss of photosynthesizing plant life eventually reduces or eliminates a number of various herbivores that depend on these plants for food. Some of the herbivores are food sources for a variety of carnivores. Further, since the majority of the human population lives in tropical and subtropical regions, vast numbers of these dependent herbivores and carnivores are people.

Human intervention in the rain forests, and in the biosphere as a whole, has severely threatened biological diversity and accelerated the rate of extinction of numerous species of organisms. Biologist E. O. Wilson has estimated that, at the present rate of rain-forest clearing, from 0.2 to 0.3 percent of all species in the forests are lost each year. Assuming a combined population of two million species, the loss resulting from deforestation could be as much as 4000 to 6000 species a year. Wilson further points out that this loss of species through intervention by deforestation may be 10,000 times greater than the naturally occurring extinction rate.

We began our look at energy transactions in Chapter 7 with a discussion of photosynthesis and cellular respiration, the basic processes that sustain life through the conversion and storage of energy. In this chapter, we have traced the flow of energy as it moves through the ecosystem. It should be evident that photosynthesis, cellular respiration, and energy flow are interrelated and ultimately dependent on sunlight, the energy source of life itself.

Energy flow sums up the use of energy by tying together cell function, organismal physiology, and ecosystem interactions. Sunlight drives the light-dependent reactions of photosynthesis and is converted to electrical energy and then stored as chemical energy in NADPH and ATP. This stored energy powers the light-independent reactions that synthesize carbohydrates and other organic molecules. These energy nutrients enter the cells of organisms and are metabolized through cellular respiration; the energy released is stored in the phosphate bonds of ATP. The flow of energy continues as energy moves from producers to consumers in the food chain. Every one of these energy interactions within a population affects another local population. In addition, the upset of energy flow also affects population interaction. Energy flow, then, is a determining and limiting factor in the structure and function of all ecosystems.

Chapter Summary

1. Ecology is the study of the interactions living organisms have with each other and with their nonliving environment. These interactions occur within ecosystems, which consist of communities and their nonliving environment.

   A food chain is a sequence of organisms, each serving as an energy source for the next group of organisms in the chain. Energy flow in an ecosystem begins with the producers, which use only a tiny percentage of the light energy striking Earth. Following producers in the food chain are primary, secondary, tertiary, and quaternary consumers. The final link in food chains consists of decomposers—bacteria and fungi.

   Key Terms: *energy flow, p. 163*; *ecosystem, p. 163*; *food chain, p. 167*; *producers, p. 168*; *consumer, p. 168*; *decomposer, p. 168*

   Concept Questions: Correlate the processes of photosynthesis and cellular respiration with energy flow in an ecosystem.

   What is the application of the first and second laws of thermodynamics in a food chain?

2. Food webs are interconnected, complex food chains that indicate the various sources of food available to an organism and the various levels at which an organism may feed. According to the second law of thermodynamics, as energy flows through a food chain, some of this energy escapes as heat. Approximately 10 percent of the energy at each level in the chain is available to the next level; this is called the 10 percent rule. Because of this rule, feeding directly on plants is ten times more energy efficient than feeding on meat.

   The amount of energy available to an ecosystem determines the biomass; producers, then, usually have the greatest biomass, while the terminal consumers have the least. The availability of energy ordinarily limits the number of links in a food chain to four, rarely five. Energy flows through ecosystems; therefore, a continuous outside source of energy is required. Nutrients and the atoms of which they are made cycle within ecosystems and are used over and over again.

   Key Terms: *food web, p. 169*; *biomass, p. 171*; *omnivore, p. 171*

Concept Questions: How does the 10 percent rule affect the functioning of an ecosystem? What is the significance of this effect?

Assume a new group of primary consumers was introduced into a stable food chain. What effect might this group have on the functioning of an ecosystem?

Explain why the number of energy levels in a food chain is normally limited to four. By constructing a hypothetical food chain, determine if you can extend the energy levels to five.

3. Conservation of matter is made possible by various cycles of inorganic substances. Some of the major cycles are the carbon, oxygen, and nitrogen cycles.

   Carbon dioxide, along with water vapor, is responsible for the "greenhouse effect." There are various arguments concerning the effects of carbon dioxide and other greenhouse gases on global warming.

   Key Terms: *conservation of matter, p. 171*
   greenhouse gas, p. 173
   nitrogen-fixing, p. 174
   legume, p. 175

   Concept Questions: What would be some of the general effects of an interruption of the nitrogen cycle? Of the carbon cycle? What is the specific role of decomposers in an ecosystem? What would be the general effect on an ecosystem if the decomposers were destroyed?

4. Eutrophication can result from the introduction of large amounts of nitrogen and organic compounds into an ecosystem. Eutrophication is a natural process, but it can be accelerated by pollution. Water is continually recycled through the living and nonliving worlds by evaporation and condensation.

   Potentially toxic substances moving through a food chain tend to accumulate and increase in concentration. This often results in the terminal consumers experiencing the most damaging effects.

   Key Terms: *eutrophication, p. 176*
   runoff p. 176
   biodegradable, p. 177
   magnification, p. 178

   Concept Questions: Explain how it might be possible to prevent a clear mountain lake from becoming excessively eutrophic.

   Using an example, explain the relationship between energy flow and eutrophication.

5. Pollution is contamination of the environment, and includes water and air pollution. Water pollution results primarily from dumping of toxic chemicals and oil. Air pollution is caused primarily by chemicals released from the burning of coal and oil. Biodegradable compounds can be decomposed to carbon dioxide and water by natural processes and thus are less harmful to the environment than are nonbiodegradable compounds.

   A mounting environmental problem is the destruction of rain forests. The harmful effects of deforestation on energy flow illustrate the results of severely altering a physical environment.

   Key Terms: *fossil fuel, p. 179*
   ozone, p. 179
   deforestation, p. 180

   Concept Question: What do you think are some of the possible effects of air and water pollution on the flow of energy through an ecosystem?

PART FIVE

Life Feature 3: Sensation and Response

Chapter Nine

Acting and Reacting

Introduction

The tennis ball clears the net, flying at you at a speed in excess of 44 meters per second (100 miles per hour). From the time the ball leaves the racket 24 meters (78 feet) away, you have just over half a second to view the ball, determine the height of the bounce, move into position, and prepare your racket for the return shot. Outside the tennis court, a scorpion lies in wait, eyeing a feeding grasshopper. The scorpion lunges, locks its pincer around the grasshopper, and thrusts its stinger into the insect's abdomen. The actions and reactions of humans, scorpions, and all other living organisms bring us to the third of our basic life features: **sensation and response.** Your perception of the tennis ball, mental decisions, and muscular movement involve the coordinated interplay of hundreds of millions of **neurons,** or nerve cells, the structural units of a nervous system. Similarly, the lunging, grasping, and stinging of prey by the scorpion are coordinated by a vast network of neurons. Bacteria, protists, plants, and many other organisms, however, do not possess neurons, but sense and respond by means of various chemicals and sensory organs. In fact, eukaryotic nerve cells may have evolved from chemical receptor molecules such as those found in the cell walls of bacteria. These molecules allow bacteria to sense chemicals in their environment and to orient their movements toward or away from the chemical stimulus.

Neurons usually bring about their sensation-response effects extremely rapidly, with some vertebrate neurons transmitting messages at speeds exceeding 100 m/s (225 mph). The slowest conducting neurons in vertebrates transmit at a speed of about 0.5 m/s (1 mph). Neurons form an elaborate communications system by integrating and coordinating all body systems, including the nervous system itself.

In the plant kingdom, responses to stimuli usually are associated with growth and development of the plant. Such responses result primarily from the effects of plant hormones and from changes in turgor pressure within the plant. A **hormone** is a chemical compound produced in one part of an organism and transported to another part where it brings about a certain response or effect. Plant hormones primarily affect cell division, growth through cell elongation, and development of different cell types. Some plant hormones also are referred to as growth regulators.

Water entering a plant cell by osmosis exerts pressure against the cell wall. The force exerted by water within the cell is called *turgor pressure.* This is the pressure that maintains the leaves and other plant structures in a firm, upright position.

Plant Responses

Unlike many animals, plants have no discrete nerves or brains, but still respond to different stimuli, such as light, temperature, gravity, and touch (Fig. 9.1). Two principal types of plant responses are **turgor movements** and **tropisms.** Turgor movements involve an increase or decrease in turgor pressure as water enters or exits the cells. The opening and closing of stomata is a common turgor movement in many plants. Stomata are bordered by a pair of photosynthetic guard cells that swell and contract as their turgor pressure changes. As shown in Fig. 9.1A, when water enters the guard cells by osmosis, the cells swell and the stoma opens; when water is lost from the guard cells, they contract and the stoma closes.

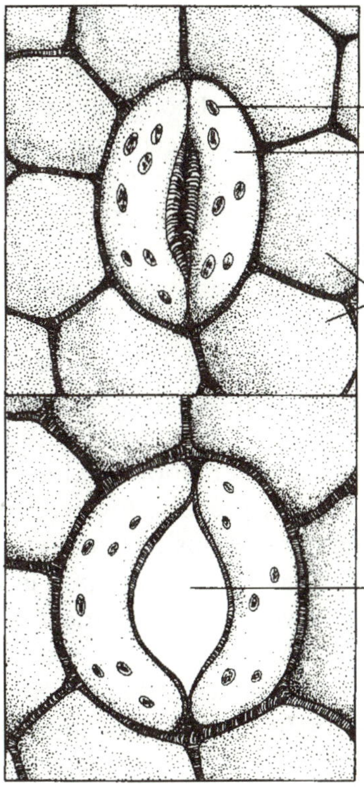

Figure 9.1A The closing and opening of a leaf stoma—an example of a turgor movement.

Two interesting turgor movements in plants are "sleep" movements and leaf closing. Sleep movements, such as those of the so-called "prayer plant", result from the expansion and contraction of pairs of certain large cells at the base of the leaves. In the daytime, expansion of one pair of the large cells and contraction of the other pair cause the leaves to spread apart. Apparently, the plant senses light by means of a receptor pigment present in the leaves. At night, when the pigment is not stimulated by light, expansion and contraction of the respective cells are reversed, and the leaves become pressed together in their "praying" position. The large cells expand or contract as a result of turgor pressure from osmosis, essentially the same mechanism that occurs in the opening and closing of stomata.

The closing of leaflets, such as those of a species of *Mimosa,* is triggered by touching the leaflets. The closing mechanism is similar to that of sleep movements, involving changes in turgor pressure within the leaflets.

As illustrated in Fig. 9.1B, tropisms are responses characterized by a bending of a part of the plant body toward or away from a stimulus. A response in the direction of the stimulus is said to be *positive;* a response away from the stimulus is *negative.* Tropisms usually require the influence of one or more plant hormones. Plant shoots and roots, for example, exhibit **gravitropism,** a response to gravity that involves one or possibly two plant hormones. If placed horizontally or at an angle, a plant shoot eventually will curve upward, an instance of negative

Figure 9.1B Soybean stems bending away from the pull of gravity, an instance of negative gravitropism.

gravitropism. Research has demonstrated that a hormone called *auxin* accumulates in the lower side of a horizontal shoot and not in the upper side. Auxin, which *stimulates* cell elongation in shoots, increases the growth rate of cells on the lower side of the shoot. The shoot cells, then, elongate faster on the lower side than on the upper side and the shoot bends upward.

In roots, however, auxin *inhibits* cell growth. A root placed on its side will bend downward, an instance of positive gravitropism. As in shoots, auxin, and perhaps another plant hormone called *abscisic acid,* accumulates on the lower side of the root, inhibiting growth in that area. In this case, growth is more rapid in the upper side of the root, and the root bends downward. Although these hormones are known to affect roots in the manner described, the complete mechanism of root gravitropism remains unclear.

Another response in many plants is **phototropism,** a bending or turning of the plant toward a light source. Phototropism also involves the hormone auxin, which accumulates on the side of the shoot away from the light. More rapid cell elongation on this darker side causes the plant to bend in the direction of the light source, an example of positive phototropism.

Auxin, abscisic acid, and several other plant hormones, including *cytokinins, ethylene,* and *gibberellic acid,* are involved in an array of additional functions, including the falling of leaves, seed germination, development of chloroplasts and other plastids, pigment formation, and flowering. Regulation of these and other plant functions is a complex process, involving different hormones and other factors, such as light and temperature, usually acting in a coordinated interplay.

Fairly recent research indicates that plants possess impulse conduction pathways somewhat similar to those of animals. Basically, this means that a plant is capable of reception of stimuli, as we have noted, and that this stimulus is converted into a chemical that induces one or more responses in the plant. Auxin, as we have seen, acts as the stimulus for cell elongation in shoots. Auxin apparently triggers activation of "messengers"—possibly chemicals or ions—within the cells. The messengers then induce the shoot cells to elongate. This pathway is suggestive of the mechanism of neuron function in animals, as we will see.

Evolution of Nervous Systems and Receptors

Basically, an animal nervous system is a group of cells, some of which may be specialized for detecting environmental changes (stimuli), some process and coordinate these changes in preparation for a response, and some cells bring about the response. A nervous system, then, carries out the sensory-integration-response mechanism in many animals; that is, the nervous system senses stimuli, interprets the stimuli in higher nerve centers, and then causes the appropriate response. This mechanism is more evident in the nervous systems of more complex animals, principally the vertebrates, but in many simpler animals, such as sponges, cnidarians and various worms, the mechanism is reduced to only sensation and response. Accordingly, these organisms are capable only of simple reflex behavior. A **reflex** in these animals is an automatic, unlearned response performed without prior experience. Incoming sensory information cannot be processed by these animals because there are no integrating centers to coordinate the information and provide some alternate response.

Millions of years before the evolution of an animal with a discrete nervous system, it was necessary for survival of the earliest bacteria and the single-celled protists that they be able to sense and respond to their environment (Fig. 9.2). Research on their modern-day descendants has demonstrated the presence of specialized sensory receptors in some of the membranes of the cells and in the cell wall of these microorganisms. These receptors allow bacteria and protists to orient their movements with respect to food, heat, chemicals, light, and other environmental factors.

In bacteria and protists, a movement oriented toward or away from a stimulus is called a **taxis,** an example of which is shown in Fig. 9.2. A taxis may be *positive,* which occurs when an organism moves toward a stimulus, such as light or food; or the taxis may be *negative,* which occurs when an organism moves away from a noxious substance. The single-celled protozoan *Paramecium,* for example, displays an avoidance reaction when it encounters a noxious chemical. On detecting the chemical, *Paramecium* stops, backs up, and orients its body in a direction away from the chemical. This is an example of negative chemotaxis, that is, movement away from a harmful chemical. Many bacteria also are capable of negative chemotaxis.

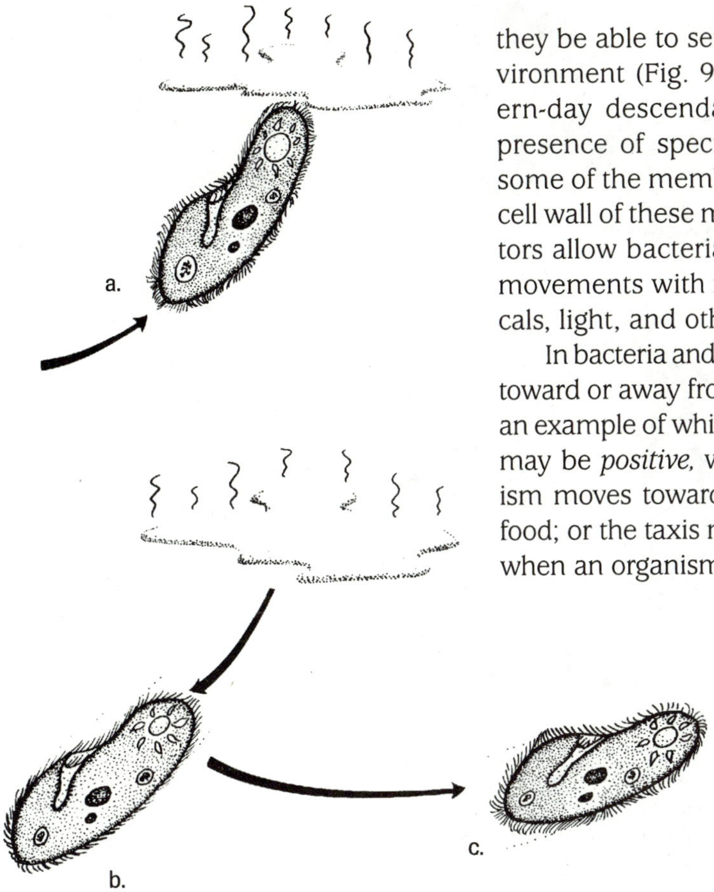

Figure 9.2 Avoidance behavior: negative chemotaxis in *Paramecium.* (a) When confronted with a noxious chemical, (b) the paramecium backs off and (c) swims forward at a different angle.

Some blue-green bacteria and photosynthetic protists display a positive phototaxis, or movement toward a light source. The flagellated protist *Euglena,* for example, has a tiny light receptor organ embedded in the membrane of the flagellum. The receptor organ allows *Euglena* to orient itself toward light, the energy source it needs for photosynthesis.

Animal Nervous Systems

All animal groups except the sponges possess some type of nervous system, although some sponge cells have sensory receptors (Fig. 9.3). Cnidarians, such as hydras, have a diffuse nerve net that runs throughout the body, but there is no control center, or brain. A *nerve net* is a system of nerve cells that acts together with sensory and contractile cells to produce movement in an organism. Although effective, movement in hydras is limited to reflex actions resulting primarily in feeding responses and changes in body shape. In other cnidarians, particularly jellyfish, body movements are better coordinated than those in hydras, mainly because jellyfish require coordinated movements for swimming.

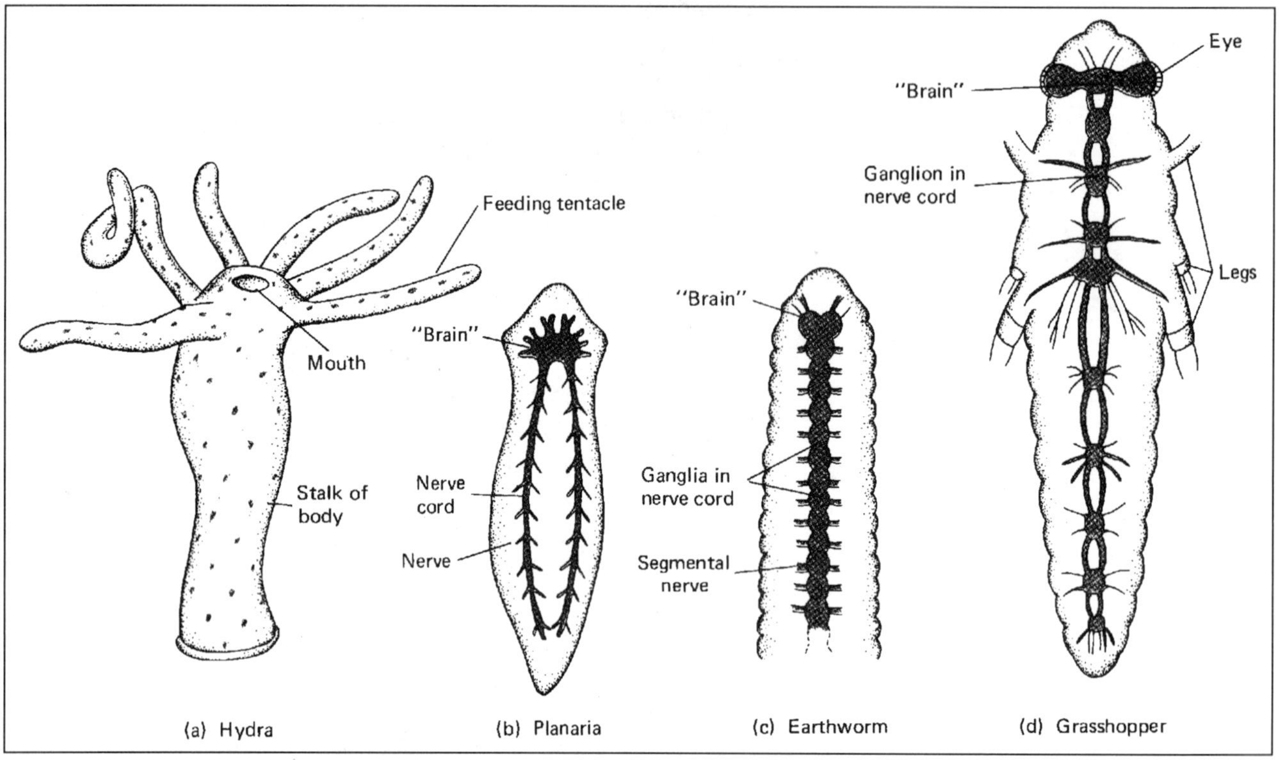

Figure 9.3 Nervous systems in lower animals. (a) *Hydra* has a simple nerve net and lacks a brain or ganglia. (b) The planarian possesses simple ganglia and shows some cephalization. In the (c) earthworm and (d) grasshopper, ganglia are more numerous and there is a greater degree of cephalization.

Evolution of the nervous system among animals more advanced than cnidarians reveals a progression toward development of an anterior, or front, end in which there is a concentration of nerve cells that form a brain. Fig. 9.3 illustrates this progression in selected animals. Development and specialization of the anterior, or head, end of the nervous system is called **cephalization.** This is first evident at the level of some of the flatworms. Many of these organisms possess a "ladder" type nervous system consisting of pairs of longitudinal nerve cords interconnected along their length by lateral cords. The longitudinal cords arise anteriorly from a brain. In roundworms such as nematodes, the brain is in the form of a ring that encircles the intestine. A network of nerves and elongated nerve cords extend anteriorly and posteriorly from the brain.

Annelids possess an anterior brain and a long single or double ventral nerve cord extending the length of the body. Lateral nerves extend from the cord into each body segment. Except for the cephalopods, the nervous system of most mollusks consists of a nerve ring around the esophagus and two pairs of nerve cords that extend posteriorly from the nerve ring. Nerves arising from the nerve cords and nerve ring supply the sensory organs, foot muscles, and other structures. Mollusks such as squids and octopuses possess well-developed brains that encircle the esophagus. These cephalopods also possess highly specialized sense organs, such as image- forming eyes, in the head region. A complex network of nerves supplies the tentacles, muscles, and eyes. Accordingly, these animals are capable of highly coordinated and complex responses.

Most arthropods exhibit highly coordinated body movements, elaborate behavioral patterns, and possess sound, odor, chemical, touch, and visual receptors. These characteristics are correlated with the high degree of cephalization in these animals. The head of the common grasshopper, for example, is admirably supplied with sensory organs, as you can see in Fig. 9.4.

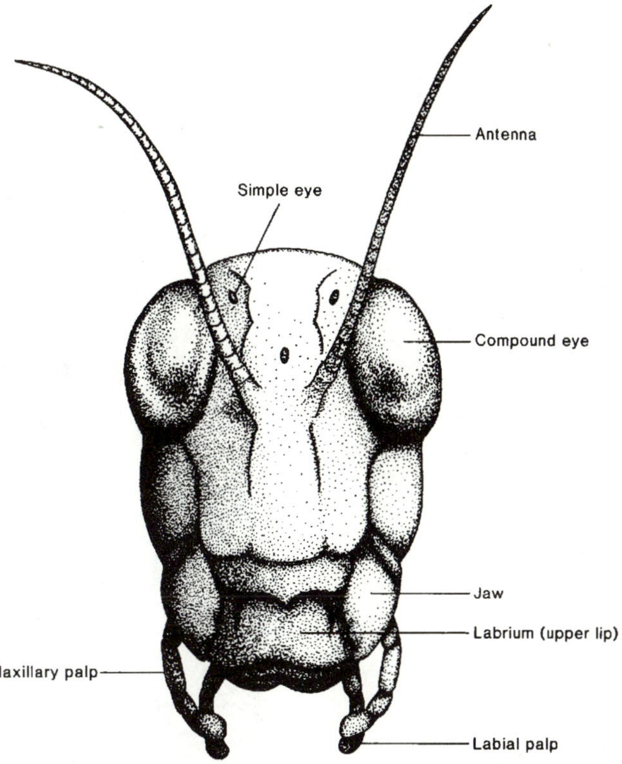

Figure 9.4 Anterior view of a grasshopper head. the antennae are touch receptors, and the simple eyes are light-dark sensitive. The palps contain taste receptors.

The basic plan of the nervous system in some echinoderms such as starfish, brittle stars, and sand dollars consists of a central circular nerve ring from which several nerves radiate outward. In other echinoderms, the circular nerve ring is replaced by a small, solid mass of neurons.

The nervous systems of many vertebrates are quite similar. Among the more advanced vertebrates—birds and mammals—evolution of the brain has resulted in an increase in the size of the brain relative to body size. In fishes, amphibians, and reptiles, however, the brain is much smaller in proportion to body size. The brains of a fish, frog, and snake of equal weight would be approximately the same size, but a mammal such as a rabbit of the same weight would have a much larger brain. Most of this increase in brain size is attributable to development of the forebrain, or anterior end of the brain.

The Vertebrate Nervous System

The nervous system of vertebrates consists of the brain, spinal cord, and associated nerves, all of which consist in part of different types of **neurons** (Fig. 9.5). Vertebrate neurons vary somewhat in size and shape, but all neurons share a fundamental architecture. The enlarged portion of a neuron is the **cell body,** which contains the organelles found in other eukaryotic cells. Extending from the cell body are one or more branched **dendrites,** processes that conduct the nerve impulse *toward* the cell body. Also extending from the cell body is a single **axon,** or nerve fiber, that conducts the nerve impulse *away* from the cell body. Axons range from only a few millimeters to over a meter in length. Although a neuron has only one axon, the tip of some axons may branch considerably. Thus, within a neuron, a nerve impulse travels from the dendrite(s) to the cell body to the axon.

Depending on their function, neurons are designated as *sensory neurons, motor neurons,* and *interneurons* (Fig. 9.6). Sensory neurons conduct nerve impulses from sensory receptors, such as heat, cold, and pain receptors in the skin, toward the *central nervous system* (brain and spinal cord). Motor neurons conduct impulses away from the

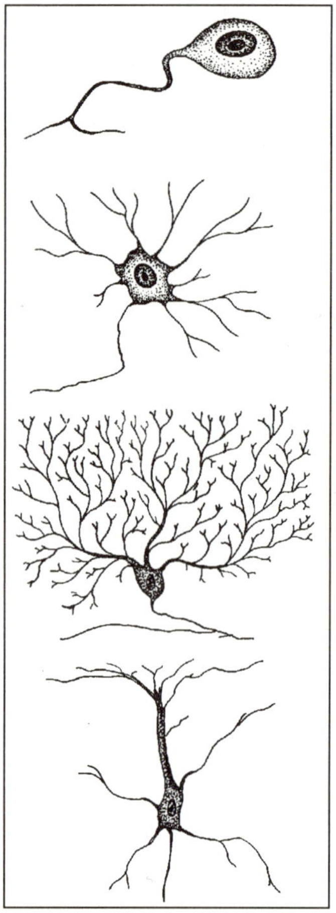

Figure 9.5 Examples of the diversity of neuron types in higher vertebrates.

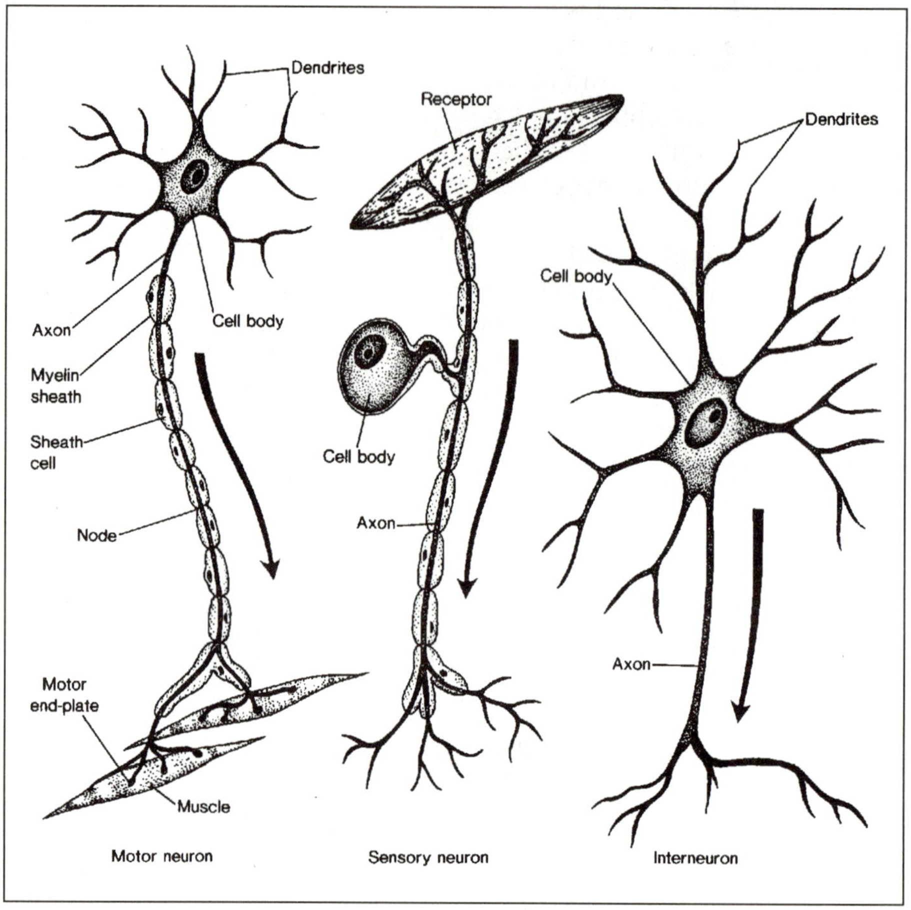

Figure 9.6 Neurons can be separated into three major groups—sensory, motor, and interneurons—according to their function.

central nervous system, out to a muscle or gland. A muscle or gland stimulated by a motor neuron is referred to as an **effector organ,** that is, an organ that causes a certain effect.

Motor impulses result in movement, either contraction of a muscle or secretion by a gland. Interneurons, which are found only in the central nervous system, are positioned between sensory and motor neurons, where they connect and integrate sensory and motor signals.

A **nerve** is a compact bundle of neuron fibers, or branches, that arises from the brain or spinal cord and extends into the muscles, organs, or glands of the body. The brain and spinal cord, as mentioned, constitute the **central nervous system** (CNS), and the various nerves make up the **peripheral nervous system** (PNS). The PNS consists of spinal nerves, cranial nerves, and nerves of the **autonomic nervous system** (ANS).

The Nerve Impulse

When it is said that a dendrite or an axon transmits impulses, what exactly is being transmitted? How does the nervous system relay information that communicates with the entire body? Basically, a nerve impulse can be thought of as an electric signal that depends on the flow of ions across the plasma membrane of the neuron (Fig. 9.7). You

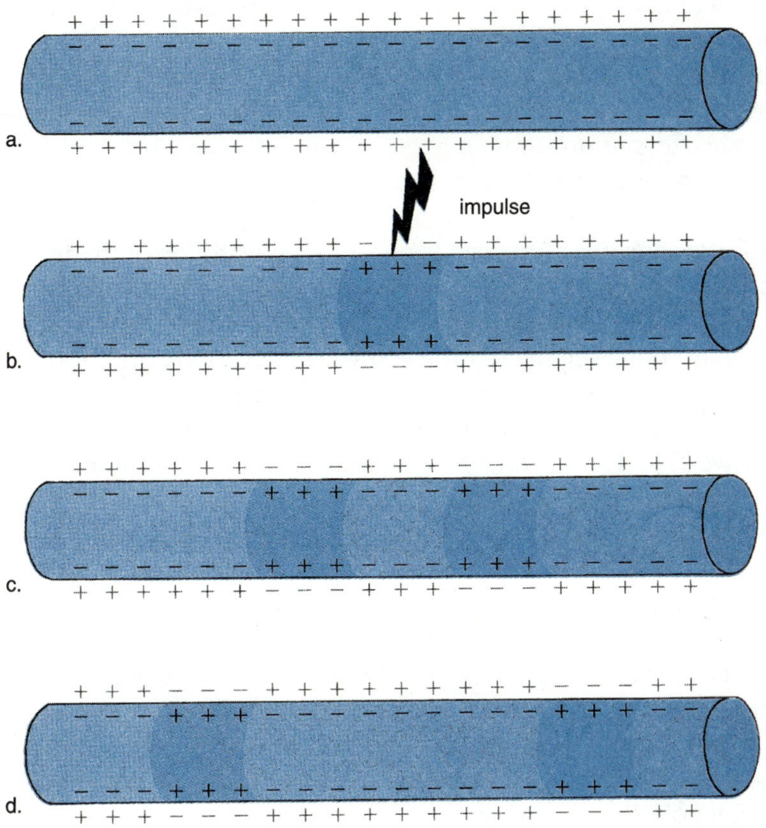

Figure 9.7 Conduction of the nerve impulse. (a) A resting (unstimulated) nerve fiber. The fiber is positively charged outside relative to the inside; i.e., there is a higher concentration of positively charged sodium ions (Na^+) outside than positively charged potassium ions (K^+) inside the fiber. (b) A stimulus is applied to the fiber (center), initiating the nerve impulse. Na^+ ions rush into the fiber reversing the polarity. (c) The reversal in polarity at the point of the stimulus stimulates adjacent points along the fiber in both directions. Within a fraction of a second after the influx of sodium ions, some of the potassium ions rush to the outside of the fiber. (d) The outward movement of the potassium ions continues as the impulse reverses the polarity along the length of the fiber.

In the next fraction of a second, the ion pumps restore the original resting state of the fiber by pumping the sodium ions out of the fiber and the potassium ions back in.

will recall from Chapter 4 that an ion is an atom having a positive or negative electrical charge. The ions essential for transmission of a nerve impulse are sodium ions (Na^+) and potassium ions (K^+). In a resting, or unstimulated, neuron, there is a high concentration of potassium ions inside the cell, while sodium ions are more concentrated in the extracellular fluid outside the cell. The unstimulated plasma membrane of the neuron is relatively impermeable to both of these ions. The relative concentrations of sodium and potassium ions on either side of the plasma membrane establish a difference in electrical charge inside and outside the neuron. In fact, all living cells have a difference in electrical charges on either side of their plasma membranes. The cytoplasm just inside the membrane has a negative charge relative to the extracellular fluid outside the membrane. Thus the nerve cell membrane and other cell membranes are said to be *polarized.*

The **polarity** of a membrane is altered when sodium ions move across the membrane into the cell and potassium ions move outward, as shown in Fig. 9.7. This movement occurs when there is a change in the permeability of the membrane to these ions. Neuron membranes possess gatelike channels through which sodium and potassium ions are exchanged. There are specific sodium channels and specific potassium channels. Any environmental factor that has the capability of changing the permeability of the membrane is called a **stimulus.** Heat, cold, pressure, pain, and other environmental effects are potential stimuli. When a stimulus is applied, the neuron membrane immediately becomes more permeable to sodium ions, the sodium channels open, and these ions cross the membrane from the extracellular fluid and enter the cell. This increased permeability causes *depolarization* of the membrane. The interior of the cell is less negative than before the sodium ions diffused inward. Milliseconds later, the membrane becomes impermeable to sodium ions, but permeable to potassium ions and these ions diffuse across the membrane through the potassium channels and out of the cell.

The diffusion of potassium ions out of the cell is part of *repolarization* of the membrane, and results in restoring the original charge across the membrane before it was stimulated; that is, the interior of the cell is again negatively charged with respect to the outside of the cell. At this point, however, the ions have reversed locations, with sodium on the inside of the cell and potassium on the outside. Movement of the ions into and out of the cell and the resulting electrical change across the membrane constitute the nerve impulse. We will leave the ions in their reversed locations for a moment and explain how the nerve impulse travels along a nerve fiber.

When sodium ions first diffuse into a cell at a given point (depolarization), these ions also diffuse laterally and stimulate an adjacent point on the membrane, which, in turn, becomes depolarized by the entrance of sodium ions. This depolarization stimulates a point adjacent to it a little farther down the membrane and this stimulation continues down the length of the fiber. Thus, movement of the nerve impulse is a series of depolarizations along the nerve fiber. This is analogous to a line of dominoes falling. Tipping the first domino (the original stimulus) tips the next domino, which tips the next one, and so on down the line to the last domino. As successive depolarizations travel down the nerve fiber, repolarization (the outward diffusion of K^+) follows immediately as a second wave of activity.

Now, how is the sodium returned to its original position outside of the fiber and potassium returned to the inside? In Chapter 5, we described the mechanism of active transport as the energy-requiring transport of a solute across a cell membrane by carrier molecules from a region of low solute concentration to a region of higher concentration. Membrane carrier molecules are also called **pumps** and some of them transport one type of molecule or ion into the cell, while transporting another type out of the cell. In other words, one type of molecule or ion is exchanged for another across the membrane. All cells, including neurons, have an exchange carrier system called the **sodium/potassium (Na^+/K^+) pump,** which is illustrated in Fig. 9.8. There may be thousands of such transport molecules in a single membrane. Following repolarization, the sodium-potassium pump transports sodium ions back across the membrane to the extracellular fluid and potassium ions back to the inside of the cell. The original resting state of the membrane is then restored and the neuron is ready to fire again.

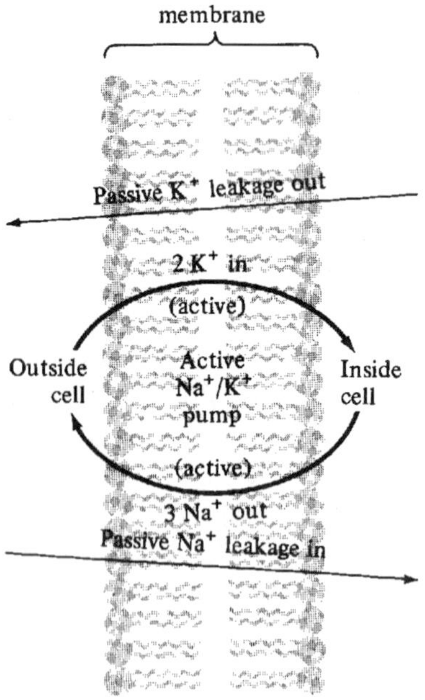

Figure 9.8 The sodium-potassium pump. This active transport mechanism in the plasma membrane by which sodium is secreted and potassium absorbed is partly responsible for maintenance of the resting potential of neurons.

Neuron Communication: The Synapse

Neurons communicate with other adjacent neurons, muscle cells, and glands at a junction called a **synapse** (Fig. 9.9). A nerve impulse is passed from the axon of one neuron to the dendrites or cell body of an adjacent neuron through the synapse between the two cells. There are, however, axon to axon synapses. A synapse between a neuron and a skeletal muscle is called a *neuromuscular junction.* Synapses may involve either electrical or chemical junctions, with chemical synapses being much more common. Cells involved in electrical synapses are physically adjacent so that the nerve impulse flows between the cells without interruption. Electrical synapses are found in some invertebrates, and particularly in the smooth and cardiac muscles of humans and other vertebrates. For many years scientists assumed that all nerve impulse conduction between cells was electrical. Subsequent research and the advent of the electron microscope, however, revealed microscopic gaps between most synapsing cells. These cells, then, are not physically connected and the nerve impulse cannot be transmitted directly from one cell to the other.

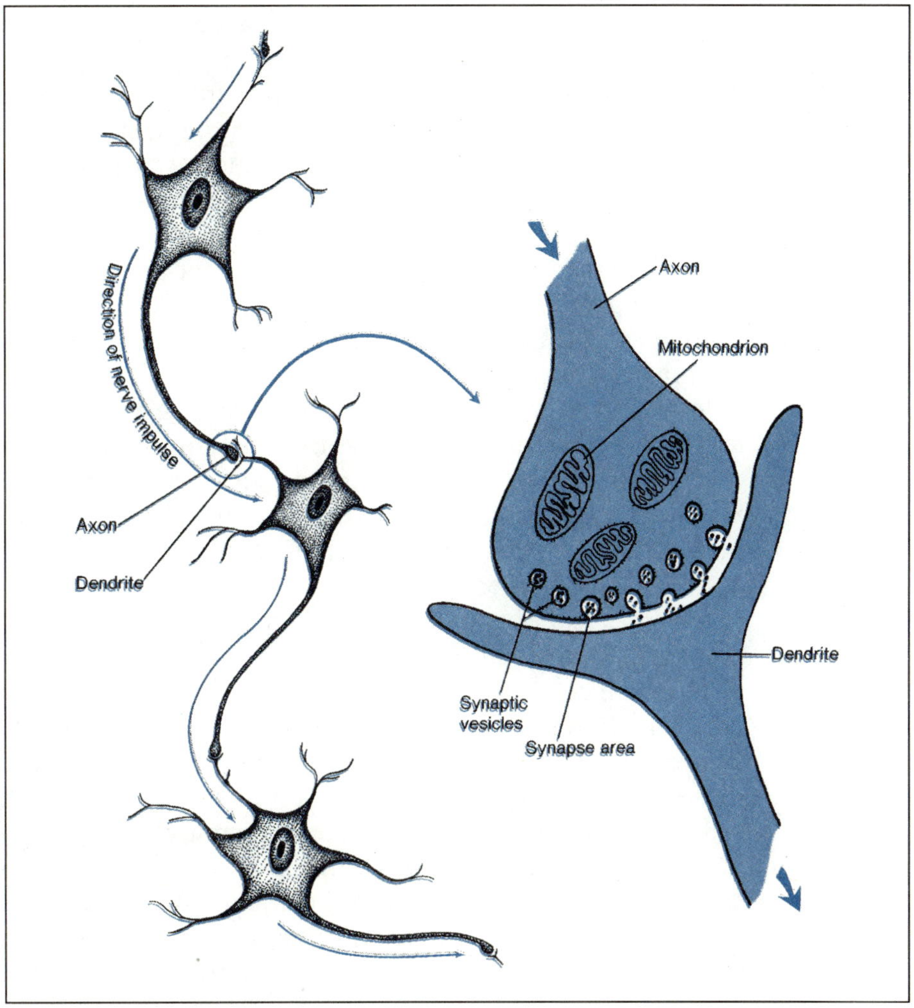

Figure 9.9 Diagram of a neuron-to-neuron synapse. The tips of the axon of one neuron and the dendrite of the adjacent neuron do not make physical contact, but are separated by a microscopic space. A nerve impulse arriving at the axon tip causes the synaptic vesicles to release a neurotransmitter. The neurotransmitter diffuses across the narrow space and stimulates the dendrite of the adjacent neuron. The dendrite transmits the impulse toward the cell body of this neuron. After the dendrite is stimulated, the neurotransmitter is broken down rapidly by deactivating enzymes. This prevents the dendrite from being stimulated continually.

As is usually the case in science, discovering a new fact or making a new observation opens the door to a multitude of new questions. In this instance, the initial and critical question from which all related questions later followed was: If the synapsing cells do not make contact, how does the nerve impulse span the gap between them to relay the signal? As early as 1921, research indicated that transmission across the synapse may involve chemicals. Subsequent research verified this finding and the theory of electrical synaptic transmission for all cells except those of muscle was abandoned.

Chemical synapses, as shown in Fig. 9.9, utilize several chemicals generally known as **neurotransmitters.** Located at the terminals of axons are thousands of tiny membranous vesicles, or sacs, that contain a specific neurotransmitter. When the nerve impulse reaches the tip of the axon in a neuron-to-neuron synapse, the neurotransmitter is released from the sacs and spills into the tiny gap between the two cells. On diffusing across the gap, the neurotransmitter stimulates special receptors on the surface of the cell membrane of the dendrite or cell body on the other side of the synapse. This depolarizes the membrane, setting up a nerve impulse that continues through this neuron.

Some neurotransmitters, however, do not stimulate neurons to conduct a nerve impulse but *inhibit* conduction of the impulse. This is important, for instance, in preventing contraction of one skeletal muscle while another is contracting. If an animal encounters a painful stimulus, such as stepping on a thorn, the resulting withdrawal reflex results from contraction of one leg muscle and inhibition of another in the same leg. If both leg muscles contracted, movement of the leg would be impossible. In another example, the human brain and pituitary gland produce a group of chemicals known as *endorphins*. Some of these are thought to function as inhibitory neurotransmitters that block the transmission of pain signals.

With the discovery of neurotransmitters and their actions, a new question logically arose. What happens to the neurotransmitter after it acts on a neuron or some other cell? If the neurotransmitter remains in the gap between cells, it would continue to stimulate or inhibit the cell and make nervous control of the body impossible. This normally does not happen partly because of the action of various **deactivating enzymes** located in the gap and in the membrane of the stimulated cell. In some cases, these enzymes break down the neurotransmitter immediately after it has acted on a cell. Portions of the degraded neurotransmitter molecule reenter the axon terminal where they are used in the synthesis of new neurotransmitter molecules. In other cases, the neurotransmitter is not completely degraded; the fairly intact molecule is transported back into the axon to be used again. Removal of the neurotransmitter from the synaptic gap allows the membrane of the affected cell to return to its resting state.

Various chemicals and drugs affect the activity of the neurotransmitters and deactivating enzymes by causing either excessive stimulation or slowing of the nervous response. Some of the nerve gases and phosphate pesticides are known to block some deactivating enzymes. The result is continued transmission of the nerve impulse, causing strong contraction and paralysis of skeletal muscles. Amphetamines or "uppers" stimulate the release of certain neurotransmitters, resulting in excessive transmission of the nerve impulse. The effect is that the nervous system becomes extremely sensitive and more easily excited than normal. The various tranquilizers exert their effects by inhibiting the nerve impulse at the synaptic junction. This results in the well-known calming effects of these drugs. Other substances, including depressants (alcohol and anesthetics), stimulants (caffeine and nicotine), and toxins (botulin and snake venom), also influence the transmission and regulation of the nerve impulse.

Spinal and Visceral Reflexes

It is of no small consequence that a horsefly irritating the rump of a plowhorse doesn't remain there for long. Instantaneous reaction to the irritation by the horse is an example of one of an animal's major protective and homeostatic mechanisms—the **reflex** (Fig. 9.10). We have mentioned that a reflex is an automatic, unlearned response to a stimulus performed without conscious thought or previous experience. Reflexes that result in contraction of skeletal

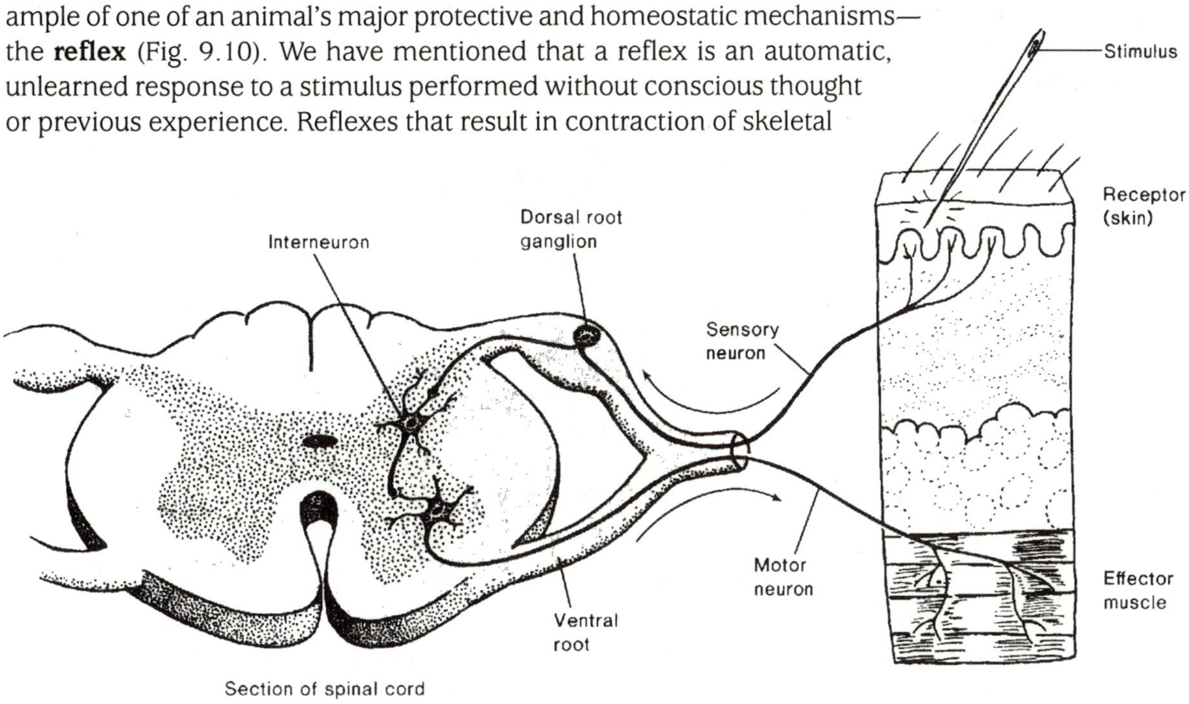

Figure 9.10A Basic components of a reflex arc. Almost all spinal cord reflexes involve three neurons: sensory, motor, and interneuron.

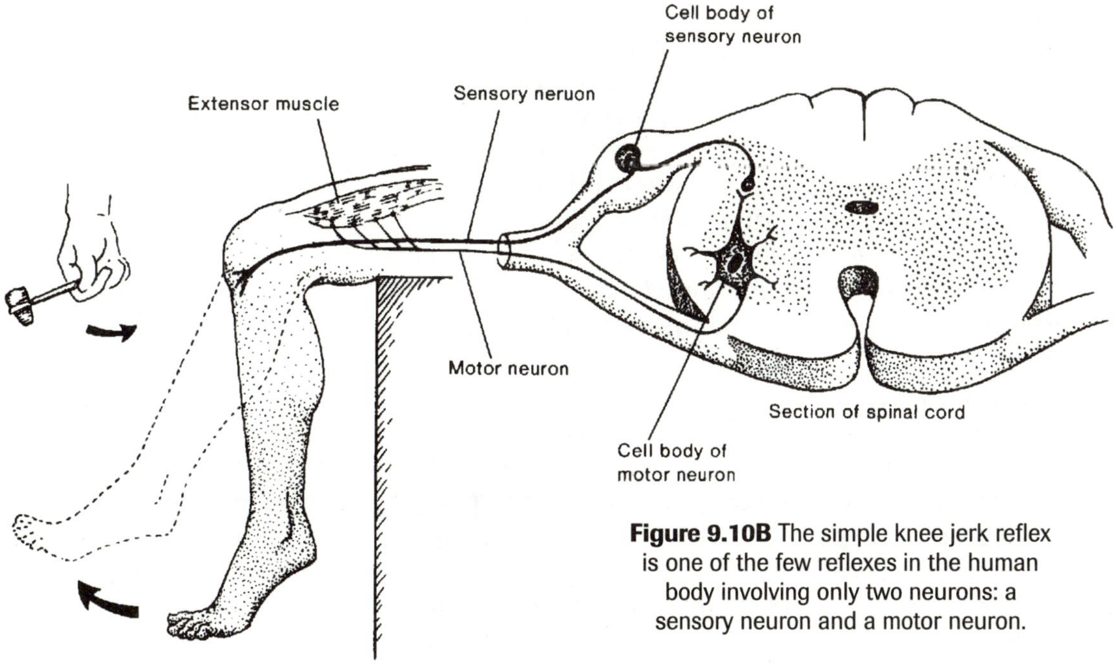

Figure 9.10B The simple knee jerk reflex is one of the few reflexes in the human body involving only two neurons: a sensory neuron and a motor neuron.

muscles are called *spinal reflexes;* reflexes that cause smooth muscle and cardiac muscle contractions are called *visceral reflexes.* **Viscera** is a general term for the body's internal organs, which contain smooth muscle.

A **reflex arc** involves a sensory receptor, a sensory neuron, usually interneurons, a motor neuron, and a responding effector organ, usually a muscle or gland. The nervous pathway from the sensory receptor through the neurons and finally to the effector organ constitutes a reflex arc. A reflex arc is the basic functional unit of the vertebrate nervous system. A reflex, then, is the result of a nerve impulse moving through a reflex arc. Reflexes not only result in certain body movements through skeletal muscle contraction, but also result in controlling heart rate, blood pressure, respiratory rate, body fluid pH, glandular secretion, and other involuntary functions.

Reflexes are important in determining the general condition of the nervous system. A normal reflex is a good indication that the part of the nervous system associated with the reflex is functioning appropriately. On the other hand, if a particular part of the nervous system has been injured, an abnormal or absent reflex may aid in locating the injured site. In adult humans, for example, stroking the sole of the foot elicits a plantar reflex, which normally involves a downward movement of the foot and toes. If certain nerve pathways in the spinal cord are injured, however, the response is upward movement of the big toe and fanning out of the other toes.

The Central Nervous System

The Brain

We have mentioned that the vertebrate nervous system consists of the central nervous system (CNS) and the peripheral nervous system (PNS). The central nervous system is comprised of the brain and spinal cord, and the peripheral nervous system is comprised of all the sensory and motor nerves. In our discussion of the central and peripheral nervous systems, we will emphasize the structure and function of the human nervous system; the nervous systems of humans and the more advanced vertebrates, as mentioned, are quite similar.

Evolution of the vertebrate brain has progressed from the small, thin structure of fish and amphibians to the highly convoluted, enlarged brain of birds and mammals (Fig. 9.11). The most striking feature of vertebrate brain evolution is development of the **cerebrum,** the largest and most anterior division of the brain. The outermost layer of nerve cells of the cerebrum is the **cerebral cortex,** which is concerned with learned behavior. In fish and most amphibians, however, the cerebral cortex is little more than a thin layer of cells mainly concerned with smell. Reptiles possess a larger cerebrum and more complex cerebral cortex than fish or amphibians. In mammals, however, the cerebrum is quite large, and the cerebral cortex is thick and highly convoluted. The trend in the evolution of the vertebrate brain, then, has been toward increasing dominance of the cerebrum.

The human brain weighs about 1.5 kg (3 lbs.), accounts for approximately 2 percent of body weight, and contains an estimated 100 billion neurons. The brain is the only known entity in the universe capable of contemplating itself. Generally, the brain is composed of three major divisions: the *forebrain,* the *midbrain,* and the *hindbrain.* It is the increased development of the forebrain that has separated humans from their vertebrate relatives. Each of the major divisions of the brain exerts control over a variety of physiological activities. For the most part, the divisions do not function separately, but

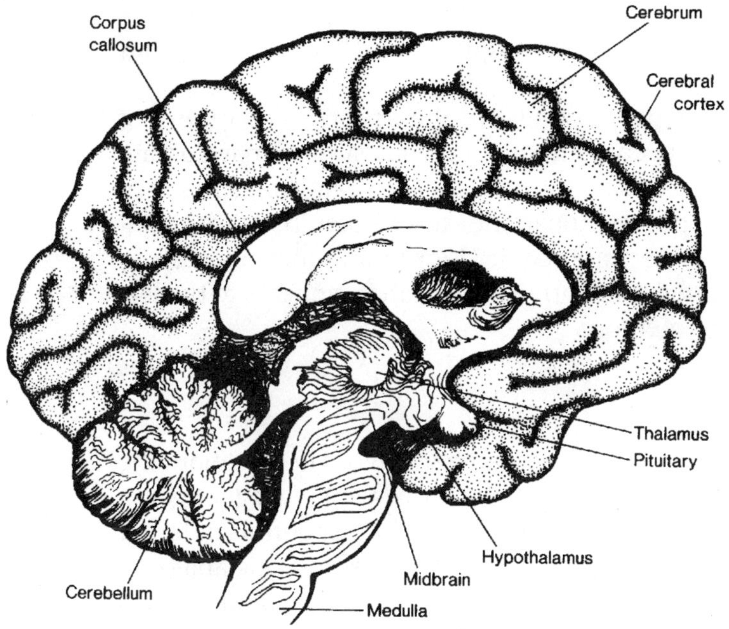

Figure 9.11A Major areas of the human brain (left hemisphere).

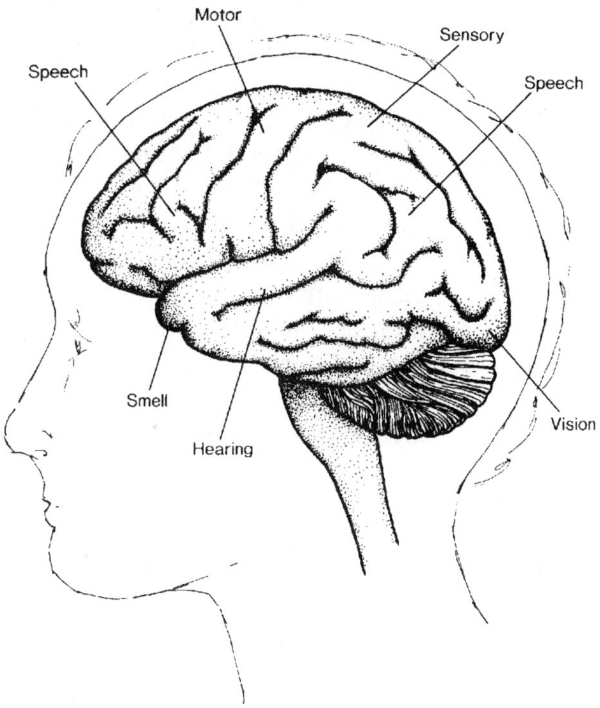

Figure 9.11B General areas of the brain responsible for detection and control of the major senses.

interact in a coordinated fashion to influence each other as well as all other parts of the nervous system. Two of the major structures of the forebrain are the cerebrum and the hypothalamus.

FOREBRAIN

Cerebrum

The cerebrum, which is composed of a right and left cerebral hemisphere, is the largest portion of the brain and the seat of reason and memory storage. As shown in Fig. 9.11B, various physiologic activities involving speech, vision, hearing, and other functions are analyzed and processed in certain areas of the cerebrum. The cerebrum also controls much of the voluntary movement of the body, while control of movement by the midbrain and hindbrain is subconscious.

Although we know very little about the underlying mechanisms of higher brain functions such as memory, emotions and thought, it appears that these phemomena may result from both physical and chemical changes in the brain itself. The processing and storage of memories, for example, are thought to involve changes in the shape of sensory neurons, particularly in the tips of axons or dendrites. In some manner, this alteration in structure may facilitate transmission of impulses across the synapse. A synapse, or a series of synapses, becomes more capable of transmitting a signal impulse if that signal passes across the synapse a number of times. Originally, of course, the signal comes from some outside stimulus, such as the sight of a painting. Viewing the painting for a sufficient period of time sends the same signal impulses repeatedly across the synapses. Eventually, the brain can generate the same impulses across the synapses even though the painting is not being viewed at the time. Thus, the person has a perception of the painting in the form of a memory.

The higher functions of the brain apparently involve the actions of a variety of excitatory and inhibitory transmitters. The entire human nervous system may contain as many as 30 different types of these chemical substances. The brain stores many of these, including acetylcholine, norepinephrine, dopamine, and serotonin. Research indicates that some of these neurotransmitters are required for brain functions such as memory and emotions. The release of norepinephrine, for instance, appears to be necessary for the imprinting of permanent memory.

Lateralization

One of the more fascinating areas of human brain research involves lateralization, the concept of a "split brain." This research, initiated by Nobel Prize winner Roger Sperry, Robert Myers, and others, revealed that each hemisphere of the brain controls different functions. The right hemisphere controls our perception of space and our ability to manipulate objects so that they are oriented or spaced properly in relation to each other. The left hemisphere is specialized for speech, language, and logical thought. In fact, there are two critical speech areas in the left hemisphere. One area controls the speech muscles of the lips, tongue, jaw, and throat; the other area has to do with understanding speech and with arranging words so that they are grammatically usable and meaningful. After you hear or see a word that you want to speak, the stimulus signal then goes to the area concerned with understanding speech and then to the area that controls the speech muscles.

The two hemispheres of the brain are connected by nerve fibers, which allow each side to communicate with the other and to integrate their information. Because of these fibers, the right hemisphere controls muscular activity in the left half of the body and the left hemisphere controls muscular activity in the right half of the body. The localization of mental functions in only one hemisphere, however, is not as well defined as originally thought. Experimental evidence indicates that a specific task is not necessarily the exclusive responsibility of only one hemisphere. Basically, then, lateralization confirms that there are specific areas in each cerebral hemisphere that are responsible for controlling different functions. It does not confirm that one hemisphere is "dominant" over the other to the extent that it determines one's capabilities and personality.

Hypothalamus

A second major area of the forebrain is the **hypothalamus,** which is located near the floor of the forebrain just above the pituitary gland. The hypothalamus consists of several masses of neurons that communicate with virtually all parts of the nervous system. Through incoming nervous stimuli and changes in the concentration, temperature, pH, and volume of the surrounding capillary blood, the hypothalamus continually monitors body function. The hypothalamus regulates the heart, blood pressure, body temperature, thirst, excretion of water, appetite, pleasure, pain, sleep and wakefulness, and sexual functions. It manufactures two of the chemical messengers called hormones that are passed to the pituitary gland, which releases these hormones into the bloodstream. In addition, the hypothalamus produces releasing factors, which are hormones that control the release of other hormones from the anterior portion of the pituitary gland. The hypothalamus and its role in maintaining homeostasis are discussed further in Chapter 11.

Midbrain

The second major division of the human brain is the *midbrain*. The midbrain is involved in visual reflexes and the subsequent reflex movements of the head known to vaudevillian enthusiasts as the "double take." The midbrain also integrates reflex movements of the head in response to sudden loud sounds, such as firecrackers, sirens, and screams of biology students. In addition, the midbrain helps control posture and serves as a relay center between the forebrain and the hindbrain. As mentioned, the midbrain is not under voluntary control, but exerts its effects subconsciously.

Hindbrain

Medulla Oblongata

The remainder of the subconscious activities of the body are controlled mostly by the hindbrain. Two of the more prominent structures of the hindbrain illustrated in Fig. 9.11A are the **medulla oblongata** and the **cerebellum.** The medulla oblongata is the posterior portion of the hindbrain and is continuous with the upper portion of the spinal cord. It functions as a conduction pathway for sensory and motor impulses between the spinal cord and the higher brain centers. It also is the center for several vital reflexes that, through input from the hypothalamus and other centers, regulate heart rate, breathing, blood vessel diameter, coughing, sneezing, swallowing, and vomiting. These basic and primitive reflexes are necessary for normal body functioning and, indeed, for life itself. A karate chop to the base of the brain not only is unpleasant, but can be dangerous if, for example, the respiratory center is damaged and breathing is interrupted.

Cerebellum

In the human brain, the cerebellum is second in size to the cerebrum. The cerebellum is a coordinating center for the maintenance of muscle tone, balance of the body, posture, and the normal fluid motion of skeletal muscle. It receives and integrates nerve impulses from the cerebrum, muscles, tendons, joints, and the inner ear, resulting in smooth and coordinated body movement. Damage to the cerebellum can result in uncoordinated movements, such as the inability to touch your nose with your finger, a staggering walk, jumbled mouth movements when attempting to talk, and other outward signs of impaired muscle control.

Enkephalins and ß-Endorphin

Another of the fascinating discoveries related to the nervous system concerns a group of brain chemicals known as neuropeptides. These are short-chain peptides, ranging from two to thirty-nine amino acids in length. Many of these have been known for years and include some of the hormones found in the pituitary gland and hypothalamus. However, in the mid-1970s, two types of neuropeptides—the enkephalins and ß-endorphin—that exhibited opiate-like activity were discovered. Their significance to overall brain function is just beginning to be understood.

Both of these chemicals closely resemble the opiate narcotic morphine. It is known that certain regions of the brain contain specific receptor sites that bind morphine and other similar drugs. When bound to the receptors, morphine and other opiates have the effect of relieving pain and producing euphoria, or a "high." It has been found that these same receptors also bind enkephalins and ß-endorphin. Thus, these naturally occurring brain chemicals also act to suppress pain and produce a feeling of well-being.

Although it is not certain, it is thought that enkephalins and ß-endorphin may act as inhibitory transmitters, thereby blocking the transmission of pain signals.

When a drug addict takes any artificial opiate, such as morphine or heroin, the brain eventually stops producing the natural ones. An attempt at withdrawal is agonizingly painful because the brain is deprived of any source of opiate. Recovery from drug addiction can occur only when the brain resumes production of its own opiate-like chemicals—the enkephalins and ß-endorphin.

The Spinal Cord

The other major structure of the central nervous system is the *spinal cord,* which is surrounded by the bony vertebral column (Fig. 9.12). The cord extends posteriorly from the medulla oblongata to the first vertebrae of the lower back. The 31 pairs of spinal nerves and some of the nerves of the autonomic nervous system originate in the spinal cord. We will consider the structure and function of these nerves in the following section on the peripheral nervous system.

Both spinal reflexes and visceral reflexes involve the spinal cord. Nerve impulses coming into the cord through sensory neurons may be relayed to various areas of the brain. The impulses are relayed along bundles of nerve fibers called *nerve tracts* that often extend the entire length of the cord. From the brain, other nerve tracts carry motor impulses back down the cord. These impulses are relayed out of the cord through motor neurons to the

Figure 9.12 (a) The central nervous system (CNS). The brain and spinal cord are protected within the bony skull and vertebrae, respectively. (b) Section through the spinal column. Each spinal nerve passes from the cord through an opening between the vertebrae. Sympathetic nerves (autonomic system) extend from the sympathetic ganglia and pass to various internal organs. Intervertebral discs are pads of cartilage that prevent the vertebrae from rubbing together. (c) Cross section of the spinal cord showing the position of the sensory and motor neurons within the cord and a spinal nerve.

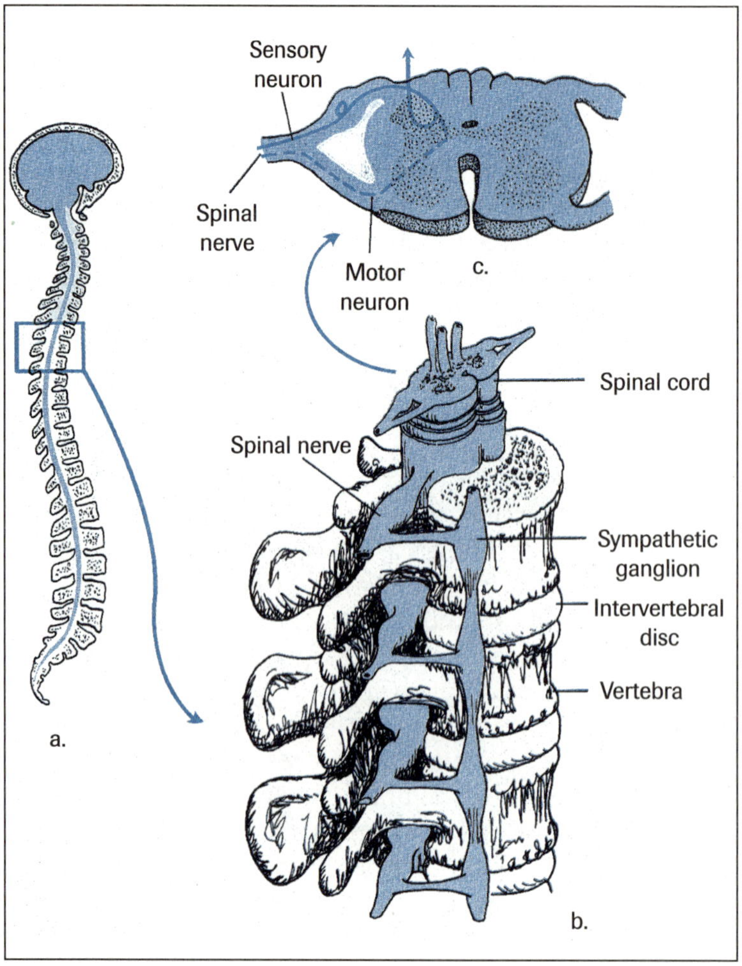

appropriate effector organ. At the junction of the spinal cord and medulla oblongata, most of the nerve fibers on the left side of the medulla cross over and run down the right side of the spinal cord; similarly, most of the fibers on the right side of the medulla cross over and run down the left side of the cord. Because the spinal cord, medulla oblongata, and cerebrum are connected by nerve tracts, this crossing-over explains why the right cerebral hemisphere controls the left side of the body and the left cerebral hemisphere exerts similar control over the right side of the body. With this arrangement, a blow to one side of the head can cause paralysis or other complications on the other side of the body.

The brain and spinal cord are covered by three separate layers of tissues called **meninges** (Gk. *meninx,* meaning "membrane"). Spaces between the meninges are filled with *cerebrospinal fluid,* a clear liquid derived from blood plasma. The meninges and cerebrospinal fluid cushion and protect the central nervous system against hard blows and jarring. Infection of the meninges by certain bacteria or viruses causes an inflammatory condition called *meningitis.* This is a potentially serious condition because the infection can spread to the vital tissues of the central nervous system. Another unfortunate condition—the alcoholic hangover—may also result from irritation of the meninges of the brain.

The Peripheral Nervous System

Spinal Nerves

The peripheral nervous system consists of spinal, cranial, and autonomic nerves, all of which arise from the brain and spinal cord. Nerve impulses entering or leaving the central nervous system travel along one or more of these nerves. In humans, each of the 31 pairs of spinal nerves contains both sensory and motor fibers discussed earlier. A spinal reflex arc involves the sensory neuron and motor neuron of a single spinal nerve. In addition to spinal reflexes, spinal nerves control voluntary muscle movement in the body. The motor neuron of a spinal nerve originates in the spinal cord and terminates in skeletal muscle or various other tissues of the body. The sensory neuron of a spinal nerve originates in muscle or other tissue and terminates in the spinal cord. The sensory and motor nerves communicate by synapsing with an interneuron in the spinal cord or brain.

Cranial Nerves

The 12 pairs of cranial nerves originate on the undersurface of the brain and extend into the head, neck and internal organs. Cranial nerves are involved in a wide range of functions, including vision, hearing, tasting, smelling, equilibrium, chewing, crying, salivation, and movements of the eyes, facial muscles, and tongue. Reptiles, birds, and mammals have 12 pairs of cranial nerves, while fishes and amphibians have 10 pairs.

Autonomic Nervous System

Everyone is familiar with the pounding of the heart brought on by the sound of a footstep in the dark and with the accompanying onset of "goose bumps", a rush of blood to the face, sweating palms, and a sudden surge of energy when we are startled or frightened. These and many other involuntary reactions of the body are called forth automatically without previous thought. The division of the peripheral nervous system re-

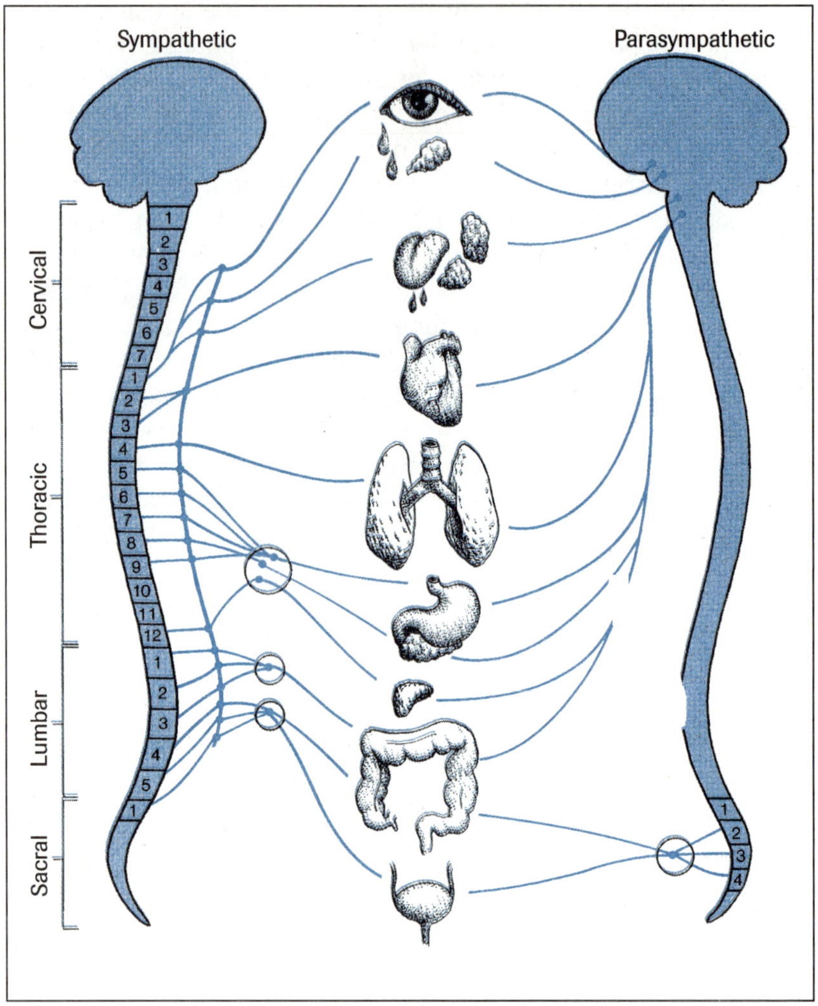

Figure 9.13 The autonomic nervous system (ANS). The sympathetic nerves leave the spinal cord at the levels of the numbered vertebrae and pass through a chain of ganglia on both sides of the cord (the other chain not shown here). Axons of the sympathetic nerves exit the ganglia and terminate in various internal organs. Parasympathetic nerves extend from the brain and sacral vertebrae to ganglia near the internal organs. From the ganglia, short parasympathetic neurons pass into the organs. Most of the internal organs, therefore, receive *double innervation,* i.e. two nerves, one from the sympathetic division and one from the parasympathetic.

sponsible for these reactions is the *autonomic nervous system.* The word *autonomic* means "self-governing." The autonomic nervous system supplies nerves to cardiac muscle of the heart, to smooth muscle of internal organs and blood vessels, and to most of the glands of the body (Fig. 9.13). These nerves are responsible for the **visceral reflexes** of the body. An increase in blood pressure, for example, initiates a reflex that stimulates nerves to the heart, resulting in a slower, less forceful heartbeat and a lowering of blood pressure. In addition, nerve impulses to the blood vessels cause expansion, or dilation, of the vessels, which also reduces blood pressure. Conversely, if the blood pressure falls too low, these visceral reflexes are reversed and the blood pressure rises.

The higher brain centers, too, affect visceral reflexes. Emotions such as anger or rage, for instance, can raise blood pressure through nervous stimulation of the hypothalamus and other centers by input from the cerebrum. From the various brain centers, nerve impulses are relayed along autonomic nerves to the heart and blood vessels as described. As we will now discuss, these autonomic nerves are separated into two distinct divisions.

Sympathetic and Parasympathetic Divisions

Structurally, the autonomic nervous system consists of two sets of nerves: one set belongs to the *sympathetic* division and the other set belongs to the *parasympathetic* division. Autonomic nerves exit the spinal cord along with the spinal nerves. Most of the organs, blood vessels, and glands receive two autonomic nerves, one from each division. The two nerves supplying a particular structure have opposing effects; that is, if one nerve has an excitatory effect, the other nerve exerts an inhibitory effect. Consider, for example, the effects of autonomic nerves on the heart. When you are frightened or angry, the sympathetic nerve to the heart is stimulated, resulting in an increase in the rate and force of the heartbeat. Stimulation of the parasympathetic nerve, in contrast, slows the heart rate and decreases the force of the heartbeat.

The sympathetic division, however, does not always increase the activity of an organ nor does the parasympathetic division always inhibit or slow the action. In the digestive tract, stimulation of the smooth muscle walls by the parasympathetic nerves causes increased muscular contraction, whereas sympathetic stimulation relaxes and slows the muscle. This, then, is opposite to the effects on the heart. With their opposing effects, the sympathetic and parasympathetic divisions act together to regulate and maintain *homeostasis,* the relatively constant internal environment of the body. Although the two divisions complement each other, the sympathetic division is usually active during periods of stress and the parasympathetic division influences the more routine involuntary functions of the body.

Sensory Receptors

Sensory receptors are structures that inform an organism of changes in its external and internal environment. These receptors are specialized to respond to a variety of stimuli such as chemicals, light, temperature, and pain. The evolution of sensory receptors in the animal kingdom has generally followed a path of increasing complexity. Some cnidarians, particularly various jellyfish, possess simple light receptor cells that allow these organisms to respond to light. Planarians usually have two light-sensitive eyes, although some species may have two or three pairs or a great number of eyes. These organisms also possess chemical receptors for detecting food, and sensory cells that orient the animals to the direction of water currents.

The nervous system of nematodes and some other roundworms is more complex than that of the flatworms. Sensory receptors for touch and chemical detection are distributed on the head and elsewhere on the body surface of many roundworms.

Evolution of the invertebrate nervous system reaches its peak in the cephalopod mollusks such as the squids and octopuses. This high degree of development is associated with the carnivorous and locomotor features of these animals. Cephalization is quite pronounced in cephalopods and the sense organs, particularly the eyes, are highly developed. The image-forming eye of cephalopods is strikingly similar to that of vertebrates and possibly is capable of detecting color. Various mollusks are supplied with touch, light, and chemical receptors.

Various annelids have two, three, or four pairs of eyes used for light detection; other annelids, however, possess two large eyes that may be capable of object detection. Receptors for touch, food detection, and orientation of the body are found in many annelids.

Arthropods display a high degree of cephalization, which is correlated with their vast array of sense organs. Most arthropods have eyes, some of which form images and detect color and some function only in light detection. The sense organs of arthropods include receptors for sight, touch, taste, sound, and smell.

Echinoderms possess only a few sensory organs, such as light-sensitive eye spots at the tips of the arms of starfish and light-sensitive cells in the tube feet of sand dollars. The epidermis of most echinoderms is supplied with sensory organs for detecting touch, temperature, and pressure.

Human Special Senses

Perception of a sensation, or sensory input to the brain, requires a stimulus, a receptor organ, a nerve impulse to the brain, and an interpretation of that stimulus by the brain. Spinal reflexes are not conscious sensations because they do not require interpretation by the brain. It is through perceptions that we and many other animals are made aware of our environment.

Cephalization has resulted in expansion of the brain accompanied by the evolution of highly specialized sense organs. The special senses of vision, hearing, equilibrium, taste, and smell in humans and many other vertebrates are the result of this evolution. These senses require complex receptors that greatly heighten sensory input to the brain, which interprets the sensation as a perception of the environment. Sensory input from these receptors requires certain cranial nerves that relay each sensation to specific areas of the brain where the sensation is interpreted.

Vision

The human eye is a remarkable organ, providing acute three-dimensional and color vision. Fig. 9.14 shows that the eyeball is a spherical structure composed of two chambers, both of which contain a fluid that gives shape to the eyeball and keeps it from collapsing. The large posterior chamber contains a transparent, jellylike substance called the *vitreous humor.* The eyeball is partially surrounded by three layers of specialized tissues: the *sclera,* the *choroid,* and the *retina.* The outer layer is called the **sclera,** or "white of the eye," and is a tough, protective covering. The **cornea** is the anterior part of the sclera and is transparent, allowing light to enter the eye. The middle tissue layer, the choroid, contains a rich network of tiny blood vessels. The third, and innermost, layer is the **retina,** which contains two types of highly specialized photoreceptor cells: the *rods* and the *cones.* Rods are responsible for vision in dim light, and cones function for color distinction and vision clarity.

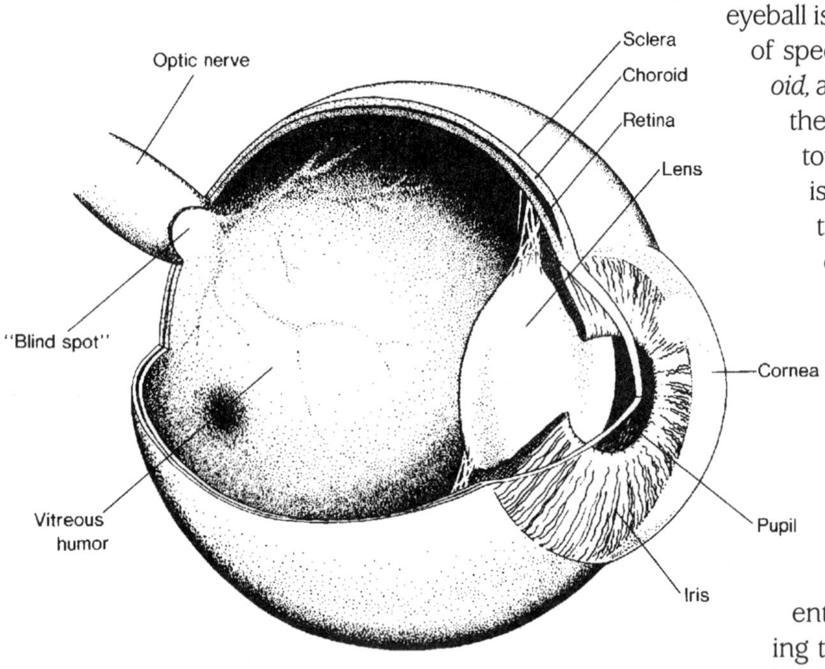

Figure 9.14 Basic structure of the human eye.

Immediately behind the transparent cornea is the **iris.** We are describing the color of the iris when we speak of brown eyes, blue eyes, or green eyes. In the center of the iris is the *pupil,* a small opening through which light passes to the interior of the eyeball. The size of the pupil, which is controlled by the autonomic nervous system, depends on the degree of constriction of two layers of tiny muscles in the iris. When we get excited or angry, sympathetic nerves to the iris constrict the outer layer of muscles, causing the pupils to dilate. Bright light entering the eye stimulates parasympathetic nerves that contract the inner layer of iris muscles, causing the pupils to constrict. In the dark, the parasympathetic nerves are less stimulated and the pupils dilate. The effect of pupil constriction and dilation is familiar to anyone who has walked out of a darkened theater into bright sunlight or vice versa.

The **lens** is behind the iris. The lens bends, or refracts, incoming light rays so that they all focus at the same point on the retina at the back of the eye. This is important in perceiving a clear image. Axons from special nerve cells at the back of the retina converge to form the optic nerve, a sensory nerve that passes into the brain.

Functionally, the eye is much like a camera. Light rays from an object pass through the transparent cornea into the pupil, and through the lens. The lens focuses the image of the object on the retina. The image is inverted, but the brain interprets it as being right-side up. Thus, our visual interpretation of what we see is not mediated by the eye but by a specific area of the brain. The visual area is in the back of the brain where the neurons associated with sight terminate. A blow to the back of the head may cause you to see "stars" because these neurons are highly stimulated by the blow.

There also is a chemical aspect of vision involved in stimulating the retina. The rods and cones contain light-sensitive pigments that are synthesized in part from vitamin A and break down when exposed to light. The cones actually contain three different kinds of pigments that are sensitive to the colors red, green, and blue. As these pigments break down, the rods and cones are stimulated and begin to transmit signals to the brain via the optic nerve. As light intensity is reduced, the breakdown process becomes reversible and the light-sensitive pigments are reformed.

The eye can detect a great variety of colors. This is possible because red, green, and blue colors are mixed and interpreted in different combinations by the brain. Color blindness results when one or more of the three types of cones are nonfunctional. People with the most common form of color blindness are missing red cones, and cannot distinguish red from green. If all of the three types of cones are missing—an extremely rare condition—the individual sees only black, white, or gray. The incidence of color blindness among males is about 10 percent, while less that 1 percent of females are color blind.

As a result of an abnormally shaped eyeball or malfunction of the lens, light rays passing through the lens may not come together to focus at the same point on the retina (Fig. 9.15).

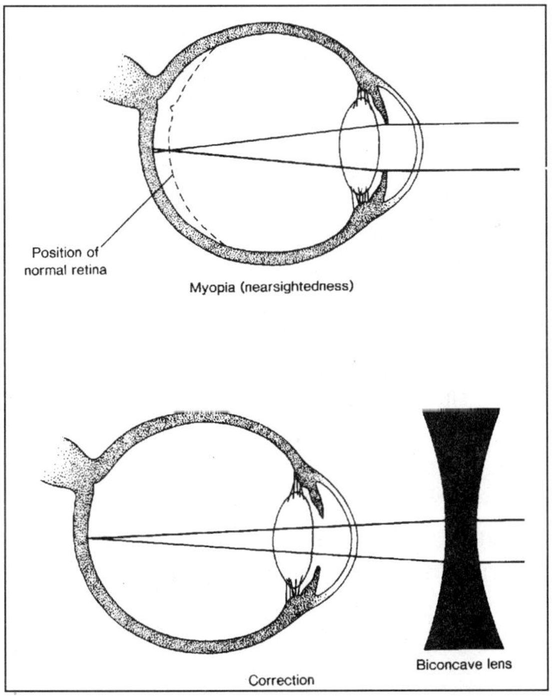

Figure 9.15A Myopia, or nearsightedness, can result from an eyeball that is too long or from a change in the shape of the lens, so that the image focuses in front of the retina. Biconcave lenses extend the focal point backward on the retina where it should be.

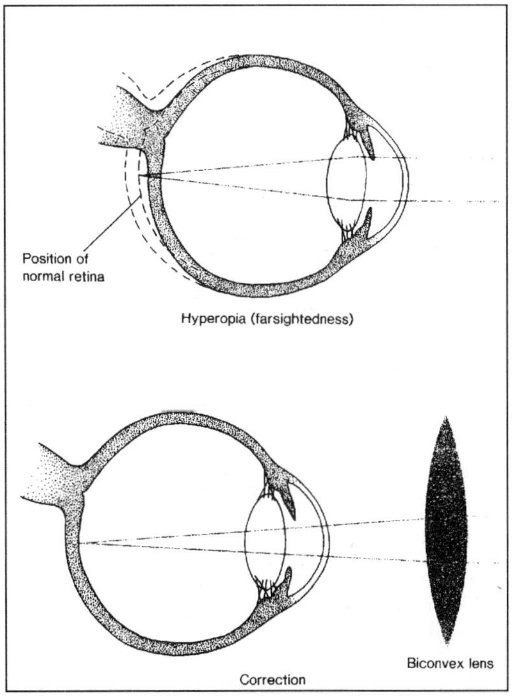

Figure 9.15B Hyperopia, or farsightedness, can result from an eyeball that is too short or from a change in the shape of the lens, so that the image focuses behind the retina. Biconvex lenses shorten the focal point forward on the retina where it should be.

If, for example, the eyeball is too long, light rays focus on a point in front of the retina. An elongated eyeball results in *nearsightedness,* the ability to see distinctly only at short distances. If the eyeball is too short, in which case the image is focused behind the retina, the result *is farsightedness.* With this condition, a person can see distant objects clearly, but near objects appear out of focus. A relatively common vision problem caused by a defective cornea and lens is *astigmatism.* Astigmatism literally means "without a spot" and results in a distorted visual image. Each of these conditions may be corrected with eyeglasses that bend incoming light rays so that the image is focused exactly on the retina.

Hearing

The sense of hearing involves structures in the outer ear, middle ear, and inner ear, as illustrated in Fig. 9.16. The outer ear consists of the fleshy appendage on the side of the head and the ear canal, which extends from the outer ear internally to the *eardrum.* The eardrum is a thin membrane that separates the ear canal from the middle ear. Sound waves enter the ear canal and cause the eardrum to vibrate. These vibrations are relayed in the middle ear by three tiny interconnected bones: the *hammer, anvil,* and *stirrup.*

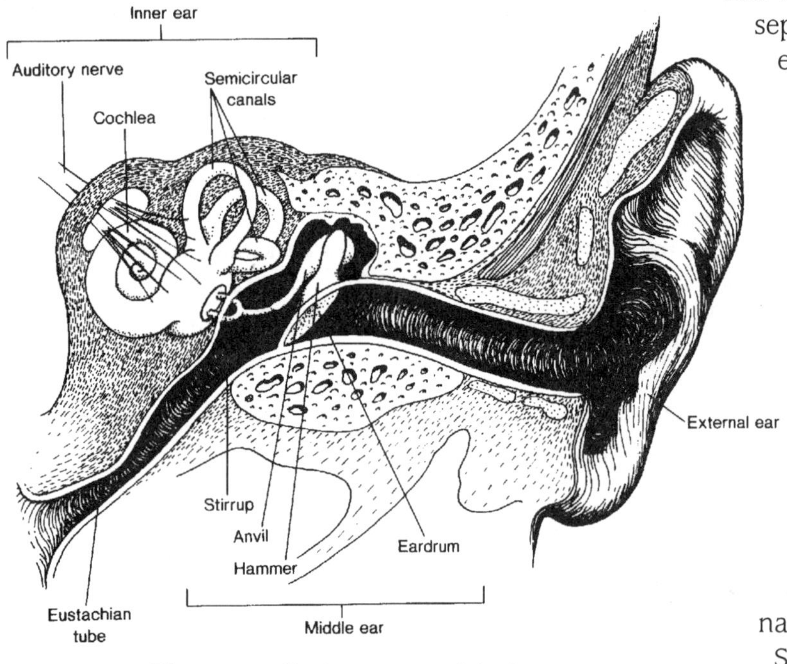

Figure 9.16 Basic structure of the human ear.

The **cochlea,** an inner ear structure shaped like a snail's shell, contains the major organs of hearing; it also contains fluids involved in conducting sound vibrations. The functional hearing unit in the cochlea is the *organ of Corti,* which contains receptor cells that send nerve impulses along a sensory nerve that terminates in the brain.

Sound waves passing through the hammer, anvil, and stirrup increase in loudness as they approach the cochlea. Vibrations of a small membrane set up movements of the fluid within the cochlea. Fluid movements in turn stimulate the receptor cells in the cochlea, and nerve impulses are transmitted to neurons that carry the impulses to the sensory nerve and into the temporal areas on either side of the cerebrum. It is in these areas, then, that the brain receives and interprets incoming sound waves. Our ability to distinguish between sounds, such as different voices, results from stimulation of different portions of the temporal areas in the cerebral cortex.

The cavity of the middle ear has a small opening that leads into the *eustachian,* or *auditory tube.* This passageway connects the middle ear with the throat. Changes in air pressure, such as those that occur in an airplane at 30,000 feet, do not cause damage to the eardrum. The eustachian tube allows air to enter or leave the middle ear so that pressure inside and outside the ear is equalized. Thus, by swallowing or yawning, you attempt to change the pressure in the middle ear to correspond to the air pressure around you. Unfortunately, the eustachian tube also is one passageway through which microorganisms cause infections in the middle ear.

Equilibrium

Normally, you can distinguish between up and down and between forward and backward linear motion. You can also tell whether or not your head is rotating; that is, undergoing angular motion. The sense organs responsible for orienting the body with respect to gravity and for detection of angular and linear acceleration and deceleration are located in the inner ear. Attached to the cochlea are the *utricle, saccule,* and three *semicircular canals,* as shown in Fig 9.17A. All three organs contain tiny hair cells that stimulate or inhibit sensory neurons associated with the cells.

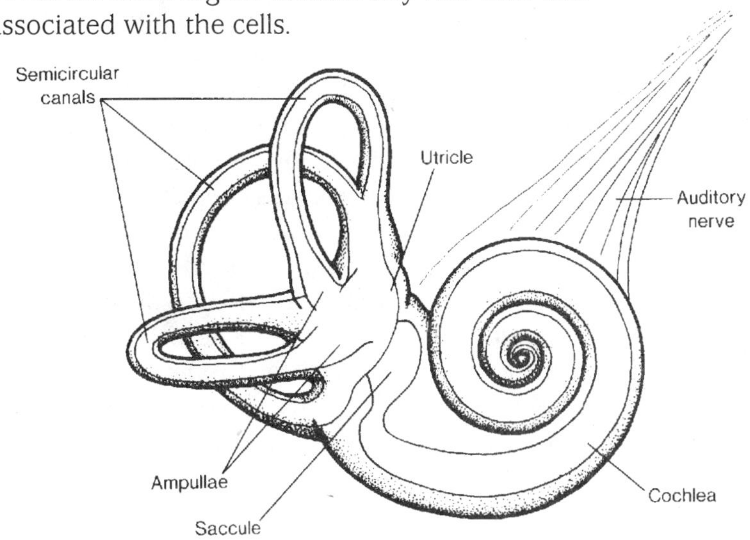

Figure 9.17A The cochlea and equilibrium apparatus of the inner ear. The cochlea functions in the sense of hearing and is not involved in equilibrium. Fluid within the semicircular canals moves when the head is rotated, stimulating hair cells that send nerve impulses to the brain. In this manner, the body is informed of the orientation of the head. The utricle and saccule contain otoliths that pull or press on hair cells that inform the body, via nerve impulses to the brain, of its position with respect to gravity.

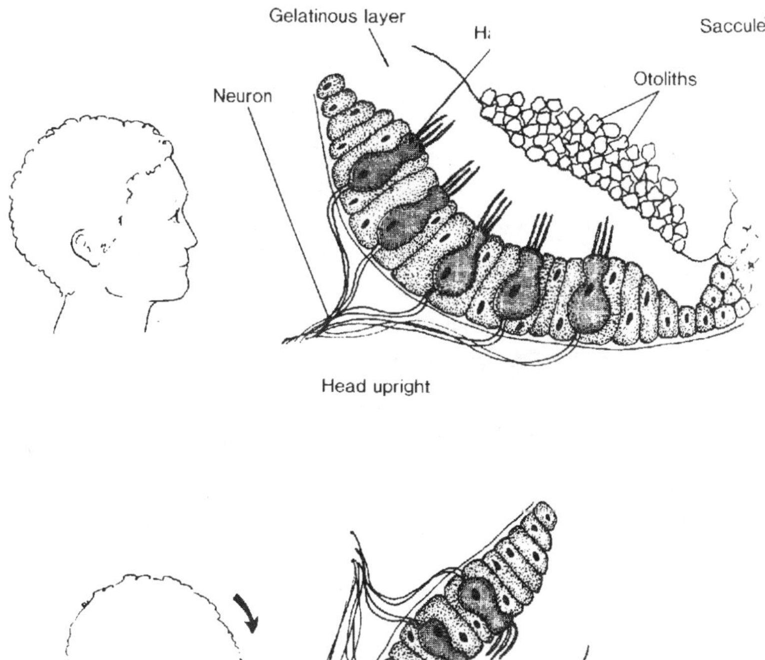

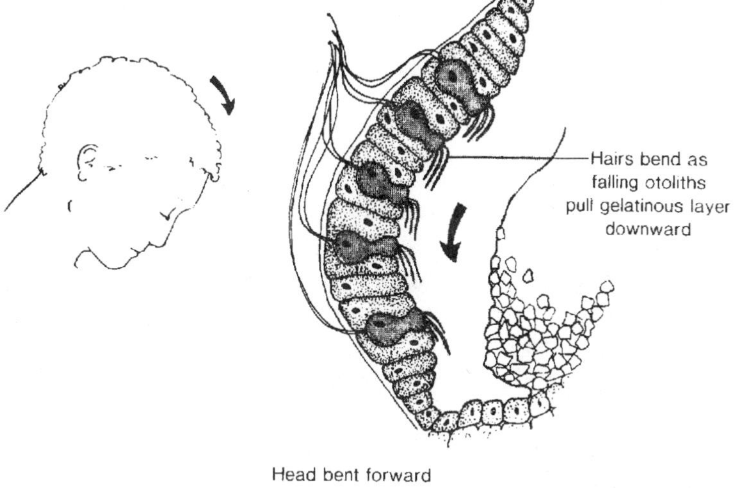

Figure 9.17B Detailed structure of the utricle. The otoliths are embedded in a gelatinous layer, lying on the surface of hair cells. Extending from the hair cells are neurons that lead to the brain. When the head is bent, the weight of the otoliths pulls on the gelatinous layer, bending the sensory hairs. Impulses from the hair cell neurons are transmitted to the brain, where the position of the head is interpreted. The same mechanism applies if the head is upright, bent backwards, sideways, or if the person is upside down.

The utricle ("small bag") and saccule ("little sac") contain receptors that inform the brain of both the position of the head with respect to gravity and linear acceleration or deceleration. More simply, these organs tell the brain if you are upside-down, right-side up, or if you are moving forward or backward.

Cytoplasmic extensions of the hair cells inside the utricle and saccule are embedded in a sticky, gelatinous substance that contains tiny crystals of calcium carbonate. These crystals are called **otoliths,** or "ear stones". If you are standing erect with your head upright, the weight of the otoliths compresses downward on the hair cells. If you bend forward at the waist, as if to touch your toes, the weight of the otoliths bends the hair cells forward. When you throw your head backward, the hair cells bend in that direction. Finally, if you are standing on your head, the weight of the otoliths pulls on the hair cells. Similarly, forward and backward thrusts of the head when the neck is not bent force the hair cells to sweep forward or backward. In each of these instances, the brain is informed of the position of the body with respect to gravity.

The semicircular canals function in detecting rotation of the head and angular motion. The three canals are oriented at right angles to each other, enabling the brain to detect movement in any direction. The canals contain a clear fluid and small crests of tissue to which hair cells are attached. Leading from each hair cell are sensory nerve fibers that pass to a branch of a cranial nerve. This nerve ultimately leads to the cerebellum. When you turn your head, the crests of tissue are displaced by movement of the fluid within the canals. This stimulates the hair cells, which transmit nerve impulses to the cerebellum. The cerebellum is informed of the movement of the head and, if necessary, to adjust the position of the body. This adjustment is necessary, for example, if you are running and suddenly change direction.

Another well-known experience is the feeling of dizziness after spinning around a few times. Once you stop spinning, the fluid in the semicircular canals continues to move back and forth like water being carried in a flat pan. This movement continues to stimulate the hair cells. which relay the impulses to the brain. You may have stopped spinning, but your brain doesn't think so. The dizziness gradually subsides after the fluid comes to rest.

In each of the organs of equilibrium, stimulation of the hair cells, as indicated in Fig. 9.17, sends nerve impulses along a branch of a cranial nerve to the cerebellum, medulla oblongata, and spinal cord. The cerebellum coordinates movements of the skeletal muscles of the head, neck, arms, and legs brought about by stimulation of motor neurons in the spinal cord. Neurons in the medulla oblongata transmit impulses to one of the cranial nerves that controls eye movements, allowing the eyes to track the movements of the body. Thus, through coordination of skeletal muscle and eye movements, the body is able to maintain balance and avoid injury.

Taste and Smell

Taste and smell are senses that result from chemical stimuli, as illustrated in Fig. 9.18. Sensory receptors for taste are located in *taste buds* found mostly on the tongue. There are four basic tastes humans can detect: sweet, sour, salty, and bitter. Our ability to detect a number of other tastes, however, probably results from a combination of two or more of the basic tastes. The various substances we taste must be in solution. This is one of the functions of saliva in the mouth. Nerve impulses from stimulated taste buds are transmitted along cranial nerves to areas on the lateral sides of the cerebrum where the tastes are interpreted.

The sense organs of smell are found in epithelial tissue, lining the upper portion of the nasal cavities. Neurons called **olfactory cells** lie within the epithelium and are the

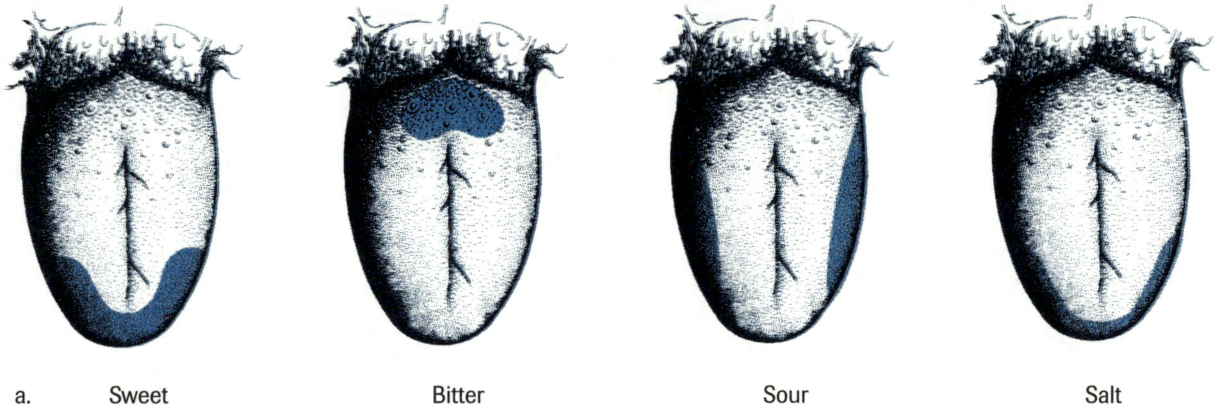

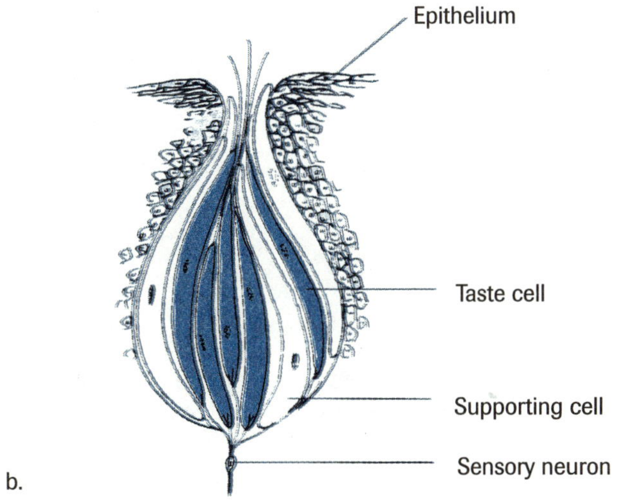

Figure 9.18 (a) The distribution of taste buds on the tongue. The four basic taste sensations are sweet, bitter, sour, and salty. (b) A taste bud of the tongue. Taste stimuli pass along the sensory neuron and then through cranial nerves to the brain, where the taste is interpreted.

actual sensory receptors of odors. When stimulated by a chemical that has an odor, the olfactory cells transmit nerve impulses along a cranial nerve to the anterior portion of the cerebrum for interpretation. We have millions of olfactory cells, which are capable of detecting thousands of different odors. It is interesting that we remember odors quite well and recognize them if we smell them at a later time. This applies with equal emphasis to dirty sneakers or Chanel No. 5.

Chapter Summary

1. In all animals except sponges, sensation and response result from the functioning of neurons, or nerve cells. In plants, sensation and response occur primarily through the influence of hormones and turgor pressure. The major plant responses are turgor movements and tropisms.

   Some bacteria and protists possess membrane receptors that allow orientation movements toward or away from a stimulus. Examples of such movement include phototaxis and chemotaxis.

   Key Terms: *hormone, p. 184*
   turgor movements, p. 185
   tropism, p. 185

Concept Questions: What are the general differences between a tropism and a taxis? What are the general functions of these responses?

2. Evolution of the animal nervous system reveals the progressive development of the head, or anterior, end of most organisms. The nervous systems of most vertebrates are quite similar; in birds and mammals, brain size relative to body size is much greater than that in fishes, amphibians, and reptiles.

   Key Terms: *reflex, p. 186*
   taxis, p. 187
   cephalization, p. 188

   Concept Question: How does the lack of a brain affect the sensation-response mechanism in some of the simpler invertebrates, such as cnidarians?

3. The nervous system of vertebrates consists of the central nervous system and the peripheral nervous system. These systems are comprised mainly of sensory neurons, motor neurons, and interneurons.

   The structural unit of the nervous system is the neuron, which consists of a cell body, dendrites, and an axon. Within a neuron, the nerve impulse travels from dendrites to the cell body to the axon.

   Key Terms: *neuron, p. 189*
   sensory, p. 189
   motor, p. 189
   effector organ, p. 190
   nerve, p. 190

   Concept Questions: A basic concept in biology is the interrelationship between structure and function. Relate this concept to the neuron, or nerve cell.

4. An unstimulated neuron has a high concentration of potassium ions in its cytoplasm; fluid outside the neuron has a high concentration of sodium ions. This results in the neuron membrane being polarized, a condition that is altered when the membrane is stimulated and the sodium ions and potassium move in opposite directions across the membrane. Movement of sodium ions and potassium ions into and out of the neuron creates the nerve impulse, which is transmitted along the length of the nerve fiber as successive depolarizations of the membrane. The sodium potassium pump is involved in restoring the original polarity, or resting state, of the membrane.

   Neurons communicate by synapses, which may be electrical or chemical. Transmission of a nerve impulse across a chemical synapse involves chemicals called neurotransmitters that are released by the axon. The neurotransmitter stimulates the membrane of the adjacent neuron or muscle fiber. A neurotransmitter may be broken down by a deactivating enzyme or it may be transported intact back into the axon.

   Key Terms: *polarity, p. 191*
   membrane pump, p. 192
   synapse, p. 193
   neurotransmitter, p. 194
   deactivating enzyme, p. 194

   Concept Question: What is the general relationship between the electrical and chemical aspects of neuron function?

5. Reflexes are designated as either spinal reflexes or visceral reflexes; the nervous pathway from the sensory receptor to the effector organ is called a reflex arc, the functional unit of the nervous system.

   Key Terms: *reflex, p. 195*
   viscera, p. 196

   Concept Question: You have just stepped on a tack with your bare feet. In terms of nervous system function, relate the mechanism of your response.

6. The human brain is divided into the forebrain, midbrain, and hindbrain. Two of the major areas of the forebrain are the cerebrum and the hypothalamus. The cerebrum is the center of reasoning, memory storage, and control of voluntary movement of the body. The cerebrum also analyzes and processes most of the information coming into the central nervous system.

   According to the lateralization concept, each of the two cerebral hemispheres controls different body functions. The right hemisphere largely controls spatial and manipulative functions; the left hemisphere controls speech and logical thought.

   The hypothalamus controls and monitors many involuntary functions of the body. It also manufactures hormones and produces releasing hormones that regulate the release of other hormones from the pituitary gland. The midbrain is a center for visual and head reflexes; it also helps control posture and is a relay center between the cerebrum and the hindbrain. Two major areas of the hindbrain are the medulla oblongata and the cerebellum. The medulla oblongata conducts nerve impulses between the spinal cord and higher brain centers, and is a center for several vital reflexes. The cerebellum coordinates skeletal muscle movement and controls balance of the body.

   Key Terms: *cerebrum, p. 196*
   cerebral cortex, p. 196
   hypothalamus, p. 198
   medulla oblongata, p. 199
   cerebellum. p. 199

   Concept Questions: What is the basic relationship between the cerebrum and the hypothalamus? Do you think it would be possible for the medulla oblongata to function without input from the cerebrum? Explain.

7. The spinal cord gives rise to the spinal and autonomic nerves. Nerve impulses coming into the spinal cord can be relayed to the brain. Both spinal and visceral reflexes involve the spinal cord.

   The human peripheral nervous system consists of spinal nerves, cranial nerves, and autonomic nerves.

   The autonomic nervous system controls the involuntary functions of the body. Autonomic nerves are responsible for visceral reflexes. The sympathetic and parasympathetic divisions of the autonomic nervous system have opposing effects on a body organ; that is, if the sympathetic system is excitatory, the parasympathetic system is inhibitory and vice versa. This opposing action is a major factor in regulating and maintaining homeostasis.

   Key Terms: *meninges, p. 200*
   visceral reflex, p. 201

   Concept Questions: Cite a specific example of the opposing effects of the autonomic nervous system. Why is this opposition necessary?

8. Various sensory receptors are found in both invertebrates and vertebrates. In most vertebrates, the special senses—vision, hearing, equilibrium, taste, and smell—involve complex sensory receptors and certain cranial nerves, which relay sensations from the sensory receptors to specific areas in the brain for interpretation.

Structures of the human eye are specialized for directing light rays to the retina, where nerve impulses are relayed along the optic nerve to the brain. The special photoreceptor cells in the retina are the rods and cones. Cones contain pigments that are sensitive to the colors red, blue, and green.

Hearing in humans involves structures in the outer ear, middle ear, and inner ear. The functional unit of hearing is the organ of Corti, located in the cochlea. Nerve impulses in the ear result from stimulation of hair cells by fluid movement in the cochlea. The impulses are relayed to the brain along a cranial nerve. Pressure changes in the ear are relieved by the auditory or eustachian tube.

The sense of equilibrium is detected by the utricle, saccule, and semicircular canals in the inner ear. The utricle and saccule detect the pull of gravity and the effects of linear motion through the shifting of otoliths. The semicircular canals detect rotation of the head and angular motion through the movement of fluid within the canals. Shifting of the otoliths in the utricle and saccule and movement of the fluid in the semicircular canals stimulate hair cells that relay impulses along a cranial nerve to the cerebellum, medulla oblongata, and spinal cord.

Taste and smell are chemical senses. The four basic tastes are sweet, salty, sour, and bitter. Taste receptors are taste buds, which transmit impulses along cranial nerves to the cerebrum for interpretation. The receptors for smell are olfactory cells, which transmit impulses along a cranial nerve to the cerebrum.

Key Terms: *cornea, p. 204*
retina, p. 204
iris, p. 204
lens, p. 204
cochlea, p. 206
hammer, p. 206
anvil, p. 206
stirrup, p. 206
otoliths, p. 208
olfactory cells, 208

Concept Questions: Give the anatomical locations of the major structures of the human eye. How is the eye functionally like a camera?

Basically, how are sound vibrations conveyed from the outer ear to the brain?

What are the functional differences between the semicircular canals and the utricle and saccule?

You are on the edge of a high cliff. You lose your balance and almost go over the edge, but manage to right yourself before falling. What general physiological mechanisms have just saved your life?

Plant and Animal Function

Chapter Ten

In discussing the organizational levels of life in Chapter 6, we noted that an arrangement of different tissues forms an organ, and that organs function together to make up a body system. The basic plant tissues are arranged to form roots, stems, leaves, and reproductive structures of higher plants. In the animal kingdom, epithelial, connective, muscle, and nervous tissues constitute the organs of most animals, especially vertebrates. Organs usually function as parts of a system, such as the nervous, respiratory, circulatory, and digestive systems of various animals. In plants, we can refer to the root system, transport system, or reproductive system. Each of these levels, as we indicated in earlier chapters, reflects the functioning of individual cells.

Tissues, organs, and systems not only make up the structure of the individual plant or animal, but also function in concert to sustain the life of the organism. The functioning, or physiology, of the organizational components that maintain the entire organism is in part an expression of sensation-response and homeostatic mechanisms. This chapter examines the physiology of some of these organizational levels in various organisms.

Plant Function

Nutrients

Water and specific organic and inorganic nutrients are required by all plants. Some nutrients are required in large amounts and others in small or even trace amounts. Most plants must acquire seventeen different elements, which are essential to the structure and function of hormones, enzymes, chlorophyll, membranes, and other cellular constituents. Elements required in relatively large amounts are called **macronutrients,** and include carbon, hydrogen, nitrogen, oxygen, phosphorus, sulfur, potassium, calcium, and magnesium (Fig. 10.1). Plants also require small amounts of **micronutrients,** which include boron, chlorine, copper, iron, manganese, molybdenum, nickel, and zinc. These same macronutrients and micronutrients are required by most algae and fungi as well. Plants take in these elements, except for carbon, hydrogen, and oxygen, in the form of naturally occurring inorganic ions referred to as *minerals.* For example, some of the positively charged ions are K^+ (potassium), Mg^{2+} (magnesium), Ni^{2+} (nickel), and Zn^{2+} (zinc); negatively charged ions include NO_3^- (a form of nitrogen), Cl^- (chlorine), SO_4^{2-} (a form of sulfur), and $H_2BO_3^-$ (a form of boron). Carbon, hydrogen, and oxygen are acquired mainly from air and water.

Plant minerals are dissolved in a watery solution in the soil and are absorbed by the plant through its root system (Fig. 10.2). Without adequate water, minerals cling to the soil particles and cannot be absorbed by the plant. Quite often, a plant absorbs elements, such as silicon and aluminum, that are present in the soil, but are not required as nutrients. The movement of water into the root cells is aided by the uptake of min-

Figure 10.1 Essential Plant Nutrients

Element	*Molecular Form Absorbed*	*Source*	*Major Functions*
Macronutrients (usually 0.1% or more of dry weight of plant)			
Carbon	CO_2	Air	Major element in organic molecules; photosynthesis
Oxygen	O_2	Air	Major element in organic molecules; cellular respiration
Hydrogen	H_2O	Soil	Major element in organic molecules
Nitrogen	NO_3^- or NH_4^+	Soil	Major element in proteins, nucleic acids, and chlorophyll
Potassium	K^+	Soil	Principal positive ion inside cells; control of stomatal opening and closing; enzyme activation
Phosphorus	$H_2PO_4^-$ or HPO_4^{2-}	Soil	Major element in nucleic acids, phospholipids, and electron carriers in chloroplasts and mitochondria
Calcium	Ca^{2+}	Soil	Component of adhesive compounds in cell walls; important in control of membrane permeability; enzyme activation
Magnesium	Mg^{2+}	Soil	Component of chlorophyll; enzyme activation; ribosome stability
Sulfur	SO_4^{2-}	Soil	Component of proteins and many coenzymes
Micronutrients (usually less than 0.01% of dry weight of plant)			
Iron	Fe^{2+} or Fe^{3+}	Soil	Needed for synthesis of chlorophyll; component of many electron carriers
Chlorine	Cl^-	Soil	Required for photosynthesis
Manganese	Mn^{2+}	Soil	Required for photosynthesis; enzyme activation
Molybdenum	MoO_4^{2-}	Soil	Required for nitrogen metabolism
Copper	Cu^{2+}	Soil	Enzyme activation; component of electron carriers in chloroplasts
Boron	BO_3^{3-} or $B_4O_7^{2-}$	Soil	Involved in sugar transport
Zinc	Zn^{2+}	Soil	Enzyme activation; protein synthesis; hormone synthesis

Reprinted from *Biology: Life on Earth,* Fourth Edition, by Gerald and Teresa Audesirk, 1996, Prentice-Hall, Inc.

eral ions. A higher concentration of solute (ions) within the root cells than in the soil establishes a concentration difference that allows *osmosis*—the inward diffusion of water—to occur. The ions also have an effect on the permeability of plant membranes. A deficiency of calcium ions, for instance, appears to increase the permeability of membranes, allowing some of the other ions to leak to the outside of the plant.

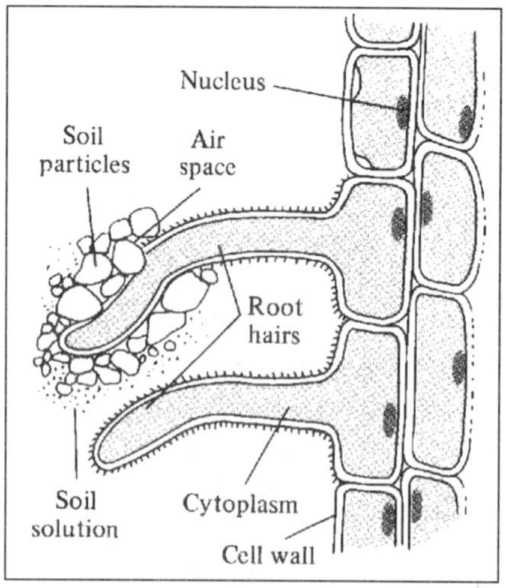

Figure 10.2 Minerals, dissolved in solution in the soil particles, enter the plant through the root hairs. The accumulation of minerals in the root facilitates the uptake of water by osmosis.

Dissolved minerals may enter the root cells by simple diffusion across the cell membranes or by active transport. Diffusion involves the passive movement of ions across the root membrane. The ions diffuse from where they are in greater concentration to where they are in lesser concentration—either into or out of the cells. Being a passive process, diffusion does not require an input of energy by the cells.

Most ions and other solutes, however, encounter difficulty in diffusing across a membrane because they must move in a direction opposite to the normal flow of diffusion; that is, they must move from where they are in lower concentration to where they are in higher concentration. To do this, the solutes are assisted by transport protein molecules embedded in the root cell membrane. Some of these proteins actively transport the solutes across the membrane, and other proteins function as selective gates that allow some ions to pass through, but not others.

Another active transport mechanism involves the use of hydrogen ions (H^+). Recall from Chapter 7 that ATP is generated in photosynthesis and cellular respiration by the diffusion of hydrogen ions across the thylakoid and mitochondrial membranes. In root cells, there is a reverse of this process in that the energy of ATP is used to pump hydrogen ions to the outside of the cells. This creates a high concentration of hydrogen ions in the solution surrounding the cells, and also causes the inside of the cells to be negatively charged relative to this solution outside. Because hydrogen ions are in high concentration outside the root, they tend to diffuse back toward the inside of the root. This inward diffusion of hydrogen ions represents a source of energy for the transport of solutes into the root cells. Many negatively charged minerals, for example, are carried along with the positively charged hydrogen ions as they diffuse into the root cells. This mechanism is known as **cotransport;** that is, the transport of one substance coupled, or linked, to the transport of a different substance. In addition, the negatively charged interior of the root cells aids in the uptake of minerals by attracting other positively charged ions, such as Ca^{2+} (calcium) and K^+ (potassium).

Figure 10.3 A yellow pitcher plant.

Various plant species that exist in nutrient-poor environments have evolved adaptations for acquiring the nutrients they need for normal growth. One of the more exotic adaptive methods of obtaining nutrients by plants is displayed by various insect-trapping species Pitcher plants, for example, possess specialized, often brightly- colored, leaves that resemble vases of various shapes (Fig. 10.3). Insects attracted by the bright colors or by sweet nectar at the base of the pitcher often fall into a pool of water and digestive enzymes at the pitcher's bottom. The inner walls of the pitcher are slippery, making it virtually impossible for the insects to climb out. Even if they make it to the top, the insects are confronted with a barricade of stiff, downward-pointing hairs. The trapped insects eventually drown, and their soft body parts are digested by the plant enzymes. Cells lining the inner surface of the pitcher then absorb the digestive solution. Pitcher plants are found in various regions of North America and in tropical areas of the Pacific Ocean.

Figure 10.4 Venus Flytrap.

The Venus flytrap probably is the most familiar of the insect-trapping carnivorous plants (Fig. 10.4). The flytrap leaf is hinged in the middle and bears stiff spines along each edge. Three tiny trigger hairs are located on the surface of each half of the leaf. When an unwary insect brushes against the hairs, the leaf snaps shut, trapping the insect in a spiny prison. Leaf cells secrete digestive enzymes that digest the soft body parts, which are then absorbed. The combination for closing the leaf is specific: one hair must be touched twice or two hairs touched once each. Closing of the flytrap leaf requires a considerable quantity of energy in the form of cellular ATP.

Research has indicated that the specific nutrients acquired from the digested insects are amino acids, the building blocks of proteins. Enzymes secreted by insect-trapping plants break down an insect's proteins into their constituent amino acids, which are absorbed by the leaf cells for use by the plant. Because insect-trapping plants are photosynthetic, insects are not major sources of food and energy. But they apparently serve only to provide the plants with nitrogen, phosphorus, and other minerals when the soil is nutrient deficient, an example, as mentioned, of an adaptation to a nutrient-poor environment.

The Need for Water

Water enters plant root cells by osmosis across the root cell membrane. This inward movement of water results from two conditions that exist within the xylem. First, as mentioned, osmosis occurs when roots transport minerals into the xylem. This transport creates a high solute concentration within the xylem, which causes water in the soil to move inward across the root cell membranes and into the xylem. This creates an upward pushing force in the roots called **root pressure.** Second, the evaporation of water from the leaf stomata, a process called **transpiration,** reduces the pressure in the root by removing water from the root xylem (Fig. 10.5). This lower water pressure within the xylem establishes a driving force for the absorption of soil water into the root cells. As the low water pressure develops within the xylem, the root pressure rapidly disappears.

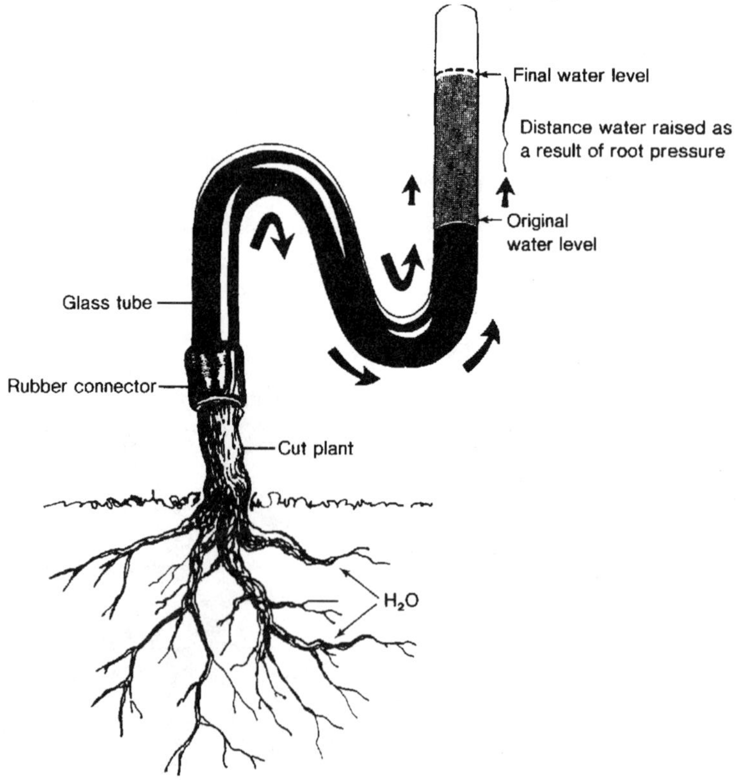

Figure 10.5 Demonstration of root pressure. A curved glass tube, partially filled with water, is sealed to a cut plant by a rubber connector. Fluid uptake by the roots raises the level of water in the tube until the downward pressure of the water equals the root pressure.

The uptake of water and minerals is greatly heightened by the thousands of root hairs that extend from the root epidermis. Root hairs enormously increase the root's surface area and also increase the amount of soil making contact with the root. Once inside the roots of vascular plants, water and minerals move laterally along various routes and then enter the xylem to be distributed upward throughout the stems and leaves.

The amount of water required by many plants is all but incredible. From the xylem, water enters small spaces in the leaves and then evaporates through the stomata into the atmosphere. Some species of trees may lose up to 379 liters (100 gallons) of water in this manner every day. If you had the same water requirements as some plants, you would have to drink nearly 38 liters (10 gallons) of water daily. Evaporation of water from the plant epidermis by direct heat of the sun helps cool hot-climate plants, and adequate amounts of water prevent plasmolysis and wilting.

Another critical function of water is its role in gas exchange. The membranes of living cells normally are bathed in a thin film of water, which is necessary for the passage of oxygen and carbon dioxide across the membranes. These gases do not diffuse efficiently across dry surfaces because the lack of water adversely affects the structure of the cell membrane. In plants, this means the leaf cells that admit carbon dioxide must be moist.

Water Transport

Although plants lose a large volume of water through evaporation, water, as mentioned, is replaced by osmosis or absorption through the roots. Root pressure is one of the factors involved in the upward movement of water in the xylem of only a few plants. In most plants, root pressure is of minor importance, and most plants do not generate root pressure at all. Water in the xylem above the roots exerts a downward pressure by virtue of its weight and the pull of gravity that would seem to be sufficient to force water back *out* of the roots. This is not the case, however, since root pressure in some plants is great enough to support water in the xylem and to push it forward and upward into the plant. But if the roots are killed or deprived of oxygen, root pressure ceases. This indicates that root pressure is not simply a passive process, but one that requires energy. The source of that energy is ATP provided by cellular respiration in the root cells.

Root pressure may account for the upward movement of water to the top of some very short plants, but it is not sufficient by itself to raise a column of water to the top of other plants that may be up to 91 meters (300 feet) tall, such as some California redwoods. Root pressure is measured in units called **atmospheres;** one atmosphere is equivalent to the pressure of air at sea level. One atmosphere of pressure can raise a column of water to a height of about 10 meters (32 feet). Measurements in several species of plants indicate that root pressure seldom is greater than 2 atmospheres, which is enough pressure to raise water only three or four feet because of resistance in the xylem, from the weight of water and from the pull of gravity.

This leads us to the current theory of water transport, which is based on the process of *transpiration,* the loss of water by evaporation from the stomata on the leaf surface. This theory, illustrated in Fig. 10.6, proposes that water is "pulled" through the plant, entering through the roots and escaping through the leaves. As long as transpiration continues, water will enter the roots and continue moving upward through the plant. This theory is based on the unique cohesive, adhesive, and evaporative properties of water. As a result of the hydrogen bonding that links water molecules, there is cohesion between the water molecules. As water molecules in the leaf cells are lost by evaporation through the stomata, these molecules draw the water molecules below them upward. These molecules, in turn, pull on the water molecules below them, and so on down through the plant to the roots. There is, then, an unbroken column of water extending from the leaves downward to the roots. As pointed out, water from the soil enters the roots partly because transpiration creates a water pressure in the xylem lower than that of the water in the soil.

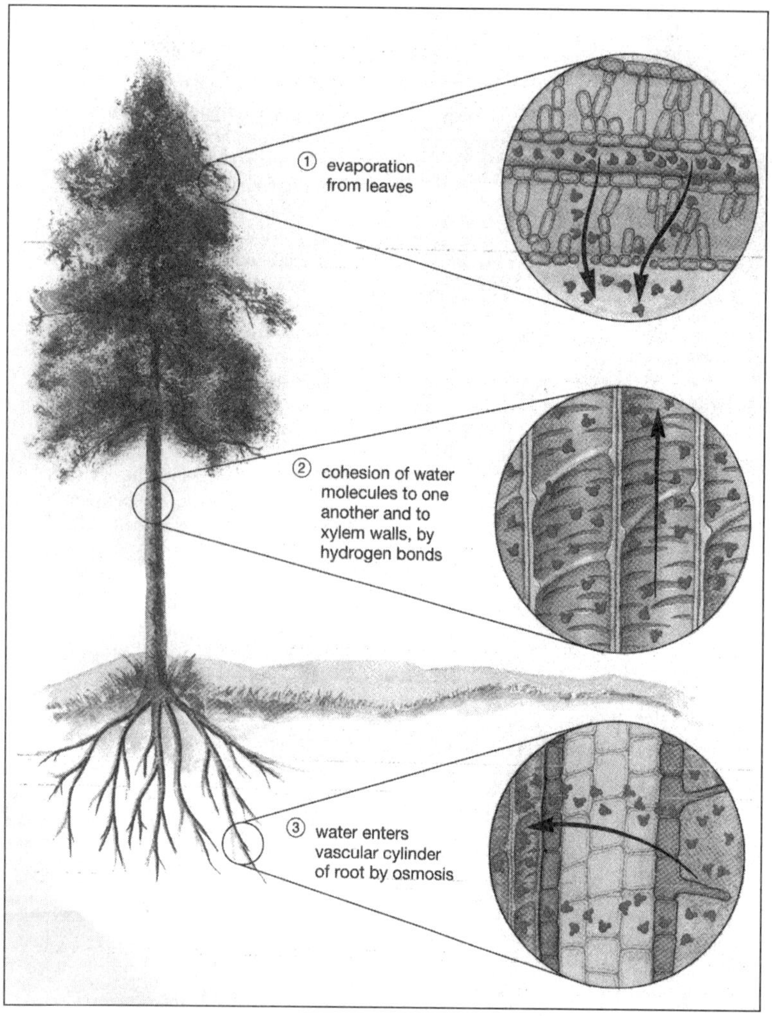

Figure 10.6 The cohesion-tension theory of water flow from root to leaf in xylem. (1) Water molecules evaporate out of the leaves, and other water molecules replace them from the xylem of the leaf veins. (2) Within the xylem, hydrogen bonding holds nearby water molecules together so firmly that the column of water behaves very much like a rope. The top of the "water rope" is pulled up by evaporation, and the rest of the rope comes along as well, all the way down to the roots. (3) As the molecules of the water rope retreat up the xylem in the roots, the decreased water concentration within the root xylem and the surrounding extracellular space causes water to enter from the soil water by osmosis, thus steadily replenishing the bottom of the rope. Reprinted from *Biology: Life on Earth,* Fourth Edition, by Gerald and Teresa Audesirk, 1996, Prentice-Hall, Inc.

In addition to cohering to each other, water molecules also adhere to the inner walls of the xylem. This attraction of water molecules for each other and for other molecules adds support for the water column. With all of these forces acting together, water can be transported upward throughout the plant. It should be emphasized, however, that the key to the upward movement of water in plants is transpiration. As long as there is a loss of water from the leaf cells, a tension, or "suction" effect, is created that allows water to move upward from the roots to the leaves. In turn, the pull of evaporation resides in the strong cohesiveness between individual water molecules. Behind the entire process is the energy of sunlight that drives evaporation, and, in a few small plants, the energy of ATP that maintains root pressure.

Food Transport

Food, the organic products of photosynthesis, is transported within vascular plants by a system of phloem tissues. The transport process is called **translocation.** Most of the solution, or sap, transported by phloem consists of the disaccharide *sucrose,* along with small amounts of amino acids and various hormones. Translocation may occur in almost any direction within the plant and follows what is called a "source-to-sink" direction. A *source* is a location, predominantly photosynthesizing leaves or storage tissue, where food is available for transport. A *sink* is the location, such as roots or developing tissues, to which the food is transported.

Translocation within the phloem of angiosperms has been explained in part in terms of the *pressure-flow,* or *bulk-flow theory,* as seen in Fig. 10.7. According to this theory, sucrose first is loaded into the phloem tubes by passing from its source, such as photosynthesizing leaf cells, through a pathway of cells that leads to the phloem. In some plant species, sucrose accumulates in the phloem by cotransport involving diffusion of hydrogen ions, as occurs in mineral uptake by the roots. The result is a very high concentration of sucrose in the phloem. This causes water from the surrounding xylem to be drawn into the phloem by osmosis, creating high turgor pressure at the source end of the phloem. At the sink end, the depletion of sucrose as it is being used by the tissues results in a loss of water and a lowering of turgor pressure. There is established, then, a pressure difference between the phloem cells at the source end and those at the sink end. This pressure difference causes the bulk, or mass, flow of sucrose all the way from the source to the sink.

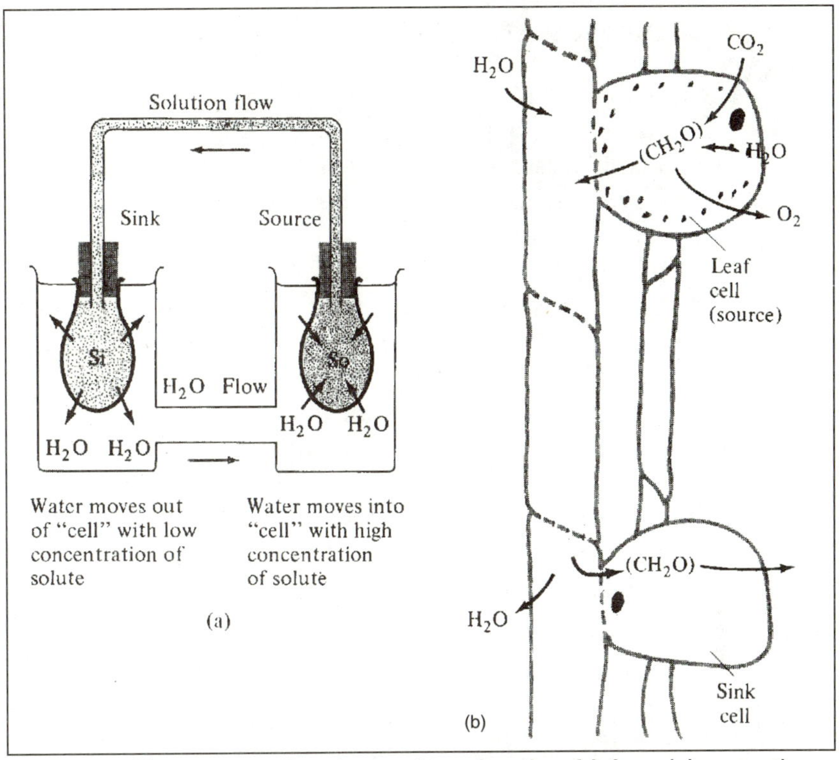

Figure 10.7 The pressure-flow theory of translocation. (a) A model suggesting how translocation occurs. Where sucrose is being produced at a source (on the right), such as leaves, osmotic pressure will be high and water will enter the phloem from the xylem by osmosis. Where sucrose is being used or stored (the "sink" at the left), osmotic pressure will fall and water will leave to reenter the xylem. The resulting difference in hydraulic pressure produces a flow of sucrose-laden fluid from the source of the sink. (b) The process depicted as it occurs in the plant. Leaves are a principal source of sucrose (CH_2O), and roots and fruits are sinks.

The pressure-flow theory satisfies many of the questions concerning translocation in angiosperms, but it may not be the complete story. For now, however, the pressure-flow theory is reflective of the best available evidence.

Animal Function

Compared to plants, animals, in general, are capable of greater movement by virtue of their muscular and nervous tissues. With their high metabolic rates, many animals require rapid delivery of oxygen to these tissues and to all metabolizing cells. Animals are heterotrophs and many of them have evolved complex systems for the physical and chemical breakdown of nutrients and the elimination of wastes. The physiological mechanisms of animals, like those of plants, are adaptations to a particular mode of existence. In this section, we will consider the physiological mechanisms of animals in relation to movement, circulation, respiration, and digestion.

Invertebrate Function

All animals display some type of movement which, like that in plants, results from an innate sensory-response mechanism. Even the sponges, which lack nerves, are capable of limited responses to stimuli. Sensation and response are carried out by the individual body cells. So limited are the movements of sponges that the ancient Greeks considered them to be plants. It was not until 1765 that sponges were clearly established as animals.

Sponges, which are aquatic, are dependent on the surrounding water for oxygen and food. As shown in Fig. 10.8, oxygen enters the body cells by diffusion from the water flowing through the body pores. Carbon dioxide, the waste product of cellular respiration, diffuses out of the cells into the surrounding water. In addition, food particles in the water are filtered out by specialized cells and some of the particles are phagocytized. Waste products, including ammonia, are swept out of the cells by the water currents.

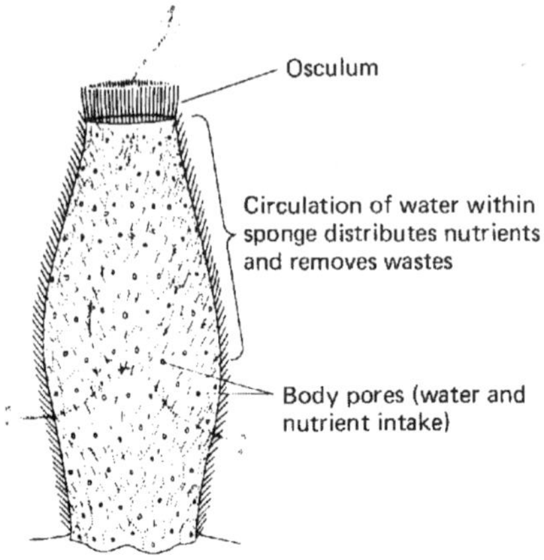

Figure 10.8 In sponges water and nutrients enter the spongocoel through tiny body pores; within the sponge, water serves as a circulatory fluid to distribute the nutrients and to remove wastes through the "mouth", or osculum.

In comparison to sponges, cnidarians, such as hydras and jellyfish, are capable of considerably more complex body movement. Most cnidarians possess specific cell types that contain a contractile fiber. These cells allow some of the animals to extend, contract, and bend over; among the jellyfish, the contractile cells are used for swimming. Hydras can inch along like worms and also can turn somersaults.

Respiratory gas exchange in cnidarians occurs across the body surface by diffusion. Oxygen in the surrounding water diffuses into the body cells, while carbon dioxide diffuses from metabolizing body cells into the water. Cnidarians feed on a variety of crustaceans and fish, which are captured by stinging cells located primarily in the tentacles. After being forced into the mouth opening by the tentacles, food is digested by enzyme-secreting cells within the hollow body cavity, although cnidarians have no discrete digestive system. Wastes are released from the body cavity into the surrounding water.

Almost all invertebrate groups above the level of the cnidarians possess well-developed muscles, which are similar to the smooth and skeletal muscles of vertebrates; discrete circulatory and digestive systems first appear in some of the roundworms. The body wall of nematode worms consists only of **longitudinal muscles** running the length of the body. Consequently, nematodes are capable only of thrashing movements. In many other invertebrates, the muscular system consists of a layer of circular muscle in addition to longitudinal muscle. Contraction of **circular muscle** lengthens a body structure, while contraction of longitudinal muscle shortens it. In the more advanced invertebrates, such as some annelids, mollusks, and arthropods, some muscles run at various angles, providing more effective muscle control and a broad a range of movement. These muscles act in conjunction with the longitudinal and circular muscles.

Animal circulatory systems are designated as either *open* or *closed,* as can be seen in Fig. 10.9. In an **open circulatory system,** blood is not confined totally within vessels, but flows through open spaces, or sinuses, during part of its circuit through the body. In this manner, blood directly bathes the internal organs and then reenters the blood vessels. All of the arthropods and most mollusks have open circulatory systems. In a

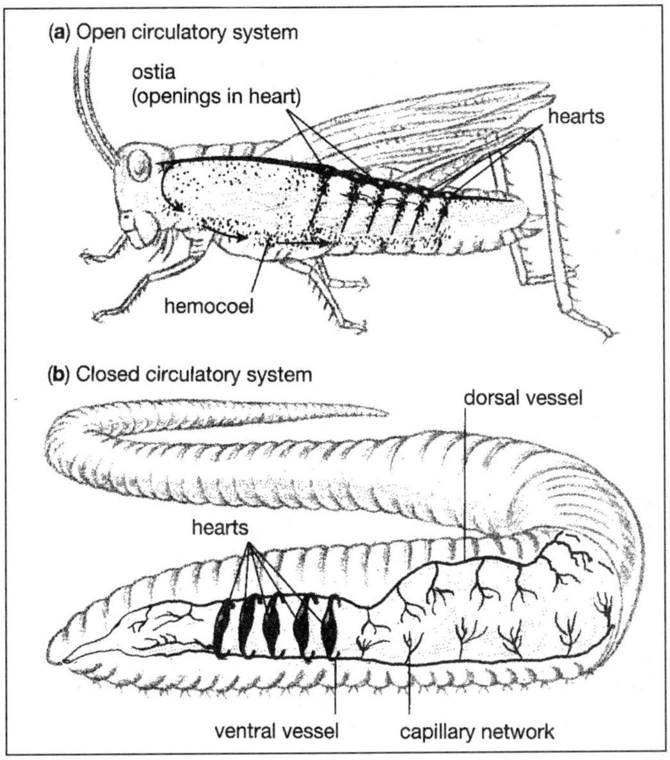

Figure 10.9 Open and closed circulatory systems (a) In the open circulatory system of insects and other arthropods, a series of hearts pumps blood through vessels into the hemocoel, where blood directly bathes the organs. When the hearts relax, blood is sucked back into the vessels through openings called ostia, guarded by one-way valves. When the hearts contract, the valves are pressed shut, forcing the blood to travel out through the vessels returning to the hemocoel. (b) In a closed circulatory system, blood remains confined to the heart and the blood vessels. In the earthworm, five contractile vessels serve as hearts and pump blood through major ventral and dorsal vessels from which smaller vessels branch off. Reprinted from *Biology: Life on Earth,* Fourth Edition, by Gerald and Teresa Audesirk, 1996, Prentice-Hall, Inc.

closed circulatory system, blood remains within blood vessels throughout its circuit in the body. Substances are exchanged between the blood and body cells by passing across the blood vessel walls. Many invertebrates and all vertebrates have a closed circulatory system.

Associated with animal circulatory systems—both open and closed—is a pumping structure, or *heart*. Some animals, such as squids, octopuses, and the vertebrates, have a single heart, while various worms and the insects have several hearts. In some cases, these "hearts" are major blood vessels that pump blood by contracting rhythmically. Blood cells that carry oxygen are found in many invertebrates, even at the level of some of the more primitive worms.

Some invertebrates, including sponges, cnidarians, and flatworms, exchange respiratory gases by diffusion across the body wall; no separate respiratory organs are present. Many other aquatic invertebrates, however, have distinct respiratory structures called *gills* that extract oxygen from the surrounding water and release carbon dioxide. Gills are found in such diverse invertebrates as marine worms, clams, crayfish, and squids. Most spiders possess modified gill-like structures called *book lungs,* which consist of thin stacked plates. Gas exchange occurs by diffusion between the blood circulating in the plates and the air between the plates.

Most terrestrial arthropods respire using a *tracheal system,* a network of small internal tubes that branches throughout the body. Oxygen enters the system by diffusing through openings on the body wall and is distributed directly to the tissues. Carbon dioxide released from metabolizing cells diffuses in the opposite direction through the tubes and out of the body. Some aquatic insects have evolved unusual methods for obtaining oxygen, such as forming air bubbles on the body and breathing the trapped oxygen when they dive underwater. Examples of the respiratory structures in various invertebrates are shown in Fig. 10.10.

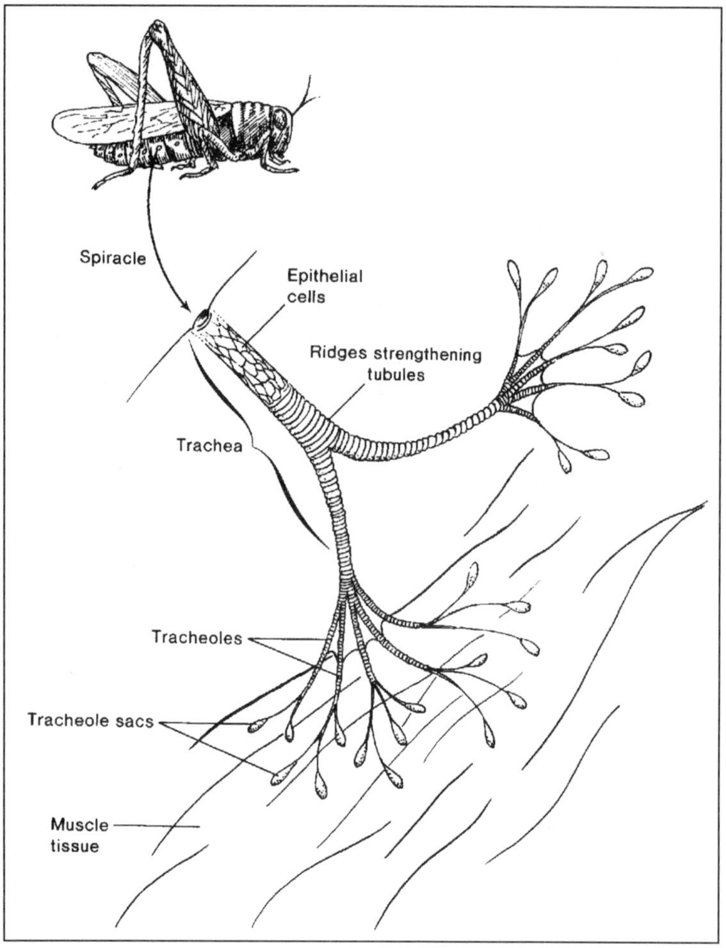

Figure 10.10A Insect tracheal system. Within the body tissues, oxygen and carbon dioxide are exchanged across the alveolar sacs.

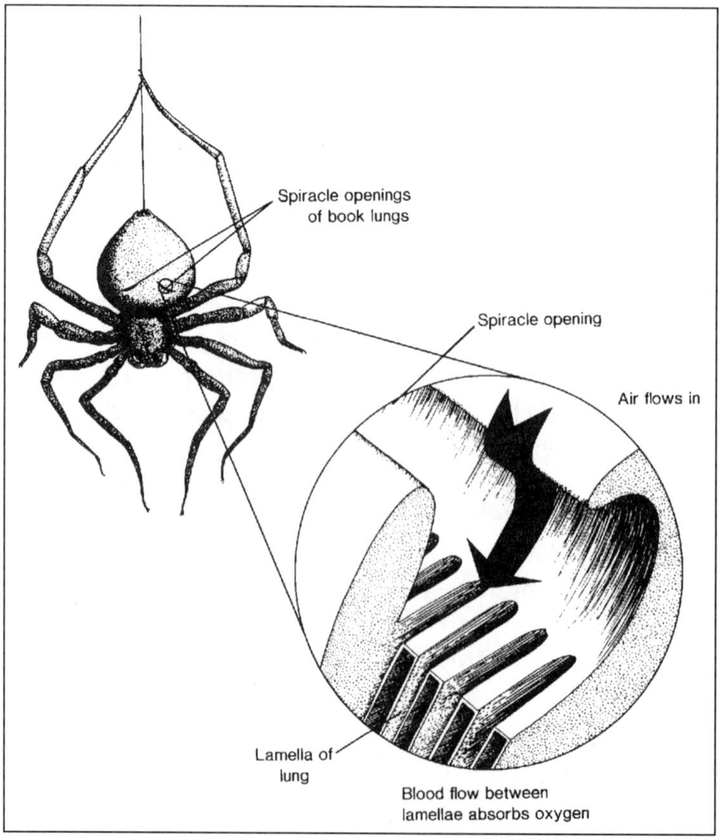

Figure 10.10B A spider, showing the location and basic structure of its book lung. The lamellae, or plates, greatly increase the surface area of the lung.

Digestive function among invertebrates ranges from the phagocytic food-trapping cells of sponges to the complete digestive systems of invertebrates above the level of flatworms. Complete digestive systems have a mouth and anus. Some flatworms have a digestive cavity similar to that of cnidarians, and food is broken down by enzyme-secreting cells; in flatworms, however, the digestive cavity branches within the body. This branching is the beginning of a trend evident in the digestive systems of more advanced animals. Branches of a digestive system increase the surface area available for the absorption of food. The digestive systems of some invertebrates are elongated and coiled within the body cavity. This extensive length also provides additional surface area for absorption. One group of flatworms—the parasitic tapeworms—do not have a digestive system, but simply absorb the products of digestion from the intestines of their hosts.

Within the digestive tract, food is broken down physically into smaller fragments by the churning action of the muscular wall and, in some invertebrates, by grinding organs. In addition, food is broken down chemically by digestive enzymes that hydrolyze large organic molecules into smaller ones. Proteins, for example, are hydrolyzed to polypeptides and then to amino acids. This is necessary because large molecules usually cannot be absorbed by the body cells. Undigested food residue is eliminated through the anus or through other body openings.

Vertebrate Function

In the vertebrates, the muscular, circulatory, respiratory, and digestive systems are well-developed. There are, however, variations among the vertebrates with regard to some of the organs and structures in each of these systems. The structure of muscle is virtually identical in almost all vertebrates.

The vertebrate heart is a muscular organ that pumps blood. In the fishes, the heart has only two chambers—one **atrium** and one **ventricle** (Fig. 10.11). Amphibians and reptiles, however, have a three- chambered heart consisting of two atria and one ventricle. In the reptiles, the ventricle is partially separated by a muscular wall. In one group of reptiles—the crocodiles—the ventricle is completely separated into two chambers. All birds and mammals have a four-chambered heart with two atria and two ventricles. A four chambered heart is more efficient in separating oxygen rich and oxygen-poor blood, a condition more favorable to active organisms with high metabolic rates and delicately balanced homeostatic mechanisms.

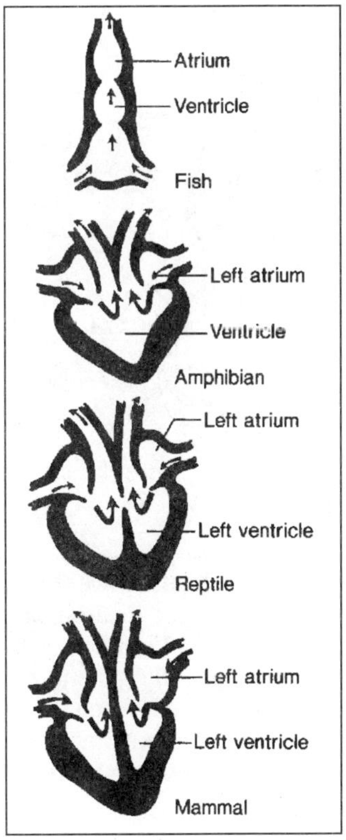

Figure 10.11 Structure of different vertebrate hearts (arrows indicate direction of blood flow). Note that the fish heart has only two pumping chambers, an atrium and a ventricle. The amphibian heart has three chambers, and most reptilian hearts have four chambers, although the ventricles are only partially divided. The heart of mammals (and birds) is completely separated into four chambers.

Agnathans, fishes, and some amphibians respire by means of gills; all other vertebrates possess lungs. The development of lungs altered the route of blood circulation from that occurring in vertebrates with gills. Vertebrates with lungs have a double circulation; that is, blood is pumped to the lungs (pulmonary circulation) where it is oxygenated and then throughout the body (systemic circulation). In the gill circulation, blood flows throughout capillaries in the gills, across which oxygen and carbon dioxide are exchanged in the surrounding water.

The digestive systems of different vertebrates are fairly similar except for a few structural differences, such as types of teeth, length of the intestines, and shape and number of stomachs, all of which are mostly geared to maximizing the efficient digestion and absorption of food. The nonparasitic lampreys, however, lose the digestive tract as adults and die a few months later.

With these provisions in mind, let us look more closely at the muscular, circulatory, respiratory, and digestive systems of vertebrates by emphasizing mammalian systems, with particular emphasis on humans.

Muscular System

We pointed out in Chapter 6 that muscle is one of the major tissues of vertebrates in addition to epithelial, connective, and nervous tissues. Individual muscle cells are called **fibers,** which are capable of contracting, or shortening, their length. Following contraction, a **muscle fiber** returns to its original length. Muscle contraction results in motion, body support, and heat production. Vertebrates, as well as many invertebrates, possess three types of muscle tissue: *skeletal, smooth,* and *cardiac.*

Skeletal muscle is voluntary, with the exception of spinal reflexes, and functions primarily in movement and support of the body. Skeletal muscle forms most of the bulk of the body, accounting for about 40 percent of body weight in humans. Smooth muscle is mostly involuntary and is found in internal organs, intestines, blood vessels, and other tubular structures, such as the esophagus, ureters, oviducts, and bile duct. Cardiac muscle is also involuntary and is found only in the heart. The more than 600 muscles in the human body constitute an extremely large cellular mass that produces a high proportion of body heat. Muscular activity, such as exercise or shivering, can increase heat production immensely.

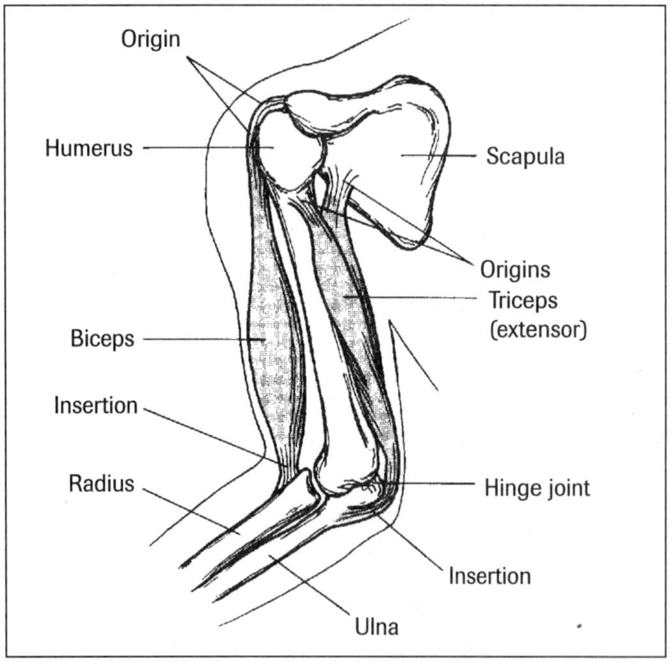

Figure 10.12 Arrangement of skeletal muscles across a joint (elbow). The *origin* of a muscle refers to its fixed, or immovable, end; the *insertion* is the movable end. Contraction of the biceps pulls the insertion toward the origin, lifting the forearm. During this action the triceps muscle is relaxed. When the triceps contracts, the biceps relaxes, and the forearm is extended. The skeletal muscles originate and insert on bones by means of *tendons.*

As illustrated in Fig. 10.12, movements of the arms, legs, neck, and other structures are possible because associated skeletal muscles extend across a *joint,* such as the shoulder, hip, knee, or elbow. A skeletal muscle is attached to separate bones on either side of a joint by tendons. A **tendon** is a tough connective tissue that binds muscles to bones. One end of a muscle is fixed, or immovable, and the other end is movable. As an example of skeletal muscle movement, the biceps and triceps muscles of the upper arm act together to move the forearm. Contraction of the biceps pulls the movable end of the muscle that is inserted on a forearm bone toward the immovable end, which originates at the shoulder, lifting the forearm. During this action, the triceps muscle is relaxed. The triceps also originates in the shoulder area and inserts on the forearm. When the triceps contracts, the biceps relaxes, and the forearm is extended. Many other skeletal muscle groups in the body have similar opposing muscles that provide for flexion and extension of body parts.

Skeletal muscle motion anywhere in the body requires stimulation by a specific nerve fiber. An otherwise healthy skeletal muscle ceases to function if its nerve supply is lost. Human skeletal muscles are controlled almost exclusively by the 31 pairs of spinal nerves that branch off either side of the spinal cord (Chapter 9). Smooth and cardiac muscle receive input from the autonomic nervous system, which controls most of the involuntary activities of the body. Many smooth muscle fibers, however, do not receive direct input from motor neurons and may contract in response to hormones, stretching, or the concentrations of oxygen and carbon dioxide.

Muscle Contraction

As discussed in Chapter 9, each skeletal muscle fiber is supplied with a neuron (axon) branch, which terminates on the fiber at a neuromuscular junction (Fig. 10.13). The tip of the axon releases a neurotransmitter that diffuses across the synapse located between the axon and the muscle fiber. Receptors on the muscle fiber bind the neurotransmitter, which stimulates the muscle cell membrane and the muscle fiber contracts. A deactivating enzyme that is released into the synaptic space breaks down the neurotransmitter molecules and the muscle fiber relaxes.

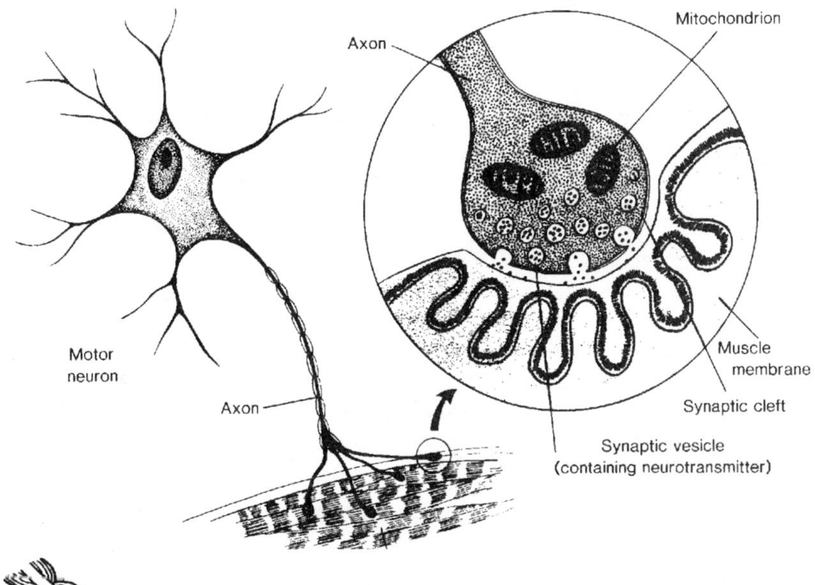

Figure 10.13A The neuromuscular junction, a type of synapse. The tip of the axon is separated from the muscle cell membrane by a microscopic space, or cleft. As the impulse from the motor neuron arrives at the tip of the axon, neurotransmitter, released from the synaptic vesicles, diffuses across the cleft and stimulates the membrane. In a fraction of a second, deactivating enzymes in the cleft break down the neurotransmitter, preventing further stimulation of the muscle.

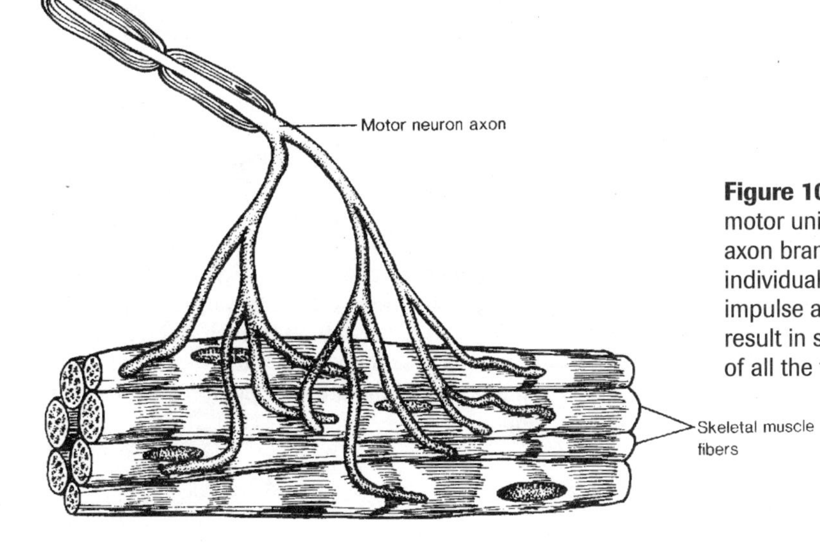

Figure 10.13B Diagram of a motor unit. The motor neuron axon branches to supply each individual muscle fiber. A nerve impulse along the neuron will result in simultaneous contraction of all the fibers in the unit.

Muscle shortening and lengthening once was assumed to be a simple stretching and relaxing phenomenon not unlike the action of a rubber band. But research, aided by the electron microscope, X-ray diffraction, and radioactive tracer techniques, has disproved this assumption. The electron microscope revealed the filamentous structure of muscle, and X-ray diffraction made possible the determination of the three-dimensional structure of muscle protein. A radioactive tracer is a radioactive isotope that can be introduced into a compound and followed by a scanner as the compound is distributed and metabolized in the body. The distribution and metabolism of various compounds involved in muscle function have been analyzed using radioactive tracers.

Individual muscle fibers consist of microscopic **actin** and **myosin** protein filaments (Fig. 10.14). The thin actin filaments and the considerably thicker myosin filaments are arranged lengthwise in a muscle so that they partially overlap. This arrangement is the key to the contraction and relaxation mechanism of a muscle. The myosin filaments possess tiny **cross-bridges,** oarlike protuberances that make connections with the actin filaments. During contraction, the cross-bridges pull the thin filaments past the thick ones, shortening the muscle. This action is somewhat like rowing a boat. The cross-bridges

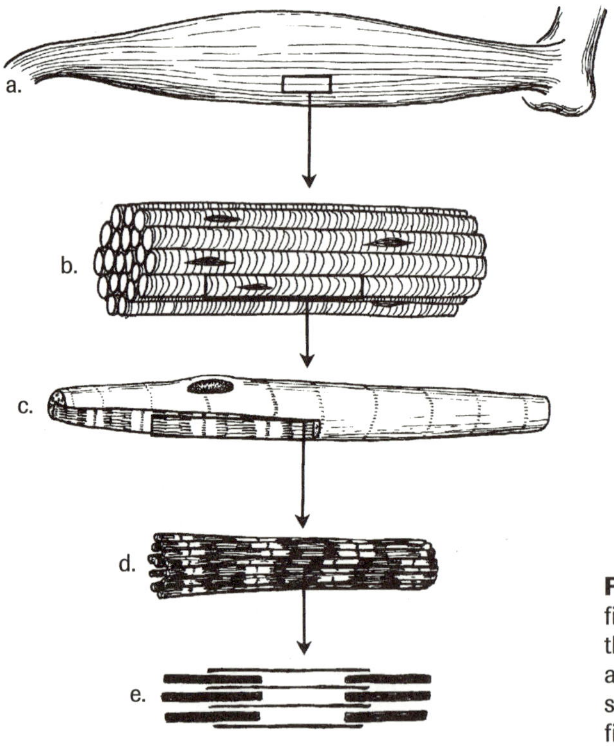

Figure 10.14A (a) Striated (skeletal) muscle. (b) A section of striated muscle showing a group of muscle fibers. (c) A single striated muscle fiber. (d) A section of a single fiber showing a group of myofibrils. (e) Arrangement of protein myofilaments (actin and myosin) in a single myofibril.

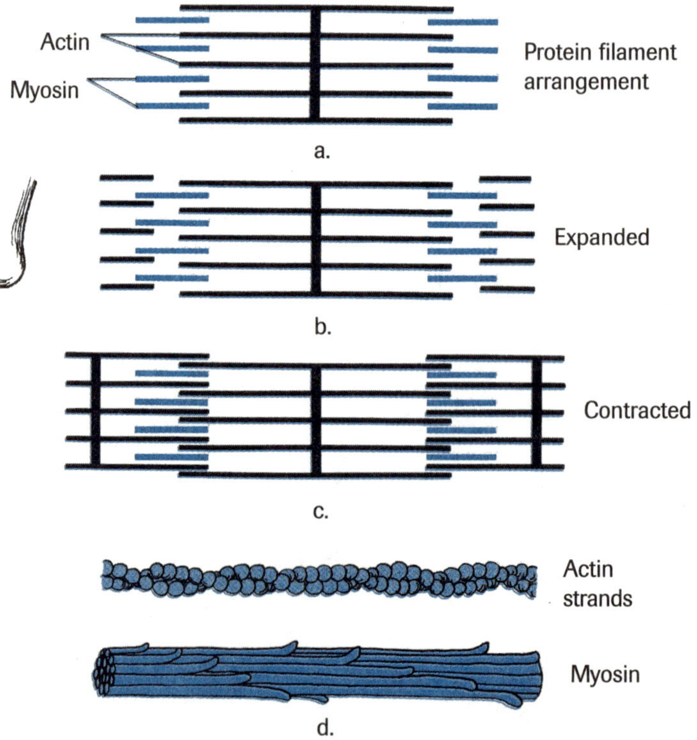

Figure 10.14B (a) Basic arrangement of myosin and actin filaments in a myofibril of skeletal muscle. (b) Appearance of the filaments when the myofibril is expanded, or relaxed, and (c) when the myofibril is contracted . According to the sliding filament theory of muscle contraction, myosin filaments possess tiny cross-bridges that hook onto the actin filaments at specific sites. By bending inward, the cross-bridges pull the actin filaments past the myosin filaments. The cross-bridges then release, bend outward and reattach to the actin, and pull again. (d) Strands of actin and myosin filaments. Note the cross-bridge projections of the myosin filament.

attach to the actin filaments, pull, and then release. This sequence is repeated, causing the muscle to continue to contract. Both the actin and myosin filaments remain the same length throughout the contraction. When the muscle relaxes, the same filaments slide past each other in the opposite direction, lengthening the muscle.

The energy at work in muscle contraction and relaxation is supplied by ATP. It is ATP that powers the pulling action of the cross-bridges, and ATP is required to detach the cross-bridges at the end of their pull. In addition, the energy of ATP is necessary to actively transport calcium ions (Ca^{2+}) into and out of the muscle cell cytoplasm. Calcium ions are necessary for contraction of a muscle; when calcium ions are removed, the muscle ceases to contract.

Types of Skeletal Muscle Fibers

A most interesting attribute of skeletal muscle is its speed of contraction. Skeletal muscle fibers are grouped into **slow-twitch fibers** and **fast-twitch fibers.** Slow-twitch fibers contract relatively slowly and do not fatigue readily. They are supplied by numerous capillaries and mitochondria and have a high concentration of **myoglobin,** an oxygen-carrying protein found in muscle cells and similar to blood hemoglobin. Myoglobin, however, has a higher affinity for oxygen than does hemoglobin and thus transports more oxygen to the muscle cells. Cardiac muscle cells also are rich in myoglobin. Slow-twitch fibers are also referred to as *red fibers.* These are the fiber types found in dark meat, such as turkey or chicken drumsticks.

The thicker fast-twitch fibers are supplied with fewer capillaries and mitochondria, and contain very little myoglobin. These fibers also are called *white fibers* the type found in the white meat of a chicken breast. Fast-twitch fibers contract rapidly and fatigue quickly. This is because the relatively small number of mitochondria in fast-twitch muscle fibers limits the amount of ATP available for energy. Some human skeletal muscles may have an intermediate fiber type that is also the fast-twitch type.

Slow-twitch fibers are advantageous in athletic endurance events, such as long-distance running. This is because slow-twitch fibers, with their extensive supply of capillaries and mitochondria, have a high aerobic capacity. The great number of mitochondria provide the relatively large amount of cellular energy necessary for the prolonged activity of slow-twitch fibers. This type of muscle fibers predominates in the postural muscles, such as those of the back and thigh. On the other hand, fast-twitch fibers are more effective in performing rapid muscle movements, such as running sprints or eye movements. These fibers are adapted to respire anaerobically.

We inherit our muscle types, which explains in part why some people excel at certain athletic events while others who have trained just as hard do not. It has been found that world class marathoners, for example, have a higher percentage of slow-twitch fibers than the average person; Olympic sprinters have a greater number of fast-twitch fibers. This explains the scientific basis for the old saying that you can't make a racehorse out of a plowhorse.

The *size* of muscle fibers increases with exercise, but the *number* of fibers does not. Enlargement of muscle fibers occurs through **isotonic** exercise or **isometric** exercise. During isotonic exercise, such as weight lifting, the muscles shorten, and the tension on the muscles remains fairly constant throughout the contraction. In isometric contraction, which results when one pushes against an immoveable object, for instance, the muscles do not shorten, but the tension increases. In our everyday activities, the muscles of the body contract both isotonically and isometrically. Walking or running primarily involves isotonic contractions, whereas standing still requires prolonged tensing of the leg muscles, an instance of isometric contraction.

Repeated, *forceful* use of skeletal muscles is required to increase their size and strength. As the muscle fibers increase in size, they are capable of contracting more strongly. On the other hand, muscles that are not used tend to *atrophy* (waste away);

consequently, they decrease in size and strength. Muscles also atrophy if their motor neurons are severed or are otherwise nonfunctional.

Skeletal muscles continually receive a few impulses from the nervous system, which maintains a slight tension or contraction of the muscles. This is referred to as **muscle tone,** and it keeps the muscles from becoming flabby or completely relaxed and enables them to respond quickly if the need arises. However, during sleep the impulses subside, and the muscles become totally relaxed.

As we grow older, the muscle fibers in all of our muscles decrease in number, and once they die, they cannot be replaced by new fibers. Instead, the muscle fibers are replaced by a tough connective tissue. Soon after death, the body becomes stiff owing to partial contraction of the skeletal muscles, a condition called *rigor mortis.* The condition persists until the dead body begins to decompose.

Circulatory System

In all vertebrates, body cells must receive an almost continuous supply of oxygen, nutrients, water, and other vital materials. In addition, metabolic wastes must be transported to structures or organs for release from the body. The transport of vital nutrients and wastes is an essential function of the circulatory system (Fig. 10.15).

In humans and other vertebrates, the circulatory system consists of the heart, the blood vessels, and the blood. An adjunct to the circulatory system is the *lymphatic system,* consisting of various *lymph vessels* and *lymph nodes.*

Blood Vessels and Blood

Arteries, veins, and capillaries constitute the mammalian internal blood transport system (Fig. 10.16). **Arteries,** by definition, carry blood *away* from the heart; **veins** carry blood *toward* the heart. Neither of these vessel types would be of much use were it not

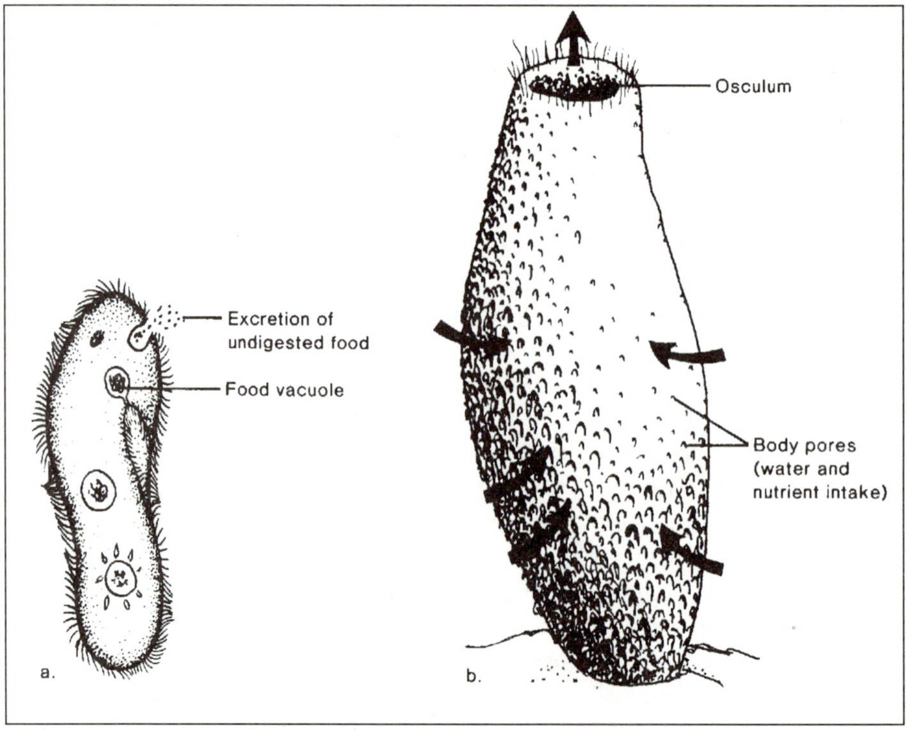

Figure 10.15 Examples of circulation in lower organisms. (a) In *Paramecium,* food taken in from the surrounding medium is enclosed in food vacuoles. Movement of the cytoplasm distributes the vacuoles throughout the interior of the organism. Undigested food is released from the vacuoles through pores to the outside. (b) In sponges, water and nutrients enter the spongocoel through tiny body pores; in the sponge, water serves as circulatory fluid to distribute the nutrients and to remove wastes through the osculum, or mouth. (Organisms not drawn to scale.)

for the **capillaries,** the tiny blood vessels that connect arteries to veins. Respiratory gases, nutrients, hormones, ions, and other substances transported in the blood move across the exceedingly thin-walled capillaries to enter living cells (Fig. 10.17). Similarly, intracellular substances exit the cells by crossing the capillary walls to enter the circulating blood. The capillary network in the body is so profuse that you cannot cut yourself anywhere without drawing capillary blood. No doubt a significant slice of the Kleenex fortune is attributable to erratic shavers who nick themselves at least six or seven times a week.

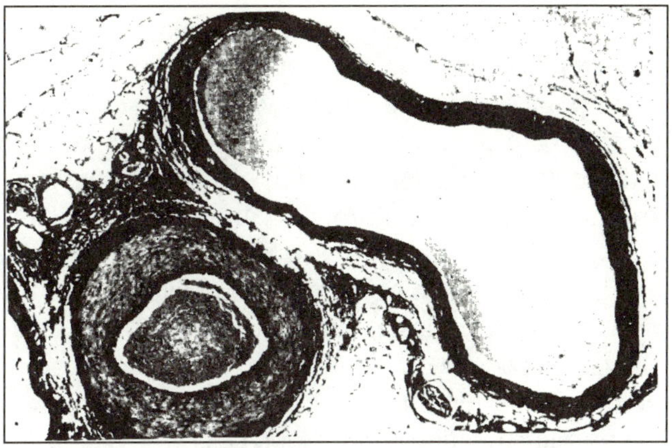

Figure 10.16 Section through an artery and adjacent vein (artery at left, vein at right).

Arteries normally deliver oxygenated blood efficiently because this blood has the pumping force of the left ventricle behind it. By the time blood reaches the veins, however, this force is greatly diminished. Consequently, venous blood traveling back to the heart tends to flow sluggishly and back up and pool. This is particularly true in the lower extremities of the body. To prevent backflow of blood, many veins are equipped with one-way valves that are forced shut when blood backs up in the veins (Fig. 10.18).

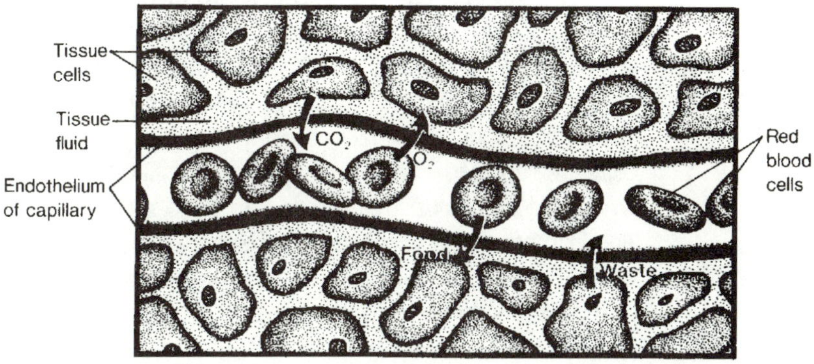

Figure 10.17 Slow movement of the blood through the thin-walled capillaries allows oxygen and food to diffuse into the tissue cells, while carbon dioxide and other wastes diffuse from the tissue cells into the blood.

Venous blood often must overcome the pull of gravity on its way to the heart. The most effective method for promoting venous blood flow is the squeezing action of skeletal muscles

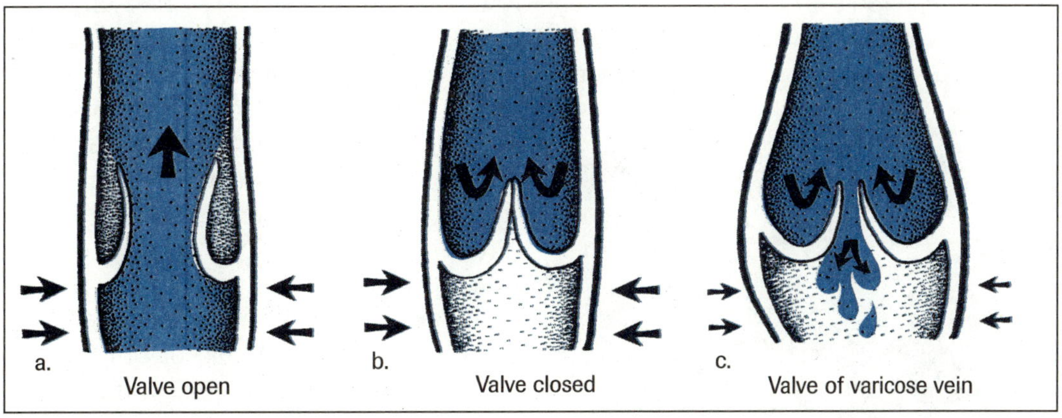

Figure 10.18 Valve action within a vein. (a) Blood flowing through the vein forces the valve open. Movement of blood through veins is greatly assisted by contraction of skeletal muscles. (b) The backward flow of blood (as from the pull of gravity, for example) forces the valves closed, preventing any further backflow. (c) Varicose veins become swollen to the extent that the valves do not close, and blood tends to back up in the veins until it encounters a normal valve lower down that stops the backflow. Faulty valve function in the veins decreases venous blood return to the heart.

on the veins. Prolonged standing or bed rest, however, reduce this pumping action and blood tends to accumulate and pool in the veins. Sometimes this causes veins to stretch to the point where the valves will not close. This can produce *varicose veins* that often become painful and swollen. Varicose veins in the rectum are called *hemorrhoids.* One of the most beneficial effects of physical activity, such as walking, is the efficient circulation of venous blood brought about by contraction of skeletal muscles.

We have mentioned that the essential function of the vast circulatory network in the vertebrate body is to transport and deliver blood. **Blood** is a liquid tissue and one of the most remarkable substances in the animal body. An adult human of average size has about 5 liters of blood. Blood is composed of a liquid **plasma** and the various blood cells, including **red blood cells** (erythrocytes), **white blood cells** (leukocytes), and **platelets** (thrombocytes). The primary function of red blood cells is oxygen transport, while white blood cells contribute to the body's defense against disease. Red blood cells contain the protein *hemoglobin,* which binds oxygen and transports it throughout the body. Platelets, which are tiny cell fragments, act in concert with other factors in the blood to bring about clot formation.

The primary site of blood cell production is bone marrow, although one type of white blood cell, the *lymphocyte,* also is produced in lymphoid tissue, such as the spleen, tonsils, and lymph nodes. Normally, blood cells have a very brief life span in the body and must be continually replaced by new cells arising from the bone marrow and lymphoid tissues. Red blood cells persist about four months in human males and for a slightly shorter time in females. Red blood cells are replaced at a rate of about 2.5 million cells per *second!* Cells in the liver, spleen, and lymphoid tissues destroy old or damaged red blood cells and break down their hemoglobin. The iron atoms and amino acids released from the hemoglobin can be recycled to make new proteins and red blood cells. Most white blood cells survive in the circulation only a few days, and platelets do not last much over a week before they are destroyed.

As blood flows through the circulatory vessels, it exerts pressure on the inner walls of the vessels, much like water in a garden hose. Blood pressure is measured by the force of blood in the arteries. This pressure varies with one's age and sex. Blood pressure is measured in units called *millimeters of mercury* (mm Hg). An average human blood pressure is 120 mm Hg/80 mm Hg. The 120 mm Hg, called the **systolic pressure,** indicates that when the ventricles contract, they exert a pressure in the two great arteries of the heart—the aorta and pulmonary artery—that can raise a column of mercury to a height of 120 mm. The 80 mm Hg, or **diastolic pressure,** represents the pressure in the arteries when the ventricles relax. This pressure, then, can raise a column of mercury to a height of 80 mm. If blood pressure rises significantly above the normal range, a person may suffer from *hypertension.* Various factors, such as kidney disease, arteriosclerosis (hardening of the arteries), and genetics appear to be partly responsible for hypertension, but it is not always evident as to what produces the condition.

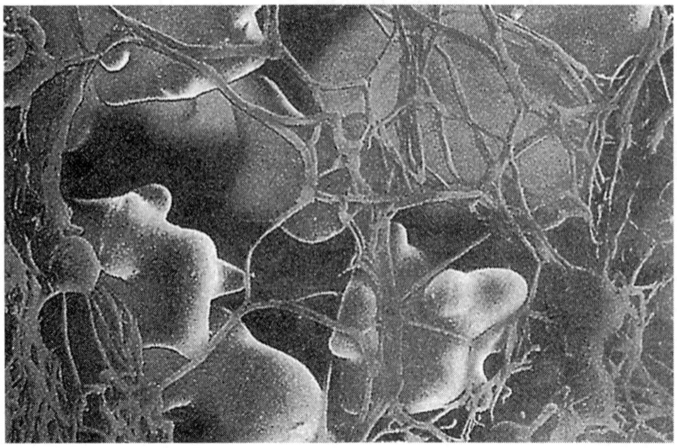

Figure 10.19 Red blood cells trapped in a mesh of fibrin strands, forming a blood clot.

From our discussion, it is clear that blood is a vital substance and that its loss would mean certain death. In many less developed animals, prevention of blood or fluid loss is accomplished simply by reflex contraction of the muscles and circulatory vessels when they are cut. In humans and many other more complex animals, various substances in the blood, together with the blood cells, form a plug or clot that prevents blood loss (Fig. 10.19). This crucial mechanism quickly closes the bleeding nicks and cuts that occur in our daily lives. As mentioned, platelets are essential to normal blood clotting. A deficiency in the normal number of circulating platelets can result in bleeding from the skin and internal organs.

Severe damage to a blood vessel requires coagulation of the blood, resulting in a blood clot. Blood clotting is an extremely complex pathway, requiring a series of steps that involve at least a dozen protein coagulation factors found in blood plasma, along with platelets, calcium ions, and vitamin K. Essentially, a blood clot involves the enzymatic conversion of a plasma protein called **fibrinogen** to **fibrin,** a threadlike protein that traps blood cells and platelets, forming the clot. Clotted blood that forms a "scab" consists of a network of fibrin strands in which the blood cells and platelets are trapped.

Heart Structure and Blood Circuit

The mammalian heart is described generally as a muscular, our-chambered, double pump (Fig. 10.20). Its sole function is to pump blood. The heart has nothing to do with emotions; you cannot love someone with "all your heart" or play ball with a "lot of heart" or "follow your heart instead of your head." You love and feel and think with your brain. Still, it is doubtful that embroidered brains will be gracing the covers of Valentine cards in the foreseeable future.

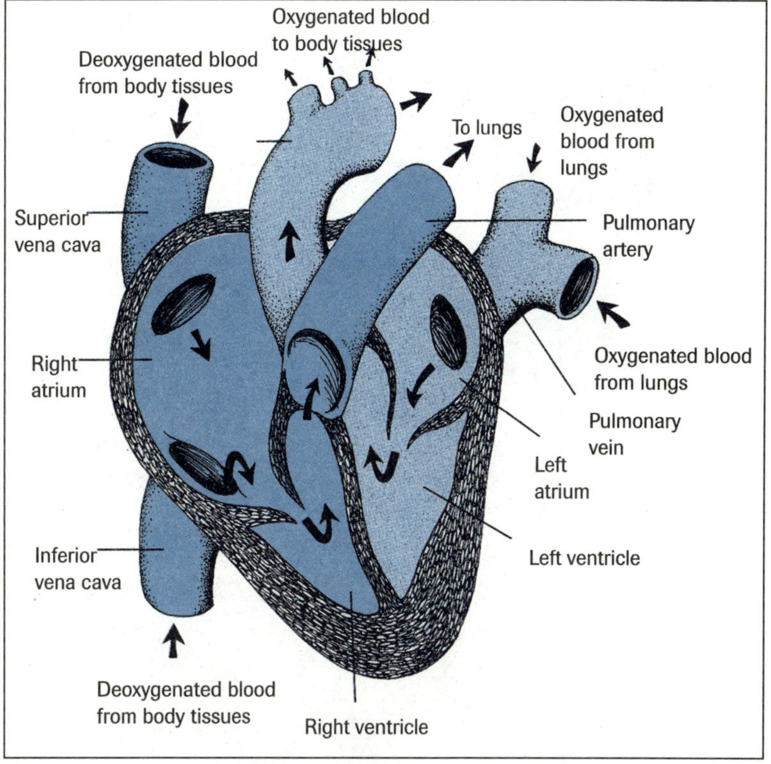

Figure 10.20 The four-chambered heart of birds and mammals.

The human heart is about the size of a fist, and lies between the lungs in the midline of the chest cavity. The four chambers of the heart include an upper right and left **atrium** and lower right and left **ventricles.** Four heart valves near these chambers keep blood moving in one direction. Normally, the valves close tightly, restricting the backflow of blood. For various reasons, the valves may fail to close completely or may be constricted. Blood leaking backward through a defective heart valve produces sounds known as *heart murmurs.* An estimated 10 percent of the United States population have heart murmurs, but many of these people do not suffer any serious effects.

All deoxygenated, or venous, blood is returned to the right atrium of the heart by two major veins—the **superior vena cava,** which collects venous blood from the upper portion of the body, and the **inferior vena cava,** which collects venous blood from the lower portion of the body.

Before entering the venous system, blood cells exchange most of their oxygen to the body cells for the waste product carbon dioxide. From the right atrium, deoxygenated blood (blood low in oxygen but high in carbon dioxide) passes through a valve into the right ventricle. When the ventricle contracts, this valve is forced shut and blood is ejected through a second valve into the **pulmonary artery.** This artery exits the heart and divides into a right and left trunk, each of which carries venous blood to a lung and into the lung capillaries. The pulmonary artery is the only artery in the adult that transports deoxygenated blood.

In the lungs, the blood cells release carbon dioxide, which diffuses from capillary blood into the lungs to be exhaled. At the same time, oxygen from air inhaled into the lungs diffuses outward into the capillary blood. Thus, the blood now is rich in oxygen, but contains a very low concentration of carbon dioxide. This oxygenated blood leaves the lungs and returns to the left atrium of the heart through the **pulmonary veins.** These are the only veins in the adult that transport oxygenated blood. From the left atrium, blood passes through a valve into the left ventricle. Contraction of the left ventricle closes this valve and forces blood out through yet another valve into the **aorta.** It is the aorta and its multitude of branching arteries that deliver oxygenated blood to the body capillaries surrounding every living cell.

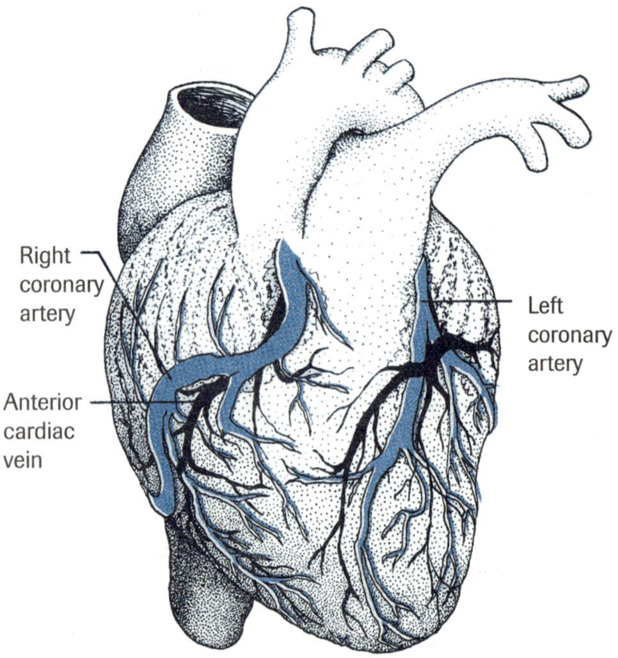

Figure 10.21 Coronary circulation that supplies oxygenated blood to the heart muscle via the arteries, and returns deoxygenated blood to the right atrium via the veins.

Thus, the double pump action of the heart refers to the right ventricle, which pumps blood to the lungs, and to the left ventricle, which pumps blood throughout the remainder of the body.

Oxygenated blood is conveyed to the heart muscle itself by the right and left **coronary arteries,** which branch off the aorta (Fig. 10.21). Subdivisions of these arteries penetrate into all parts of the heart. Deoxygenated blood, saturated with carbon dioxide from metabolizing heart muscle, collects in the **coronary and cardiac veins,** which return venous blood to the right atrium, along with blood entering from the inferior and superior venae cavae.

The human heart usually averages about 70–72 beats per minute, although somewhat faster or slower rates are within a normal range. As pointed out in Chapter 9, the heart rate is controlled by

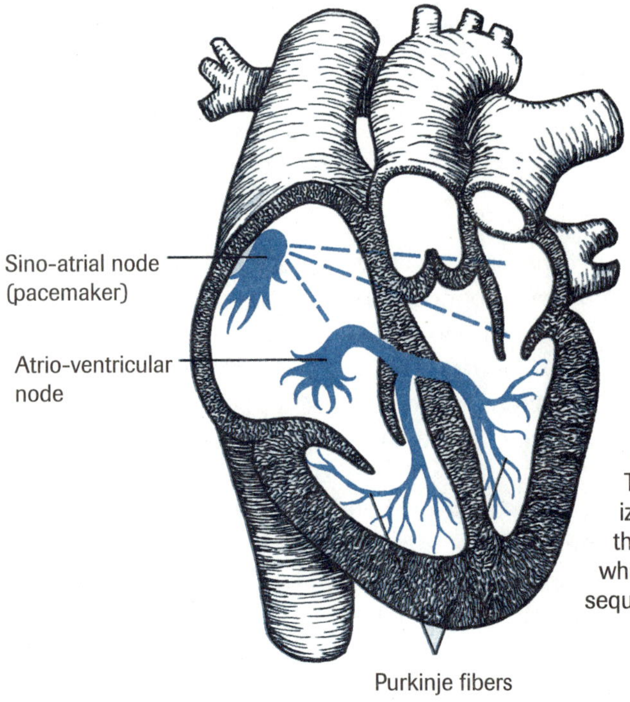

Figure 10.22 The conduction system of the mammalian heart. Electrical impulses originate in the specialized muscle cells—the sino-atrial, or SA, node—in the right atrium. The impulses then spread through the atria, which contract simultaneously while the ventricles are relaxed. From the atria, impulses reach the atrio-ventricular, or AV, node near the junction of the right atrium and right ventricle. The AV node sends impulses out along a system of specialized muscle cells—the Purkinje fibers—that ramify throughout the two ventricles. The ventricles then contract simultaneously, while the atria are relaxed. When the SA node fires again, the sequence is repeated.

nerves from the brain. Cardiac cells have an intrinsic rhythm and can contract when isolated from the body. In the functioning heart, however, the rhythm must be regulated by sympathetic and parasympathetic nervous input in association with the **sinoatrial node,** or the pacemaker, of the heart. You will recall that sympathetic stimulation increases the rate and force of the heartbeat, while parasympathetic stimulation decreases the rate and force. The pacemaker, as shown in Fig. 10.22, is a small, specialized mass of muscle tissue located in the right atrium and initiates each cardiac contraction; that is, it is the site where the normal heartbeat originates. The sinoatrial node sends out electrical impulses that spread to the atria, causing them to contract in unison. The impulses stimulate another node, the **atrioventricular node,** located near the floor of the right atrium. At the atrioventricular node, the impulse is delayed about 0.15 seconds and then the impulse is relayed downward to the ventricles, which contract in unison. The delay at this node allows the atria time to empty their blood into the ventricles before the ventricles contract.

Diseases of the heart have the dubious distinction of being the number one cause of human death in the United States. The most common forms of these diseases are heart attacks, primarily those involving the coronary arteries. When a coronary artery or one of its branches is blocked by a blood clot or cholesterol, blood cannot reach the area of the heart supplied by the vessel. Consequently, cardiac muscle cells in that area cannot metabolize and eventually die.

Arteriosclerosis, or hardening of the arteries, contributes to heart attacks by preventing normal contraction and expansion of arteries, forcing the heart to work harder. Some blood clots in the circulatory system can enter arteries leading to the brain. If the clot restricts blood flow to the brain, the result may be a *stroke.* A stroke occurs when there is an interruption of blood supply to a portion of the brain. The affected neurons begin to die within minutes; if severe, strokes can cause paralysis, loss of speech, and various mental disorders.

The Lymphatic System

The lymphatic system is an adjunct to the blood circulatory system (Fig. 10.23). Blood capillaries, because of their thin walls and the outward pressure of blood, tend to leak. Fluid derived from plasma, along with small blood proteins, leak out of the capillaries into the surrounding tissues. A network of lymph capillaries absorbs this fluid and protein mixture and returns it through large lymph ducts to veins just below the neck. In this manner, lost fluid and blood proteins are returned to the circulating blood. Once the mixture of fluid, proteins, and surrounding tissue fluid enters the lymph capillaries, it is called **lymph.** If the absorption mechanism fails, fluid collects in the body tissues, producing *edema,* or swelling. If edema occurs in the lungs, breathing and respiratory gas exchange can be severely impaired.

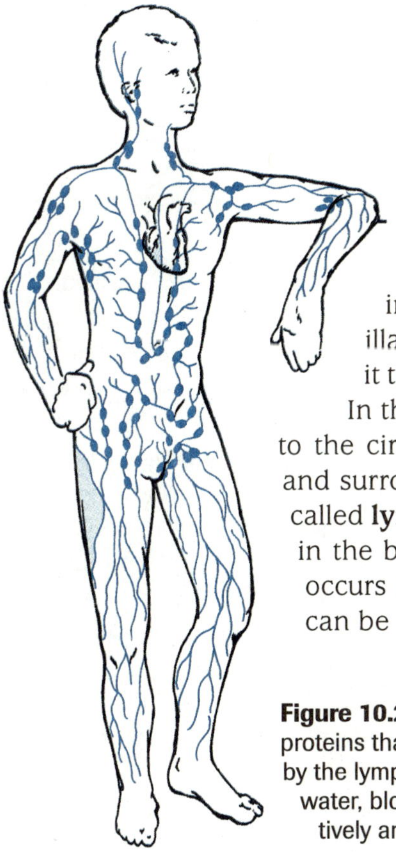

Figure 10.23 The human lymphatic system. Plasma water and small blood proteins that leak from the blood capillaries are returned to the circulation by the lymphatic system. Once they enter the lymph capillaries, the plasma water, blood proteins, and some of the surrounding tissue fluid collectively are called *lymph.*

Lymph vessels, like some veins, possess one-way valves that keep lymph moving in one direction. Lymph is moved along mainly by skeletal muscle contraction in the same manner as in veins.

As lymph flows within the lymph vessels, it passes through small oval masses of tissue called **lymph nodes** that filter out harmful materials and destroy bacteria. This bacterial destruction is carried out by specialized white blood cells called **lymphocytes** and by phagocytic cells. Production of lymphocytes is a critical function of the lymphatic system. We will return to lymphocytes in more detail during our discussion of immunity in Chapter 11.

At times, lymph nodes are overrun with infectious microorganisms and become swollen and painful. This occurs, for example, in the tonsils, which are lymphoid organs located in the throat.

The human body contains a vast network of lymph capillaries that extends throughout most tissues. Special types of lymph vessels called *lacteals* are found in the lining of the small intestine where they absorb and transport digested lipids out of the digestive tract. Two large lymphoid organs in humans are the *thymus,* located between the lungs just above the heart; and the *spleen* located behind the stomach (Fig. 10.24). In the infant, the thymus is involved in immune system functions and the spleen, in people of all ages, filters blood, produces some lymphocytes, and destroys old red blood cells.

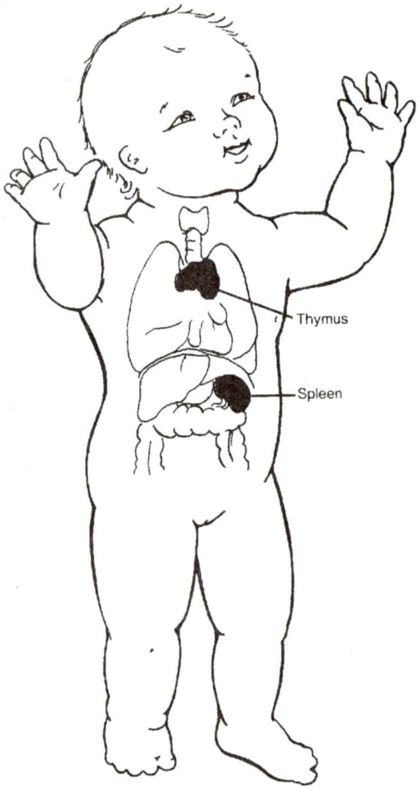

Figure 10.24 Location of the human thymus and spleen.

Respiratory System

We have commented previously on the oxygen requirements of living cells and on the need for removal of carbon dioxide. Because of constant cellular activity, the vertebrate body cannot function very long without oxygen. In humans, loss of consciousness results if brain cells are deprived of oxygen for four or five minutes; after about ten minutes, oxygen deprivation results in death.

Terrestrial vertebrates, including most amphibians, reptiles, birds, and mammals, have evolved lungs that act in conjunction with the circulatory system to deliver oxygen and remove carbon dioxide (Fig. 10.25). The respiratory system of humans and other mammals consists of air-conducting passageways, along with sites for respiratory gas exchange in the lungs (Fig. 10.26). Air is inhaled through the nostrils where dirt particles are filtered from the air by hairs before the air enters the nasal cavities, where it is moistened and warmed. Air then passes into the **pharynx,** or throat, down into the **larynx,** also called the voice box, and into the **trachea,** or windpipe. The trachea branches into a right and left **bronchus,** each of which leads into a lung where these two bronchi divide and subdivide to form numerous small **bronchioles.**

Located at the terminals of the tiniest respiratory bronchioles are the **alveoli,** about 350 million of which are found in each lung of humans. Alveoli are the sites of respira-

Figure A.1 Evolutionary adaptation: A white willow Ptarmigan blends in with the snow. © Tony Stone Images

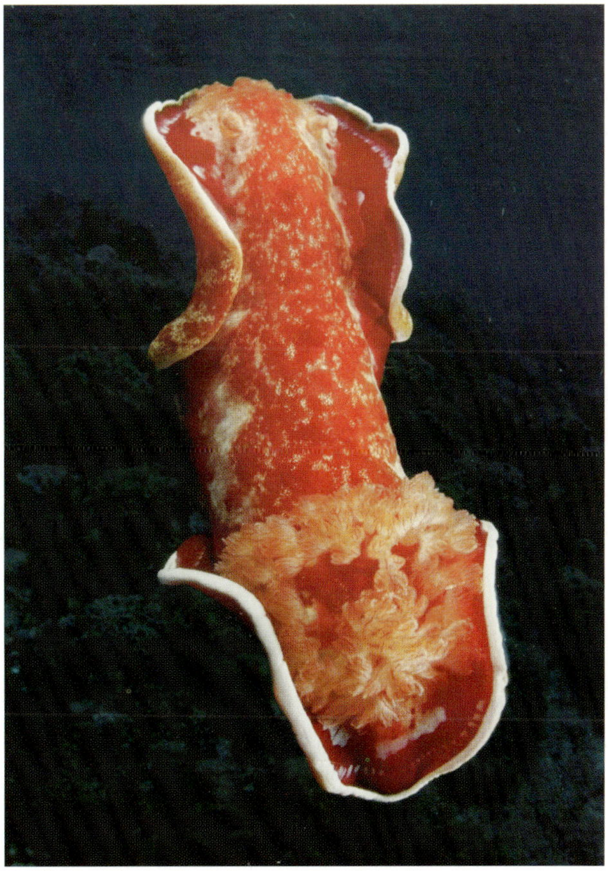

Figure A.2 A reddish Spanish dancer Nudibranch is a mollusk that lacks a shell. © Dave B. Fleetman/Tom Stack & Associates

Figure A.3 A female ringnecked snake with her eggs. © Jim Merli/Visuals Unlimited

Figure A.4 The tough exoskeleton of this spiny Lobster is made of chitin.
© Andrew Wood/Photo Researchers, Inc.

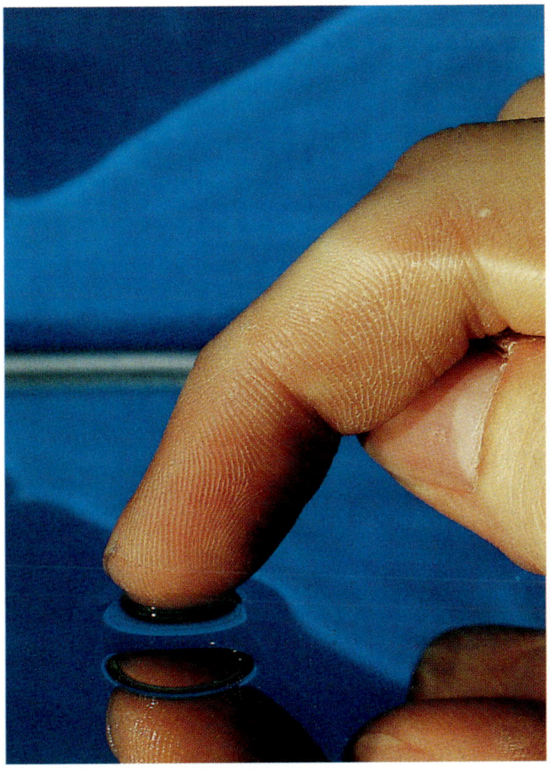

Figure A.5 A demonstration of the surface tension of water; this property of water allows insects to walk on its surface. © Visuals Unlimited

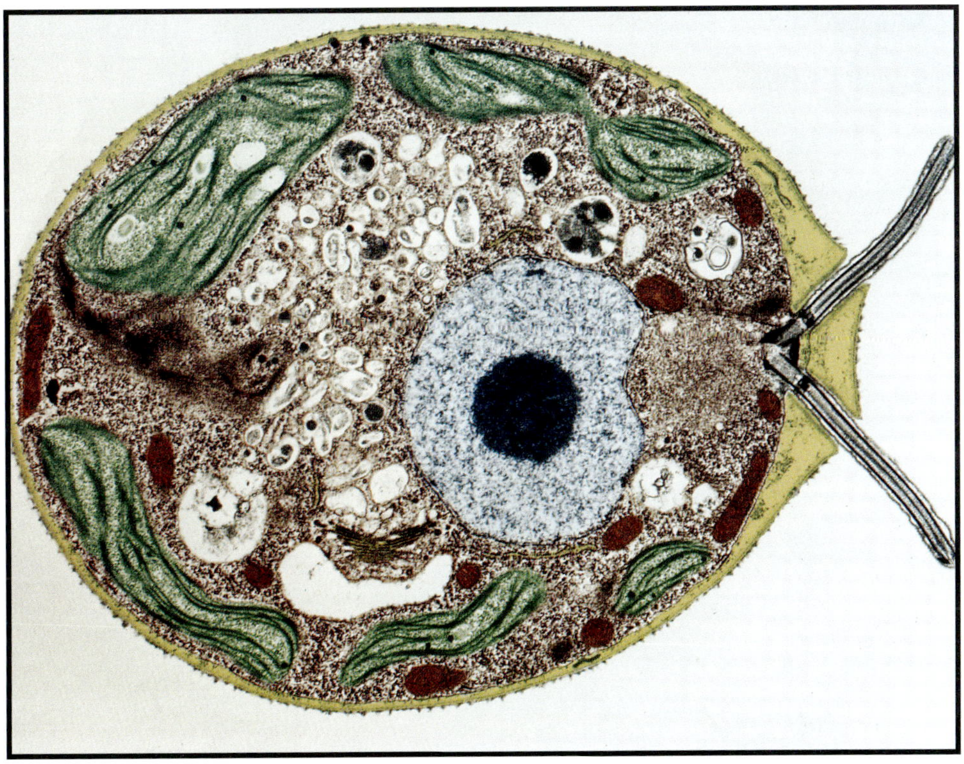

Figure A.6 An electron micrograph of a single-celled protist; paired flagella extend on the right, and large green chloroplasts can be seen in the cytoplasm. © Tony Stone Images

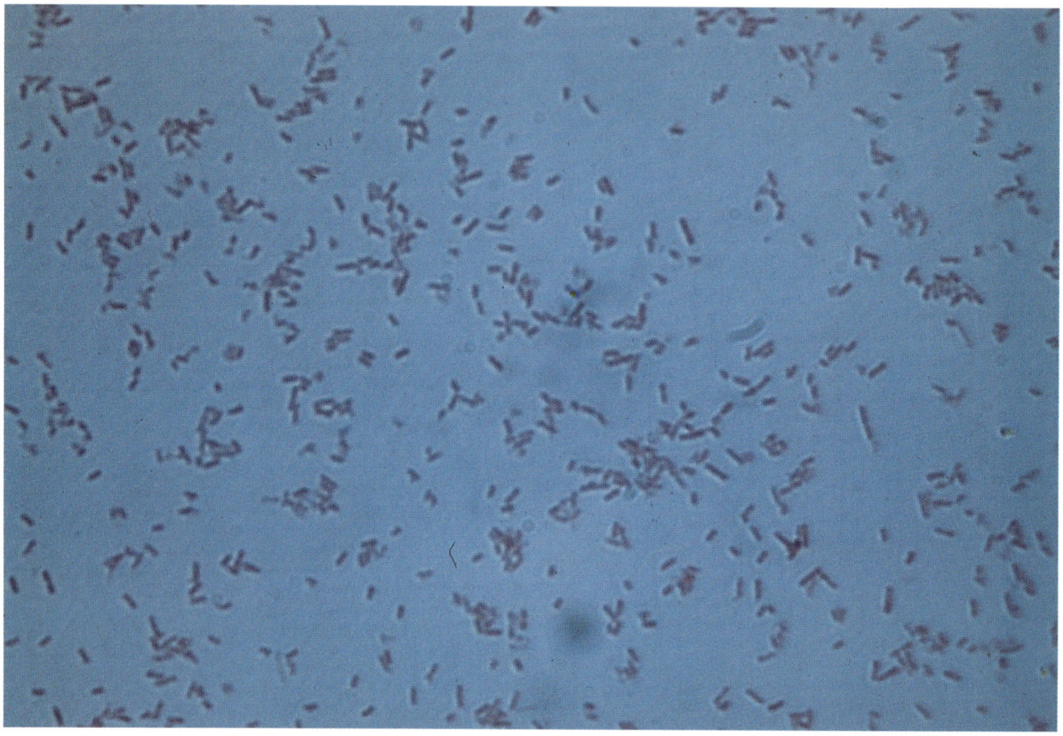

Figure A.7 A group of rod-shaped *Salmonella* bacteria. Species of *Salmonella* cause such conditions as typhoid fever, inflammation of the intestinal tract, and food poisoning. © Visuals Unlimited

Figure A.8 A cicada emerges from its old casing, or exoskeleton. © Animals Animals/Earth Scenes

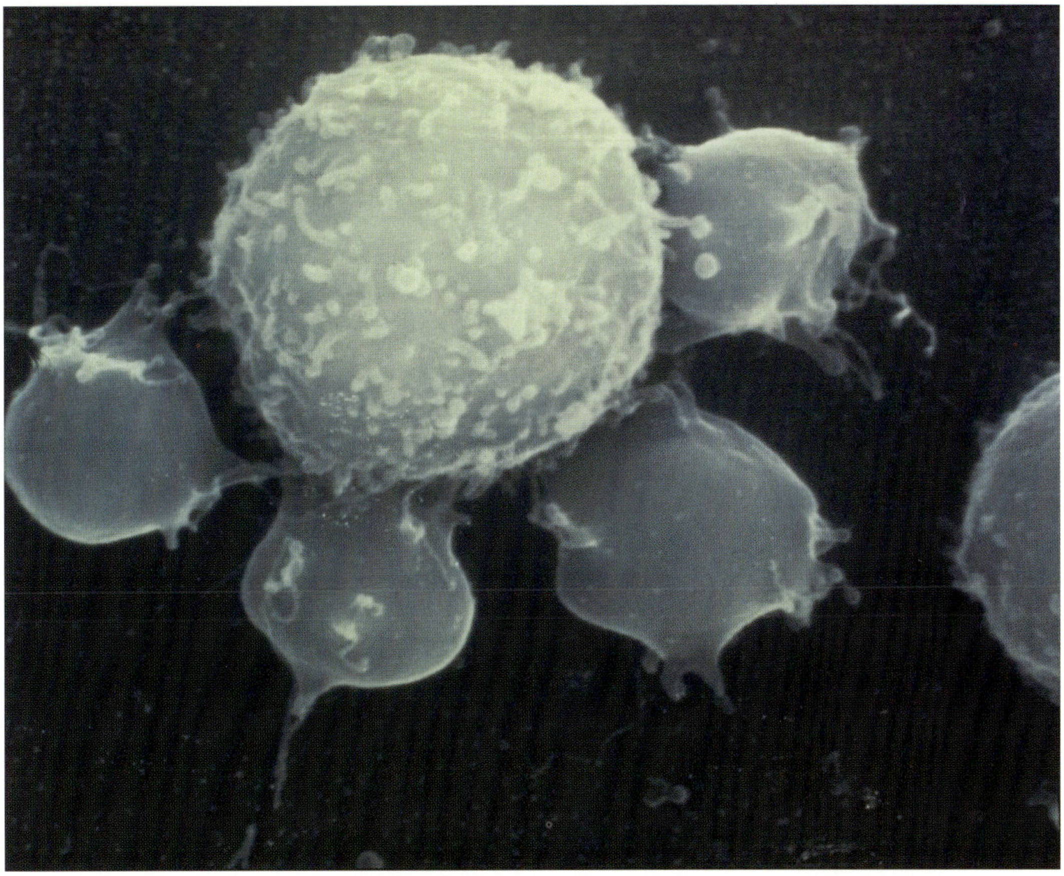

Figure A.9 A spherical white blood cell (lymphocyte) attacks four smaller grey bacteria.
© Benjamin Koziner/Phototake NYC

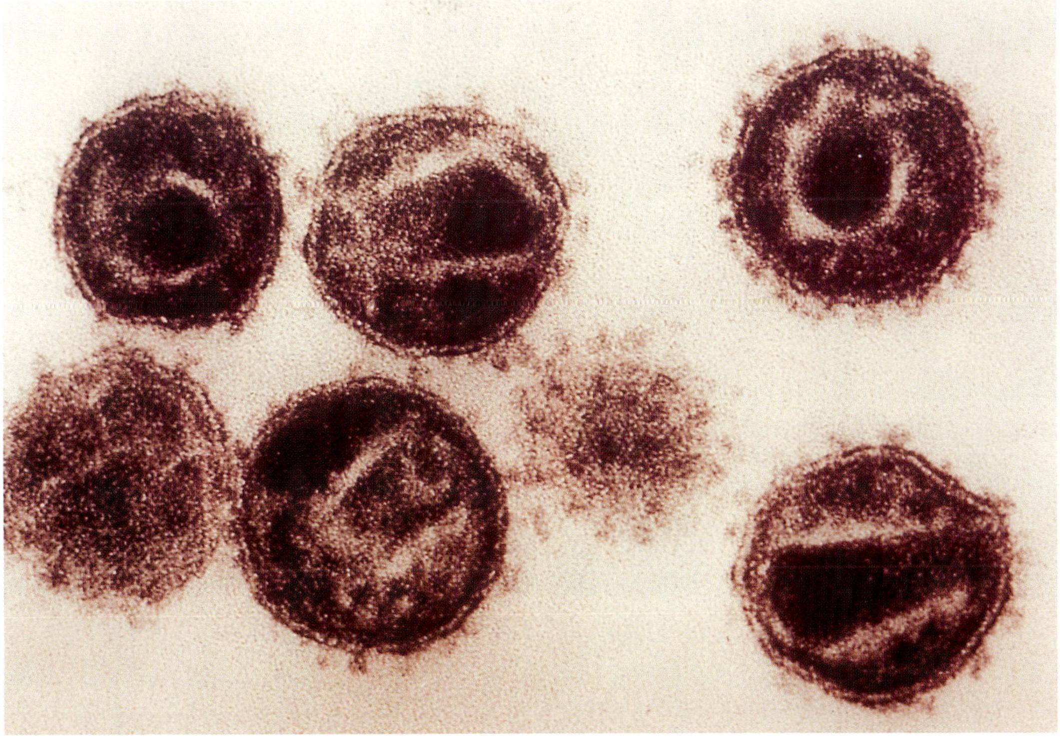

Figure A.10 Human immunodeficiency viruses (HIV), the causative agents of AIDS.
© Visuals Unlimited

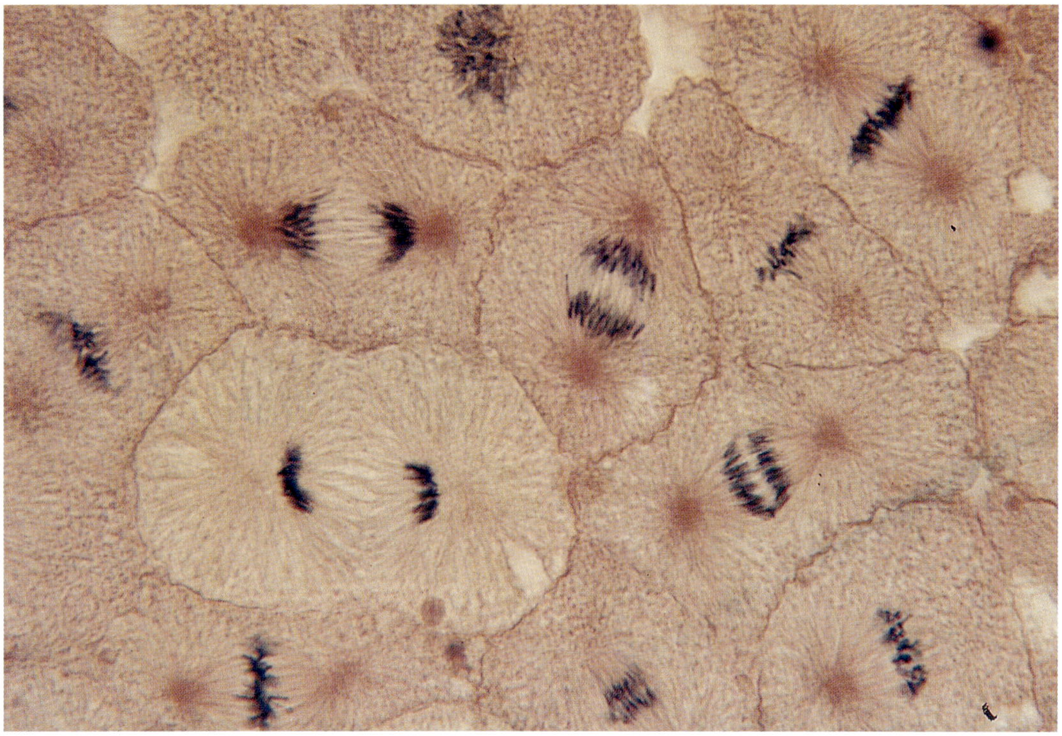

Figure A.11 Animal mitosis; phases prophase through telophase can be seen in the photo. © Visuals Unlimited

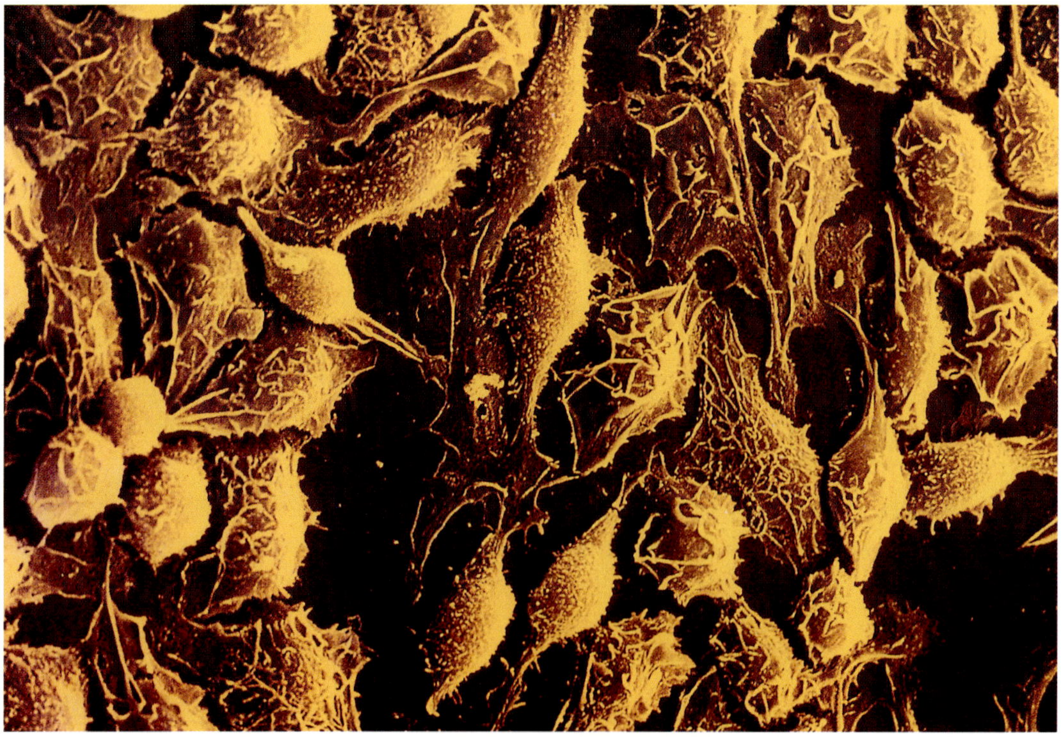

Figure A.12 Human cancer cells. © Visuals Unlimited

Figure A.13 Fruit flies *(Drosophila)*. Compare the head, eyes, body and wings of the normal fly on the left with those of the mutant fly on the right. © University of Hawaii/CCRT

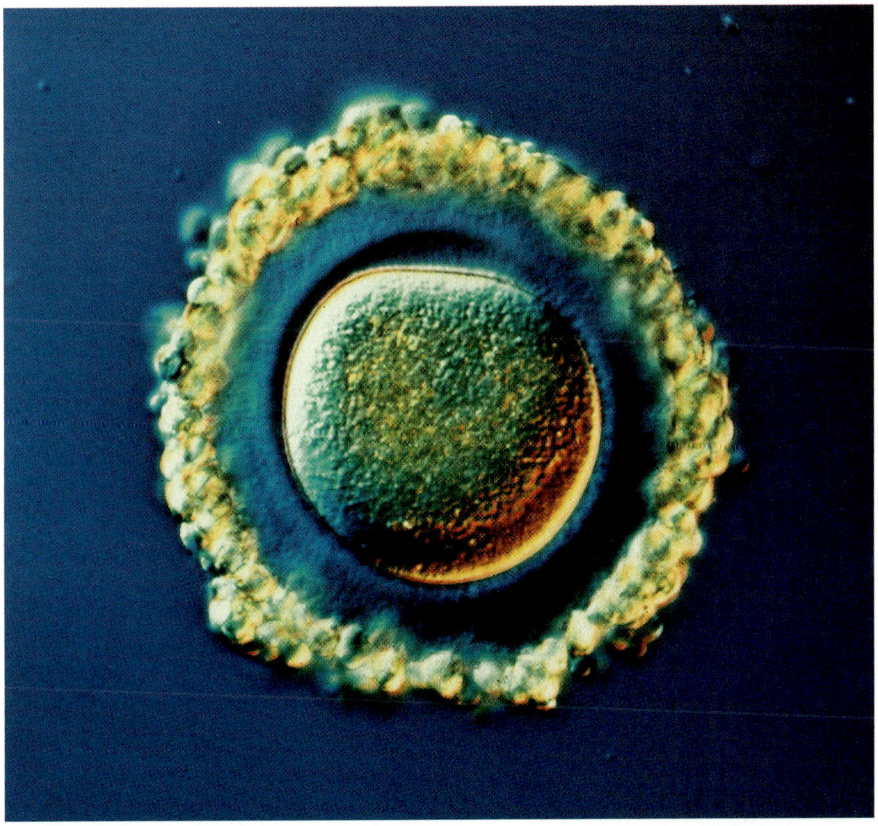

Figure A.14 A human secondary oocyte, or "egg cell," shortly after ovulation (release of oocyte from the ovary); the oocyte is surrounded by a protective layer of cells. © Lennart Nilsson/Bonnierforlagen AB

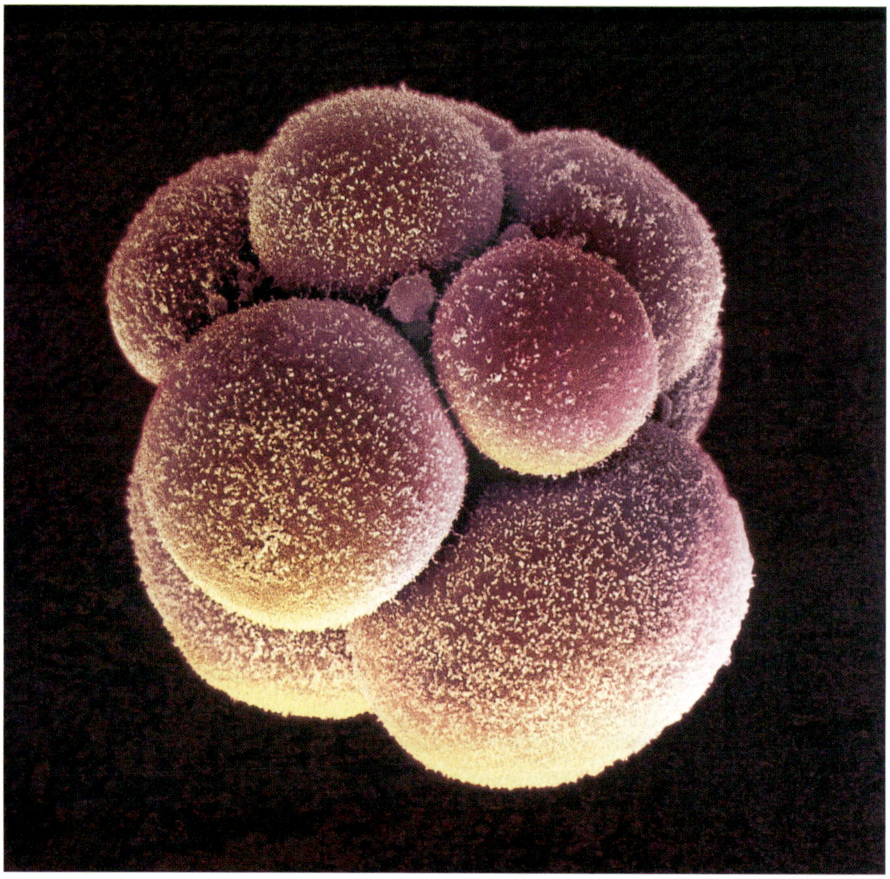

Figure A.15 An early multicellular human embryo undergoing cleavage.
© Tony Stone Images

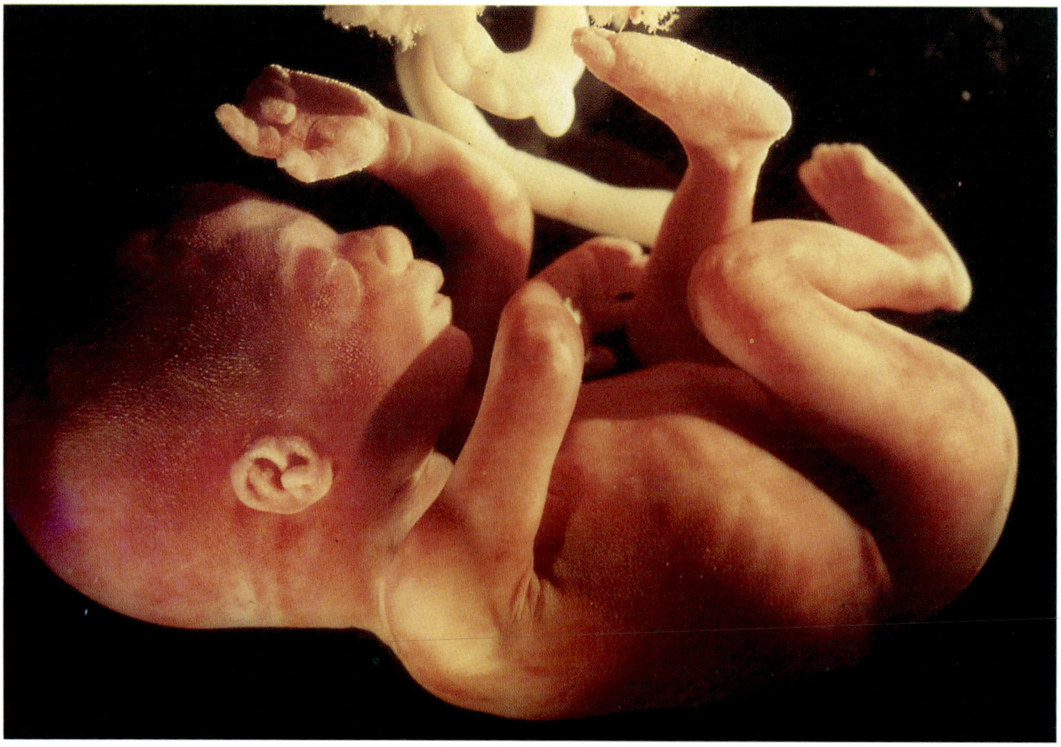

Figure A.16 A human fetus at age 16 weeks. © Animals Animals/Earth Scenes

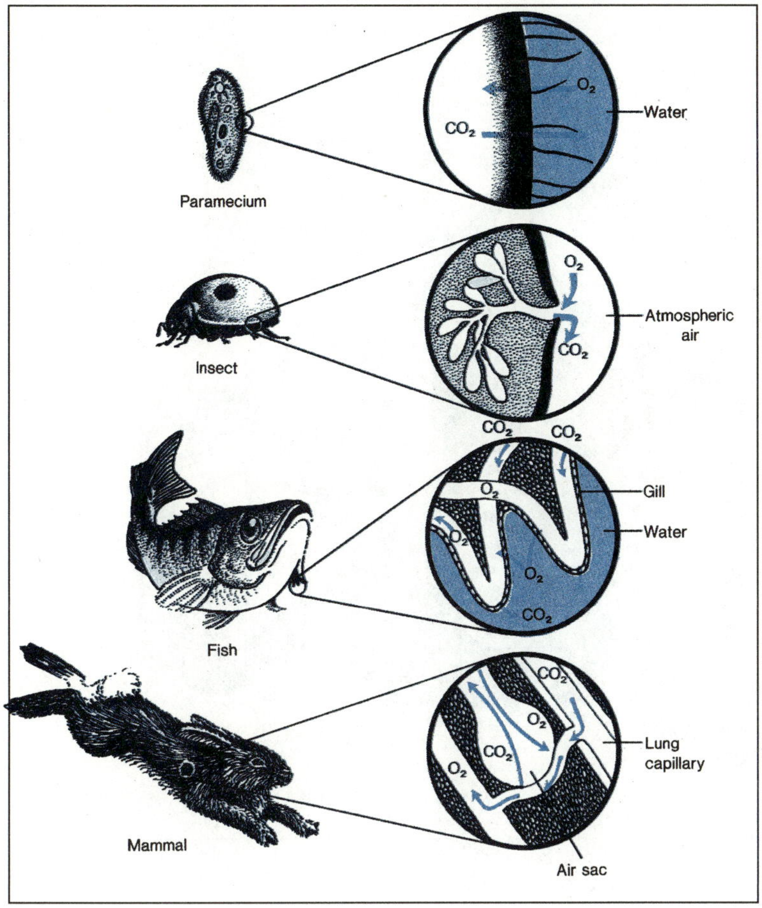

Figure 10.25 Respiratory activity in a variety of organisms. The paramecium exchanges carbon dioxide and oxygen by diffusion directly through the cell membrane. Oxygen diffuses into the tracheal system of the insect and is distributed to the body cells. Carbon dioxide, a by-product of cellular metabolism, diffuses from the cells, into the tracheal system and out of the body. The gills of a fish extract oxygen from the surrounding water. The oxygen is delivered via the blood to the body cells. Carbon dioxide from the cells diffuses into the blood, which returns to the heart. Blood is then pumped to the gills, where the carbon dioxide diffuses out into the water and oxygen again diffuses inward. Oxygen and carbon dioxide in the lungs of a mammal are exchanged across tiny air sacs surrounded by lung capillaries. Oxygenated blood is delivered to the heart for distribution to the body, and the carbon dioxide in the air sacs is exhaled.

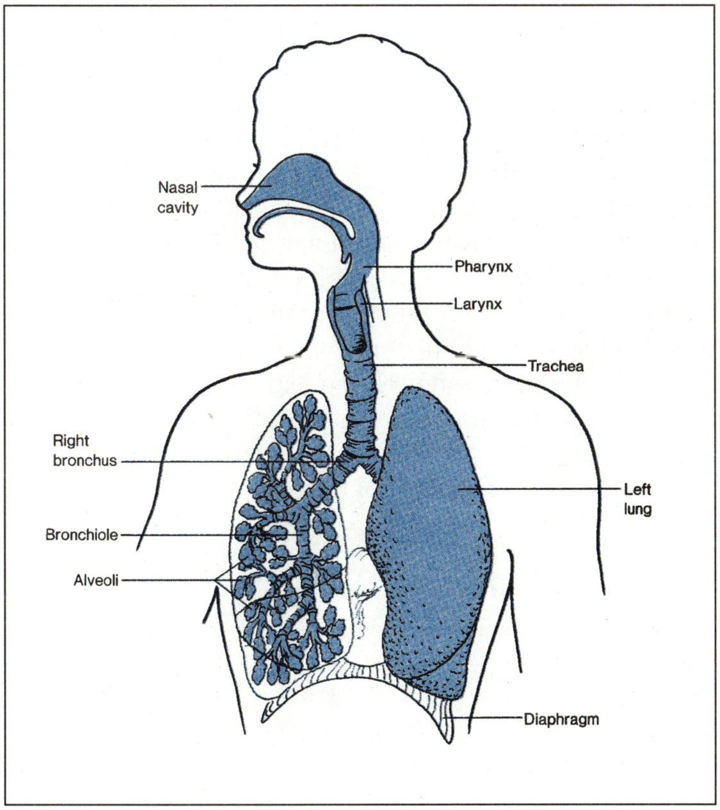

Figure 10.26 Major structures of the human respiratory system.

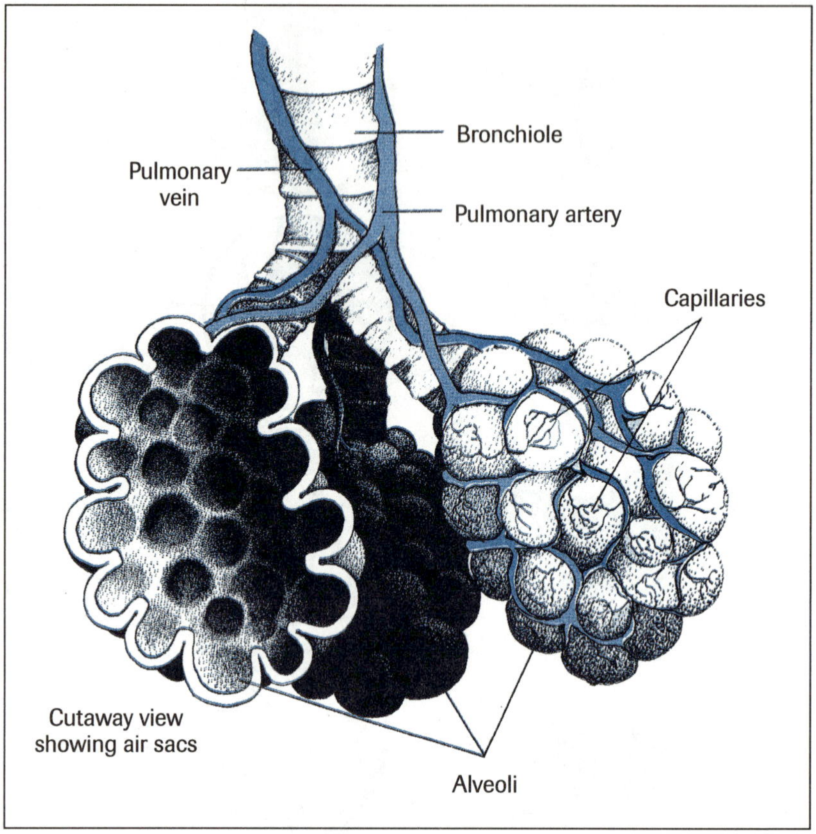

Figure 10.27 Diagram showing the relationship between the lung alveoli (air sacs) and the pulmonary capillaries. The pulmonary artery branch transports deoxygenated blood to the alveoli, and the pulmonary vein branch transports oxygenated blood from the alveoli toward the heart.

tory gas exchange in the lungs. They are microscopic sacs only one cell layer thick, and are surrounded by a profuse network of blood capillaries. The tremendous number of alveoli greatly increases the surface area of the lungs, a condition essential for the effective diffusion of gases between alveoli and the surrounding capillary blood.

Lung alveoli have a tendency to collapse because of surface tension created by a film of watery fluid inside the alveoli. Collapse of the alveoli normally is prevented by a phospholipid substance called **surfactant** secreted by alveolar cells. Surfactant lowers the surface tension of the fluid, allowing the alveoli to remain open. Some premature babies lack adequate surfactant in their alveoli. The alveoli, then, tend to collapse, making lung inflation extremely difficult. Many babies with this condition, known as *respiratory distress syndrome,* can be kept alive by ventilators and by administration of surfactant taken from animal lungs. Another common problem experienced by premature babies is incomplete development of the respiratory system; the body cells, therefore, cannot receive adequate oxygen. Moreover, surfactant in the alveoli is not produced in the normal human fetus until about the eighth month of development.

Mechanics of Breathing

In mammals, the act of breathing in and out depends on movements of the **diaphragm,** a broad sheet of muscle separating the chest and abdominal cavities, and on muscles located between the ribs. Contraction and relaxation of these muscles change the volume of the chest cavity, allowing air to enter and exit the lungs (Fig. 10.28). When you inhale, your rib muscles contract, which expands the rib cage, and the diaphragm contracts downward. This increases the volume of the chest cavity and lowers the air pressure in the lungs below the pressure of the air outside the body. Air moves from regions of higher pressure to regions of lower pressure and thus is drawn through the mouth or nostrils and down into the alveoli. On exhaling, the rib muscles relax, the diaphragm rises, and the volume of the chest cavity is reduced. This compresses the lungs and increases the air pressure in the chest cavity above that of atmospheric air. As a result, air is forced out of the alveoli into the respiratory passageways and exhaled through the mouth or nostrils.

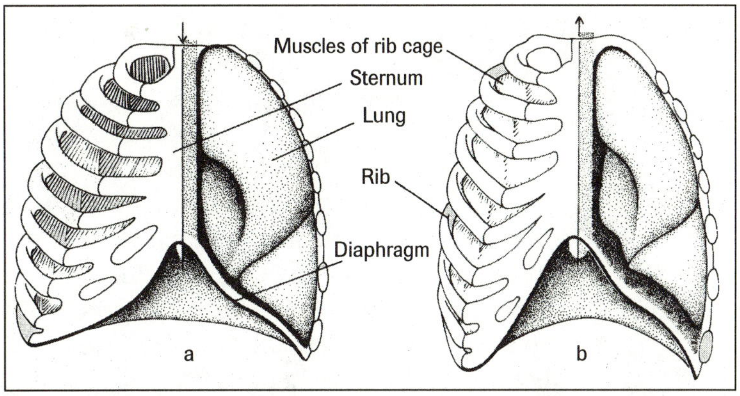

Figure 10.28 The mechanism of breathing. (a) Inhalation. The diaphragm drops and the chest cavity expands, allowing air to enter the lungs. (b) Exhalation. The diaphragm rises and the chest cavity is compressed, forcing air out of the lungs.

Increasing the air pressure in the chest cavity is the basic principle behind the *Heimlich maneuver,* a technique for removing obstructions from the trachea of a choking victim. The Heimlich maneuver involves applying strong upward pressure with the fist just below the rib cage. This action compresses the air in the chest cavity and lungs, forcing air up through the trachea to dislodge the obstruction.

Respiratory Gas Exchange

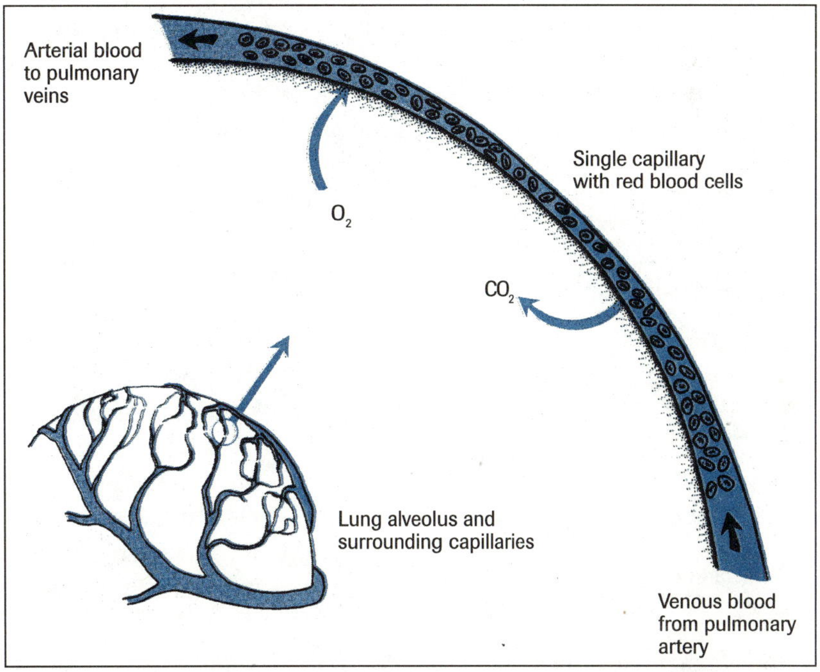

Figure 10.29 Respiratory gas exchange in the mammalian lung. Venous blood, entering the lung from the pulmonary artery, has a high concentration of carbon dioxide (CO_2). Consequently, the CO_2 diffuses *into* the alveolus where the concentration of CO_2 is considerably lower. On the other hand, the concentration of oxygen (O_2) in the blood from the pulmonary artery is low, while the O_2 concentration in the alveolus is high. Thus, O_2 diffuses *out* of the alveolus into the capillary blood.

As indicated in Fig. 10.29, inhaled air delivered to the alveoli contains oxygen at a higher pressure and concentration than the oxygen in the capillary blood surrounding the alveoli. This is because the circulating blood gave up most of its oxygen to the body cells before returning to the lungs. Oxygen in the alveoli, therefore, diffuses out of the alveoli into the capillary blood and is cycled via the pulmonary veins to the heart to be pumped throughout the body. As this newly oxygenated, or arterial, blood flows through the body, it releases oxygen, which diffuses into every living cell. As we described in Chapter 7, this is the oxygen required to carry out cellular respiration.

Oxygen transported in the blood diffuses into the body cells because its pressure and concentration are greater than that of the minimal amount of oxygen inside the cell. Cells utilize oxygen for metabolism about as fast as it enters the cells. As the cells metabolize, the carbon dioxide concentration increases, and this gas begins to diffuse out of the cells into the blood that moments before had released its oxygen. Carbon dioxide diffuses out of body cells because it has a higher pressure and concentration than the carbon dioxide in the capillaries surrounding the body cells. This now deoxygenated, or venous, blood is returned to the heart and pumped to the lungs.

On reaching the lung capillaries, venous blood gives up its carbon dioxide, which diffuses across the capillary walls into the alveoli. Carbon dioxide diffuses in this direction because it has a higher pressure and concentration in the lung capillaries than it does in the alveoli. When you exhale, the carbon dioxide is forced upward through the respiratory passageways and expelled from the body. On the next inhalation, this entire breathing and gas exchange cycle is repeated. On average, you will inhale and exhale some 20,000 times a day to maintain this life-sustaining exchange.

Regulation of Breathing

When you are sitting quietly, intently studying biology, you probably breathe between ten and twenty times a minute, the normal range for most adults. But this rate can increase up to twentyfold during strenuous exercise. Breathing, like most physiologic activities of the body, is regulated by input from the nervous system. The diaphragm and rib muscles are supplied with nerves from the brain and spinal cord. The automatic rhythm of breathing is controlled by neurons located in the medulla oblongata at the base of the brain (Fig. 10.30). When stimulated, these neurons, collectively known as the **respiratory center,** send impulses along the nerves leading from the brain and spinal cord. The major stimulating factor acting on these neurons is the concentration of carbon dioxide in the blood. The neurons respond to carbon dioxide stimulation by sending impulses to the diaphragm and rib muscles. Higher levels of carbon dioxide increase the rate and depth of breathing, which occurs when you exercise vigorously. The high levels of carbon dioxide result from increased cellular metabolism during exercise. Reduced levels of carbon dioxide lessen the stimulatory effect on the neurons and the breathing rate slows. This is what happens as you "catch your breath" after a strenuous workout.

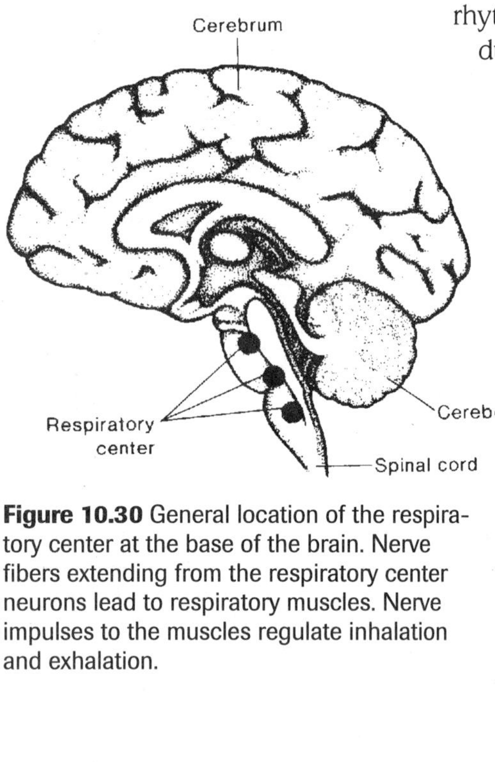

Figure 10.30 General location of the respiratory center at the base of the brain. Nerve fibers extending from the respiratory center neurons lead to respiratory muscles. Nerve impulses to the muscles regulate inhalation and exhalation.

Taking rapid deep breaths, or *hyperventilating*, releases so much carbon dioxide that the respiratory center is depressed and breathing may stop. The results of prolonged hyperventilation are constriction of blood vessels in the brain and dizziness. The traditional yet effective remedy for hyperventilation is breathing into a paper bag. By rebreathing exhaled air that contains a higher concentration of carbon dioxide than the surrounding air, the blood level of carbon dioxide is brought back up to normal and the respiratory center again sends signals to the breathing muscles. If you hold your breath for too long, reduced oxygen to the brain can lead to dizziness and unconsciousness. Should you pass out, however, the high level of carbon dioxide that was held in the blood stimulates the respiratory center and normal breathing resumes. Breath-holding is a favorite ploy of overcoddled children who have discovered its terrifying effect on their elders. For the reason described, however, it is not possible to commit suicide by holding your breath.

Digestive System

Digestion is the process by which food is broken down mechanically and chemically into simpler molecules that can be used by the body. Most of the carbohydrates, lipids, and proteins consumed by humans and other vertebrates are too large to be absorbed into the blood and must be broken down into their smallest constituent units. Large carbohydrates, then, are ultimately reduced to monosaccharides; lipids to free fatty acids, glycerol, and monoglycerides (fat molecules with two of the three fatty acids removed); and proteins to their constituent amino acids. In vertebrates, food is digested within the tubular **digestive tract,** which extends from the mouth to the anus. In addition, digestion depends on secretions of three types of accessory glands: the *salivary glands* in the mouth, the *pancreas,* and the *liver* (Fig. 10.31).

After being chewed and swallowed, food is propelled slowly through the digestive tract by rhythmic waves of contraction and relaxation by the muscular walls of the tract (Fig. 10.32). This action, called **peristalsis,** also serves to break down food mechanically; that is, to break down pieces of food into physically smaller particles. Peristalsis aids in mixing food with the secretions of the digestive tract for more efficient chemical digestion. The secretions involved in the chemical digestion of foods are the digestive *enzymes* secreted by specialized cells throughout much of the digestive tract. You will recall from Chapter 7 that an enzyme is an organic molecule, usually a protein, that catalyzes a specific chemical reaction.

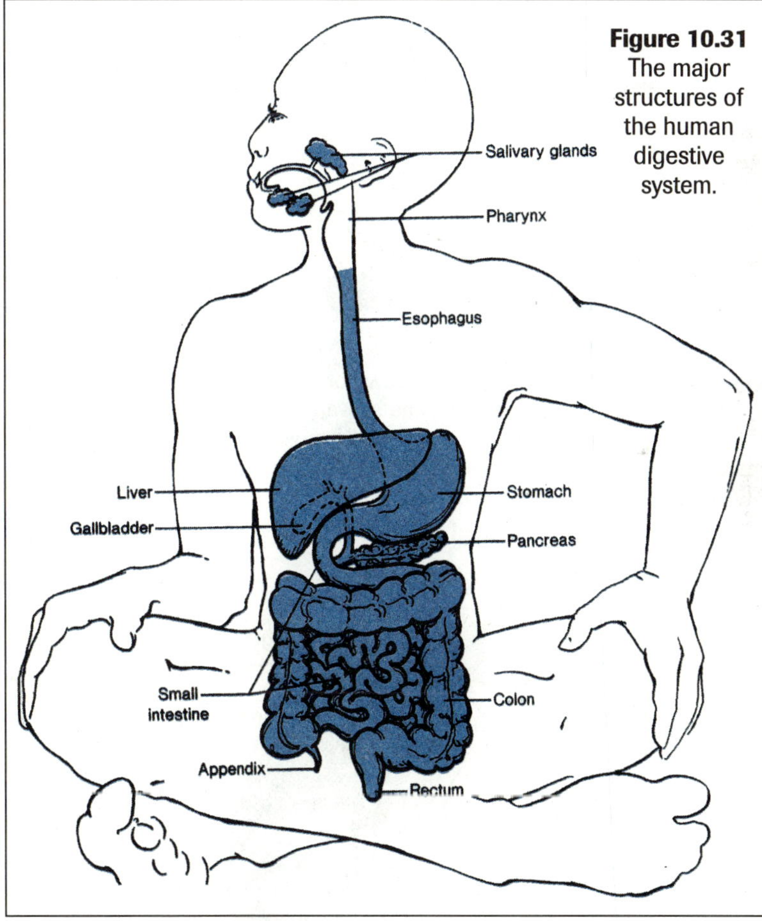

Figure 10.31 The major structures of the human digestive system.

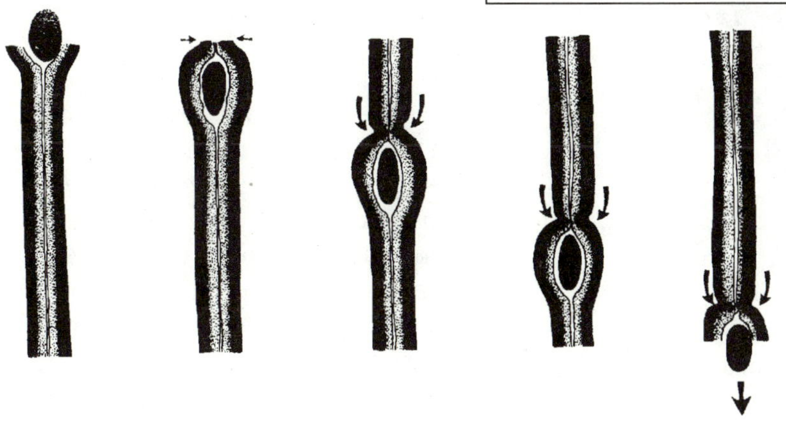

Figure 10.32 Peristalsis in the human digestive tract. The alternative contraction and relaxation of the smooth muscle wall of the tract is controlled by the autonomic nervous system. A wave of contraction forces the food mass downward; this action is followed by a moment of relaxation, and then another wave of contraction.

Digestion of Food

Assume you have consumed your favorite triple-decker cheeseburger. The cheeseburger contains all three major organic nutrients: the bun is carbohydrate, consisting of the polysaccharide starch; the meat patties are protein; and the slice of cheese is lipid. Chewing mixes the food with saliva, which is secreted by three pairs of salivary glands in the mouth. Saliva is a mixture of watery *mucus* and the starch-digesting enzyme **salivary amylase.** Salivary amylase begins the digestion of carbohydrates by breaking down starch into several smaller molecules, including the disaccharide *maltose* (Fig. 10.33). There are no lipid- or protein- digesting enzymes in the mouth, although both nutrients are broken down physically. Salivary amylase, as with all enzymes, functions most effectively at a certain pH. The optimum pH for salivary amylase function is about 6.7.

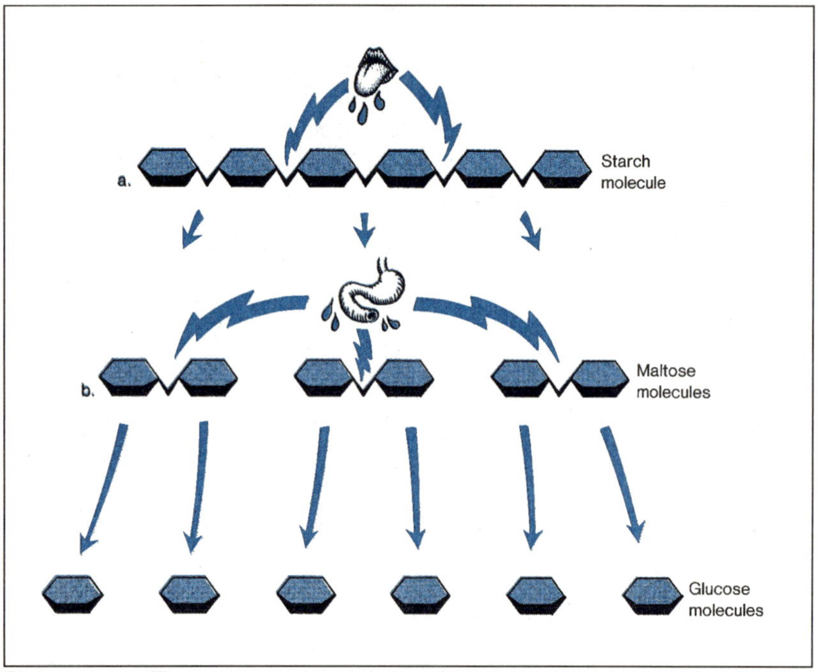

Figure 10.33 The enzymatic breakdown of starch in the human digestive system. (a) In the mouth, the enzyme salivary amylase acts on the polysaccharide starch, breaking it down into separate molecules of (b) the disaccharide maltose. Enzymes secreted into the small intestine break down the maltose molecules into individual molecules of (c) the monosaccharide glucose.

From the mouth, the liquefied cheeseburger passes into the pharynx, or throat, down the esophagus, and into the stomach. At the junction of the esophagus and stomach, the muscular wall of the esophagus forms a constriction called the **cardiac sphincter** that normally prevents regurgitation of the stomach contents into the esophagus (Fig. 10.34). Sometimes, however, this cannot be prevented and the regurgitated acid contents from the stomach produce *heartburn,* although it is your esophagus that is burning, not your heart. Repeated regurgitation of acid from the stomach may result in erosion of the lining of the esophagus and development of *esophageal ulcers.*

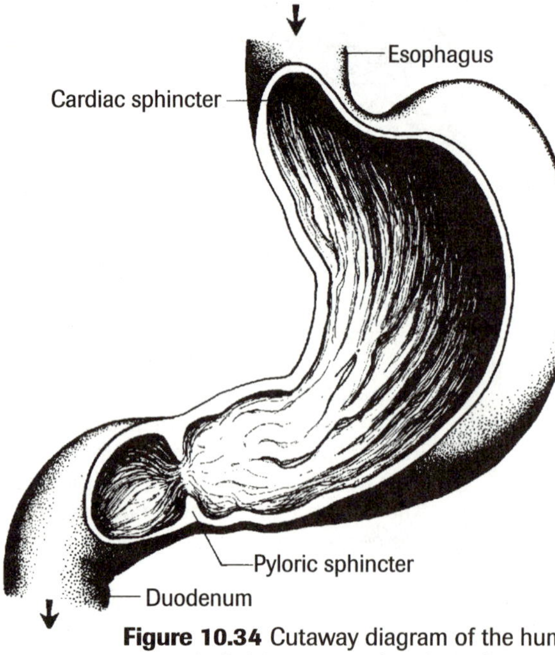

Figure 10.34 Cutaway diagram of the human stomach. Sphincter muscles close the upper entrance and lower exit regions of the stomach.

Cells in the stomach secrete several different products, including *mucus, hydrochloric acid* (HCl), and the protein-digesting enzyme **pepsin.** Actually, this enzyme is secreted in its inactive form called *pepsinogen,* which is converted to pepsin in the presence of hydrochloric acid. Because of hydrochloric acid, food entering the stomach is exposed to an extremely acid environment having an average pH of about 2.0. This is the most effective pH for the action of pepsin, which breaks down proteins into small polypeptide fragments. Stomach secretions increase under the influence of the hormone *gastrin,* which is secreted by the stomach and the upper portion of the small intestine (duodenum) in response to the presence of food in these organs.

The stomach, the muscles of which are protein, does not digest itself because the enzyme pepsin, as mentioned, is secreted in an inactive form, and is converted to its active form only when HCl concentration is sufficient. In addition, stomach cells secrete a thick layer of mucus that protects the stomach lining. Erosion of the stomach lining resulting in *gastric ulcers* apparently is caused primarily by a species of bacteria. Hypersecretion of pepsin and HCl also contribute to ulceration in the stomach.

Very little lipid and no carbohydrate digestion takes place in the stomach. Salivary amylase from the mouth is inactivated by the acid environment of the stomach.

The degraded cheeseburger, now mixed with mucus, enzymes, and acid, moves from the stomach into the small intestine. This ghastly mixture is referred to as **chyme.** The junction of the stomach and small intestine forms another tight constriction called the **pyloric sphincter** that regulates the movement of chyme into the small intestine and prevents backflow into the stomach. The upper 25 cm (10 in.) of the small intestine is the **duodenum.** The next section is the **jejunum,** approximately 1 m (3 ft.) long, followed by the remaining section of the small intestine, the **ileum,** which is about 2 m (6–7 ft.) long. The duodenum is connected through ducts to the pancreas, which secretes digestive enzymes and an alkaline bicarbonate solution into the duodenum (Fig. 10.35). The bicarbonate solution helps neutralize the acid chyme as it exits the stomach. This can raise the pH value in the duodenum to around 8.0. If chyme is not neutralized or if the duodenum secretes inadequate mucus, stomach acid can erode the lining of the duodenum, producing *duodenal ulcers.*

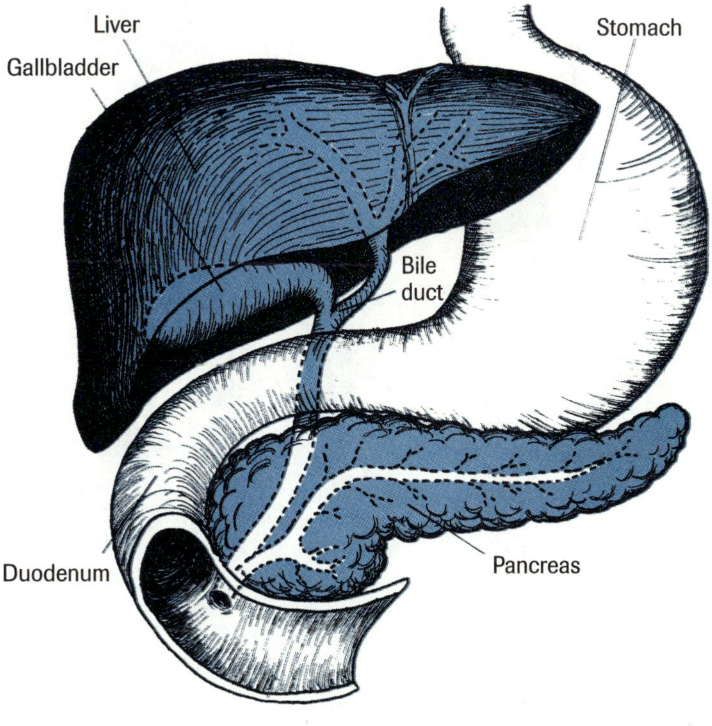

Figure 10.35 Anatomical relationship of the liver, gallbladder, stomach, duodenum, and pancreas. Note that the bile duct and pancreatic ducts empty into the duodenum.

Secretion of the pancreatic digestive enzymes is regulated by hormones released from the small intestine and by neural stimulation. These enzymes hydrolyze all three major nutrients. The enzyme **pancreatic amylase** continues digestion of starch, while several **proteases,** or protein-digesting enzymes, break down polypeptides from the stomach into shorter peptides and amino acids. Lipid digestion requires the enzyme **pancreatic lipase,** along with **bile,** which is produced by the liver. Bile is not an enzyme, but a substance that coats fat droplets, causing them to break apart into very tiny droplets. This action, known as **emulsification,** makes it easier for enzymes to remove fatty acids from lipid molecules. Bile is stored temporarily in the gall bladder and is then secreted through the bile duct into the duodenum.

As a result of pancreatic lipase activity, the original slice of cheese on the cheeseburger now exists in the duodenum as free fatty acids, glycerol, and monoglycerides. This essentially completes lipid digestion. Digestion of carbohydrates and proteins is completed later by enzymes secreted by cells in the intestinal wall. The cheeseburger bun ends up as free glucose molecules, and the meat patties become individual amino acids. The three major nutrients, then, have been broken down into their smallest constituent units and are ready to be absorbed. In addition, the cells lining the small intestine contain *nucleases,* enzymes that break down the nucleic acids DNA and RNA.

Absorption of Food

Before the body can be nourished, food must be absorbed from the small intestine into the circulating blood, which delivers the food to every living cell. Absorption of food occurs mostly in the jejunum, although some absorption also occurs in the duodenum and ileum. The small intestine also absorbs most of the ions, water, and other fluids taken into the body. As shown in Fig. 10.36, the lining of the small intestine is highly folded and contains microscopic fingerlike projections called **villi** that extend from the membrane lining the intestine. The folds and villi greatly increase the surface area of the intestinal lining, which significantly increases the efficiency of absorption. Each villus contains a rich network of capillaries and a lymphatic vessel called the **lacteal.** Monosaccharides and amino acids are taken up by villi cells and absorbed directly into the villi capillaries, but fatty acids and monoglycerides are absorbed by the lacteals.

Following absorption, monosaccharides and amino acids are transported in the blood to the liver, from which they eventually are released into the general circulation. After a carbohydrate rich meal, the liver may store excess glucose in the form of the polysaccharide glycogen. Should the body need glucose at any time, it can draw on the stored glycogen. Fatty acids and monoglycerides are taken up by the villi cells and then reconstructed into lipid molecules. These molecules are coated with a protein found in the villi cells and the lipid-protein complexes then enter the lacteals. The combined lipid-protein molecules move out of the lacteals through lymphatic vessels that eventually deliver the molecules to the circulating blood. The blood distributes

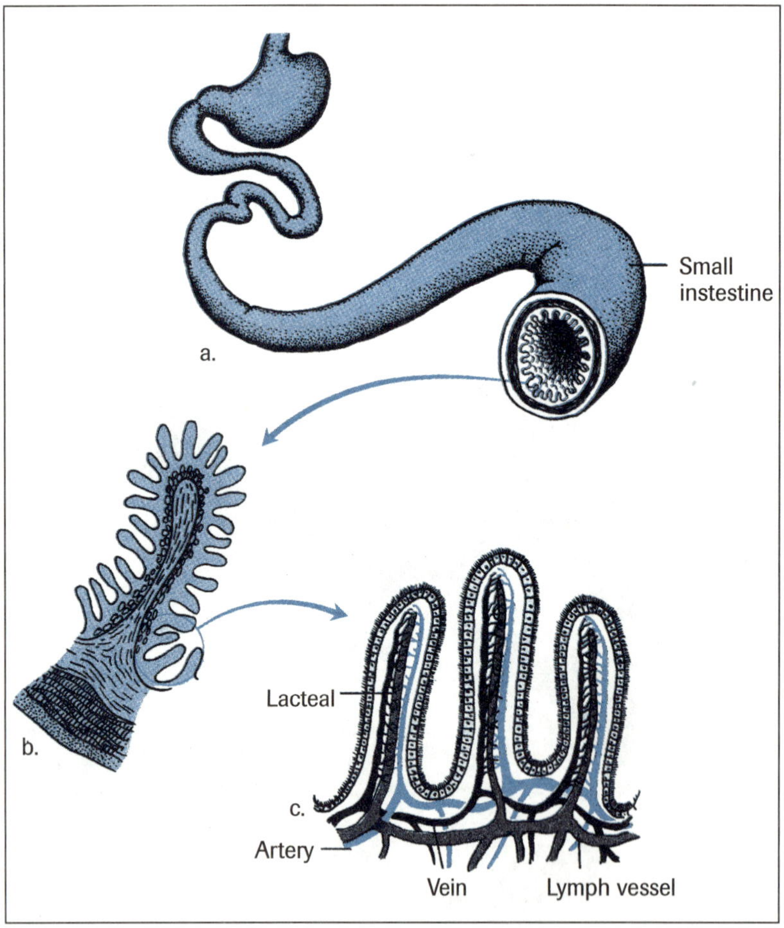

Figure 10.36 (a) Cross section through the human small intestine, showing its folded lining. (b) A single fold covered by fingerlike villi. (c) Three villi and their blood supply. In the core of each villus is a *lacteal* that aids in the absorption of digested lipids. The tiny hairlike extensions surrounding each villus are *microvilli.* Villi or microvilli enormously increase the absorptive surface area of the inner wall of the small intestine.

the absorbed monosaccharides, amino acids, and lipid compounds throughout the body.

The final extension of the digestive tract is the large intestine, or *colon,* which is about 1.5 m (5 ft.) in length. In Fig. 10.37 you can see that the large intestine originates as a closed pouch, the **cecum,** in the right lower abdomen. The terminal end of the ileum joins the large intestine just above the cecum. Attached to the cecum is the worm-like *appendix*. The appendix contains lymphatic tissue, but its function, if any, is unknown. Inflammation of the appendix is called *appendicitis;* the greatest danger from appendicitis is rupture of the appendix with release of bacteria and wastes that can cause inflammation of the lining of the abdomen.

The large intestine terminates at the *rectum* in the pelvic region and the lower few centimeters of the rectum form the *anal canal,* which opens to the outside through the *anus.* There are no digestive enzymes secreted by cells in the large intestine and, hence, no further digestion of the major nutrients.

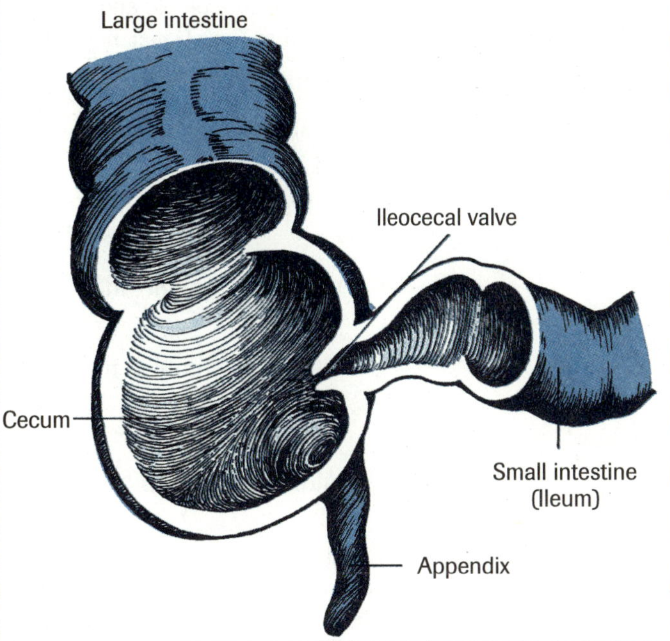

Figure 10.37 Junction of the large and small intestines. Undigested wastes enter the large intestine from the ileum via the ileocecal valve. The valve prevents wastes from backing up into the small intestine.

The primary functions of the large intestine are absorption of water and ions not taken up by the small intestine and the storage and expulsion of undigested wastes. In addition, the large intestine harbors certain species of bacteria that perform several functions. Some bacteria, for example, break down the polysaccharide cellulose, which humans cannot digest in the same manner as other foods. Bacteria also synthesize vitamin K, a factor necessary for normal blood clotting. A person taking excessively large doses of antibiotics to treat an infection may experience internal bleeding and prolonged blood clotting times because the antibiotics may kill the colon bacteria as well. Vitamin K deficiency often can be countered by taking the vitamin in tablet form. Other vitamins, including B_{12}, thiamine, and riboflavin, also are synthesized by bacterial action in the large intestine. Vitamin B_{12} is required for normal blood cell formation and for nucleic acid metabolism. Thiamine (vitamin B_1) deficiency causes muscle weakness and nerve malfunction. Riboflavin (vitamin B_2) is necessary to maintain normal epithelial tissue in the skin and in the lining of the intestine, blood vessels, and other structures.

Nutrition and Diet

Proper nutrition involves the intake of a sufficient quantity and balanced proportion of nutrients so that the body's metabolic needs are met. Carbohydrates, lipids, and proteins are generally referred to as macronutrients, that is, nutrients required by the body in large quantities. Nutrients required only in relatively small amounts are called micronutrients. The micronutrients necessary for normal metabolism are classified as vitamins and minerals. Vitamins are organic compounds, only a few of which can be synthesized by the body; consequently, most vitamins must be obtained in the diet. In humans the necessary water-soluble vitamins include the large group of B complex vitamins (B_1, B_2, B_6, B_{12}, and so on) and Vitamin C; as mentioned previously, the fat-soluble vitamins are A, D, E, and K. Table 10.1 lists some of the vitamins and the symptoms resulting from their deficiency in the body.

Table 10.1 The Major Water-soluble and Fat-soluble Vitamins and Some of the Symptoms Resulting from Their Deficiency in the Body

	Vitamin	Symptoms of Deficiency
Water-soluble:	Vitamin B_1 (thiamine)	Impaired function of CNS; weakened heart muscle; gastrointestinal disorders
	Vitamin B_2 (riboflavin)	Scaly skin; cracking of skin at the angles of the mouth
	Vitamin B_6 (pyridoxine)	Inflammation of the skin; gastrointestinal disturbances; convulsions
	Vitamin B_{12} (cobalamin)	Pernicious anemia; occasional loss of nervous sensations and development of paralysis
	Vitamin C (ascorbic acid)	Prolonged wound healing; cessation of bone growth; fragile capillaries (scurvy)
	Niacin	Inflammation of the skin; gastrointestinal disturbances; mental disorders
	Folic acid	Impaired growth; anemia
Fat-soluble:	Vitamin A	Failure of growth; scaly skin; impaired function of reproductive system; nightblindness
	Vitamin D	Bone deformities; rickets in children
	Vitamin E	Impairment of growth; male sterility
	Vitamin K	Prolonged blood clotting time

Minerals are inorganic compounds, usually in the form of ions, that the body requires in relatively small amounts. These include sodium (Na^+), chloride (Cl^-), potassium (K^+), calcium (Ca^{++}), magnesium (Mg^{++}), iodine (I^-), zinc (Zn^{++}), and several others. Each of these is required in varying amounts. Some of the minerals, along with their functional roles, are given in Table 10.2.

Energy Needs. Each of the major macronutrients—carbohydrates, lipids, and proteins—contains chemical energy that is released during cellular metabolism. Each gram of carbohydrate or protein contains approximately 4 Cal (Kcal), and each gram of lipid contains 9 Cal. The amount of energy needed by the body varies with size, sex, and extent of physical activity. The maintenance of good nutrition for a 70-Kg (154-lb) adult male requires about 2,500 to 3,000 Cal per day; a 54-Kg (120-lb) female requires about 2,000 to 2,200 Cal per day. During pregnancy and nursing, the daily requirement for females includes an additional 300 to 500 Cal. These caloric requirements, of course, may be much higher if someone is engaged in a vigorous daily exercise program.

In terms of good health, one of the major concerns is the effect of caloric intake on body weight. If the number of calories taken into the body is equal to the number expended, the body weight remains constant. When more calories are consumed than given off, the excess food is stored as fat, and the individual *gains* weight. Since each gram of lipid contains 9 Cal, the body stores 1 gram of lipid for every 9 Cal of *excess* food energy consumed. On the other hand, one *loses* weight when the number of calories expended is greater than the number taken into the body. This is because the stored fat is "burned" and is not replaced by an equivalent amount of food energy.

Table 10.2 Some of the Minerals and Their Functional Roles in the Human Body

Mineral	Functional Role
Sodium (Na)	Maintenance of osmotic pressure of body fluids; necessary for muscle contraction and conduction of nerve impulses
Chlorine (Cl)	Maintenance of osmotic pressure of body fluids; important in acid-base balance
Potassium (K)	Aids in maintaining osmotic pressure within cells; necessary for muscle contraction and conduction of nerve impulses
Calcium (Ca)	Constituent of bones and teeth; necessary for blood coagulation, muscle contraction, and conduction of nerve impulses
Magnesium (Mg)	Necessary for normal muscle and nerve function; associated with production of ATP
Phosphorous (P)	Constituent of bones and teeth; component of DNA, RNA, and ATP
Iodine (I)	Necessary for the synthesis of thyroid hormones
Iron (Fe)	Constituent of hemoglobin, myoglobin, and respiratory enzymes
Copper (Cu)	Constituent of many enzymes; necessary for hemoglobin synthesis and production of melanin
Zinc (Zn)	Constituent of many enzymes; necessary for normal wound healing

Chapter Summary

1. Plants require various nutrient elements, which are acquired in the form of inorganic ions, or minerals. Minerals are taken up from the soil by diffusion and active transport through the root cells. Insect-trapping plants can acquire nitrogen and other nutrients through absorption of amino acids from digested insect protein.

   Water enters root cells by osmosis, which occurs in response to high solute concentration in the root or reduced water pressure in the xylem. Root hairs greatly increase the surface area of the root and aid in the uptake of water and minerals.

   Plants lose water by evaporation through stomata, a process called transpiration. Water transport in plants results primarily from transpiration. Solar energy drives transpiration, and, in a few plants, the energy of ATP maintains root pressure. Plant surfaces must be moist to facilitate effective gas exchange.

   Translocation, the transport of food in the phloem of vascular plants, follows a source-to-sink direction, and has been explained in terms of the pressure-flow theory. The major organic nutrient transported is the disaccharide sucrose.

   Key Terms: *macronutrient, p. 213* · *micronutrient, p. 213* · *cotransport, p. 215* · *root pressure, p. 216* · *transpiration, p. 216* · *translocation, p. 218*

Concept Questions: What factors facilitate the uptake of nutrients by plants? What physiologic and environmental influences might counter this uptake?

It has been reported that the tallest tree currently standing is a species of California redwood that is 121 meters (362 feet) tall. Explain how water in the tree is transported to this great height. Explain how sucrose in leaf cells of a plant reaches the growing cells in the root of the plant. In the early morning, water droplets often appear on the tips of grass blades. What processes do you think might be responsible for this phenomenon?

2. All animals except sponges and cnidarians have distinct muscular systems. Definite circulatory and digestive systems are first found at the level of roundworms. Animal circulatory systems are either open or closed. Some invertebrates and all vertebrates have a closed circulatory system. Some invertebrates exchange respiratory gases by diffusion across the body wall. Respiratory organs in some invertebrates include gills, book lungs and tracheal systems. Digestion in sponges occurs within phagocytic cells; in cnidarians and some flatworms, digestion is accomplished by enzyme-secreting cells within a hollow body cavity. Other invertebrates and all vertebrates possess a complete digestive system consisting of mouth and anus.

   Key Terms: *longitudinal muscle, p. 220*
   circular muscle, p. 220
   open circulatory system, p. 220
   closed circulatory system, p. 221

   Concept Questions: Compare generally the similarities and differences in the exchange of respiratory gases in cnidarians and insects. What do you think might be the comparative advantages and disadvantages of open and closed circulatory systems?

3. Fishes have a two-chambered heart, while amphibians and most reptiles have a three-chambered heart; birds and mammals have four-chambered hearts. Agnathans, fishes, and some amphibians respire by means of gills; reptiles, birds, and mammals possess lungs. The digestive systems of different vertebrates are similar.

   The three types of vertebrate muscle tissues are skeletal, smooth, and cardiac. Muscle contraction results in motion, body support, and heat production. Skeletal muscles that move the bones of the body extend across a joint. Control of muscle activity usually requires input from the nervous system. Muscle contraction and relaxation involve the sliding action of thick and thin protein filaments that comprise the muscle fibers. Skeletal muscle fibers may be slow-twitch or fast-twitch.

   Key Terms: *atrium, p. 223*
   ventricle, p. 223
   muscle fiber, p. 224
   actin, p. 226
   myosin, p. 226
   cross-bridges, p. 227
   myoglobin, p. 227
   slow-twitch fiber, p. 227
   fast-twitch fiber, p. 227

   Concept Question: You have been awarded a degree in muscle physiology. A runner complains to you that he cannot break three hours in the marathon, but can run the 100-meter dash in 11 seconds flat. Inform him of the characteristics of skeletal muscle and ease the exasperation behind his complaint.

4. The circulatory system of mammals consists of the heart, blood vessels, blood, and lymphatic system. Blood circulatory vessels include arteries, veins, and capillaries; their essential function is to transport blood. In addi-

tion, capillaries function in the exchange of substances between blood and the body cells. Blood is a liquid tissue consisting of plasma and blood cells. Blood pressure is the force of blood against a blood vessel wall.

The mammalian heart is a four-chambered pump that delivers deoxygenated blood to the lungs and oxygenated blood to all living cells. Heart rhythm is regulated by nerves of the autonomic nervous system in association with the pacemaker of the heart.

The lymphatic system produces lymphocytes and consists of lymph vessels, lymph, and lymph nodes. The system picks up plasma fluid and blood proteins that leak from blood capillaries. Lymph nodes filter lymph and contain lymphocytes and phagocytes that destroy bacteria and other foreign matter.

Key Terms: *artery, p. 228*
vein, p. 228
capillary, p. 228
blood, p. 229
blood cells, p. 229
systolic pressure, p. 230
diastolic pressure, p. 230
superior vena cava, p. 231
inferior vena cava, p. 231
pulmonary artery and veins, p. 231
aorta, p. 232
coronary arteries and veins, p. 232
lymph, p. 233
lymphocyte, p. 234

Concept Questions: The vertebrate heart has evolved from the single ventricle of fishes, amphibians, and most reptiles to the completely separated paired ventricles of birds and mammals. In a single ventricle, oxygenated and deoxygenated blood are mixed to some extent, while in the paired ventricles, oxygenated blood is confined to the left ventricle and deoxygenated blood is confined to the right ventricle. What might be the effects of this mixing and separation of blood on the functioning of animals?

As an eminent veterinarian, you inform an owner that his dog has a blood clot in its right shoulder and that this presents the possibility of a future stroke in the animal. How do you explain this to the owner?

Why is the lymphatic system considered an adjunct to the blood circulatory system?

5. The vertebrate respiratory system functions in supplying and removing respiratory gases in the body. Alveoli are the sites of respiratory gas exchange in the lungs. Gas exchange occurs across capillaries by diffusion; this exchange occurs between the lungs and the blood, and between the blood and body cells. The major factor regulating the respiratory rate is the level of carbon dioxide in the blood.

Key Terms: *pharynx, p. 234*
trachea, p. 234
bronchus, p. 234
bronchiole, p. 234
alveoli, p. 234
surfactant, p. 236
diaphragm, p. 236
respiratory center, p. 238

Concept Questions: Trace the route of oxygen from the atmosphere to a cell in your big toe, and explain what happens to carbon dioxide in the same cell.

6. In the digestive tract, food is broken down mechanically and chemically into forms that can be absorbed into the blood. Digestive enzymes, found in the mouth, stomach, pancreas, and small intestine, break down food chemically. Absorption of foods occurs in the small intestine. Carbohydrates and proteins are absorbed directly into the circulating blood, but lipids are absorbed by lacteals in the intestinal villi.

The large intestine absorbs some ions and water, but does not secrete digestive enzymes. It also stores and expels digestive wastes. Bacteria in the large intestine break down cellulose and synthesize various vitamins.

Key Terms: *peristalsis, p. 239*; *salivary amylase, p. 240*; *pepsin, p. 240*; *pancreatic amylase, p. 241*; *pancreatic lipase, p. 241*; *protease, p. 241*; *bile, p. 241*; *emulsification, p. 241*; *villi, p. 242*; *lacteal, p. 242*; *cecum, p. 243*

Concept Questions: Why is digestion referred to as a physical and chemical process? Describe the general roles of the mammalian muscular, circulatory, and lymphatic systems in the processes of digestion and absorption of food.

Part Six

Life Feature 4: Homeostasis

Chapter Eleven

The Balance Inside

Concept and Significance of Homeostasis

This chapter and the following chapter will explore the life feature of *homeostasis,* a term derived from the Greek *homeos,* "similar" and *stasis,* "standing." Homeostasis refers to the ability of an organism to maintain relatively constant physiological conditions in response to a changing external environment. It also refers to the maintenance of stability and the resistance to change in communities and ecosystems, more complex levels of organization. Accordingly, it is actually impossible to separate homeostasis from the other life features because it is concerned with the functional maintenance of all levels of organization in the living world. This chapter deals with homeostasis of individual organisms; homeostasis in communities and ecosystems will be considered in Chapter 12.

Evolution has provided adaptive mechanisms that allow organisms to survive in extremely different environments and under varying physiological conditions. In dry deserts, salty waters, subfreezing temperatures, or in any other environment, the interior, or cellular environment, of an organism must remain relatively constant. This principally involves regulating the composition of the fluid bathing every living cell. Within certain limits, organisms normally are able to make the necessary internal adjustments automatically if their environment or physiology is altered. Inability to make these adjustments results in malfunction or death.

In previous chapters we noted that organisms must exchange materials across their cell membranes, regulate water balance, maintain a relatively constant pH, control the cellular concentration of ions, and synthesize ATP through cellular respiration. If all physiological activities result in a balance between nutritional needs and elimination of wastes, the organism is said to be in a **steady state** condition. This, as we will see, is the essence of homeostasis.

Negative Feedback

Regulation of homeostasis requires that an organism be able to detect any change in its internal environment and to make an appropriate response. This sequence requires some type of *sensory receptor,* which may be located in the cell membrane of single-celled organisms or within specific structures of multicellular organisms. Information received by the sensory receptor is relayed in some organisms to a *control center,* which

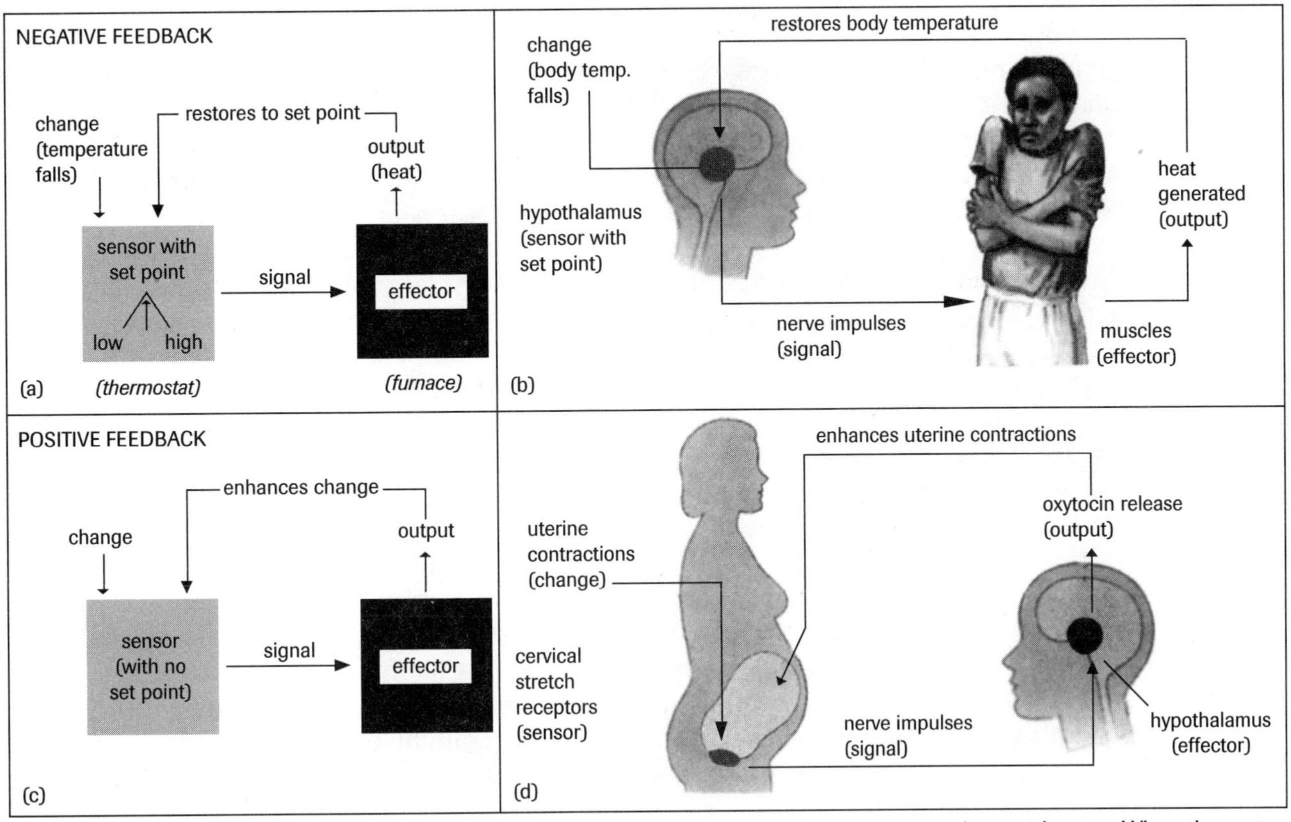

Figure 11.1 Negative and positive feedback (a) Negative feedback controls the temperature in most homes. When the system changes (the house cools below a specific temperature), the change is sensed by a sensor with a specific set point (the thermostat, set to 68 F). The sensor sends a signal to an effector (the furnace) whose output (heat) returns the system to its set point. The signal is turned off, which turns off the effector. (b) Body temperature is both set by and sensed by neurons in the hypothalamus (the sensor with a set point). A drop in body temperature causes the hypothalamic neurons to send a signal to effectors such as muscles, which begin shivering. Shivering generates heat, which helps restore body temperature to its set point, shutting off the signal, and causing shivering to cease. (c) In positive feedback, a change in the system is sensed by a sensor with no set point, which then sends a signal to an effector. The output of the effector causes an even greater change in the system, rather than attempting to return it to previous conditions. (d) At the onset of labor, uterine contractions are sensed by receptor neurons in the cervix, which send a signal to the hypothalamus that results in secretion of a hormone (oxytocin) from the posterior pituitary gland. The hormone, in turn, stimulates more and stronger uterine contractions, ultimately leading to the expulsion of the baby and placenta from the uterus.Reprinted from *Biology: Life on Earth,* Fourth Edition, by Gerald and Teresa Audesirk, 1996, Prentice-Hall, Inc.

regulates a response carried out by an *effector*. This pathway maintains homeostasis in organisms primarily through a mechanism known as **negative feedback** (Fig. 11.1).

A simple way to explain negative feedback is to compare it to the principle of a thermostat on a home air conditioner. If the thermostat is set at 20°C (68°F), the air conditioner clicks on when the room temperature rises above the temperature setting. When the room temperature reaches 20°C, the thermostat shuts off the air conditioner. This maintains a nearly constant temperature in the room. The mechanisms of cause and effect are opposites; that is, an elevated temperature causes the air conditioner to turn on and effect a lower temperature. Because the effect is opposite to the cause, the mechanism is said to be negative. In other words, negative feedback shuts off or decreases the effect of a process that otherwise would continue to magnify in effect. In our example of the air conditioner, the temperature of the room would continue to rise unless countered by the thermostat control mechanism. In living organisms, the mechanism of negative feedback acts to detect an alteration in the body's internal environment and responds to that change by returning the internal environment to its original condition.

Many animals also have internal "thermostats" that regulate numerous physiological processes. A familiar example is the maintenance of human body temperature at 37°C (98.6°F). As discussed in Chapter 9, several masses of neurons, or nerve cells, located in the forebrain above the pituitary gland form the hypothalamus. These neurons communicate with other parts of the nervous system through interconnecting neurons. Some of the neurons in the hypothalamus are sensitive to temperature changes in the blood flowing through arteries in the area. If the blood temperature is above normal, the neurons send impulses to the sweat glands and capillaries in the skin. The sweat glands release perspiration and the capillaries dilate, or expand, which allows heat to escape from the surface of the skin. This tends to cool the body and return the temperature back to within its normal range. This is the mechanism at work when you exercise vigorously on a hot day.

If your blood temperature falls below normal, as it tends to do in winter, impulses from the hypothalamus inhibit sweating, constrict skin capillaries, and promote shivering by the skeletal muscles. Lack of sweat and constriction of skin capillaries restricts heat loss, while shivering increases heat production as the muscles metabolize more rapidly. Negative feedback not only affects temperature regulation in many animals, but it also controls water balance, the amount of sugar in the blood, the concentration of ions in the body fluids, the production of blood cells, and virtually every other physiologic activity of the body.

Positive Feedback

Another regulatory mechanism in organisms is positive feedback. In **positive feedback,** a stimulus causes a response that further increases the effect of the stimulus, instead of decreasing it as is the case with negative feedback. Positive feedback occurs, for example, during a muscle cramp. Muscle cramps may result from dehydration and loss of sodium ions, restriction of blood flow, a change in pH, or damage to a muscle. As the affected muscle contracts, sensory impulses are sent to the spinal cord. The impulse is relayed to a motor neuron that sends impulses from the cord to the muscle, causing it to contract further; this contraction sends sensory impulses back to the spinal cord where a motor impulse again is sent to the muscle, causing it to contract even further. This sequence is repeated until the muscle is in a state of tight contraction we call a cramp. The effect of the nerve stimulus, then, is to increase muscle contraction instead of decreasing it. The muscle eventually relaxes when the cause of the stimulus has been corrected or removed.

Positive feedback is less common than negative feedback in regulating homeostasis and does not necessarily maintain the body in a steady state. This is not to say that positive feedback is unimportant. In many instances, positive feedback is critical for the normal functioning and maintenance of an organism. In Chapter 10, for example, we discussed the digestion of food in the stomach. The hormone gastrin is secreted by stomach cells in response to partially digested proteins in the stomach. Gastrin then circulates in the blood and returns to the stomach where it stimulates the secretion of acid and enzymes by the stomach cells. The acid and enzymes break down additional proteins which in turn stimulate release of more gastrin which stimulates more secretion of acid and enzymes.

Using our example of muscle contraction as another illustration, the uterine muscles of female mammals undergo increasingly stronger contractions during the birth process. The head of the fetus stretches the cervix at the opening of the birth canal, stimulating contraction of the uterine muscles, which push the fetus against the cervix. This further stretches the cervix, causing more contractions, which continue to push the fetus, and so on. After birth, the cervix is no longer stimulated and the positive feedback sequence ceases.

Quite often, positive feedback is associated with illness or malfunction of the body. Stress, for example, increases tension, which constricts blood vessels, which raises blood pressure, which damages the heart and kidneys. Cancer, which involves an uncontrollable increase in cell number, is another example. For the most part, our emphasis will be on the mechanisms of negative feedback, which normally maintain the body in its steady state condition.

Homeostasis at the Cellular Level

In a single-celled organism, such as a bacterium, all homeostatic regulation is confined to only one cell. While there is no separate control center in unicellular organisms, they maintain homeostasis by the same basic principle of sensation and response operative in multicellular organisms. Specialized protein molecules embedded in the bacterial wall or membrane function as sensory receptors that detect stimuli in the external environment. These receptors respond to stimuli by bringing about movement, speeding up cellular metabolism, activating enzymes, or transporting substances. These same processes may be inhibited or slowed in accordance with environmental conditions. As an example, assume that a cellular substance produced in one area of the cell cytoplasm diffuses to another area in the cytoplasm where it is metabolized by the cell. Normally, the production, diffusion, and metabolism of the cellular substance occur at about the same rate so that the cell is in a steady state condition. In other words, the rates of production and metabolism of the substance are in balance; the substance, therefore, does not tend to accumulate in the cell as it would if the rate of production were too fast or if the rate of metabolism were too slow. If the production rate of the cell substance suddenly increased, however, it would diffuse more rapidly through the cytoplasm and would tend to accumulate at the site where it is metabolized. This, in turn, would also tend to increase the rate at which the substance is metabolized. In time, the cell will compensate, within limits, for the increase in the production of the substance to the extent that the rates of diffusion and metabolism of the substance will equal the rate at which it is produced. This establishes a new steady state and illustrates the tendency of living cells to maintain a homeostatic condition.

To maintain a steady state condition, there must be a balance among all the processes affecting the well-being of the cell. Cellular metabolism, for example, depends on the transport of organic nutrients into the cell, and on the activation of enzymes necessary to catalyze the reactions of metabolism. As we have mentioned, the rate of cellular metabolism cannot be too slow or too fast. If too slow, metabolism cannot meet the energy demands of the cell; if too fast, the end products of metabolism may accumulate in the cell, disrupting the functioning of other cellular processes. Single cells are confronted with the same basic problems of existence—including the maintenance of homeostasis—as are the most complex multicelluar organisms.

Regulating Homeostasis: Interplay of the Vertebrate Endocrine and Nervous Systems

In Chapter 9, we examined the role of the vertebrate nervous system in carrying out the sensation-response mechanisms of the body. It was pointed out that the nervous system, through its intercommunicating network of neurons, provides almost instantaneous response to external and internal stimuli. In vertebrates, as well as many other animals, the nervous system is indispensable to the maintenance of homeostasis.

The other system that functions in concert with the nervous system to exert overriding control of most body functions is the *endocrine system,* which consists of several organs and glands that release chemical messengers called hormones into the bloodstream (Fig. 11.2). A **hormone** is released in one part of the body and transported in the blood to its target organ, a body structure on which the hormone exerts a specific physiological effect. Some hormones, however, affect many or most parts of the body. Some of the effects of hormone stimulation include muscle contraction, glandular secretion, cell growth, blood cell production, and release of other hormones.

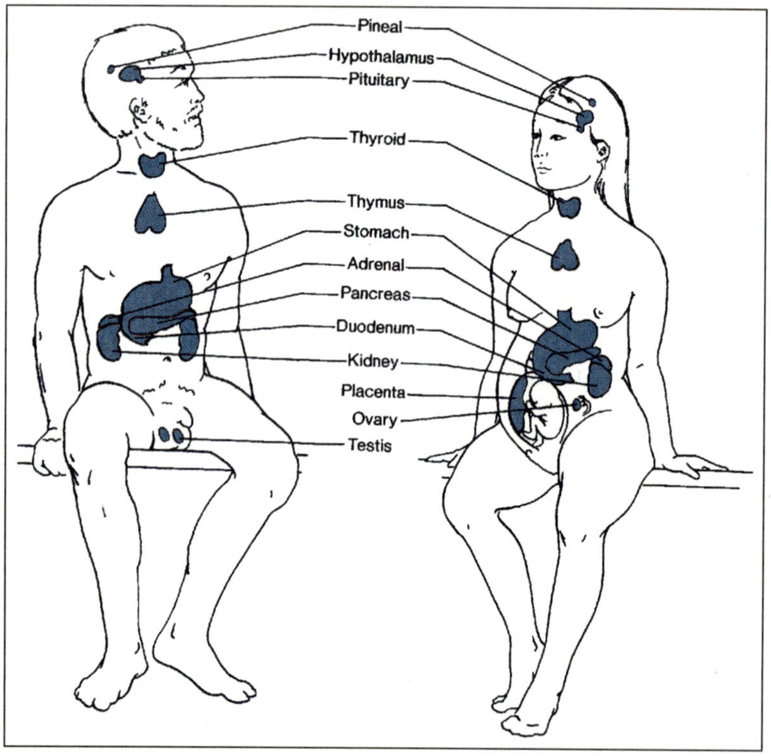

Figure 11.2 Location of the human endocrine organs.

The interaction between the endocrine system and nervous system is reciprocal; that is, hormones stimulate neurons, causing them to fire, and neurons stimulate endocrine organs, causing the release of hormones. Hormones also are released in response to chemical stimuli, including other hormones. One human endocrine organ—the hypothalamus—produces hormones that stimulate the pituitary gland to secrete nine important hormones. Two hormones—antidiuretic hormone and oxytocin—secreted by the pituitary gland are produced by the hypothalamus itself. Through the reception of chemical and nervous stimuli from throughout the body, the hypothalamus plays a critical role in the maintenance of homeostasis.

Regulatory Effects of the Pituitary Gland and Hypothalamus

As shown in Fig. 11.3, the human pituitary gland is suspended by a small stalk under the floor of the forebrain just below the hypothalamus. The gland is divided into an anterior lobe and a posterior lobe. Nine known hormones are secreted by the pituitary

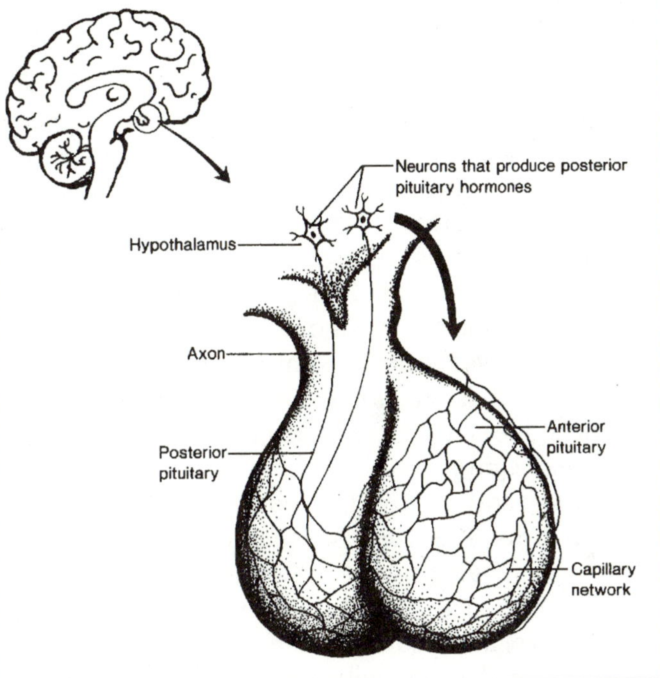

Figure 11.3A The pituitary gland lies beneath the hypothalamus on the underside of the brain. Secretion of the *anterior* pituitary hormones is controlled by releasing and inhibitory factors in the hypothalamus. The factors enter the capillaries and are transported in the blood to the anterior pituitary. Hormones of the *posterior* pituitary are produced by neurons in the hypothalamus. The hormones travel along the axons of the neurons and are absorbed into the capillary blood.

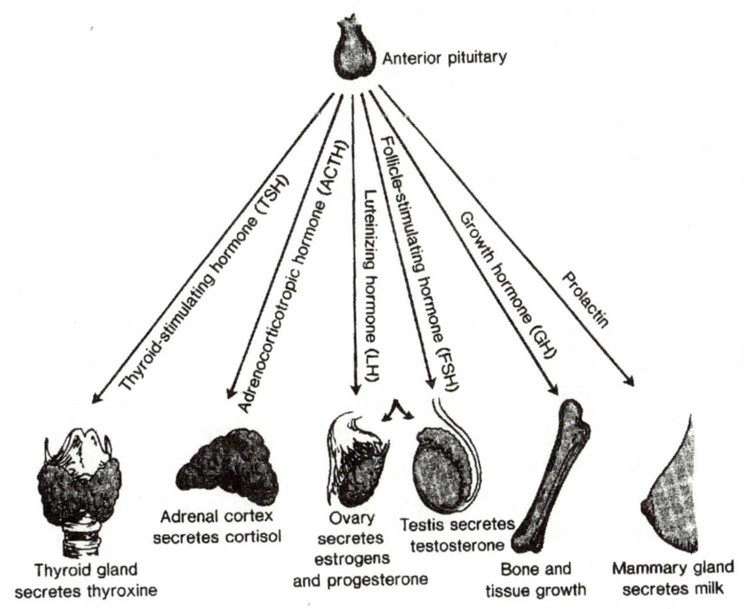

Figure 11.3B Hormones secreted by the anterior pituitary gland. The target organs stimulated by each pituitary hormone are indicated.

gland, seven by the anterior lobe and two by the posterior lobe. Neurons in the hypothalamus produce releasing and inhibiting hormones that cause the anterior pituitary either to release one of its hormones into the blood or to inhibit release of the hormone. In turn, release of hormones from both the hypothalamus and pituitary gland is controlled by negative feedback from secretions of other endocrine organs and target organs. Antidiuretic hormone and oxytocin of the posterior pituitary gland are produced by neurons in the hypothalamus. Secretion of these hormones is controlled by sensory impulses coming into the hypothalamus.

Other Endocrine Organs and Hormones

In addition to the organs stimulated directly by secretions of the pituitary gland and hypothalamus, other organs of humans and some other vertebrates also have an endocrine function. The *pineal gland,* located in an area above the hypothalamus, consists of cells that synthesize the hormone *melatonin,* which inhibits production of the pigment melanin in skin cells. In some animals, melatonin also inhibits reproductive function by slowing the maturation of reproductive organs. Whether melatonin exerts this function in humans is unclear. Production of melatonin is influenced by light entering the visual pathways; nerve branches from these pathways enter the pineal gland. Melatonin production is highest in darkness and lowest in daylight.

The kidneys release the hormone *renin* when the blood pressure or volume decline. Renin secretion stimulates formation of certain blood proteins that cause the blood vessels to constrict and the kidneys to retain water, preventing a further decline in blood volume. The kidneys also release the hormone *erythropoietin,* which stimulates bone marrow to increase red blood cell production (Fig. 11.4). Erythropoietin is released in response to a lowered concentration of oxygen in the body. This occurs when the blood volume is low or it can occur at high altitudes where there is less oxygen that at lower altitudes. Additional red blood cells in the circulation increase delivery of oxygen to the cells.

An endocrine organ discussed later in this chapter is the *thymus,* which is located above the heart. It produces several hormones, the major one being *thymosin,* which is involved in immune system function. The human thymus is quite large in children, but is less than one-third as large in older adults.

A group of fatty acids known as *prostaglandins* often are referred to as "local hormones" because they act on the tissues that release them. Virtually all human body tissues produce prostaglandins, which have a broad range of effects. For example, prostaglandins stimulate contraction of the uterus, raise or lower blood pressure, reduce stomach secretions, inhibit release of female sex hormones, and stimulate pain receptors. Aspirin reduces fever, pain, and inflammation because of its ability to block the synthesis of prostaglandins; that is, in the absence of prostaglandins, many of the body's pain receptors are not stimulated.

Other hormones and hormonelike substances include the digestive hormones discussed in Chapter 10, and the endorphins mentioned in Chapter 9.

It is evident that homeostasis in the vertebrate body is highly dependent on a complex system of chemical communication. To illustrate the chemical regulation of homeostasis, let's consider three contrasting examples—the adrenal glands, the pancreas and the thyroid gland.

Figure 11.4 Control of red blood cell production by negative feedback. A deficiency of oxygen in the body tissues stimulates the release of an enzyme from the liver or kidney. The enzyme acts on one of the blood proteins to release the hormone *erythropoietin.* Erythropoietin stimulates the bone marrow to increase its production of red blood cells. With the increase in the number of red cells in the circulating blood, there is increased oxygen transport and a rise in the level of oxygen in the body tissues. The rise in oxygen concentration inhibits further release of erythropoietin, and red blood cell production by the bone marrow declines.

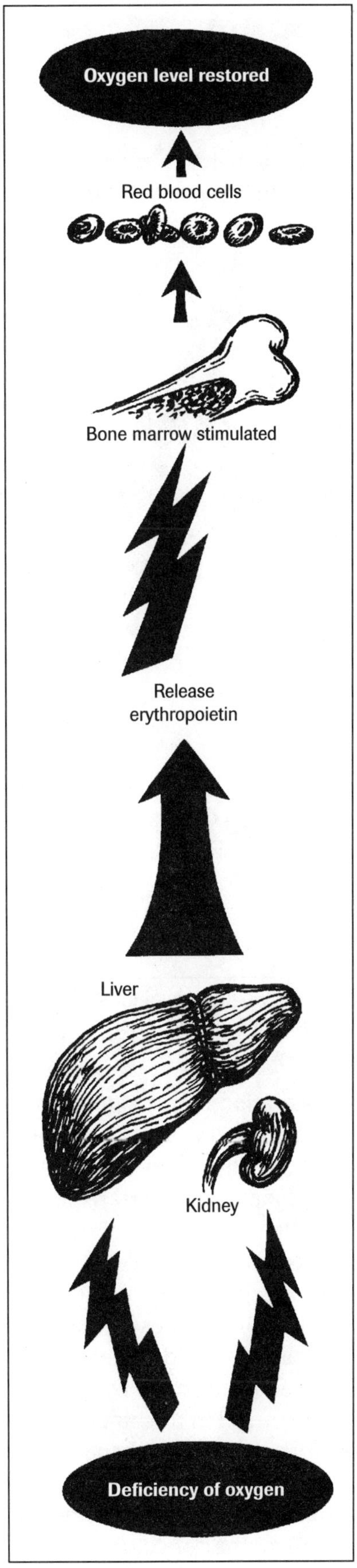

Function of the Adrenal Glands

Mammals have two adrenal glands, one sitting atop each kidney embedded in fatty tissue (Fig. 11.5). An adrenal gland is separated into two functional areas: an outer *adrenal cortex* and an inner *adrenal medulla.* Both areas secrete hormones, but release of the hormones is controlled by different mechanisms. Release of some of the adrenal cortex hormones is controlled by *adrenocorticotropic hormone* (ACTH) from the anterior pituitary gland; release of hormones from the adrenal medulla results from direct stimulation by the sympathetic nervous system.

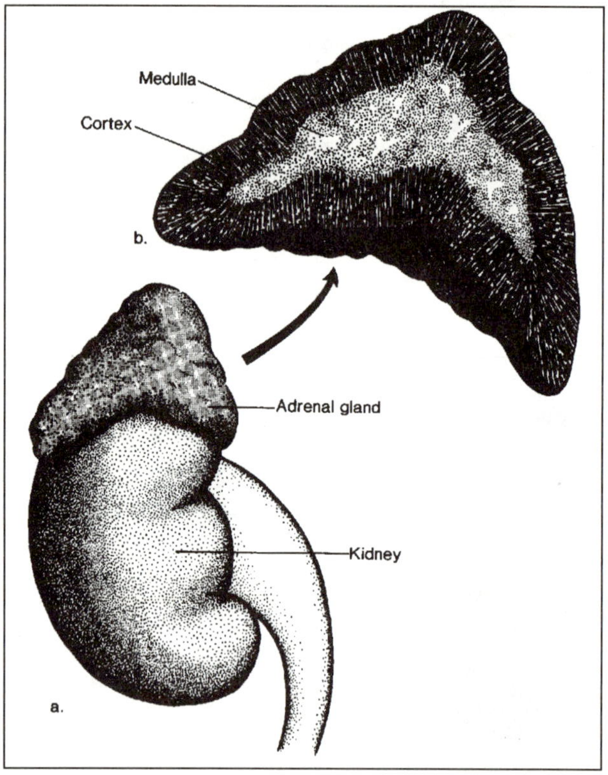

Figure 11.5 (a) Location of one of the two adrenal glands. (b) Cutaway view of an adrenal gland, showing the outer cortex region and the inner medulla.

The Adrenal Cortex

A major hormone secreted by the adrenal cortex is **cortisol**. Cortisol secretion is regulated by negative feedback involving *adrenocortocotropic hormone* (ACTH) which is released from the anterior pituitary gland. In turn, release of adrenocorticotropic hormone is controlled by *corticotropin-releasing hormone* (CRH) from the hypothalamus. The primary factor responsible for the release of cortisol is physical or mental stress. Exercise, illness, injury, and fasting are examples of stress that promote elevated levels of cortisol in the blood. Cortisol causes the release of glucose, fatty acids, and amino acids from most body cells. However, cortisol also prevents the uptake of glucose by these cells. The body then begins to use fatty acids as an energy source, while conserving glucose. Some of the glucose, however, is taken up by the liver and stored as glycogen. Cortisol also promotes the uptake of amino acids by the liver.

When a person or other mammal experiences a stressful situation, nerve signals from the higher brain centers are relayed to the hypothalamus, which begins to secrete corticotropin-releasing hormone into the blood. This hormone travels to the anterior pituitary gland where it causes the release of adrenocorticotropic hormone, which is transported by the blood to the adrenal glands where the hormone stimulates the adrenal cortex to release cortisol.

The effects of cortisol is directed toward providing defenses against the stressful situation. The decreased use of glucose for energy in preference to fatty acids conserves glucose and glycogen and thereby ensures an adequate supply of cellular fuel if the stress persists. If body tissues are damaged or depleted of vital proteins by the stressful condition, amino acids in the blood can be taken up by the liver and synthesized into new proteins.

As the level of cortisol in the blood rises to a certain point, feedback signals are sent to the hypothalamus that decrease the secretion of corticotropic-releasing hormone and also reduce the output of adrenocorticotropic hormone by the anterior pituitary gland. The adrenal cortex responds by lowering its secretion of cortisol. Through this negative feedback mechanism, the concentration of cortisol in the blood is maintained at a normal level when the body is not under stress.

Cortisol and some of the other compounds of the adrenal cortex are *anti-inflammatory* agents. These agents counteract the redness, swelling, and tissue damage associated with inflammation. Anti-inflammatory agents reduce the migration and activity of white blood cells and also inhibit the release of certain chemicals that promote inflammation in the affected area.

Derivatives of these agents are used in medicine to treat inflammatory diseases such as arthritis, allergies, and various infectious conditions.

The Adrenal Medulla

Figure 11.5 shows that the inner layer of the adrenal glands, the medulla, is composed of neurons of the sympathetic nervous system. The two major hormones secreted by the adrenal medulla are **epinephrine,** also called adrenaline, and smaller amounts of **norepinephrine,** or noradrenaline. More specifically, these are *neurohormones,* which, like neurotransmitters that are released at synapses in the nervous system, bring about their effects very rapidly. As with cortisol, these hormones are released in times of physical or emotional stress. Secretion of epinephrine and norepinephrine, however, greatly increases the ability of the body to respond to instant emergencies.

If you encounter a threatening or intense situation, your higher brain centers activate the sympathetic nervous system, causing epinephrine and norepinephrine to be released into the circulating blood. Almost instantly, the heart rate increases and blood vessels in the skeletal muscles, brain, and heart dilate—all geared to increasing the oxygen supply to these organs. This is accompanied by constriction of blood vessels and a reduced oxygen supply to less active organs. The respiratory rate quickens and more energy is made available as glycogen in the liver is broken down into glucose, raising the blood glucose level. Blood pressure rises and you become keenly alert. Overall, the rate of cellular metabolism increases. You are now prepared for what is called the "fight or flight" response. If you were to hear the sound of approaching footsteps in a dark alley, for example, your response might be to stand and fight or depart in haste. The chosen action is undertaken with an intensity far greater than normal. You can think more quickly, fight more furiously, or run much faster than would ordinarily be the case. Once the stressful stimulus is removed, most of the epinephrine and norepinephrine is excreted or metabolized and the body eventually returns to its normal physiological state. The "fight or flight" response is opposite to what could be termed the "rest or repose" response, which is the normal state when the body is not under unusual stress.

The Thyroid Gland

The thyroid gland is situated around the trachea (windpipe) just below the larynx (voice box). The gland secretes several iodine-containing hormones, the most abundant and important of which is thyroxine.

Thyroxine. The principal effect of thyroxine on the body is to increase the rate of metabolism. Thus, under the influence of thyroxine, bone growth increases, glucose is metabolized more quickly, oxygen is consumed more rapidly, the heart rate increases, food is absorbed more readily from the intestine, and so on. Thyroxine also affects overall growth of the body and appears to be necessary for the normal functioning of the growth hormone.

To illustrate the metabolic effect and feedback control of thyroxine secretion, consider the following example. Assume that you have been transported via jet from a tropical island to the Arctic Circle. As you step out of the plane into the subzero temperature, receptors in the skin send nerve impulses to the brain, informing the body of the extreme cold. The impulses

Figure 11.6A Secretion of thyroxine by the thyroid gland is regulated by negative feedback. *Thyroid stimulating hormone* (TSH) from the anterior pituitary acts on the thyroid gland to increase its secretion of thyroxine. The effect of thyroxine is to increase the rate of metabolic activities in many body cells. When its concentration in the blood increases above the required level, thyroxine feeds back on the brain, inhibiting further TSH secretion by the anterior pituitary.

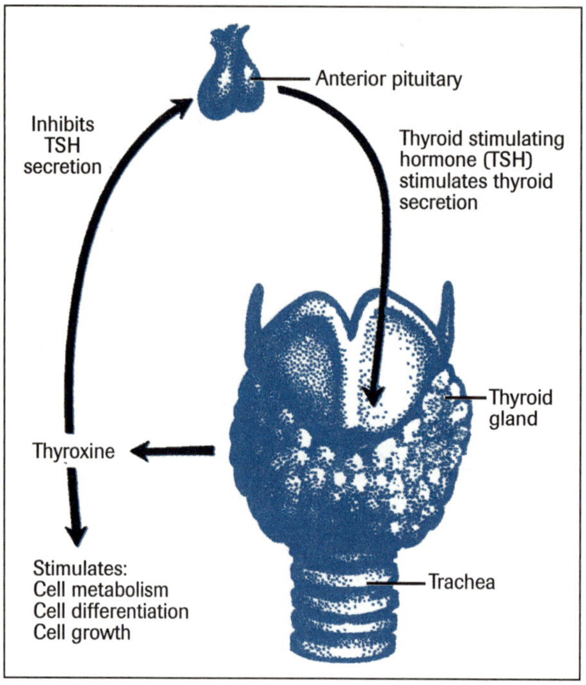

stimulate the hypothalamus, which begins to secrete **thyroid-stimulating hormone releasing factor** (TRF). In turn, TRF stimulates the anterior pituitary to increase its output of **thyroid-stimulating hormone** (TSH). This hormone circulates in the blood to the thyroid gland, which responds by releasing thyroxine into the circulation. As pointed out, thyroxine acts on body cells to increase their rate of metabolism. You will recall that metabolizing cells release *heat energy,* which is necessary to maintain the temperature of the body. Cells metabolizing more rapidly release more heat energy. In the frozen Arctic, this increased heat energy is required to *keep* the body temperature within its normal range. Thus, people who move to cold regions normally develop a higher metabolic rate.

Once the appropriate metabolic rate is set, further secretion of thyroxine is inhibited by negative feedback. The elevated level of thyroxine in the blood decreases the secretion of TRF and TSH, and the thyroid gland is no longer stimulated. In this manner, the level of thyroxine in the blood can be adjusted to maintain homeostasis (Fig. 11.6A).

Endocrine Function of the Pancreas

You will recall from Chapter 7 that the primary fuel for metabolizing cells is the simple sugar glucose. Body cells obtain glucose from the circulating blood primarily after the sugar is absorbed by the small intestine. Normally, the level of glucose in the blood remains fairly constant. This is principally due to the effects of two hormones—**insulin** and **glucagon**—secreted by the pancreas (Fig. 11.6B). The digestive functions of the pancreas were discussed in Chapter 10; in addition, the pancreas acts as an endocrine organ. Insulin and glucagon regulate the blood glucose level by negative feedback. Notice that this regulation mechanism does not involve direct influence of the pituitary hormones.

Following a carbohydrate meal, the concentration of glucose in the blood rises. Cells in the pancreas respond to the increased blood glucose level by secreting insulin; at the same time, the secretion of glucagon is inhibited. Insulin in the blood acts on the cell membranes of body cells, principally those of the liver, fat tissue, and muscles, causing them to take up glucose from the blood. Excess glucose taken up by skeletal muscle and liver cells is stored as the polysaccharide glycogen. Insulin, then, lowers the blood glucose level, allowing the body cells to take up the sugar for metabolism and store it in the form of glycogen. The action of insulin, therefore, prevents **hyperglycemia**, or high blood sugar.

During fasting or between meals, when the blood sugar level tends to fall, the pancreatic cells secrete glucagon and the secretion of insulin is inhibited. In the absence of insulin, body cells are prevented from taking up glucose in the blood. Although there is no intake of food during periods of fasting, the blood sugar level does not fall to dangerous levels. This is attributable to the effect of glucagon, which causes the release of glucose from the glycogen stored in the liver. Secretion of glucagon, then, prevents **hypoglycemia,** or low blood sugar, by promoting release of sugar from the liver.

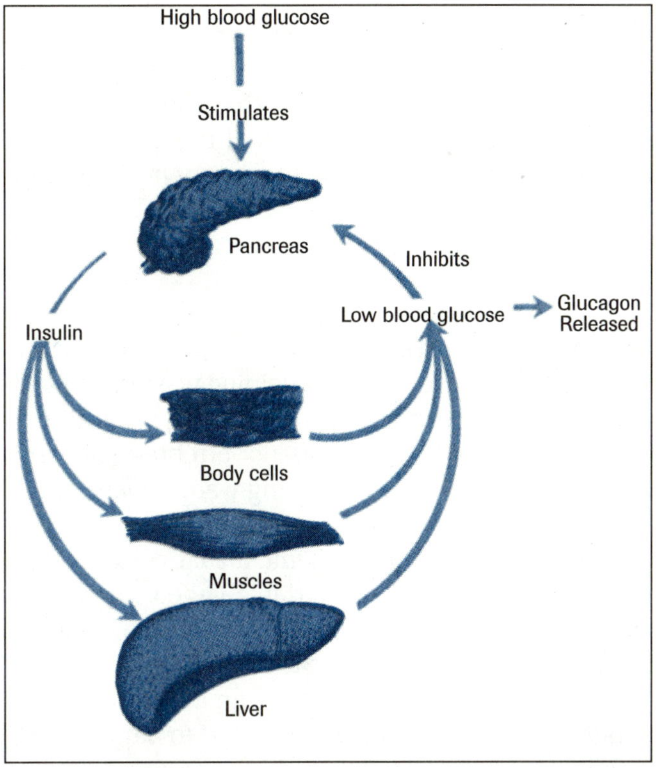

Figure 11.6B Negative feedback control of insulin secretion by the pancreas. A high level of blood glucose stimulates the pancreas to release insulin. Insulin causes the body cells to take up excess glucose from the blood and also stimulates the conversion of glucose into glycogen, which is stored in the liver and muscles. As the blood glucose level falls, the pancreas is inhibited and reduces its secretion of insulin. The falling level of glucose in the blood also stimulates the release of glucagon from the pancreas.

Thus, the secretion of insulin and glucagon are regulated mainly by the concentration of glucose in the blood. If the blood glucose level rises, insulin is released and glucagon is inhibited; if the blood level of glucose falls, glucagon is released and insulin is inhibited. This negative feedback mechanism helps prevent a serious decline in blood glucose levels during fasting, and is a further illustration of a homeostatic mechanism at work in the body.

The Pancreas and Diabetes

Hyperglycemia, or high blood sugar, is associated with the disease *diabetes mellitus.* There are two basic forms of diabetes, referred to as Type I and Type II. Type I diabetes results from destruction of the insulin-secreting cells of the pancreas. Without these cells, no insulin is produced and the level of blood glucose rises, producing hyperglycemia. Sometimes the blood glucose level increases 10 to 12 times its normal value.

In addition, the pancreatic cells secrete an abnormally high amount of glucagon. This further increases the blood glucose level. Unable to utilize glucose, the body cells switch to metabolizing fats as an energy source. Prolonged metabolism of fats releases several acidic by-products into the blood. As these acidic compounds accumulate, the pH of the blood may fall to dangerous levels, often inducing vomiting. If the blood pH, which has an average value of 7.4, falls to a value of 6.9, the central nervous system becomes depressed and the person often becomes disoriented and may pass into a coma. An individual cannot survive Type I diabetes without insulin therapy, which involves a prescribed regimen of daily insulin injections. The regimen, which involves the amount of insulin required and how often it needs to be taken each day, is geared to maintaining stable and normal blood glucose levels over as long a period of time as possible.

Type II diabetes, which most often occurs in obese individuals over 40 years of age, is characterized by normal or even elevated levels of insulin in the blood. The body cells, however, do not respond to the effects of insulin and hyperglycemia results. The cause of this lack of sensitivity to insulin is unknown, but heredity may play a role. Treatment for Type II diabetes includes weight loss and a diet that restricts foods, such as carbohydrates, which tend to increase insulin secretion.

We have seen that the endocrine system functions primarily through chemical messengers that maintain homeostasis through negative feedback. Another system having wide ranging effects on body function is the urinary system, which helps maintain homeostasis by regulating the composition of blood and tissue fluid. The urinary system of vertebrates also is influenced by regulatory hormones of the endocrine system.

Homeostasis and Excretion

All animals—both aquatic and terrestrial—are confronted with the problems of water balance and the excretion of nitrogen-containing wastes. The cellular metabolism of proteins and nucleic acids releases nitrogen, which in most animals is converted into the toxic substance ammonia (NH_3). Most aquatic vertebrates and invertebrates excrete nitrogen in this form directly into the surrounding water where the ammonia can be washed away. Many marine fishes and most terrestrial animals, such as mammals and some amphibians, however, convert ammonia into *urea*. Urea is much less toxic than ammonia and therefore does not need to be diluted in a large volume of water. Birds, insects, and some reptiles excrete nitrogen in the form of *uric acid,* which is a pasty substance from which water has been reabsorbed from the urine. The excretion of nitrogen in the form of urea or uric acid conserves water and thus is an adaptation to a terrestrial existence.

All vertebrates have kidneys which function to varying degrees in excreting water and/or nitrogen wastes. The kidneys of fishes excrete a dilute urine, but the kidneys are relatively unimportant in the excretion of nitrogen wastes, which are excreted in the form of ammonia mainly by the gills.

Various invertebrates have an internal tubular system equipped with discrete excretory organs. In flatworms and some annelids, for example, the tubules contain cilia that propel dissolved excretory products out of the body through tiny pores in the body wall (Fig. 11.7). Insects and other terrestrial arthropods possess excretory tubules that are outpocketings of the intestine. Nitrogen wastes accumulate in the tubules and pass into the intestine where they are eliminated along with the feces.

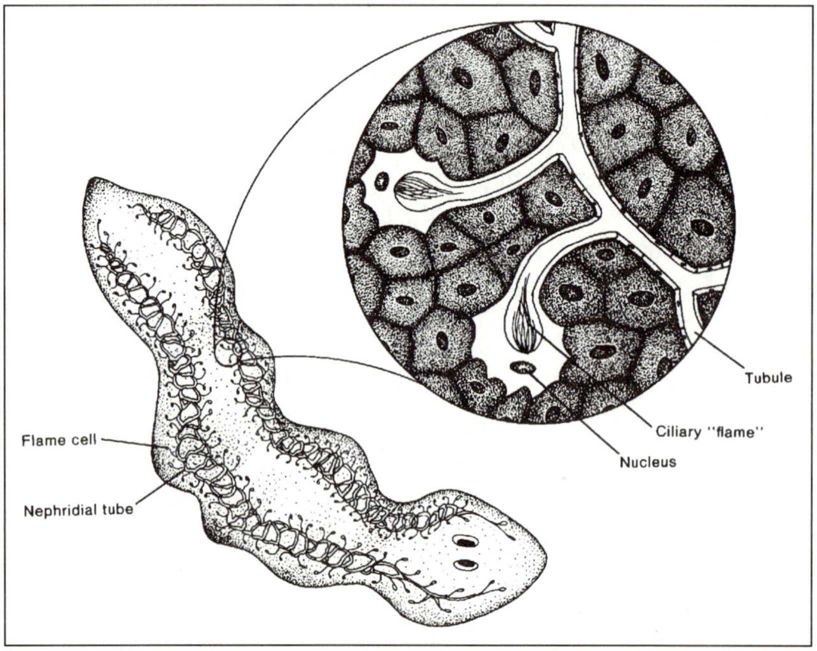

Figure 11.7 Excretory structures of a planarian, showing a close-up of the flame cells. The undulating cilia sweep water and wastes into the tubule and out the body through tiny pores.

Excessive loss or gain of water is a threat to the survival of all organisms. Normally, the amount of water entering and leaving an organism across cell membranes is in balance. We saw in Chapter 5 that too much water entering an animal cell can cause it to swell and burst, while the loss of too much water causes the cell to shrink. The cells of some aquatic animals, such as the freshwater sponges and cnidarians, are directly exposed to the water in which they live. The cellular fluids of these organisms are **hypertonic** to the surrounding water. Consequently, water tends to enter the body cells by osmosis. This is countered by cytoplasmic contractile vacuoles that repeatedly pump water to the outside of the organisms.

The cells of most organisms other than sponges, cnidarians, and a few others are bathed in an internal watery fluid. Almost all aquatic organisms live in an environment that is either hypertonic or **hypotonic** to this fluid. The body fluids of most marine invertebrates, however, are isotonic to the surrounding sea water. The hypotonic environment of freshwater organisms presents the same problem of excessive water uptake as that experienced by sponges and cnidarians. In fishes and other freshwater animals, this problem is countered by excreting a large volume of very dilute urine. Some marine fishes and other vertebrates exist in a hypertonic environment and thus tend to lose water by osmosis. These animals counter this effect by drinking large amounts of sea water and excreting the accumulated salt through salt glands.

Although terrestrial animals face the same basic problem of water balance that confronts aquatic animals, it is the loss of water and the threat of drying out that presents a major homeostatic problem for terrestrial animals. Life on land entails a loss of water by evaporation from the body surface, a condition unimportant to aquatic animals. Water is also lost through breathing, urination, and elimination of wastes. To balance this loss, terrestrial animals replace most of their water through drinking and eating. Regardless of the environment, however, the problem of water balance, as well as that of nitrogen waste excretion, has been solved in most animals through the evolution of discrete excretory structures. One of the most elegant of these structures is the mammalian kidney.

Mammalian Urinary System

The urinary system is critical to the maintenance of homeostasis. The composition of the extracellular fluid, which includes tissue fluid surrounding the cells and the liquid portion of blood, or plasma, is critical to this maintenance. The composition of tissue fluid and plasma is regulated by the kidneys through the formation of urine, which actually is blood plasma that is cleaned or "cleared" as it passes through the kidneys.

The Kidneys

The two bean-shaped kidneys lie against the posterior abdominal wall on either side of the spinal column (Fig. 11.8). Within each kidney, over a million microscopic functional units called **nephrons** clean the blood plasma and carry out the formation of urine (Fig. 11.9). Extending from each kidney is a **tubular ureter** that transports urine to the **urinary bladder** for storage. From the bladder, urine is excreted to the outside of the body through the **urethra.**

Oxygenated blood enters the kidneys through the right and left renal arteries that branch off the abdominal aorta. After being processed in the nephrons, blood exits each kidney through a renal vein that empties into the inferior vena cava. Through filtration and urine formation, the kidneys regulate fluid volume, eliminate wastes in the blood, control the concentration of ions in the extracellular fluid, and maintain the pH of the blood.

Nephrons

As you can see in Fig. 11.9, a nephron is basically a thin, elongated tubule associated with a network of blood capillaries. On entering the kidney, the renal artery divides into smaller arteries that, in turn, subdivide into smaller arteries. Very small arteries, or arterioles, formed from this branching carry blood to tiny clusters of capillaries, each called a **glomerulus.**

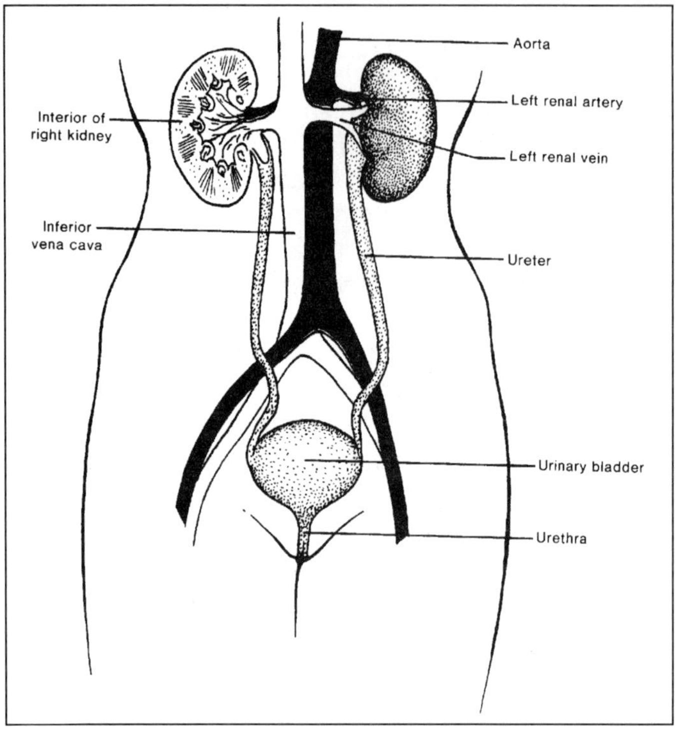

Figure 11.8 Diagram of human urinary system and associated blood vessels.

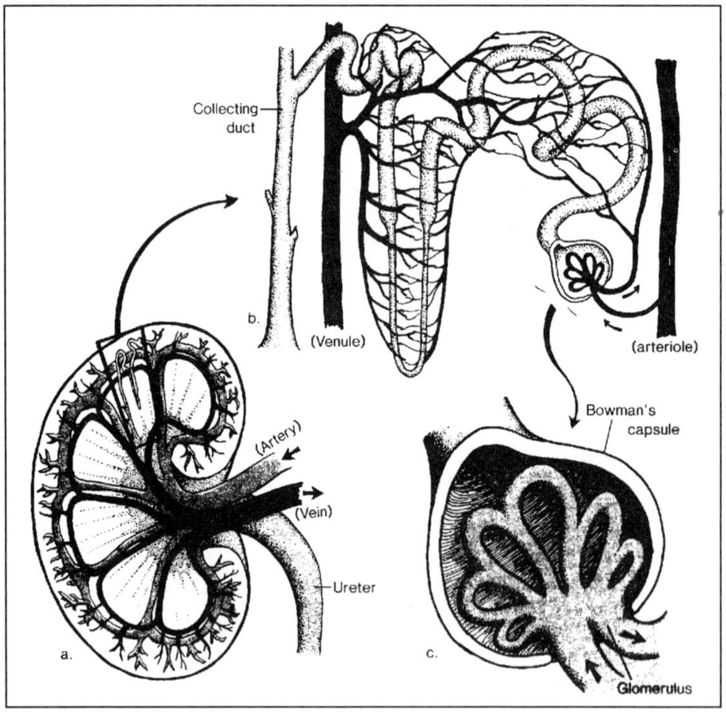

Figure 11.9 (a) Interior of human kidney showing arterial and venous blood vessels. (b) Basic structure and associated blood vessels of a nephron. (c) Enlarged view of Bowman's capsule (cut-away) and aglomerulus.

Urine Formation

The formation of urine involves three basic mechanisms: filtration, reabsorption. and secretion. Blood circulating through the capillaries of the glomerulus exerts pressure against the capillary walls. Because of the high permeability of these capillaries, water and other small blood constituents, such as ions, salts, glucose, amino acids, and urea, readily pass across the capillary walls into the renal capsule (Bowman's capsule). Larger blood constituents, such as blood cells and plasma proteins, however, cannot be filtered into the capsule and thus remain in the blood to circulate within the glomerulus and out into the capillary network surrounding the nephron. The water and small blood constituents that pass from the glomerulus into the renal capsule constitute the **filtrate.**

From the renal capsule, the filtrate moves into the proximal tubule and then into the loop of Henle, and finally into the distal tubule. As the filtrate moves along, some of the blood constituents in the filtrate are reabsorbed through the walls of the nephron tubule into the surrounding capillaries. In other words, the filtered constituents move out of the tubule and back into the blood circulating within the capillaries.

For the most part, substances needed by your body, such as glucose and amino acids, are reabsorbed and returned to the blood; waste products, however, remain in the tubule and eventually are excreted.

Some substances in the capillary blood surrounding the nephron tubule move out of the capillaries and into the tubule. Excess potassium and hydrogen ions, for example, are transported out of the capillary blood into the nephron tubule where the ions are secreted into the filtrate. Secretion, then, is the opposite of reabsorption, and serves as a mechanism whereby substances not filtered from the blood in the glomerulus can still enter the tubule to be excreted. Secretion of hydrogen ions into the filtrate aids in maintaining the blood pH. If the hydrogen ion concentration in the blood is too high, the blood pH tends to become more acidic. By removing the excess hydrogen ions from the blood and secreting them into the plasma filtrate, the nephron lowers the hydrogen ion concentration, which raises the blood pH back toward its average normal value of 7.4 and maintains the pH of urine within its normal range of 4.5-8.

Water Balance

The plasma filtrate, or urine, is mostly water. Maintaining water balance in the body is one of the critical functions of the kidneys. Both the volume of blood and the volume of tissue fluid must be regulated within certain limits. As with other components of urine, water either remains in the nephron tubule to be excreted or is reabsorbed back into the circulating blood. Whether water is excreted or reabsorbed is determined in part by the effects of *antidiuretic hormone* (ADH). A diuretic is a substance that increases the output of urine. An antidiuretic, then, *decreases* urine output. Antidiuretic hormone is produced in the hypothalamus and is secreted by the posterior pituitary gland. Receptors in the hypothalamus respond to changes in blood volume, which either stimulate or inhibit release of antidiuretic hormone. When the volume is low, as occurs with dehydration, more antidiuretic hormone is released; if excessive amounts of fluid are taken in, antidiuretic hormone release is inhibited.

Antidiuretic hormone decreases urine output by increasing the permeability of the collecting ducts to water. Thus, instead of remaining in the collecting ducts to be excreted, water passes through the walls of the ducts and is reabsorbed into the surrounding capillaries. Water is drawn out of the collecting ducts by osmosis because of a high salt (NaCl) concentration that builds up around the loop of Henle (Fig. 11.10). The rate of osmosis, however, is determined by the permeability of the walls of the collecting ducts. Thus, if antidiuretic hormone release is inhibited as a result of excessive fluid intake, water remains in the collecting ducts and moves into the ureters and urinary bladder to be excreted.

The filtrate moving from the distal tubule into the collecting duct is now called urine and contains the excess ions, urea, and other unwanted substances that will pass into the ureters and down into the urinary bladder to be excreted through the urethra. The blood circulating through the capillaries surrounding the nephron now has been cleared of these impurities and will reenter the general circulation by way of the renal vein. In a 24-hour day, human kidneys process over 1700 liters (1800 quarts) of blood, resulting in the production of nearly 1.5 liters (1.6 quarts) of urine.

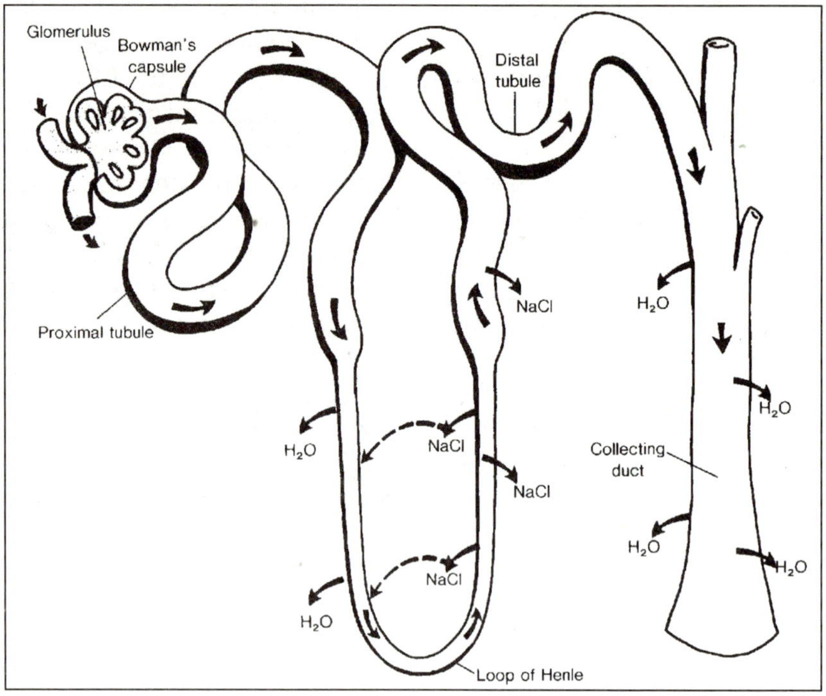

Figure 11.10 Diagram of water and NaCl movement in the mammalian nephron. Some of the NaCl leaving the ascending limb of the loop of Henle reenters the descending limb. The countercurrent mechanism creates a region of high NaCl concentration in the tissue fluids around the loop and collecting duct. Consequently, water is withdrawn from these tubules by osmosis (see text for details).

THE KIDNEYS AND REGULATION OF BLOOD PRESSURE

Blood pressure, you will recall, is the force of the blood pushing against the wall of a blood vessel. The control of blood pressure is influenced not only by the nervous system but by hormonal mechanisms as well. One of these mechanisms involves the hormone **angiotensin,** derived from one of the plasma proteins. If the blood pressure in the arteries should fall too low, the kidney responds by secreting an enzyme called **renin.** In the blood, renin splits one of the plasma proteins (angiotensinogen) to form angiotensin. This hormone is an extremely powerful *vasoconstrictor,* that is, it causes blood vessels to constrict. In particular, angiotensin constricts small arteries and thereby raises the arterial pressure back to normal. In addition, angiotensin also stimulates the release of aldosterone from the adrenal glands. Aldosterone causes the kidney to decrease its excretion of water, which elevates the blood volume. The higher blood volume also aids in raising the blood pressure back to normal.

This same mechanism occurs if blood flow to the kidney is reduced through arterial obstruction or disease. The result in this case, however, is that if the obstruction is not removed or the disease corrected, the blood pressure remains too high, and the individual develops hypertension.

Immunity: Biological Defense

Like other organisms, vertebrates are bombarded and invaded by a multitude of potentially pathogenic, or disease-causing, substances and microorganisms. Normally, the body has the ability to recognize, resist, and eliminate most of these foreign invaders through certain *immune responses.* These responses are the essence of **immunity,** which contributes to homeostasis by defending the body against infection and disease. Relatively little is known about immunity in invertebrates, but most vertebrates carry out immune responses in much the same manner as do humans.

Specific and Nonspecific Immune Responses

Immune responses are designated as either **nonspecific** or **specific.** These two types of responses usually act together to destroy the pathogenic invader. Nonspecific immunity includes external and internal factors that protect the body against invading substances without having to "know" the specific identity of the substances. Any foreign invader, including a bacterium, fungus, or pollen, can be physically barred from entering the body or attacked the very first time it is encountered. The skin, for example, is a physical barrier to microorganisms, and also secretes acidic enzymes that destroy bacteria. Other enzymes in saliva, tears, and nasal secretions, along with acids produced in the stomach and urinary tract, kill microorganisms.

In addition, there are at least eight other types of body chemicals that participate in nonspecific immune responses. For example, injured tissues release the chemical *histamine,* which increases local blood flow by dilating capillaries. The walls of the capillaries become more permeable and fluid leaks from the blood into the surrounding tissues. This increased blood flow causes the area to redden, and the leaking fluids cause swelling. This is the typical inflammatory response familiar to anyone who has banged a thumb with a hammer or suffered from a sore throat.

Other chemicals produced by various body cells include *interferons* that interfere with virus reproduction and *pyrogens* that cause fever. Both interferons and pyrogens are small protein molecules. Some evidence suggests that a mild fever may contribute to inactivating some bacteria. Finally, several cell types, primarily certain white blood cells, destroy foreign invaders by phagocytosis. Chemicals released by the infected tissues and by the invaders attract the phagocytic cells, which move to the infected area and engulf and destroy the invaders. Thus, through these different types of nonspecific immune responses, the body possesses a natural immunity against many infectious and pathogenic substances.

Specific immune responses entail recognition of a specific invader and remembering the identity of the invader to initiate a rapid defense if the invader returns. Specific immunity results from the actions of certain white blood cells called *lymphocytes* (Fig. 11.11).

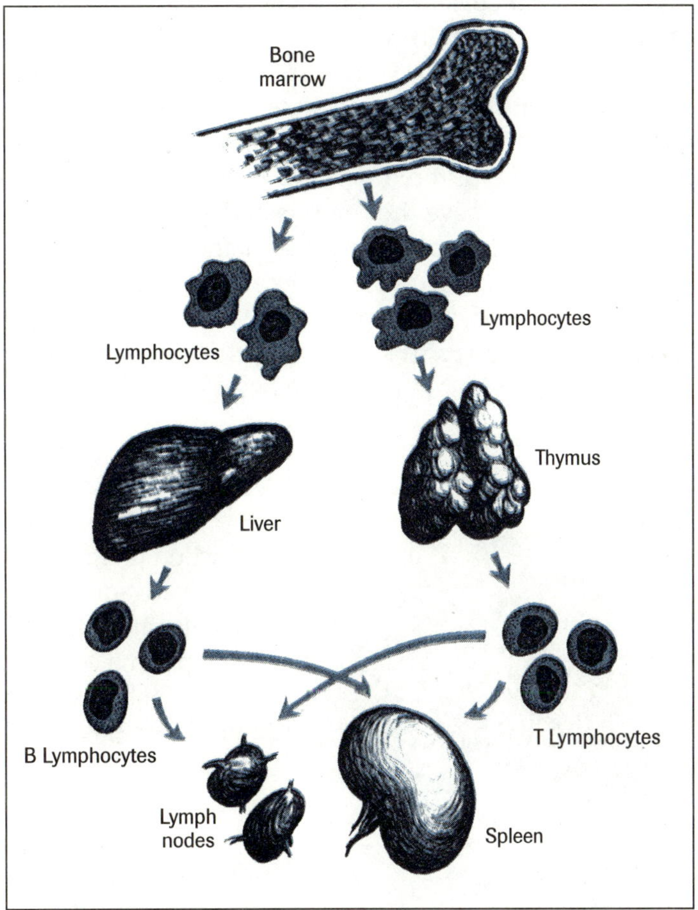

Figure 11.11 Illustration of B lymphocyte and T lymphocyte production. Some lymphocytes produced in bone marrow enter the circulating blood and move into certain organs or tissues, perhaps the liver or intestine, where the blood cells are "activated" to become B lymphocytes. Other lymphocytes from bone marrow and activated in the thymus gland to become T lymphocytes. Both the B and T lymphocytes soon enter the lymph nodes and spleen.

This type of immune response does not occur immediately when the body is invaded, but requires several days to a few weeks before taking effect. The second invasion of the same foreign substance, however, stimulates a more rapid immune response. For example, the first time a virus enters the human body it may cause damage or disease, but the second time the virus is encountered, the body "recognizes" the invader and may destroy the virus before any disease symptoms develop.

An invading microorganism or any substance that stimulates a specific immune response is called an **antigen.** More specifically, an antigen is any substance that stimulates production of **antibodies,** which are protein molecules produced by certain lymphocytes. Antibodies bind to the antigen that caused their production or to a similar antigen. In the process, the antibodies inactivate or destroy the antigen. An antigen such as a virus, bacterium, or drug may enter the body from the outside and is called a *foreign* antigen; substances produced by one's own body that stimulate antibody production are *self*-antigens. Even transplanted tissues or organs are regarded as foreign antigens by the immune system of the transplant recipient. Transplants between identical twins, however, is an exception to this because they possess similar antigens.

Most blood cells arise from certain base cells in bone marrow. Some of the cells produced in bone marrow are the previously mentioned white blood cells called lymphocytes, shown in Fig. 11.11. After leaving the bone marrow and entering the circulating blood, some of the lymphocytes circulate to the thymus gland where they are processed by the hormone *thymosin* and perhaps other secretions into **T cells.** T cells develop into one of three types of cells: killer T cells, helper T cells, and suppressor T cells. The T cells leave the thymus and enter the spleen, lymph nodes, and other lymphatic tissues. Apparently, the spleen and lymph nodes are able to produce new T cells in adults since the thymus degenerates after puberty.

Other lymphocytes are processed within the bone marrow into **B cells,** which also are carried by the blood to the spleen and lymph nodes. T cells and B cells are the major functioning cells in the specific immune responses of the body.

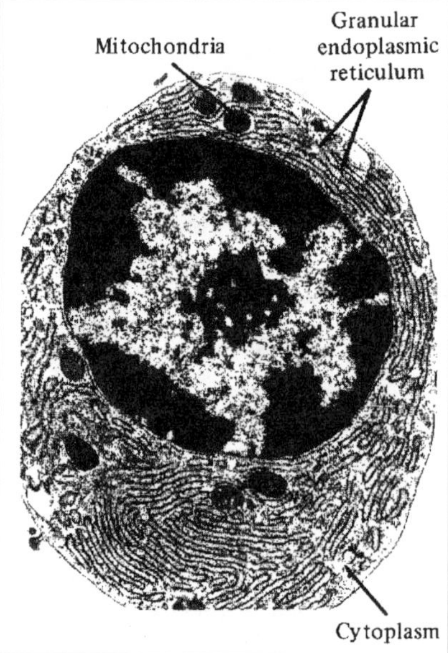

Figure 11.12 Electron micrograph of a guinea pig plasma cell. The abundant cytoplasm is packed with a granular endoplasmic reticulum and ribosomes, the sites of antibody synthesis.

Antibody-Mediated Immune Responses

B cells are the principal cell type involved in the destruction of antigens by antibodies. B cells do not attack an invading antigen directly, but are transformed into **plasma cells** that produce antibodies (Fig. 11.12). Each plasma cell produces many antibodies directed against one specific antigen. Upon encountering the appropriate antigen that stimulated their production, the antibodies bind to it, usually resulting in its ultimate destruction. Specific T cells called helper T cells mediate the activities of the B cells by stimulating division of the B cells and thereby causing production of many identical antibody-producing plasma cells. Suppressor T cells regulate this response by reducing the production of antibodies by B cells. Thus, a balance between helper T cell stimulation and suppressor T cell inhibition contributes to immune system homeostasis. Responses involving B cells and the production of antibodies constitute **antibody-mediated immunity.**

Some of the B cells remaining in the spleen and lymph nodes become **memory cells** after being exposed to an antigen. These cells have become "sensitized" to the antigen; that is, they can recognize the antigen and attack it quickly if it should invade the body a second time. Memory cells, then, constitute an exceptionally strong immune defense that may last for many years or even for a lifetime.

Cell-Mediated Immune Responses

A cell-mediated immune response involves killer T cells attacking an antigen directly. As their name indicates, killer T cells make direct contact with an antigen and destroy it. In addition, killer T cells release chemicals that activate phagocytic cells. Killer T cells are also stimulated by helper T cells, producing more T cell progeny to battle an invader. A virus residing within a human cell will cause killer T cells to attack that cell, puncturing the membranes, and destroying the infected cell. This prevents the virus from multiplying within the cell. As with B cells, some T cells remain in the spleen and lymph nodes as memory cells.

Active and Passive Immunity

Immunity to a specific antigen can be acquired in several ways. An individual who is exposed to a disease-causing antigen mounts a specific immune response, which activates T cells and B cells. The person usually develops the disease or symptoms of the disease, but on second exposure to the antigen at a later time, the person is immune and the disease does not develop. This is known as **active natural immunity.** "Active" refers to the fact that one's own immune system is involved in activating T cells and producing antibodies.

There is also **active artificial immunity,** an acquired immunity resulting from the introduction of an antigen into the body by vaccination. A *vaccine* contains inactivated or weakened antigens that cause the body to form antibodies in the same manner as active natural immunity. The Salk vaccine for polio contains an inactivated form of the polio virus, while the oral Sabin vaccine contains a live but weakened form of the virus. When injected into the body or taken orally, the virus (antigen) stimulates production of antibodies that are available thereafter to attack a live polio virus if it should enter the body.

Immunity is also acquired passively by the transfer of antibodies from a mother to her child. The antibodies may diffuse from the mother's blood across the placenta into the baby's blood or the baby may acquire the antibodies from the mother's milk. **Passive immunity** can also be acquired artificially. This is accomplished through the injection of antibodies produced in the blood of an animal, often a horse, that has been exposed to a certain antigen. The blood of the animal is withdrawn and an **antiserum** containing the antibodies is prepared. Injection of the antiserum into humans provides antibodies that the immune system could not produce or did not have time to produce for itself. In the case of a snakebite, for example, one cannot wait several days for the body to build up its own antibodies against the venom. This type of passive immunity, unlike active immunity, is short-term, usually lasting only a few days or weeks. This is because the body has not had the opportunity to "recognize" the antigen and to form antibodies and memory cells against it. Consequently, after the antiserum antibodies are disposed of by the body, there is no remaining defense against the antigen if it should reenter the body.

Autoimmunity

Unfortunately, all immune defenses of the body are not always beneficial. Situations sometimes arise in which the body cannot identify "self" from "nonself"; that is, it cannot distinguish between antigens normally found in the body and foreign antigens. This is possibly due in part to suppressor T cells being unable to prevent an immune response to self-antigens. In effect, the body's recognition system goes awry. The result is that B cells become activated and are transformed into plasma cells that begin to produce antibodies against one's own cells and tissues. The body's immune system also may attack antigens that have been present in the body for years. If a foreign antigen similar to a self-antigen invades the body, the immune system not only destroys the foreign antigen but may continue to act against the similar self-antigen. It may be that Type I diabetes, in which the insulin-secreting cells of the pancreas are destroyed, develops in this manner. In reacting against these cells and antigens, the body literally is attacking itself. The development of immunity against one's own antigens is called **autoimmunity.**

Autoimmunity can have severe effects on homeostasis. In addition to Type I diabetes, various other conditions, including rheumatic fever, inflammation of the thyroid gland, connective tissue disorders such as arthritis, and certain allergies, are considered to result from autoimmune responses.

Allergy

An allergy (Gk. *allos,* other, + *ergon,* work) can be defined as a heightened sensitivity in an individual to a substance that in normal amounts is harmless to the majority of people. Some types of allergies may occur in practically anyone, while there are other types that are genetically based and are inherited from one's parents.

Among the more common allergies are those that often cause skin eruptions in response to poison ivy, poison oak, or poison sumac. The toxic chemical antigen in these plants penetrates the skin and combines with the tissue proteins. T lymphocytes then are mobilized and become sensitized to the antigen after a period of 1 to 2 weeks. This initial exposure to the toxic chemical usually does not cause any serious tissue damage. However, if one again is exposed to the chemical following sensitization of the T lymphocytes, reddening and blisterlike eruptions appear on the skin, and chemicals are released from the T lymphocytes. This sets up a series of reactions through which toxic substances and lysosomal enzymes are released from the lymphocytes. In addition, large phagocytic cells are attracted to the affected area. The combination of toxic substances, enzymes, and phagocytes acts to destroy the invading antigen. In some cases, however, release of these products may be so extensive as to cause serious tissue damage. This type of allergy, because it does not occur on first exposure to the antigen, is called a *delayed-reaction* allergy.

Hay Fever

Hay fever, also known as *rhinitis,* or inflammation of the nose, usually occurs in the fall when various weeds pollinate. However, hay fever can be seasonal or may occur throughout the year. Some people are allergic to the plant or mold allergens released during a particular season of the year, while other people are allergic to dust, pet fur, pollen, and household fungal spores that are in the immediate environment all year long.

When inhaled, the various allergens react with certain antibodies, releasing histamine that causes increased capillary permeability. Fluids and various chemicals begin to leak into the nasal tissues, which then become inflamed and swollen. The well-known treatment for hay fever is the use of an antihistamine drug that reduces the swelling.

Asthma

Asthma, a term from the Greek, meaning "to breathe hard" or "to pant," is characterized by a narrowing of the trachea, bronchi, and bronchioles. There are several types of asthma, some of which are induced by a variety of organic, chemical, medicinal, and other allergens. Still other types may be induced by viruses or even by exercise. Some types of asthma cannot be explained in terms of allergic reactions, but appear to result from functional abnormalities of the autonomic nervous system. It even has been suggested that such abnormalities may underlie all asthma. The fundamental autonomic system effect in asthma is spasm of the bronchial smooth muscle and increased secretion of mucus. Consequently, breathing is difficult, particularly on exhalation when the bronchioles are compressed.

It should be noted that virtually every person has the capacity to develop some type of allergic disease because of the characteristic response of the immune system to a foreign antigen. It has been found that human beings can be allergic to virtually anything, including their clothes, food, furniture upholstery, cosmetics, and even each other.

Loss of Immunity: Acquired Immune Deficiency Syndrome

The disease known as *acquired immune deficiency syndrome,* or AIDS, is caused by the *human immunodeficiency virus (HIV),* which is thought to have evolved from another virus in central Africa (Fig. 11.13). A *virus* is a tiny infectious agent that can enter a cell and reproduce itself. Viruses consist of a core of either DNA or RNA enclosed within a protein coat. Some viruses also possess an outer membrane surrounding the coat. Viral infection involves the release of viral nucleic acid, either DNA or RNA, into the host cell where the nucleic acid takes control of the cell's genetic and metabolic machinery. The viral genes direct the host cell to synthesize new viral nucleic acid and proteins, which assemble into new viruses. These new viruses either destroy the host cell or cause it to malfunction. In some cases, certain viruses may remain dormant in a cell for long periods of time before undergoing this reproductive process.

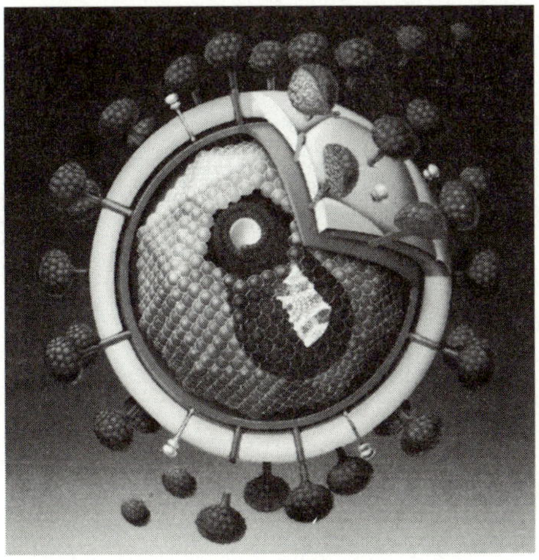

Figure 11.13 Model of the AIDS virus.

HIV, which is an RNA virus, attaches to a surface molecule found primarily on helper T cells. A few other blood and tissue cells, along with neurons, also possess this surface molecule. The virus then injects its RNA into the T cell where the RNA directs the synthesis of new HIV particles. Initially, the exposure to the HIV causes the body to produce antibodies against the virus, but eventually the immune system is overwhelmed and the virus destroys the helper T cells. With the loss of these cells, killer T cells and B cells are inactivated and the body is unable to mount a specific immune response. In addition, tissue cells infected by the HIV can carry the virus into the central nervous system where nerve cells may be damaged or destroyed.

Some individuals infected with the HIV suddenly suffer from fatigue, fever, headache, and swollen lymph nodes that may persist for up to two weeks or, more commonly, several months. This condition is referred to as *AIDS-related complex* and the infected individual becomes HIV positive. This means that the antibodies initially produced by the immune system against the HIV can be detected in the circulation.

The time between infection by the HIV and development of full-blown AIDS may vary from 5 to 10 years or more. Because of suppression of the immune system, AIDS victims are highly susceptible to infections such as pneumonia, meningitis, and yeast infections in the mouth. A rare form of cancer known as Kaposi's sarcoma (tumors of the skin, lymph nodes, and internal organs), along with various nervous system disorders, also are associated with AIDS. Although new drugs are available that slow the progress of the disease, in virtually 100 percent of the cases, AIDS is ultimately fatal.

HIV is transmitted by body fluids, such as infected blood or semen, and breast milk from mothers to their nursing infants. Transmission of the virus most often occurs between infected sexual partners or intravenous drug users who share needles contaminated with the virus. There is no evidence that the HIV can be transmitted by physical contact such as shaking hands or kissing.

At this writing there is no cure for AIDS, and treatment consists of managing the infections and administering drugs that interfere with reproduction of the HIV within the body. The drugs presently available, however, have harmful side effects, or the patient becomes resistant to them after several months. Current research on AIDS is centered on finding a vaccine that will cure the disease. Until such a vaccine is developed, however, the most effective approach in preventing the spread of AIDS is educating people concerning their sexual habits and the risk of using contaminated needles.

Blood Type

Immune responses depend on the ability of an organism to discriminate between its own cells and tissues and those of another organism. By virtue of this ability, each organism has a unique immunological identity. Your T cells and plasma cell antibodies impart this unique immunological identity to you by reacting against antigens that are foreign to your cells and tissues. Another self-identifying immunological trait is blood type.

In some people, there are specific antigens on the surface of their red blood cells. The antigens are a type of lipid or protein that determines one's blood type. The major and most commonly used group of antigens constitute the ABO system (Fig. 11.14). People with antigen A on their red blood cells have blood type A; those with antigen B have blood type B. If both A and B antigens are present on a person's red blood cells, the blood type is AB. In the absence of both antigens A and B, the blood type is O.

Figure 11.14A Summary of Blood Group Antigens and Antibodies

Blood Type	Red Cell Antigen	Plasma Antibody
A	A	*b* (anti-B)
B	B	*a* (anti-A)
AB	A and B	None
O	None	*a* and *b*

Figure 11.14B Compatibility of Blood Types

Blood Type	Can Donate Blood To	Can Receive Blood from
A	A, AB	A, O
B	B, AB	B, O
AB	AB	A, B, AB, O
O	A, B, AB, O	O

The presence or absence of a red cell antigen determines the type of blood antibodies produced in the plasma. If your blood type is A, you produce *b* (or anti-B) antibodies; you produce *a* (or anti-A) antibodies if your blood type is B. If your blood type is AB, you do not produce either antibody, and if you are blood type O, you produce both *a* and *b* antibodies. It is not fully understood why the body produces *a* and *b* blood antibodies since their production, like that of any other antibody, would require that a person be exposed to a corresponding antigen. People with type B blood, for example, could not develop *a* antibodies unless they were exposed to blood type A antigens. Since the blood antibodies are first found in the blood only two months after birth, very few people would have been exposed to blood antigens through a blood transfusion. It has been suggested that A or B antigens on bacteria or food in the intestine stimulate production of antibodies against antigens that are different from one's own antigens. In our example, a person with type B blood would produce type *a* antibodies against the A antigens on the bacteria or food.

The information on blood types, the work of the Austrian-born pathologist and 1930 Nobel Prize winner Karl Landsteiner, is invaluable in blood transfusions because incompatible blood types will cause clumping of the donor's red blood cells in the blood of the recipient (Fig. 11.15). If the donor is blood type B or AB and the recipient is type A, the *b* (anti-B) antibodies in the recipient's blood attach to the B antigens on the donor's red blood cells, causing the cells to *agglutinate,* or clump together. Agglutination also occurs if type A or AB blood is donated to a type B recipient because the *a* (anti-A) antibodies in the recipient's blood attach to the A antigens of the donor's red cells. People with type AB blood, however, can receive any other blood type because type AB blood contains no blood antibodies. The red cells of blood type O individuals do not possess blood antigens and in most cases will not undergo agglutination when transfused regardless of the blood type of the recipient.

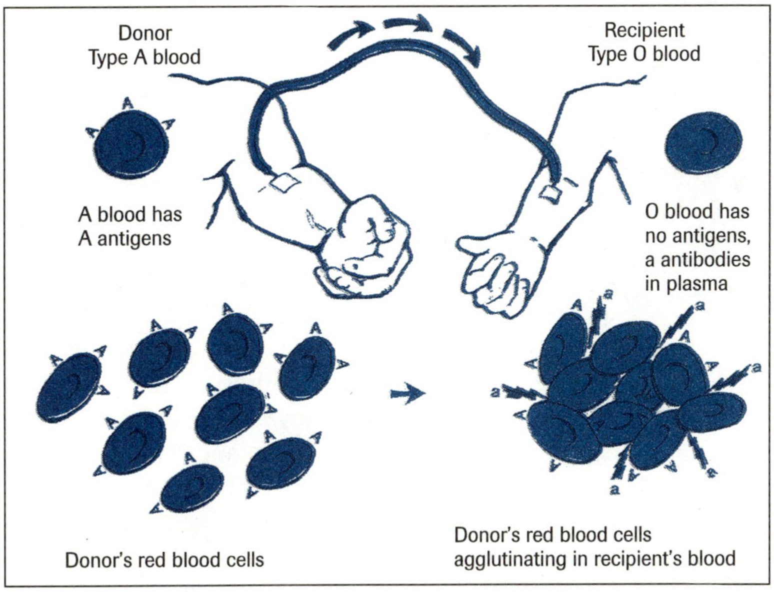

Figure 11.15 Illustration of blood type incompatibility. Transfusion of blood from a type A donor to a type O recipient results in agglutination, or clumping, of the recipient's red blood cells. This is because the *a* antibodies in the plasma of the recipient react with the *a* antigens on the red cells of the donor.

The red blood cells of some people also have another surface antigen called the **Rh factor,** named for the Rhesus monkey in which the antigen originally was found. People are designated Rh positive if they possess this antigen, and Rh negative if they do not. Thus, each blood type—A, B, AB, and O—is also designated as either Rh positive or Rh negative.

Clumping of red blood cells also can occur when Rh positive blood is transfused to a person with Rh negative blood. The Rh antigen in the donor's blood causes the recipient to produce anti-Rh antibodies that attach to the Rh antigen and clump the donor's red blood cells. If Rh negative blood is transfused to a person with Rh positive blood, however, there is no reaction because the donor's red blood cells do not have the Rh antigen. Thus, in transfusing blood, the effects of both the A and B antigens and the Rh antigen must be considered.

The Rh antigen also may present problems for expectant mothers who are Rh negative (Fig. 11.16). An Rh negative woman whose baby is Rh positive may receive a dose

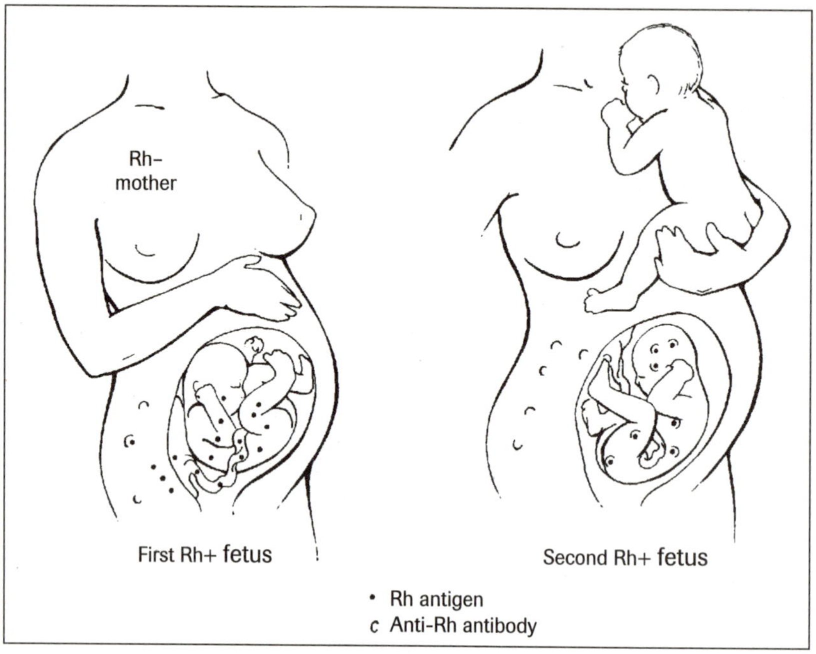

Figure 11.16 The onset of erythroblastosis fetalis. The condition most often affects an Rh positive infant whose mother is Rh negative and whose father is Rh positive. In the uterus, the blood of the fetus releases Rh antigens into the mother's blood. The mother's blood develops anti-Rh antibodies against the antigens. The antibodies later seep across the placenta into the infant's blood, where they agglutinate the infant's red cells. Because the mother's blood does not build up a high enough concentration of antibodies with the first Rh positive child, erythroblastosis fetalis is more common in the second and subsequent Rh positive children.

of Rh positive antigens during the birth process when she is exposed and sensitized to the baby's blood. In some cases, but not all, the Rh negative woman produces antibodies against the Rh antigen. These antigens can cross the placenta and agglutinate the baby's red cells. Usually, however, this is not a problem in the first pregnancy because the mother has not had enough time to build up enough antibodies to harm the baby. In future Rh positive pregnancies, antibody concentration increases and these antibodies may seep across the placenta and clump the red blood cells of her Rh positive fetus. This can result in the condition known as *erythroblastosis fetalis,* which causes anemia, jaundice, and other complications in the newborn infant. This condition usually is prevented by injecting the Rh negative woman with antibodies that inactivate the Rh positive antigens she received from the first fetus. Treatment for babies with erythroblastosis fetalis includes a transfusion of Rh negative blood and exposure to fluorescent light, which breaks down a substance called bilirubin released from the baby's destroyed red cells. It is bilirubin in the circulation and cell spaces that is responsible for the yellowish tinge of the skin called jaundice.

Chapter Summary

1. Homeostasis, the maintenance of physiological balance within an organism, usually involves a sensory receptor, control center, and effector. Regulation of homeostasis operates mainly through negative feedback and sometimes through positive feedback.

   A major homeostatic regulatory center in vertebrates is the hypothalamus. Hormone secretion by the hypothalamus and pituitary gland also is regulated by negative feedback from other organs and tissues. The endocrine system regulates homeostasis through the release of hormones, and interacts with the nervous system to maintain homeostasis. The two systems are influenced by external stimuli and by internal stimuli arising from higher brain centers and other organs.

Key Terms: *steady state, p. 248*
negative feedback, p. 248
positive feedback, p. 250
hormone, p. 252

Concept Questions: Explain why an animal usually maintains homeostasis by negative feedback rather than by positive feedback. What is the general role of the hypothalamus in maintaining homeostasis?

2. The human adrenal glands consist of an outer cortex and an inner medulla. A major hormone of the adrenal cortex is cortisol. Secretion of cortisol and some other adrenal cortex hormones is controlled by adrenocorticotropic hormone from the anterior pituitary gland. Adrenocorticotropic hormone secretion is regulated by corticotropin-releasing hormone from the hypothalamus. Secretion of the adrenal medulla hormones, epinephrine and norepinephrine, is stimulated by the sympathetic nervous system.

Key Terms: *cortisol, p. 256*
ACTH, p. 256
CRH, p. 256
epinephrine, p. 257

Concept Questions: What is the general relationship between negative feedback and response to stress? Having just been informed that you will present today's biology lesson to the entire class of 300 students, you immediately break out in a cold sweat. Blood drains from your face, your heart pounds, and there are other outward and inward signs of devastating panic. What general role does the endocrine system play in your responses?

3. The pancreas secretes the hormones insulin and glucagon. Insulin causes most body cells to take up glucose from the blood. Glucagon prevents the uptake of glucose by body cells and causes the release of glucose from glycogen stored in the liver. Insulin and glucagon act through negative feedback to regulate blood glucose levels.

   The disease diabetes mellitus results in hyperglycemia, or high blood sugar. Type I diabetes results from destruction of the insulin-producing cells of the pancreas. Type II diabetes results from a lack of sensitivity of body cells to the effects of insulin.

Key Terms: *insulin, p. 257*
glucagon, p. 257
hyperglycemia, p. 258
hypoglycemia, p. 258

Concept Questions: It has been determined that a person has a blood glucose level five times the normal value. What factors could possibly lead to this condition? What is the role of negative feedback in regulating the blood glucose level?

4. Organisms are confronted with the problems of water balance and excretion of nitrogen wastes. Aquatic organisms in a hypotonic environment tend to take in water, while those in a hypertonic environment tend to lose water. Nitrogen is excreted in various forms by different organisms. Almost all animals above the level of cnidarians have discrete excretory organs for elimination of water and nitrogen.

   The vertebrate urinary system consists of the kidneys, ureters, urinary bladder, and urethra. Blood enters the kidneys through the renal arteries and exits the kidneys through the renal veins. The functional unit of the vertebrate kidney is the nephron.

Key Terms: *hypertonic, p. 260* *glomerulus, p. 262*
hypotonic, p. 260 *filtrate, p. 262*
nephron, p. 261 *ADH, p. 263*

Concept Questions: With reference to the excretory system, explain why humans cannot drink sea water but some marine vertebrates can. How would drinking a liter of water affect kidney function in humans?

5. Immunity is the ability of the body to recognize, resist, and destroy harmful substances that invade the body. Nonspecific immune responses do not require previous recognition of or exposure to the harmful substance. Specific immune responses, which involve recognition and memory, are carried out by lymphocytes called T cells and B cells.

   Lymphocytes that become T cells are activated in the thymus, and are involved in cell-mediated immunity. B cells, activated in bone marrow, differentiate into plasma cells that produce antibodies. B cells are involved in antibody-mediated immunity. After exposure to an antigen, some T cells and B cells remain in the spleen and lymph nodes as memory cells.

   Specific immunity can be acquired by natural exposure to an antigen or by vaccination; it also can be acquired by the transfer of antibodies from mother to child or by administration of an antiserum. Autoimmunity results from the inability of the body to identify self from nonself.

Key Terms: *antigen, p. 264* *plasma cell, p. 265*
antibody, p. 264 *memory cell, p. 266*
T cell, p. 265 *antiserum, p. 266*
B cell, p. 265

Concept Questions: Homeostasis maintains the body in a steady state condition. With reference to the immune system, what does this mean? You have been infected with the creeping crud virus. Describe the general types of responses your immune system might mount to counteract the virus.

Since passive immunity acquired from an antiserum is only short term, why would one need to acquire it? How does this type of immunity differ from active artificial immunity?

6. Acquired immune deficiency syndrome is caused by the human immunodeficiency virus. The virus suppresses immune system function primarily by destroying helper T cells.

   Human blood types are determined by the presence or absence of A, B, and Rh antigens on the surface of red blood cells. Agglutination reactions occur when the antibodies of the blood recipient clump the red cells (antigens) of the blood donor.

Key Terms: *HIV, p. 267*
blood antigen, p. 269
blood antibody, p. 269
Rh factor, p. 270

Concept Questions: Could blood type A be transfused safely to a person with blood type O? Explain. Would it be necessary to consider the Rh factor in both the donor and the recipient? Why? Answer these same questions if the donated blood were type O and the recipient's blood were type A.

The Balance Outside

Chapter Twelve

Ecosystem Homeostasis

The question is sometimes asked, albeit jokingly, as to why we are not up to our knees in alligators or houseflies. After all, a single pair of houseflies could produce nearly 200 *quintillion*—200 followed by 18 zeroes—offspring in about four months if all the offspring survived. In other words, why don't alligators or houseflies increase in number to the point where they literally overrun Earth? The question, of course, could apply to virtually any group of organisms. This and many other questions relating to ecosystem stability fall under the heading of ecosystem homeostasis, or the "balance of nature"; that is, the tendency to maintain a stable external environment.

Natural ecosystems are subject to many opposing conditions, or **feedback;** that is, conditions that tend to disrupt the ecosystem and conditions that tend to protect against disruptions. The result is the maintenance of a dynamic balance and ecological stability. This means that, although all ecosystems fluctuate, a stable ecosystem eventually counters one imbalance with an opposite imbalance. For example, a given ecosystem may be subject to extreme climatic changes, fires, or drought. A stable ecosystem can adjust to these changes before the effects become severe. The essence of a stable ecosystem is that it usually can regenerate itself without being altered significantly. As a case in point, the grassy plains of Africa are home to most of the world's population of lions. Recall that a population consists of members of a species living in a certain geographical area. Lions are mostly predators, feeding on antelope and other hoofed animals that graze on the grasses of the plains. Growth of the grasses, of course, depends on adequate rainfall. During a drought, however, food for the grazers becomes more scarce and many of them become weak from hunger. This is an advantage for the lion population because the weaker animals are easier to catch. Thus, the lion population tends to increase or maintain its size, while the population of grazers decreases. With the next rainfall, however, the grasses grow again and the smaller population of grazers has adequate food because there are fewer animals feeding. The lions now must compete for food from a smaller population of grazers. This, then, tends to reduce the lion population or hold it in check, resulting in the eventual return of the grazing herds to their normal population size.

In this example, the ecosystem is stable because it is able to counter one imbalance with another through a negative feedback mechanism. A decrease in the grazer population because of inadequate food and subsequent attacks by the lions is later countered by an increase in the food supply and an increase in population size. The same feedback mechanism applies to the stability of the lion population. Such stability renders an ecosystem less likely to be destroyed and increases the probability that it will persist for a longer time.

Our intent in this chapter is to examine some of the basic processes that control and maintain homeostasis in natural ecosystems. Some of these processes often involve both positive and negative feedback and include unique relationships between organisms and their environments.

Ecology: Life Interacting

The study of the interactions organisms have with each other and with their environment constitute the science of **ecology,** which is a term derived from the Greek *oikos,* "house," and *logos,* "study of"; hence, ecology literally means "study of the house." The reference here, of course, is to the "house" of nature. Within that house, organisms are found virtually everywhere from 6100 meters (20,000 feet) up in the Himalayan Mountains to 7600 m (25,000 ft) deep in the Pacific Ocean.

The environment in which an organism lives is called its **habitat,** and the manner in which it occupies that habitat is its **niche.** A habitat is the residence or "address" of an organism in nature. For a given organism, its habitat could be a small plot of ground, a mountain stream, the intestines of an animal, or an entire continent. Niche is the total role an organism plays in nature—what it consumes, the wastes it produces, its requirements for mating, space, and temperature, and its behavioral characteristics. Niche is an organism's "profession" and determines if an organism can survive in its habitat.

Most populations are parts of communities. Communities are composed of all the populations of organisms that live within a specific geographic area. Individual organisms within the populations interact both with other organisms in their population and also with organisms in the other populations that make up the community. Because of these interactions, each population adapts or adjusts to the other, often with mutual benefits. Simultaneous evolutionary adaptation by interacting species or populations is called **coevolution.** Actually, coevolution is the general rule in nature because, as members of communities, populations change over time in relation to other populations in the community. Any change in one population has the potential to affect the entire community. Such a change, however, is not necessarily inevitable nor do organisms in a population "choose" to change. Any change in one species simply has the potential to effect a countering adjustment in another species. We can illustrate this point with an excerpt from Charles Darwin's famous book *On the Origin of Species.* Darwin wrote: "Many cases are on record showing how complex and unexpected are the checks and relations between organic beings, which have to struggle together in the same country." Darwin then presents a few examples of these "checks and relations," one of which involved insects and cattle.

In Paraguay, Darwin wrote, the cattle population never increased to the point where it ran wild. This population check was attributed to a certain fly that laid its eggs in the navels of newly born cattle. Because the flies destroyed some of the cattle, the number of flies had to be checked, or the entire cattle population would be wiped out. The fly population probably was held down by other parasitic insects. But if certain insect-eating birds were to decrease in Paraguay, the parasitic insects probably would increase. This would decrease the number of navel-inhabiting flies, and the cattle population then would increase to the point of running wild. Increased cattle ranging would affect vegetation, which would affect the population of insects, and then the population of insect-eating birds. Thus, as Darwin concluded, insects actually determined the existence of cattle in Paraguay.

Population interactions, then, are part of the answer in controlling population size, which is critical to homeostasis. Darwin further emphasized this point by noting: "Every being, which during its natural lifetime produces several eggs or seeds, must suffer destruction during some period of its life, and during some season or occasional year, otherwise . . . its numbers would quickly become so inordinately great that no country could support the product. . . . There is no exception to the rule that every organic being naturally increases at so high a rate, that, if not destroyed, the earth would soon be covered by the progeny of a single pair." Darwin himself calculated that a single pair of elephants, which are notoriously slow breeders, would have 19 million living descendants in about 750 years. Some insects, as mentioned, have the potential of producing quadrillions of offspring in a few months. Through simple fission, a single bacterial cell could produce more than one million descendants in a little over 6 hours.

Population Size

The **reproductive capacity** of a species refers to the maximum rate of population growth that would occur if all the factors affecting the species' existence were ideal. Such factors include unlimited food resources, freedom from disease, competition, or attacks from other organisms, and unlimited breeding opportunities. A population tends to increase in size because it possesses more offspring than necessary to replace members that die. Natural population growth, however, eventually tends to level off—a necessary occurrence if an ecosystem is to remain relatively stable and in homeostatic balance. Nineteen million elephants, each consuming 500 pounds of food a day, would be impossible for an ecosystem to support.

It should be noted that homeostasis also can be upset in some populations that are too small. Various species of plants and animals may fail to survive or reproduce if the population size declines beyond a certain point. A low population size in some cases may fail to allow individuals to huddle for body warmth or to be protected from drying winds or to be able to find a mate. In terms of survival, these small populations tend to benefit from a limited increase in population size. As components of all ecosystems, populations of living plants and animals are subject to certain ecological interactions that affect the size of these populations.

Regulation of Population Size

Environmental Effects

The reproductive capacity of a natural population normally is not realized because of several environmental and species interactions known collectively as **environmental resistance.** Every population has a limiting size that represents what is called the **carrying capacity** of its environment. This refers to the maximum number of organisms the environment, or available resources, can "carry," or support, for an extended period of time. Thus, there is a struggle of sorts between the reproductive capacity of a population and those factors—the environmental resistance—that act against unlimited population growth.

Such factors generally are separated into **density-independent** and **density-dependent** types. Density refers to the number of organisms living in a given area at a particular time. Depending on the population, the organisms may be dispersed in various ways. In some populations, organisms may be clumped together in a number of groups; in other populations, organisms may be dispersed randomly.

Population growth can be attributed to several factors, such as an increase in birth rate, a decrease in death rate, or additional organisms moving into the population. Observations of natural populations have revealed a predictable pattern of growth rates. If environmental resistance were at its minimum, the growth rate of a new population would be slow at first, but would become quite rapid, increasing to the extent that, within each given time period, the number of individuals added to the population is greater than the number added during the preceding period. Eventually, however, the growth rate slows because the environmental resistance increases. Finally, the size of the population reaches the carrying capacity of the environment and the growth rate levels off. The foregoing can be illustrated using a population *growth curve,* in which the number of organisms is plotted against the passage of time (Fig. 12.1). This is an idealized curve, but some populations come close to following it. More often, however, population density fluctuates to varying degrees, particularly when population density approaches the carrying capacity shown at the top of the curve. Studies of populations of microorganisms, plants and animals, including humans, have shown that their growth curves in many instances are quite similar.

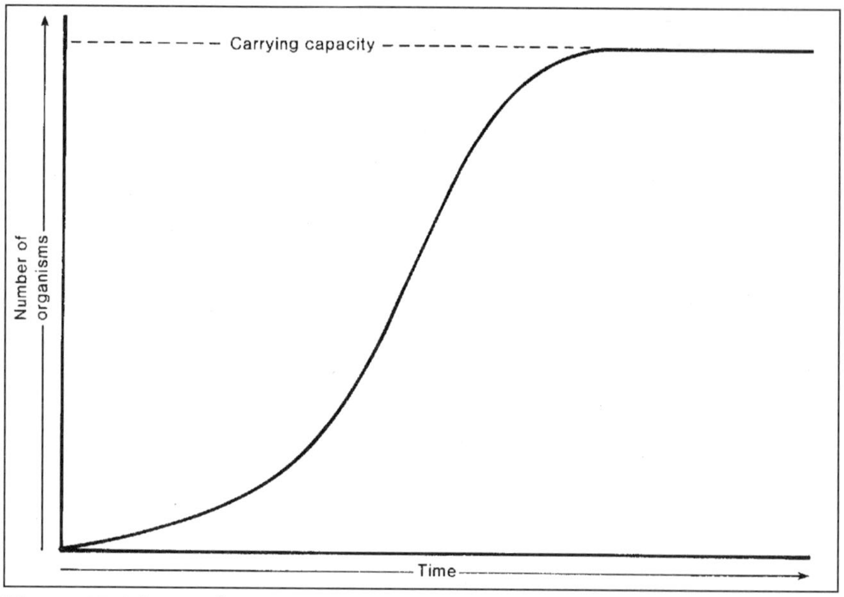

Figure 12.1 A growth curve, representative of biological populations. The curve indicates that the population grows geometrically, or exponentially, and then growth slows and levels off when it reaches the carrying capacity of the environment.

Reproduction itself is positive feedback—one organism produces another which produces another and so on. A growing population experiences increased environmental resistance, an increase in the death rate, and a lower survival rate. Reproduction, however, may increase as a result of a decrease in population size. This is an example of negative feedback and may occur because a reduction in population size may decrease the environmental resistance, which may lower the death rate and raise the survival rate. Both an increase and a decrease in the size of some populations, therefore, tend to maintain a stable population size over a period of time.

Birth and Death Rates

As mentioned, the rate of increase in population size is a function of the birth rate and the loss of members by death. The *birth rate* of a population is the number of organisms born or produced in a certain period of time, while the *death rate* is the number of organisms dying per unit time. Birth rates and death rates can be correlated with the range of ages within a population to provide information about the stability of the population in terms of future changes in size.

Virtually all populations contain members of reproductive age, members too young to reproduce, and members of post-reproductive age. Populations in which the birth rate is high and the death rate low can generally be expected to experience rapid growth. When the number of individuals in all age groups is about the same, the size of the population usually does not increase significantly. Finally, if there are relatively few members that are too young to reproduce, while the other age groups contain many members, this indicates that the birth rate is low and that the population size probably will decline.

While birth rates and death rates are determining factors in regulating population growth, there are, in turn, factors affecting birth rates and death rates. As mentioned, the sum of these factors is environmental resistance, the mechanism that normally prevents us from wading knee-deep in alligators or houseflies. In many cases, the ultimate effect to be gained from these factors is survival of the population. Let us look at some of these factors and their effects on regulating population size.

Density-Independent Factors

Climate and Geology

The primary density-independent factors acting to limit population size can be classified generally as *climatic* and *geologic* events. A given physical environment may be altered more frequently as a result of drought, storms, or floods, and much less frequently by geologic upheavals such as volcanic eruptions and earthquakes. These climatic and geologic disasters usually affect an entire population regardless of the number of individuals, whether 100 or 10,000, living in the affected area; that is, regardless of the density of the population. The result of such density-independent factors acting on a population may be that organisms starve or drown or have their habitats destroyed.

Extreme changes in temperature would also tend to affect entire populations. For example, there is a sharp decline in certain insect populations during the cold winter months. In any of these instances, a small population may be affected as readily as a large one, regardless of the density of either. The result is that the reproductive capacity is not realized over an extended period of time and population size, although it may fluctuate to some degree, is held in check.

Density-Dependent Factors

Competition

We know that most organisms cannot exist without adequate nourishment, water, light, or space. As the density of a population increases, competition for these resources becomes more and more intense and some members may simply lose out. Depending on the species, competitive activity may be the same from one day to the next. Birds that dig for worms, for instance, must compete with other birds for that food source every day. Other animals, however, engage in a contest for food or space by fighting. The winner survives and the loser must go elsewhere to compete again for food or space. In some cases, competition in the winner's sphere may disappear for a time. Competition is particularly important as a density-dependent factor in populations that are near their carrying capacity. There is little competition where resources are unlimited.

Competition for the limited resources of the environment may be confined to members of the same population. Such interactions among members of the same species are referred to as **intraspecific;** that is, within the species. Competition also occurs among members of different species and thus is termed **interspecific** competition. In some forests, for example, trees and other plants of various species may compete for light, water, and space in a manner similar to the competition displayed by members of the same population for these essential factors. The same principle applies when planting crops or a garden, in which seeds are spaced to reduce competition. In both intraspecific and interspecific relationships, competition results from the increasing density of the populations involved and serves to keep population size in check. Two other density-dependent factors considered along with competition are predation and parasitism.

Predation

Predation is the act of killing and eating another organism. Since all heterotrophs must feed on other organisms, predation is an integral part of almost all ecosystems. The relationship between predator and prey is sometimes a delicately balanced interaction. Although seemingly paradoxical, the predator-prey relationship can be beneficial to both groups, although it may not always be beneficial to an individual organism. Let us say, for example, that the density of a certain prey species increases because of favorable conditions. As is often the case, the density of the predator species feeding on the prey also tends to increase. But as a result of increased predation, the density of the prey population eventually begins to decline. With fewer prey, the density of the predator population also may decline after a time. Thus, the availability of the prey as food is a limiting factor for the predator population while predation tends to limit the density of the prey population. The predator-prey relationship, therefore, is an example of ecosystem homeostasis maintained by negative feedback.

Predators may also switch to another source of prey if the population size of the original prey source begins to decline. This may benefit both prey species. In the absence of continued predation, the size of the original prey population would tend to increase. In contrast, predation may prevent the size of the alternative prey population from exceeding the carrying capacity of its environment.

There is also another twist in the predator-prey relationship. In the absence of its natural predators, a prey population may reach such a high density that competition for food, space, and other requisites may actually result in a greater loss to the prey population than if the predators had never been removed. Such was the case in the state of Arizona during the early part of this century. A population of some 4000 deer living on a plateau near the Grand Canyon increased in number to around 100,000 in less than 20 years—a population increase that was considerably above the carrying capacity of the area. The factor responsible for this 25-fold increase was the intentional killing of thousands of wolves, coyotes, and cougars, which are the deer's natural predators. There was, then, a subsequent increase in the number of deer competing for a limited food supply. Eventually, nearly 60 percent of the herd died from starvation. This left a population of about half of what could have been supported if the predators had not been destroyed.

Parasitism

A relationship between organisms of different species is known as **symbiosis,** a term literally meaning "living together." Strictly speaking, symbiosis applies to all species within a community because they do, in fact, "live together." Symbiotic relationships may be harmful, beneficial, or neutral. Many of these relationships are examples of coevolution in which two different species adapt simultaneously. One form of symbiosis is **parasitism,** which, by definition, is an intimate relationship between two different species in which the host is harmed to a greater or lesser extent, while the parasite benefits from the relationship. In many ways, parasitism is similar to the predator-prey relationship.

Here again, as the host (prey) population increases, it becomes easier for the parasite (predator) to find and victimize members of the host population. This tends to reduce the size of the host population, which is followed by a reduction in the prey population. On the other hand, victimizing the prey is less possible if the density of this population declines. This is another example of negative feedback operating in ecological communities.

Eventually, the parasite-host relationship must reach a state of homeostatic balance. For the parasite to survive, it would be counterproductive for it to kill its host. To do so would mean extinction for the parasite. The host must be able to exist in reasonable comfort in order to provide the needs of the parasite. What is often seen in nature, then, is that as the host species becomes increasingly resistant to the demands of the parasite, there is a corresponding adaptive adjustment by the parasite. This is evident, for example, in the evolution of an enzyme-resistant body covering by species of worms that inhabit the intestines of vertebrates. This is an instance of coevolution resulting in a relatively stable relationship. It is known, however, that in some instances parasites do weaken and even kill their host. More often than not, this results from some favorable alteration in the life cycle of the parasite that greatly increases its population size. Consequently, this may lead to massive and thus fatal infestation of the host.

Parasitism is considered to have had a long evolutionary history during which the parasites have undergone modification and specialization. Apparently, adaptations to parasitism have involved the loss of organs no longer useful and the development of new ones. Some adaptations have resulted in extreme specialization to the extent that a certain parasite can exist only with one host species, while other parasites can infect several different host organisms. A certain species of plants in the morning glory family, for example, is known to grow only on flax. Some of the parasitic flatworms, however, can exist within the intestines of several different vertebrate hosts, such as pigs, dogs, and humans.

Symbiotic Relationships

MUTUALISM

As mentioned, symbiosis within a community may be harmful, beneficial, or neutral in its effects on the organisms involved. The evolution of symbiosis may have begun 3.5 billion years ago with the rise of the first single-celled prokaryotes. With the later evolution of filamentous and colonial forms, these bacterial communities could presumably have benefited from their interrelationships. Bacterial filaments, for example, exhibit division of labor in which different cell types are specialized for a particular function. We saw in Chapter 5 that the evolution of eukaryotic cells may have resulted from a symbiotic relationship in which certain species of prokaryotes took up residence within larger prokaryotes. The first known eukaryotic cells to arise from this relationship appear to have been algal-like protists, which evolved some 1.5 billion years ago. Today, many of the protists themselves exist in close symbiotic relationships.

There is, for example, a form of symbiosis called **mutualism,** which is defined as a relationship between two different species in which both species benefit. Such is the case with some flagellated protozoans that inhabit the gut of termites. The termites feed on wood, but they cannot digest the cellulose in the wood. Digestion of the cellulose is carried out by the symbiotic protozoans, which benefit from the relationship by being provided a habitat and a supply of food.

Many plants, including pine trees and orchids, live in a very close association with mutualistic fungi, which are attached to the roots of the plants. The fungi facilitate the uptake of mineral ions by the roots, and the plants, through photosynthesis, provide a source of carbohydrate for the fungi.

Mutualism is widespread in the living world and may involve relationships between many different species of organisms. In Chapter 8, for example, we discussed the relationship between legumes and nitrogen-fixing bacteria, and it was pointed out in Chapter 10 that humans harbor mutualistic bacteria in the large intestine.

Commensalism

Another common type of symbiotic relationship is **commensalism.** Roughly translated, *commensalism* means "at the table together." More exactly, commensalism is a relationship between two different species in which one member benefits and the other is unaffected. As you can see in Fig. 12.2, the so-called shark sucker, or remora fish, attaches itself to the body of a shark and gains a free ride. Moreover, when the shark feeds, the remora dines on the leftovers and returns to the shark to await the next meal. Through it all, the shark apparently is indifferent to and unaffected by its freeloading companion. Commensalism is widespread in the living world. Barnacles attach themselves to the bodies of whales, which is a significant mechanism to disperse their species; various fish and other aquatic organisms inhabit the burrows of larger sea animals; and birds build their nests in trees.

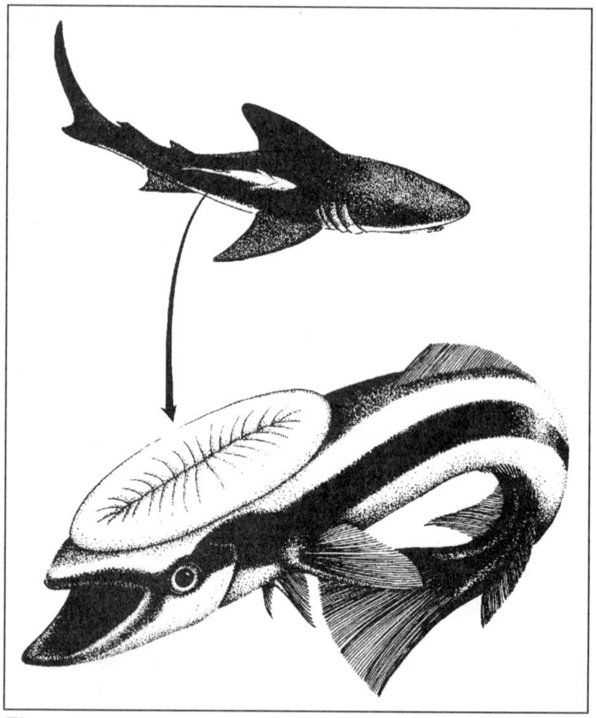

Figure 12.2 The remora fish, or shark sucker, showing its modified dorsal fin used for attachment to the body of a shark. The remora and the shark share a commensalistic relationship in that the remora benefits by being transported and provided with food, while the shark is unaffected.

Neutralism

While parasitism is harmful to one of the symbiotic members and mutualism and commensalism are beneficial, neutral relationships are neither harmful nor beneficial. Although members of the same community, a deer and a chipmunk in the forest are not in direct competition for food or other requirements. The relationship between the two animals is neutral in the sense that neither animal directly benefits or harms the other. It is probable, however, that deer and chipmunks in the same community affect each other to some degree indirectly. It is possible, for example, that the deer might accidentally destroy the chipmunk's burrow, or that the chipmunk may startle or distract a deer being chased by a predator.

Some of the symbiotic relationships designated as mutualism, commensalism, or neutralism are often difficult to determine. Just how a species in a certain relationship is affected may be open to debate. What can be said, however, is that interspecific relationships within a community can be quite varied, and that they often serve the ecological function of maintaining the stability and overall welfare of the community over the long term.

Social Interactions

Behavior and Societies

Instinctive Behavior

The behavior of many animals is the result of instinct, which is genetically determined behavior that does not have to be learned. Fundamentally, animal behavior is based on a concept called the **fixed action pattern,** which is any precise, unlearned behavioral pattern characteristic of a given species. One of the classic behavioral experiments illustrating a fixed action pattern was carried out by the French naturalist J.H. Fabre (1823–1915).

Fabre experimented with a species of wasps that preys on crickets (Fig. 12.3). The wasp paralyzes the cricket with venom from its stinger and then stores the cricket in a small burrow in the ground. Before storing the cricket, the wasp follows a precise ritual. The cricket is deposited at the mouth of the burrow and the wasp goes inside the burrow for an apparent inspection. Returning to the outside, the wasp picks up the cricket and stuffs it into the burrow. In his experiment, Fabre moved the cricket a slight distance from the burrow entrance while the wasp was inside the burrow inspecting. On emerging from the burrow, the wasp appeared disoriented by the position of the cricket, but eventually brought it back to the entrance and *again went down to inspect the burrow.* Fabre again moved the cricket; again the wasp brought it back and returned to inspect the burrow. This little scenario was repeated some forty times, with the wasp always reinspecting the burrow each time it returned the cricket to the burrow entrance.

Fabre's experiment reveals that the wasp has a chain of fixed action patterns that are highly stereotyped and apparently difficult to alter. The first fixed action pattern is the deposition of the cricket at the burrow entrance. This activity releases the next fixed action pattern, which is inspection of the burrow. Finally, the position of the cricket at the entrance acts as a stimulus to release the next fixed action pattern of stuffing the cricket in the burrow. Thus, if the cricket is not in position near the burrow entrance, the fixed action pattern sequence is

Figure 12.3 Illustrations of Fabre's experiment demonstrating a fixed action pattern (FAP). (a) after paralyzing a cricket, the wasp deposits it near the mouth of its burrow. (b) The wasp then inspects the burrow. (c) The cricket is moved farther away from the mouth of the burrow. (d) On emerging, the wasp drags the cricket back near the mouth of the burrow and (e) again inspects the burrow. Repetition of this experiment yielded the same result, with the wasp reinspecting the burrow each time the cricket was moved.

interrupted and the wasp must begin the sequence all over again. It is evident, then, that the wasp did not learn from experience, even though the experiment was repeated many times. It has been found, however, that a different species of the wasps used by Fabre will drag its prey directly into the burrow if the experiment is repeated a few times.

There are some fixed action patterns that require a certain amount of practice before the animal can perform them efficiently. Newborn chicks, for example, instinctively peck at crumbs or other small objects, but they almost invariably miss their targets. But after a few days of practice, their aim improves and the chicks, although never learning to hit the objects, come much closer to their targets. The pecking behavior eventually becomes efficient as the chicks continue to practice and mature.

There are, of course, many examples of instinctive behavior, including such familiar ones as the spinning of webs by spiders, nest building by birds, and the activities of many social animals. Whatever the form of instinctive behavior, no prior information or observation is necessary for the animal to act. In the course of evolution, such behavior has proved to promote the welfare and survival of the species involved.

Learning Behavior

Much of the behavior observed in animals cannot be attributed solely to instinct. Instinctive behavior, however, can be modified by learning, which usually develops through experience. With some animals, it is difficult to determine if a change in their behavior is truly learned or if it is the result of maturation or physiological changes. A young animal, for instance, may not be able to walk on its first try, but with maturation of its bones, muscles, and nerves, it soon accomplishes the task. Consequently, it is doubtful that walking in this case is a type of learned behavior. To illustrate learning, we will comment briefly on four types of learned behavior: habituation, conditioning, imprinting, and reasoning.

Habituation is quite simple and is characterized by a loss of response to previously experienced stimuli. More exactly, an animal becomes habituated to certain unimportant stimuli to the extent that eventually it ignores them. As shown in Fig. 12.4, a new

Figure 12.4 Habituation. (a) On first experiencing a new stimulus, an animal usually makes a particular response. (b) Eventually, the animal becomes habituated to the stimulus, and thereafter does not respond to it.

puppy or kitten in the home may, at first, withdraw from or display fear at every household sound. In time, however, the animal becomes habituated to the sounds and, finding that the sound stimuli are unimportant to its welfare, finally ceases to respond at all.

Conditioning behavior is generally of two types: *classic* conditioning and *operant* conditioning, both also known as *associative* learning. Classic conditioning is evident, for example, in the response of predators to certain insects. Some insect species are brightly colored or patterned and taste unpleasant to potential predators. After a few unpleasant encounters, the predators become conditioned to avoid many other brightly colored insects, although they may not taste disagreeable at all. The predators, then, become conditioned to the particular color stimulus and associate it with all similarly colored insects. Many animals display similar behavioral patterns in associating one stimulus with another.

Operant conditioning is based on reward or punishment resulting from a particular action. Experimentally, an animal can be induced to perform some activity for which it is immediately rewarded or it may be punished if it fails to perform correctly. This is also referred to as *trial-and-error* learning because the animal may have several options available in finding the one that is rewarded. Through trial-and-error experiments in the laboratory, animals such as ants, earthworms, planarians, and mice, have navigated various mazes to be rewarded with food. If they make a wrong turn in the maze, however, they are punished with a mild electric shock. Over time, each animal learns to avoid the disagreeable shock and seek its reward instead.

Trial-and-error learning was involved in much of what got us in and out of trouble as children. Hurling unpleasantries at your neighbor often resulted in painful atonement at the hands of your neighbor, your parents, or both. Thereafter, you knew that hurling unpleasantries was an act to avoid—at least audibly. By trial and error we all learn which acts are to be avoided and which ones result in reward. In a lighter vein, your relative proficiency at sports and games comes about to some extent by trial-and-error learning.

Early in the lives of many animals, there is a strong response and attachment to the first moving object they see. Ordinarily, of course, this object is their mother, but it has been found that these young animals will **imprint** on any moving object, including a toy, another animal, or even a human being. In baby chicks and ducklings, the critical period for imprinting is about 36 hours after hatching. Beyond this time, imprinting behavior is lost, even if the young animal encounters its own mother. Apparently, the early attachment phase is necessary for the young animal to learn to recognize the imprinted object. Something no doubt similar to imprinting occurs in human infants, who form a strong maternal bond early in life. Imprinting affords survival value for a young animal, ensuring that it remains for a time under the protective custody of its mother.

The most complex form of learning is the result of **reasoning,** also referred to as *insight* learning (Fig. 12.5). Reasoning is the ability to draw conclusions from known or presumed facts. It reaches its highest form in humans, followed by the higher primates, particularly chimpanzees and monkeys. When confronted with the problem of reaching a bunch of bananas hanging from the ceiling, a chimp will begin to stack boxes lying about the room until it can climb up and reach the food.

Other animal groups also demonstrate behavior based on the ability to reason. Jackdaws, for example, have learned to locate food found only in boxes with a certain number of circles on their covers (Fig. 12.6). If a bird had been trained to find food in a box marked with four circles, it would ignore boxes of food marked with a different number of circles. To dramatize this discriminatory ability, one jackdaw that had been

Figure 12.5 Insight (a) A dog, lacking insight in this situation, fails a detour problem on the first try. It will eventually learn by trial and error. (b) Unable to reach the bananas, the chimpanzee exhibits insight by stacking boxes beneath the bananas to extend its reach. Reprinted from *Biology: Life on Earth,* Fourth Edition, by Gerald and Teresa Audesirk, 1996, Prentice-Hall, Inc.

Figure 12.6 Learning behavior. (a) Jackdaws were trained to find food in containers marked with a certain number of circles; in this illustration, four circles. (b) When offered a container with three live worms on top, the jackdaws ignored the worms. When four worms were placed on the containers, the jackdaws immediately ate them.

trained using a four-circle box cover, was offered two boxes, one with three worms lying on the cover and one with four worms on the cover. The bird promptly ate the four worms off of one box, but ignored the three worms on the other box.

Associated with the process of reasoning in humans are factors such as concept formation and abstraction. A certain animal may be able to distinguish the color red, but it is unable to comprehend the concept of redness. Similarly, animals cannot separate the attributes of an object from the object itself. Humans may refer to a given object as "beautiful," but "beauty," as far as we know, is an abstraction impossible for an animal to make.

The behavior of organisms within a community reflects their adaptations to a particular environment. Like instinct, learning behavior is the product of evolution and serves to promote survival of a species. With its influence on adaptation and survival, behavior is yet another factor in the maintenance of ecosystem homeostasis. We can further observe this influence among organisms that interact closely within their own group; that is, within a social context.

Social Behavior

On a barren ice field in Antarctica, a rookery of emperor penguins begins the task of incubating a clutch of newly laid eggs. After laying the eggs on the ice, the females swim out to sea, leaving the males to tend the unhatched brood. While standing on one foot, the male holds an egg against his belly with another foot. For the two months' incubation period, the male cradles the egg close to his body while standing on one foot and then the other. When the chicks are ready to hatch, the females return to rear the brood and the exhausted males take to the open sea to feed.

The parental conduct of the emperor penguins is a unique example of *social behavior.* It is common knowledge that most species of animals live and interact with each other and also interact with members of other species. Animal groups include deer or zebra that run in herds, wolves that form packs, birds of a flock, or a rookery of penguins (Fig. 12.7). Within these groups, there is communication between the members that results in various types of social behavior.

Figure 12.7 A pronghorn buck herding his harem.

Communication may involve chemical stimuli, sight, or sound. Animals such as bees, moths, and ants release chemicals called *pheromones,* or "social hormones," that communicate information to other members of the same species. A female gypsy moth releases a pheromone that can be detected by a male moth several miles away, and worker ants secrete a "tracking" pheromone as they wend their way to a food source and back to the nest. Other members of the ant colony follow this chemical trail to the food source and also release the pheromone on the return trip home. In this manner, the trail stays fresh as long as the food source is available. Pheromones are secreted by many animals, especially insects and mammals.

Visual displays, particularly during courtship, are widespread in the animal kingdom. Male fireflies, for instance, flash their presence to females that flash their response in turn. After a series of such amatory luminescence, the male approaches and mating takes place. It has been noted, however, that some species of female fireflies can mimic the flashes of females of other species. When the unwary male approaches the foreign female, he is eaten. Male greater prairie-chickens have large orange throat pouches which they display and inflate at mating time. With a group of females looking on, the males first begin to strut, wings hanging at their sides, and then prance back and forth with their bills close to the ground. Adequately impressed, each female eventually chooses one of the males for her mate.

These examples of animal communication and resulting social behavior represent only a small sample of social interactions of which animal species are capable. In the following sections, we will consider a few additional examples in a more detailed fashion in an attempt to relate their effects to ecosystem homeostasis.

Societies

Individuals of the same species that interact to mutual adaptive advantage, as they usually do, form a **society.** A society is characterized generally by interbreeding within the group, division of labor among the members, and a cooperative organization of the group. Nonsocial groupings of animals are referred to as *aggregations.* For example, animals of the same species that simply gather at the watering hole or houseflies that converge on a hamburger patty are aggregating, but they do not form a society. The importance of social organization to our discussion is that it has a regulatory effect on the maintenance and size of populations. For example, some societies offer protection from predators and provide an increased opportunity of finding a mate. Such effects are critical to homeostasis.

Insect Societies

The most highly developed societies, depicted in Fig. 12.8, are found in the insects and mammals. The caste society of the common termite is a good example of an insect society. Having been around for some 200 million years, termites constitute the oldest animal society known. Division of labor among the members is quite exact within the termite colony. The *reproductives* include only the queen and king. The queen exists solely to lay eggs and the king usually remains in the chamber with her. The *work-*

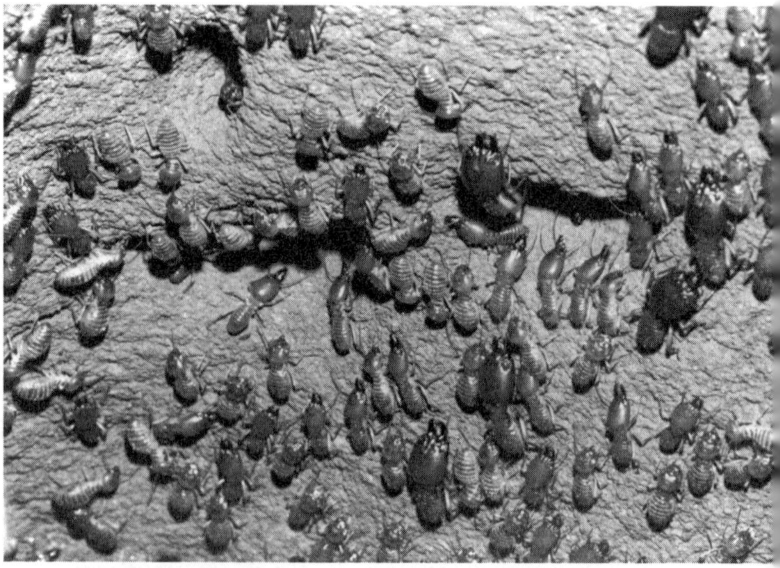

Figure 12.8 Worker and soldier termites in a termite mound in Africa. The large termites are the soldiers.

ers, which are sterile males or females, tend the colony, feed the young, and wait on the king and queen. The *soldier* termites, which possess strong jaws and a thick body covering, defend the colony against enemies. The reproductives and soldiers are fed by the workers. Eggs laid by the queen develop into either workers or soldiers, depending on the needs of the colony at the time. Although the workers and soldiers are sterile, at certain times of the year winged reproductive males and females develop and soon fly away to found a new colony.

An even more diverse division of labor is seen in honeybee societies. A honeybee colony has one queen, several hundred males (drones), and thousands of sterile female workers for a total of 20,000 to 80,000 members in the colony. The queen controls the population of the colony by secreting several pheromones, one of which, known as "queen substance," prevents the production of other queens. This pheromone is licked off the queen by the workers and passed around the hive by regurgitation. If production of "queen substance" declines, the workers begin to prepare the hive for a new queen.

When a new queen emerges from a colony, she leaves the hive on her nuptial flight, followed by the drones of the colony. She mates with several drones, accumulating enough sperm in her sperm sac to provide for the next 5 to 10 years of egg laying. The drones die almost instantly after mating. The queen returns to the hive to begin a new colony; in the meantime, the old queen and many of the workers, having been expelled from the old hive, swarm to a new location where they, too, establish a new colony.

The role of the queen honeybee is limited almost exclusively to reproduction, but the workers perform a variety of tasks. Young workers feed the larvae and prepare the hive for the eggs laid by the queen. Other workers act as defenders of the hive and store food brought in from the outside by the older workers. The sole function of the drones is to fertilize the queen. Those not involved in mating during the nuptial flight eventually are driven out of the hive or killed by the workers before the new drones emerge.

Further cooperation among the members of the bee colony is evident in the maintenance of temperature within the hive. In winter, the members huddle closely together in an effort to generate heat; in warm weather, some of the worker bees fan their wings to maintain the hive at an appropriate temperature. Temperature regulation and all the other interactions in the honeybee society require coordination and cooperation. This is accomplished efficiently because of division of labor among the members of the society. The honeybee society, like that of all social insects, resembles a single organism with each of its parts specialized for an exact function.

At this point, let us retrace our steps and employ the insects as representatives of the intraspecific and interspecific activities characteristic of all organisms in the grand scheme of ecosystem homeostasis. As members of a community, termites, honeybees, and all other organisms interact within their own groups and with other members of their community. Maintenance of the honeybee colony is dependent not only on interactions between its members, but on interactions with other species of organisms as well. By pollinating flowers, for example, honeybees obtain nectar for the hive. This interaction has a regulatory effect on the size of both populations; the honeybee colony needs the nectar to exist and pollination is necessary for reproduction in the flower population. Termites and honeybees are also subject to predation by various organisms, including other arthropods. Honeybees are fed on by scorpions and hornets, while termites serve as food for birds and assassin bugs.

Honeybees, as noted, feed on nectar from various flowers, while termites feed on wood, which is digested by symbiotic protozoa in the termites' intestines. Thus, both insect groups are involved in mutualistic relationships. Ant nests often harbor honeybees, which coexist with the ants as guests. As with all organisms, termites and honeybees interact with their nonliving environment. The effect of temperature alters the behavior of honeybees as they attempt to keep warm or cool the hive. Termites inhabit damp environments and will dry out and die if overexposed to the air.

Insects, then, are subject to the same effects of density-independent and density-dependent factors as are most other organisms. They normally do not exceed the carrying capacity of their environments because of the factors—the environmental resistance—just discussed. Thus, the population size of honeybees and termites is held in check, a necessary requisite for the maintenance of ecosystem homeostasis.

Vertebrate Societies

Dominance

Animal societies can contribute to ecological stability by limiting population size and density. Within many vertebrate societies, the group is organized into a hierarchy, which involves social *dominance.* Depending on the society, there may be one or several dominant individuals or there may be one or more dominant groups. The former instance is characteristic of chickens, in which the hierarchy consists of a "pecking order." The dominant bird in the order pecks all the others, but they will not peck back. The bird second in rank pecks all the chickens of inferior rank in the order and is itself pecked only by the one dominant chicken. This is repeated at successive levels in the hierarchy, with the unfortunate chicken at the bottom being dominant over none of the others.

Pecking orders regulate population size by limiting breeding opportunities among the animals low in the pecking order. The more dominant animals breed more frequently and leave the most offspring.

There is also a pecking order of sorts among various primate groups, such as monkeys and baboons. In these societies, groups of individual males and females may enjoy higher social status than do others. Such dominant positions may be inherited in part from dominant forebears, but age, strength, seniority, and other factors appear to be involved as well. Within the social group, dominance serves to reduce fighting and establish order. This is of significant survival value in terms of both protection and mating.

Although group life heightens the possibility of finding a mate, the dominant males and females in some societies usually select among each other, whereas animals on the lower strata often must choose among others of the same social status. This is not always the case, however, as females will often reject a high-ranking male in favor of a male with less rank in the group. In many groups, however, mating among members low in the order is less frequent than among the more dominant animals. This, too, has survival value for the group by ensuring a strong and healthy posterity.

Territoriality

Interactions among mammals and other vertebrate groups assume a variety of forms. As shown in Fig. 12.9, one common form, particularly among birds and mammals, is **territoriality,** the defense of a section of the home range against intruders of the same species and sometimes other species. An animal's home range is the area in which it normally carries out its day-to-day activities. The home ranges of some species, however, may overlap with those of another species. The home range of some birds may cover thousands of square kilometers, while that of a field mouse may include only a few dozen square meters.

Figure 12.9A Threat display by a male mandrill baboon *(Papio sphinx)* of West Africa.

Figure 12.9B Male toads, such as this *Bufo americanus,* call from shallow water.

Figure 12.9C A male mocking bird proclaiming his territory by singing from a prominent song perch.

A territory usually is within, and therefore smaller than, the home range and is the specific area defended most strongly by an animal group. More often, it is males who establish and defend a territory. Would-be intruders are discouraged by threat displays, such as bared teeth or boisterous vocalizations. Many animals establish the boundaries of their territory by marking them with glandular secretions or body wastes. Other animals mark their territories vocally by howling or singing. In addition to serving the function of protecting the feeding and mating areas, territoriality also aids in social stabilization by preventing overcrowding. As a major result, intraspecific competition is minimized. The defense of a territory against others of the same species contributes to keeping the population size in check by preventing mating among members outside the territory.

Ecological Succession

Homeostasis is characterized by stability, or a resistance to change. Many ecosystems, of course, are quite stable while others are much more susceptible to change. It may appear that a given ecosystem is a rather static, inflexible unit; lush meadows or thickets, however, often occupy an area that once was an open, barren field. Equally striking is the appearance of plant and animal life in a region that initially was a nearly lifeless stretch of sand dunes. Such dynamic changes in communities can occur in almost any area of the biosphere. These changes, ordinarily requiring considerable spans of time, come about gradually and result from both environmental and organismal influences. The process of community change over time is referred to as **ecological succession.**

Environmental effects, such as climatic changes, shifting winds, and volcanic eruptions often are important to succession because they alter the physical surroundings and thus have a direct influence on the types of plant and animal life that will inhabit the area. Succession also involves alterations of the environment caused by incoming organisms or by those already in the area. In time, a species may so alter the environment that it can no longer occupy the area. Such alterations, however, often make the area more suitable for a different species. The new species, in turn, may repeat the alteration process and eventually be replaced by another species, and so on. Essentially, then, succession involves a modification of organismal habitats brought about to a large extent by the organisms themselves. The result is the replacement of one dominant species by another.

The process of succession eventually ends with formation of a **climax community,** a community that has reached a relatively stable, or homeostatic, state. Technically, the concept of a climax community presents various problems that have led some ecologists to question the suitability of the concept. For example, a community may be affected by outside factors, such as fire, that prevent the community from reaching what would normally be its climax stage. In effect, the community is arrested at the preclimax stage. Nonetheless, we will use the concept, noting that exceptions to our points concerning climax communities are possible and do exist.

In contrast to the communities that preceded it, a climax community is often stable and self-perpetuating. This is because a climax community does not tend to upset or destroy the conditions necessary for its own survival. The central factor originally thought to be responsible for the stability of climax communities is diversity. Most ecologists today, however, doubt that increased diversity actually causes a climax community—or any other community—to be more stable. Just what is meant by "stability" has been the subject of lively debate and no consensus has been reached. Nonetheless, having more habitats, a greater number of different species, more intricate food webs, a greater biomass, and, hence, more opportunities for varied ecological niches, a climax community is less likely to be disrupted by short-term changes in the environment.

A classic example of succession is the arctic biome. A **biome** is any large, stable ecosystem. At one time, glacier movement in the area produced a surface of smooth, barren rock. As the temperature rose and fell, the rocks expanded and contracted, often breaking into fragments. Water, seeping into the crevices of the rocks, expanded as it froze, further aiding in the breakdown of the rocks. In time, wind-transported lichens gained a foothold on the surface of the rocks, establishing themselves as the first, or pioneer, stage of succession. The lichens released corroding acids that further broke the rocks into small particles, forming a base for the production of soil. With the eventual death of some lichens, their remains mixed with the rock particles and continued the soil buildup.

Once established, patches of soil made possible the invasion of wind-blown moss spores, initiating the second stage of succession. The established moss plants, through cellular secretions and finally decay, added organic enrichment to the soil. Eventually, growth of the mosses produced a thick, absorbent mat. The next inhabitants—seeds of various herbs—fell into the mat beds and germinated, sending their roots deep into the soil. Because of the cold climate, more discriminating and complex forms of plant life, such as shrubs and trees, could not survive. The community of herbs, therefore, represents the final stage of succession in the arctic biome.

In more temperate climates, additional and more complex stages of succession may occur in which various communities of grasses, shrubs, and trees become established after the lichens, mosses, and herbs have each taken hold. With each community replacing the previous one, the environment may be altered substantially. Each species adds its waste products and chemical secretions to the soil; root systems rearrange the soil, affecting aeration and water drainage. Decomposition of the bodies of dead plants and animals adds organic nutrients to the soil, aiding the growth of new and different forms of plant life.

Along with the succeeding stages of vegetation, various species of animal life capable of being supported also change. For instance, it would be impossible for the lichen community to support a large bird population. Most birds require seeds or berries or worms for food and trees in which to nest. With succession, then, come new habitats and the opportunity for varied ecological niches.

With the progressive succession of communities, an area eventually becomes modified to the extent that further stages of succession are unlikely. When this occurs, the climax stage of a community has been reached. If not affected by extreme environmental factors, either geologic or climatic, or by disease, a climax community may exist for centuries. The climax community, then, is highly homeostatic, displaying a relatively constant ecological balance. This is not to say that climax communities, or other ecosystems, are static units. Rather, they are dynamic systems that remain within relatively stable bounds.

World biomes, such as deserts and tropical rain forests, may be considered as major climax formations. The total time for succession to progress from the pioneer stage to the climax community may, in many instances, require hundreds or even thousands of years.

The Human Factor

As we have seen, homeostasis is affected by population size. A natural population normally does not exceed its carrying capacity because of environmental resistance. One of the most widely and seriously considered issues of our time centers on the possible effects of continued growth of the human population. To put the issue in perspective, we need to look at how and why we came to be as many as we are today (Fig. 12.10). A reliable account of human population size can be found only for the years after about 1650. Educated speculations of human population size extend backward from this date to the rise of agriculture, some 10,000 to 12,000 years ago. A reasonable guess is that at the advent of agriculture, the human population numbered somewhere between 5 million and 10 million people. By 4000 B.C. the population had jumped to about 90 million. In 6000 years, the human population doubled between three and four times with each doubling requiring over 1300 years. Projections indicate that by the year 2025, the human population of the world will increase to 8.5 billion. An estimated 95 percent of this growth will occur in the less developed countries, such as Bangladesh, Ethiopia, and Turkey. To appreciate these figures, consider that more than two-thirds of all the people born since A.D. 1500 are alive today.

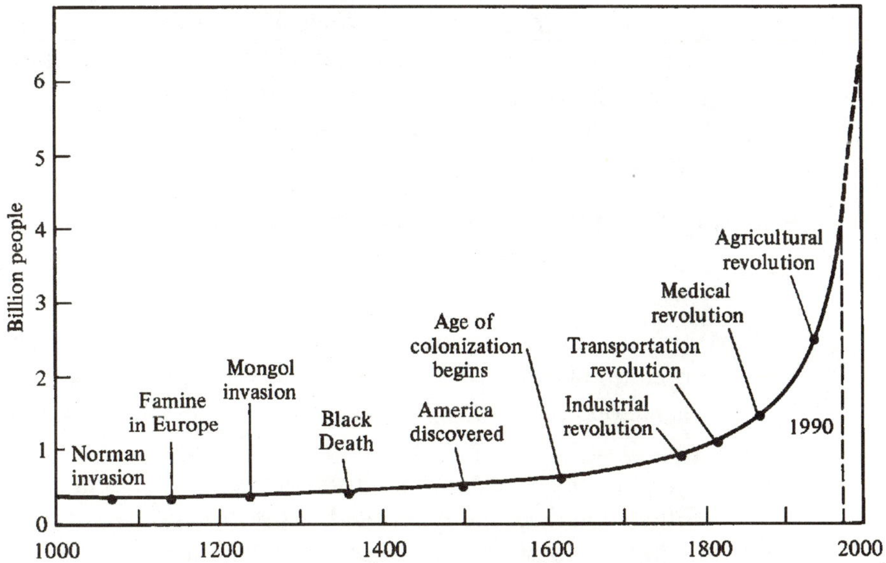

Figure 12.10 Growth curve of the world population of *Homo sapiens* from the year 1000 to 1990.

The world population currently is growing at a rate of about 1.7 percent, or 17 people per thousand people per year. This works out to a doubling time of about 41 years, which means our numbers could go from the present 5.6 billion to over 10 billion by the year 2035. Since A.D. 100, the world **population density** has gone from 1 person per square kilometer of ground (a square kilometer is 0.386 square miles) to over 30 people per square kilometer today.

An increasing population density carries with it the likelihood of increasing environmental resistance. Humans, like other living organisms, must compete for light, space, air, food, and water. With unlimited population growth, we will reach the carrying capacity of the environment. Overpopulation carries with it the increased need for food, building materials, and fuel. But in some countries, this need has resulted in severe damage to the environment. Several years ago, for example, Thailand banned all logging because of flooding caused by deforestation, and millions of acres of rain forest have been destroyed in Brazil in attempts at farming the land.

Unrestricted population growth presents us with an ultimatum: either reduce the reproductive rate or it will be reduced for us. The sociologist Nathan Keyfitz reported that in 1989 the annual population growth rate in Nigeria was 3.4 percent, which represents a doubling rate of 22 years. At this rate of growth, the population of Nigeria would equal that of the entire world in less than 150 years. Keyfitz points out, however, that this will not happen because of either a fall in the birth rate or a rise in the death rate. As has been the case in the past and in our own time, the consequences of uncontrolled growth too often are famine, starvation, war, and disease. The tragedy is that it need not be so.

What, then, can be done? Evidence indicates that birth rates and death rates decline with the advent of economic progress and dissemination of information on birth control. Declining death rates are partly attributable to medical advances and improved nutrition. Although the reasons for the decline in birth rates are not fully understood, birth control information and access to contraceptives have substantially reduced birth rates in Mexico and in China, South Korea, and other Asian nations over the past 20 to 30 years. In addition, the increasing financial independence and improved status of women in less developed countries appear to play a role.

Finally, people must be educated concerning the effects of population density on the economic progress and ecological stability not only of their own country, but on the entire world. They must understand that to achieve economic growth without destroying their environment is in part answered by lowering the rate of population growth. For those countries not already suffering, whether this will be understood in time to avert the tragic consequences remains to be seen.

Upsetting the Balance

We have observed many of the factors affecting homeostasis within an ecosystem: competition for food, water, light, and space; density-dependent and density-independent factors; birth rates, death rates, and societal behavior. These factors operate together in regulating a community. There are, of course, factors that act in opposition to this regulation. It has been demonstrated that effects such as pollution and excessive deforestation have the potential of upsetting homeostasis in a variety of ecosystems.

There are other factors less well understood, but also potentially damaging to Earth. In the last 20 years or so, there has been increasing concern over the gradual depletion of the **ozone layer,** particularly over the colder regions of Earth. It has been shown that the major contributors to ozone depletion are chemicals called chlorofluorocarbons (CFC). These are used commercially in aerosol cans, refrigerators, and as solvents. Although harmless at low altitudes, CFCs that diffuse into the stratosphere (a layer of the atmosphere about 10 to 50 kilometers above Earth at the poles) are bombarded by ultraviolet (UV) radiation, which breaks apart the chemicals, releasing free chlorine atoms. These atoms act as catalysts in converting ozone (O_3) into oxygen (O_2). Thus, ozone concentration decreases, allowing more UV radiation to reach Earth. This could damage aquatic and terrestrial vegetation and also could increase the incidence of human cancers and cataracts. In addition, other gases containing compounds such as carbon monoxide and methane may prove to have harmful effects on the atmosphere in the years to come. Carbon monoxide lowers the self-cleansing ability of the atmosphere by reducing the concentrations of molecules that act as cleansing agents. A reduction of these agents would raise the atmospheric concentrations of trace gases and thus could alter the physical and chemical properties of the atmosphere. Methane has the same effect on the atmosphere as carbon monoxide and also contributes to the greenhouse effect and the depletion of ozone. In an effort to determine the long-range effects of CFCs and other chemicals, scientists continue to monitor the concentration of these chemicals in the atmosphere.

Recalling our discussion of ecosystems in Chapter 8, food and energy move through the food chains and webs from autotrophs to consumers. Associated with this movement are the nutrient cycles that conserve matter for reuse. **Eutrophication,** or nutrient enrichment, and the intrusion of chemicals into food chains exert their effects on the functioning of an ecosystem. Eutrophication normally is a natural process, resulting in a healthy growth of plant and animal life. The runoff of sewage, animal wastes, fertilizers, and other nutrients, however, can cause a dramatic increase in the population of aquatic photosynthesizers. Eventually, the biomass exceeds the availability of oxygen, and the algae and water plants die. Decomposers break down the dead organic matter, consuming most of the remaining oxygen in the process. Without adequate oxygen, fish and aquatic invertebrates cannot survive.

This scenario has been played out in untold numbers of freshwater ecosystems throughout the world. Introduction of toxic chemicals into a food chain, you will recall, can result in the chemicals becoming more concentrated as they move through the chain. This, of course, most severely affects the carnivores at the end of the food chain.

All of the processes cited here can be going on at the same time. When we include the factors discussed in this chapter, it is apparent that regulation of natural ecosystems requires a balance among the many processes occurring within them. This is not to say that stable ecosystems are in balance all the time. Ecosystems fluctuate, compensating for an imbalance in one direction with an imbalance in the opposite direction. The key to stable ecosystems is that their homeostatic mechanisms can help an ecosystem remain within a certain range of stability and correct imbalances before they become destructive.

An Alternative Hypothesis

Having examined many of the factors affecting ecosytem homeostasis, the conclusion accepted by most biologists is that Earth has provided the physical and chemical stage on which organisms act out their ongoing play. Organisms, then, must adapt according to the limits placed on them by their environment. But what if we look at it another way and consider the proposal that life itself is the regulator of the environment? What if the activities of life maintain the physical and chemical conditions of air, land, and water? Is it possible that life, interacting with Earth, maintains the homeostatic conditions necessary for itself ? This concept, proposed by the British scientist James E. Lovelock, is known as the *Gaia hypothesis,* named for the Greek goddess of Earth.

Lovelock proposed that all life on Earth could be considered one superorganism, the parts of which regulate different environmental factors. Plants replenish the atmosphere and oceans with oxygen, and metabolizing cells continually release carbon dioxide. Microorganisms recycle vital nutrients, and plants and animals add their wastes and organic remains to the soil. Through all of their activities, according to Lovelock, organisms actually control Earth's temperature, atmospheric conditions, soil composition, and so on. Organisms interacting with Earth, therefore, maintain global homeostasis.

Taken in the extreme, the Gaia hypothesis is controversial but nonetheless significant in its emphasis on the effects interacting organisms may have on the environment. We know, of course, that activities of organisms in one area can affect organisms in a distant area. There is no question about the interrelatedness of nature. This is apparent, for instance, in the cycling of matter. Whether or not organismal activity actually regulates homeostasis of Earth, however, is an alternative hypothesis remaining to be verified.

Environmental conditions regulate the kinds and numbers of organisms that inhabit a given ecosystem. At the same time the activities of organisms modify the environment in which they live. There is, therefore, a close relationship between organisms and the environment they inhabit.

Just as the advent of photosynthesizing organisms led to the production of oxygen which provided conditions favorable for aerobic organisms and unfavorable for strictly anaerobic organisms, so are human activities modifying the world in which they live. While no one is certain how these changes will affect Earth's physical characteristics, it is certain that, as has happened before, populations of all living organisms must adapt to these changes or cease to exist.

Chapter Summary

1. Ecosystem homeostasis is affected by population size. The reproductive capacity of a population normally is not realized because of environmental resistance. Every environment can support only a limited number of organisms; that is, each has a certain carrying capacity. Environmental resistance increases as the population approaches the carrying capacity.

   The factors responsible for environmental resistance are both density-independent and density-dependent. Population size is affected by birth rates and death rates, which in turn can be affected by density-independent and density-dependent factors. The major density-independent factors are climatic and geologic events; density-dependent factors include competition, predation, and parasitism. Symbiotic relationships, including mutualism, commensalism, and neutralism, also play a role.

   Key Terms: *feedback, p. 274* · *ecology, p. 275* · *habitat, p. 275* · *niche, p. 275* · *reproductive capacity, p. 276* · *environmental resistance, p. 276* · *carrying capacity, p. 276* · *symbiosis, p. 279*

   Concept Questions: Consider the following statements: (1) Unlimited population growth tends to lower the carrying capacity of the environment. (2) Raising the carrying capacity of the environment tends to foster unlimited population growth. What do these statements mean? Do you agree or disagree with either one or both of them? Explain your answers.

Species A and species B coexist in a mutualistic relationship as part of a forest community. An epidemic eliminates species A. Based on any assumptions you wish to make about species A and B, what effects might the elimination of species A have on ecosystem homeostasis?

2. Animal behavior is based on the fixed action pattern. The types of behavior are instinctive and learning, which includes habituation, conditioning, imprinting, and reasoning. Social behavior is reflected in animal communication and in the activities of insect and vertebrates societies.

Key Terms: *fixed action pattern, p. 282*, *habituation, p. 283*, *conditioning behavior, p. 284*, *imprinting, p. 284*, *reasoning, p. 284*, *society, p. 287*, *dominance, p. 289*, *territoriality, p. 289*

Concept Questions: Do you think it would be possible for an insect society to function through learning behavior rather than by instinct? Explain.

Dominance and territoriality are stabilizing factors in many animal societies, but how do these factors contribute to homeostasis of an entire ecosystem?

3. Ecological succession is affected by climatic events and by organismal activities. Ecological succession culminates with formation of a climax community, which usually is relatively stable and self-perpetuating.

Key Terms: *succession, p. 290*, *climax community, p. 291*, *biome, p. 291*

Concept Question: Why is a major world biome, such as a tropical rain forest, considered to be a climax community?

4. The human population has increased from an estimated 5 to 10 million people at the time of the advent of agriculture 10,000 to 12,000 years ago to approximately 5.6 billion people today. The human population, like that of other organisms, is subject to environmental resistance and the carrying capacity of the environment.

Other factors affecting ecosystem homeostasis are pollution, excessive deforestation, depletion of the ozone layer, eutrophication, and toxic chemicals in food chains.

The Gaia hypothesis proposes that the collective activities of all organisms maintain global homeostasis.

Key Terms: *population density, p. 293*, *ozone layer, p. 293*, *eutrophication, p. 294*

Concept Question: Assume a toxic chemical was introduced into photosynthetic algae in a lake stocked with fish and various invertebrates. Explain how homeostasis could be upset in this aquatic ecosystem.

PART SEVEN

Life Feature 5: Reproduction

Chapter Thirteen

Protein Synthesis and Cell Division

Evolution and Significance of Cell Division

Life arises from life. In a broad sense, this statement summarizes our life feature of reproduction, which is the major focus of Chapters 13, 14, and 15. In all its forms, life begins as a single cell. Some of these life forms, such as various bacteria and protists, remain single cells throughout their existence. The single cell beginning of other life forms, however, is a prelude to multicellularity, which is the result of one cell giving rise to two cells or to trillions of cells. All organisms, however, whether unicellular or multicellular, reproduce their own kind by a precise process of cell division. As a result of this process, hereditary material within each cell is passed to descendant cells. The specific nature of this material determines the basic structural and functional characteristics of every cell in the lineage. On acquiring their hereditary makeup, living cells grow and develop to become specialized for their specific task in maintaining the life of the organism.

The rise of eukaryotic cells about 1 billion years ago was one of the critical stages in evolution. In the preceding 2.5 billion years, prokaryotic bacteria apparently were the only life forms on Earth. Bacteria, you will recall, lack most of the organelles characteristic of all eukaryotic cells. According to one theory, however, eukaryotic organelles arose from incorporation of small prokaryotes into larger cells (Chapter 5). Yet one organelle, the nucleus, most likely did not arise in this manner; it probably evolved through an infolding, or invagination, of the plasma membrane. The nucleus houses the cell's genetic information in the form of DNA, which is passed from cell to cell through cell division.

Prokaryotes

Bacteria, although lacking a nucleus, reproduce and pass on their DNA by a simple method of cell division called **binary fission** (Fig. 13.1). In this method, a circular, cytoplasmic organelle called a *chromosome* copies itself, or replicates, and the bacterial cell subsequently elongates and splits into two cells. Because the circular chromosome is composed primarily of DNA, one copy of the chromosomal DNA goes to one cell and one copy to the other.

Eukaryotes: Protists

This primitive method of cell division apparently persisted to some degree in the protists, the next group of organisms in the evolutionary scheme. Research on dinoflagellates and euglenoids, for example, indicates that both groups reproduce by cell elongation and pinching in two in a manner similar to that of bacteria. Protists, however, are eukaryotes and possess multiple chromosomes, along with parts of the cell division apparatus that are present in more complex cells. Cell division in the protists, therefore, appears to be an intermediate stage between bacterial fission and more complex methods of division.

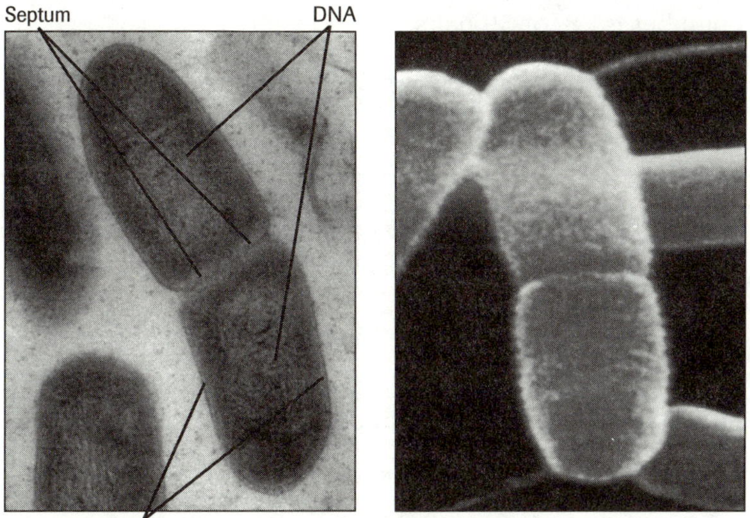

Figure 13.1 (a) Beginning of binary fission in a bacterial cell. (b) Completion of cell division.

Eukaryotes and Mitosis

The distinguishing and critical feature of eukaryotic cells is the membrane-bound nucleus. With the evolution of the nucleus, these cells held one of the keys to the subsequent evolution of all higher forms of life. That key was a type of nuclear division called **mitosis,** a complex sequence of chromosomal arrangements that, like binary fission, apportions the hereditary material (DNA) equally to every cell. Some of the primitive unicellular eukaryotes employed mitosis in asexual reproduction, producing a genetically identical lineage that persists to the present day. In turn, through mitosis and subsequent cell division, the primitive eukaryotes eventually gave rise to the first multicellular organisms a little less than 1 billion years ago.

The inherent capacity of living cells to replicate their DNA and divide to produce others of the same kind is also the basis for the processes of growth and tissue repair in the living world. The famous dictum that "all cells come from cells" is one of the cornerstones of the cell theory.

The body cells of eukaryotic organisms contain a certain number of **chromosomes** that are located in the nuclei of the cells. Eukaryotic chromosomes are linear, not circular, and consist primarily of deoxyribonucleic acid, or DNA, the hereditary material, and certain associated proteins. Each chromosome contains one DNA double helix; that is, two bonded spiral-shaped strands that wind around each other. The units of heredity called **genes** are located along the length of the helix. The proteins associated with chromosomal DNA are involved in the structure of the chromosome and help to regulate gene activity.

The number of body cell chromosomes is characteristic for each species: dogs have 78 chromosomes, mosquitoes have 6, carrots have 18, and humans have 46 chromosomes or 23 pairs of chromosomes (Fig. 13.2). The 46 human chromosomes, as well as the other examples listed, are referred to as the *diploid* number of chromosomes, designated as $2n$. The term **diploid,** meaning "double," refers to the fact that chromosomes usually come in pairs (one chromosome of a kind inherited from each parent organism). Thus, the diploid number of chromosomes for humans is 46, or 23 pairs, and for dogs the diploid number is 78, or 39 pairs.

Figure 13.2 Diploid chromosome numbers of various plants and animals.

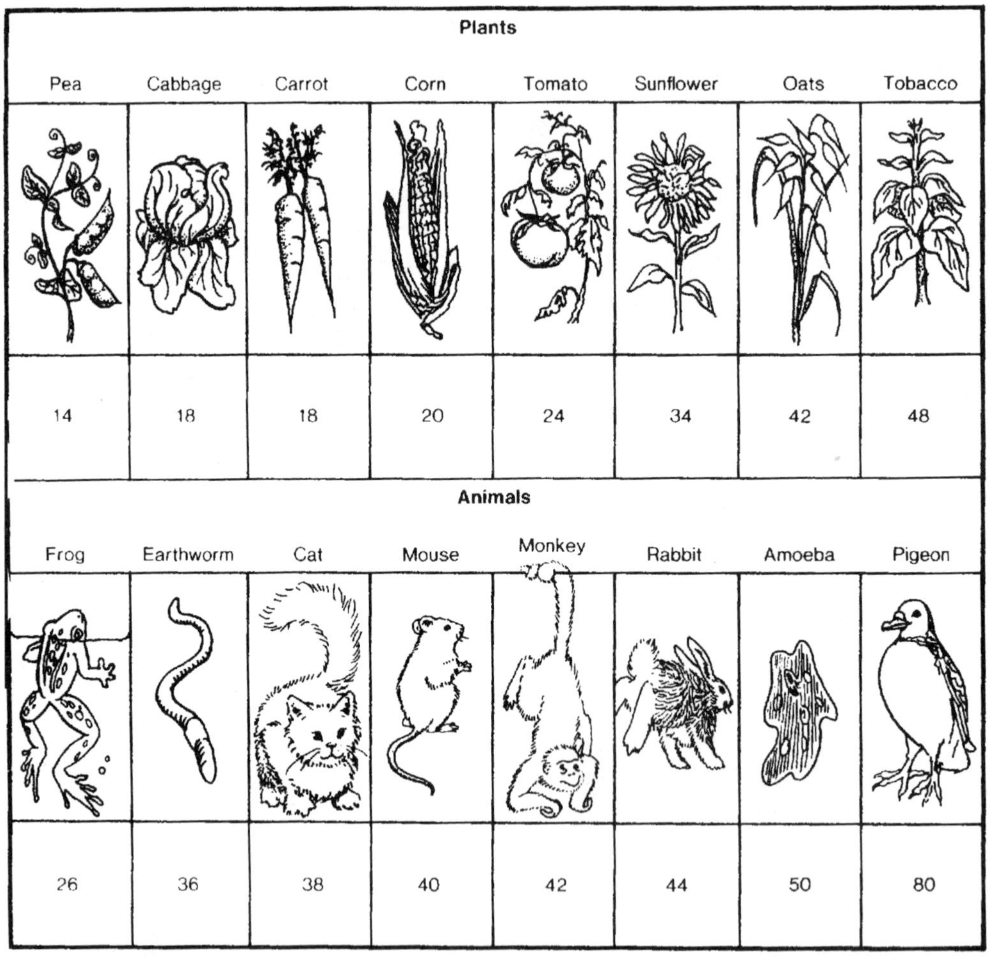

Humans have 1 pair of sex chromosomes and 22 pairs of body cell chromosomes, or **autosomes.** The sex chromosomes are designated X and Y. Normal human females have two X chromosomes (XX), one contributed by the mother and one contributed by the father. Normal males have one X and one Y chromosome (XY); the X chromosome is contributed by the mother and the Y chromosome is contributed by the father. As to the 22 pairs of autosomes, 22 individual chromosomes are contributed by the mother and 22 by the father.

The essential accomplishment of mitosis is the formation of two identical daughter nuclei; that is, the division of one nucleus into two nuclei, each having the same number of chromosomes and, hence, the same hereditary information, or DNA, as the parent nucleus. Mitosis, subsequently followed by cell division, is necessary for replacement of lost or injured body cells. If you scrape your knee, for example, the skin cells are replaced by mitotic cell division and each new skin cell is genetically the same as all the other skin cells. The same is true in regard to growth. Every body organ or structure increases in size primarily by the addition of more cells, and all the cells making up each tissue of an organ or structure are genetically identical.

Mitosis is usually followed by division of the cytoplasm and thus two cells are generated from one, but this is not always the case. Cells of some fungi and green algae, for instance, undergo mitosis without subsequent cell division. Consequently, these cells may have two or several nuclei in their cytoplasm and thus are binucleate or multinucleate. In the majority of living species, however, nuclear division is accompanied by division of the cytoplasm, also known as *cytokinesis.*

The Cell Cycle

Before a cell can divide, it must produce the components necessary to construct a new cell. These include a complete copy of the DNA, new organelles, various enzymes and other proteins. The events responsible for producing these components, along with actual cell division, constitute the **cell cycle** (Fig. 13.3). In other words, the cell cycle

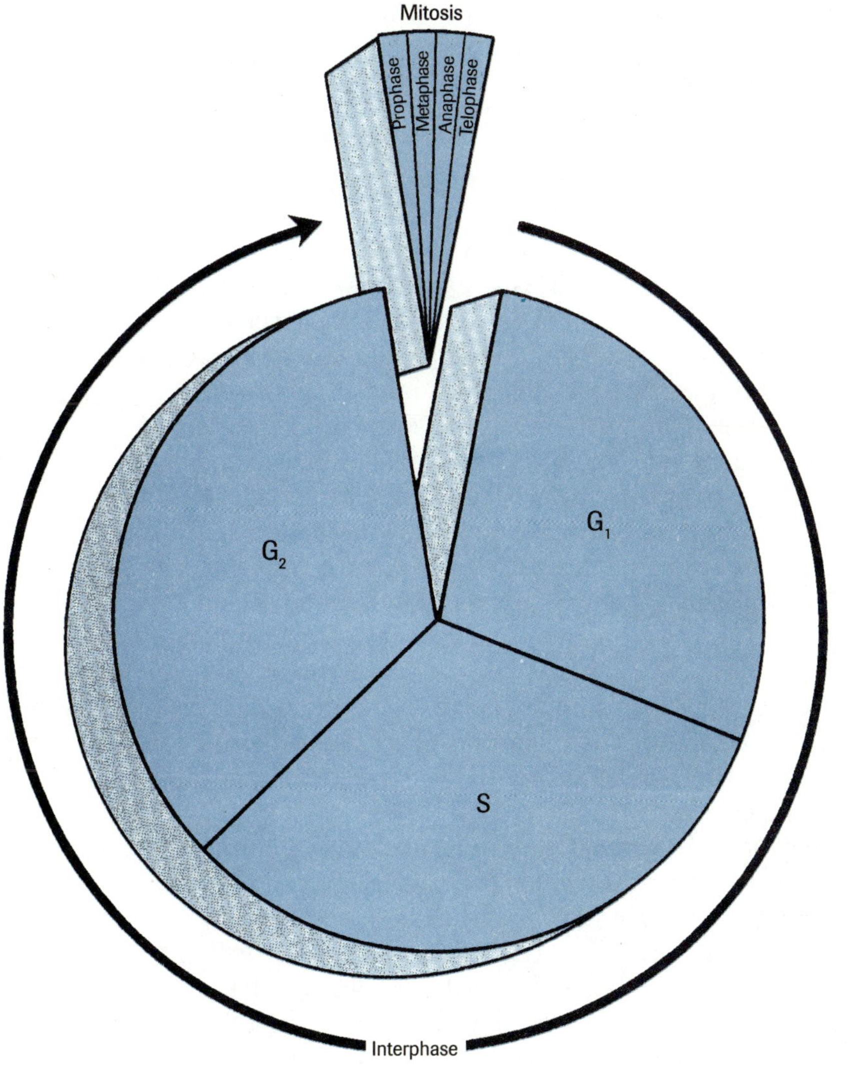

Figure 13.3 Schematic of eukaryotic cell cycle. During the G_1 phase, the cell synthesizes the enzymes and other substances required in the S phase, during which DNA is replicated. Final preparation for division, apparently involving synthesis of proteins necessary for mitosis, occurs in the G_2 phase, G_1, S, and G_2 constitute interphase.

comprises cell growth, duplication of all cytoplasmic components and DNA, and subsequent nuclear and cytoplasmic division. On production of the two daughter cells, the cell cycle begins again. The period of time a cell spends actually dividing is very short compared to the total time of the cell cycle. In some mammalian cells, mitosis is completed in about an hour, which represents only about 5 percent of the cycle. Thus, about 95 percent of the time the cells are synthesizing components that are necessary for their functioning and also preparing for division. This extended period of time between divisions is known as **interphase.** It is during interphase, then, that the cell synthesizes all the components necessary for nuclear division and formation of the two daughter cells.

Although the cell cycle is a complex process, the principle underlying it is relatively simple. Basically, the cell replicates (duplicates) its DNA and doubles its cytoplasmic components; these components are then distributed equally to each daughter cell during cell division.

The key to cell division is replication of the DNA molecule. Because DNA is the hereditary material, exact copies of it must end up in each new cell that is produced. Replication doubles the DNA content of a cell and, in eukaryotic cells, mitosis divides the nuclear DNA equally between the two resulting cells. If these two cells are to divide, each must replicate its DNA to form four cells, which in turn will form eight, and so on. Knowledge of the structure of DNA is essential to understanding its replication and central role in the making of proteins. Our task, then, is to describe the basic structure of DNA and then apply this information to its replication and subsequent role in protein synthesis. Keep in mind that these events are part of interphase of the cell cycle and are essential to the production of new cells from preexisting ones.

DNA Structure

DNA is constructed of units called **nucleotides,** each of which consists of one of four nitrogen-containing compounds called *bases.* The four bases in DNA are *adenine, guanine, cytosine, and thymine.* Bonded to each base is the five-carbon monosaccharide *deoxyribose,* from which DNA acquired its name. Bonded to deoxyribose is a phosphate group (PO_4^{-3}). Each DNA nucleotide, then, consists of a base, deoxyribose, and a phosphate group (Fig. 13.4). In the structure of DNA, the nucleotides of one strand are arranged in a linear sequence, held together by covalent bonds between the sugar of one nucleotide and the phosphate group of another nucleotide. There is formed, then, a long strand of alternating sugar-phosphate groups, with each nitrogen base extending perpendicularly along the strand.

The DNA molecule is composed of two of these polynucleotide strands. The two strands are oriented so that a base on one strand is paired only with a specific base on the other strand. In DNA, adenine (A) always bonds with thymine (T), and guanine (G) always bonds with cytosine (C). The attractive force between each complementary base pair is hydrogen bonding; these bonds, then, hold the two DNA strands together. The DNA molecule, therefore, somewhat resembles a ladder, with the base pairs A-T and G-C forming the rungs, and the alternating sugar-phosphate groups forming the sides of the ladder. As shown in Fig. 13.5, the entire molecule is twisted into a spiral, forming the now-famous double helix.

DNA nucleotides may occur in virtually any sequence along a strand. It is the specific sequence of nucleotides that dictates which proteins will be produced by a cell. For example, a section of one strand might include the sequence TGAAC. The corresponding section of the other strand would then be ACTTG. Because of their precise base pairings, the strands are said to be *complementary.* Depending on the species, a DNA molecule may consist of thousands or millions of nucleotides. What makes each organism different is its unique sequence of nucleotides in its DNA molecule.

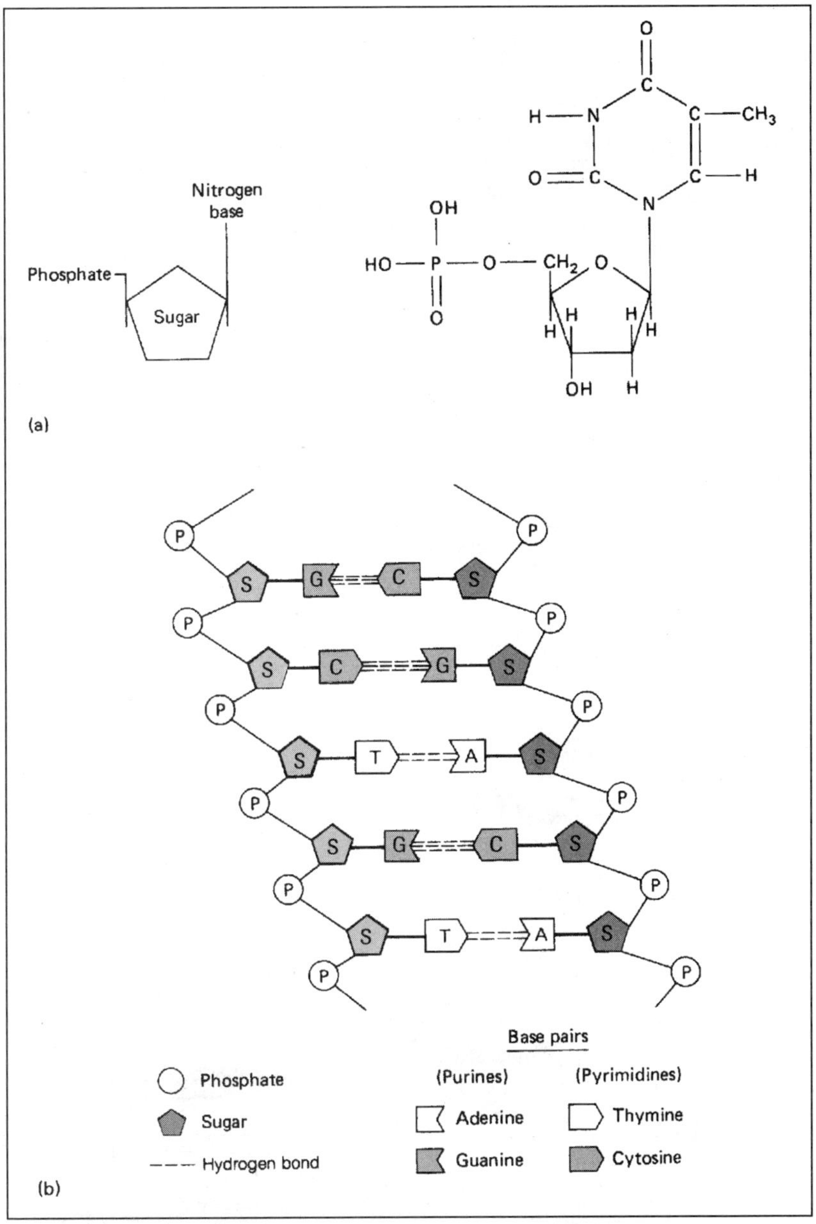

Figure 13.4 (a) General structure of a nucleotide and detailed structure of a thymine nucleotide. (b) Molecular structure of DNA, showing base pairing and orientation of sugar-phosphate "back-bone."

The structure of DNA was worked out by James D. Watson and Francis H. C. Crick in 1953. This structure is referred to, then, as the Watson-Crick model of DNA. The story behind the discovery of the structure of DNA is one of brilliant insight, drama, and inspiration. By the early 1900s, research had demonstrated that chromosomes consist mainly of DNA and proteins, but it was not known which of these components was the bearer of genetic information in the cell. But in 1944, three scientists, Oswald T. Avery, Colin MacLeod, and Maclyn McCarty, experimenting with certain bacteria, established beyond reasonable doubt that the genetic material was DNA and not protein. Thereafter, the race was on to discover the structure of DNA. Everyone in the race was aware of the information available concerning DNA, such as the fact that it was a long polynucleotide strand and that its content of adenine was equal to that of thymine and that the

Figure 13.5 Model of double helix structure of DNA. Colored areas of the strands are the sugar deoxyribose; light areas are the phosphate group.

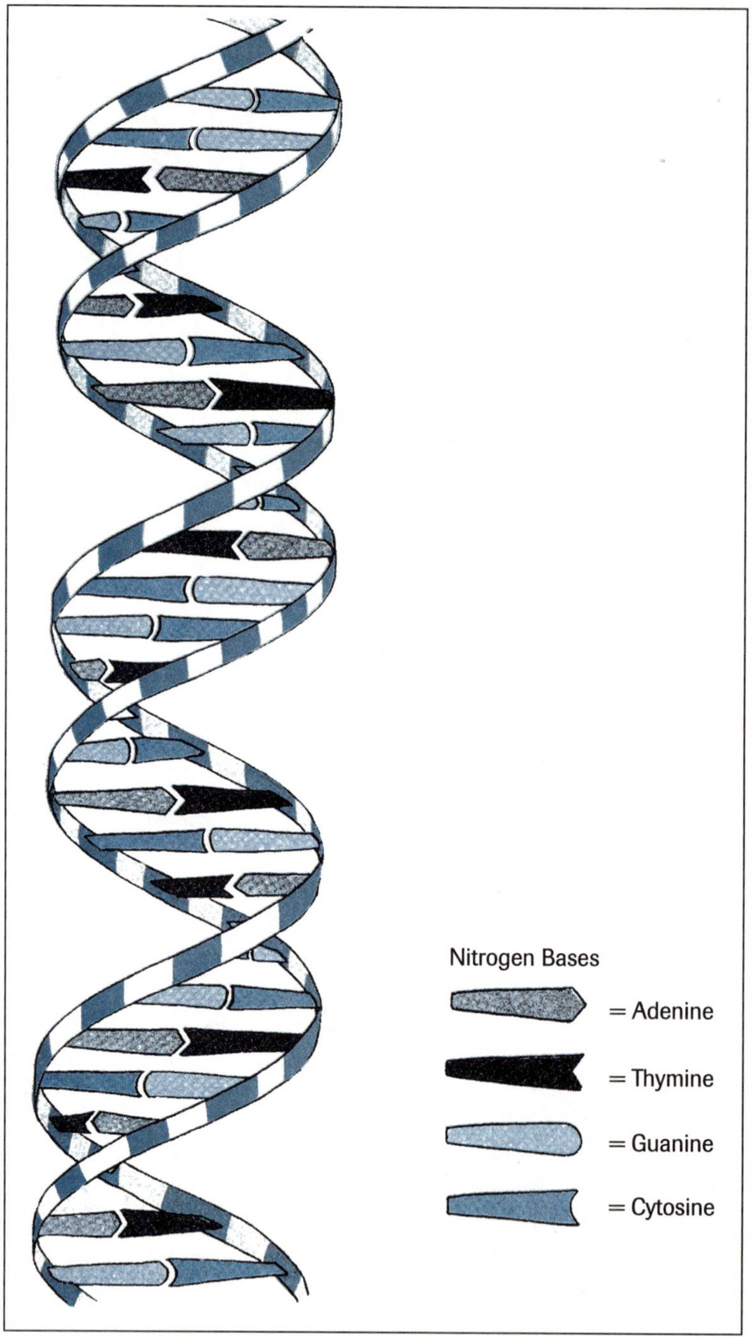

contents of guanine and cytosine were also equal. In addition, X-ray diffraction data provided by Rosalind Franklin and Maurice Wilkins indicated that DNA was helical and double-stranded. It was an inspired Watson and Crick who used these data to discover the key to DNA structure: base pairing. By aligning models of DNA strands side by side, the two investigators demonstrated that DNA consisted of repeating A-T and G-C pairs situated between two elongated strands constructed of sugars and phosphates. DNA, they concluded, was a double-stranded molecule twisted into the shape of a double helix. For their epoch-making discoveries, Watson, Crick, and Wilkins were awarded the 1962 Nobel Prize. Franklin died in 1958 and was ineligible for consideration because the Nobel Prize is not awarded posthumously.

With the structure of DNA confirmed, it was then possible to explain how the molecule could replicate itself.

DNA Replication

The replication of a DNA molecule from another DNA molecule is a complex process, requiring more than a dozen enzymes and other molecules to complete the process (Fig. 13.6). The concept of replication, however, is elegantly simple. During interphase of the cell cycle, replication begins by separation of the two strands of DNA. This involves an unwinding of the two strands of the spiral helix assisted by an enzyme. The strands unwind in opposite directions until they partially separate into two single strands.

After the DNA is "primed," or prepared for replication by another enzyme, each separated strand serves as a **template,** or "pattern" for the production of a new, complementary strand of DNA. As shown in Fig. 13.6, this begins when individual nucleotides in the cell align themselves, again with enzyme assistance, along the nucleotides of the separated DNA strands by complementary pairing. Thus, a single nucleotide containing adenine (A) would pair with a nucleotide containing thymine (T) on a DNA strand; a cytosine (C) nucleotide would pair with a guanine (G) nucleotide, and so on along the length of each DNA strand.

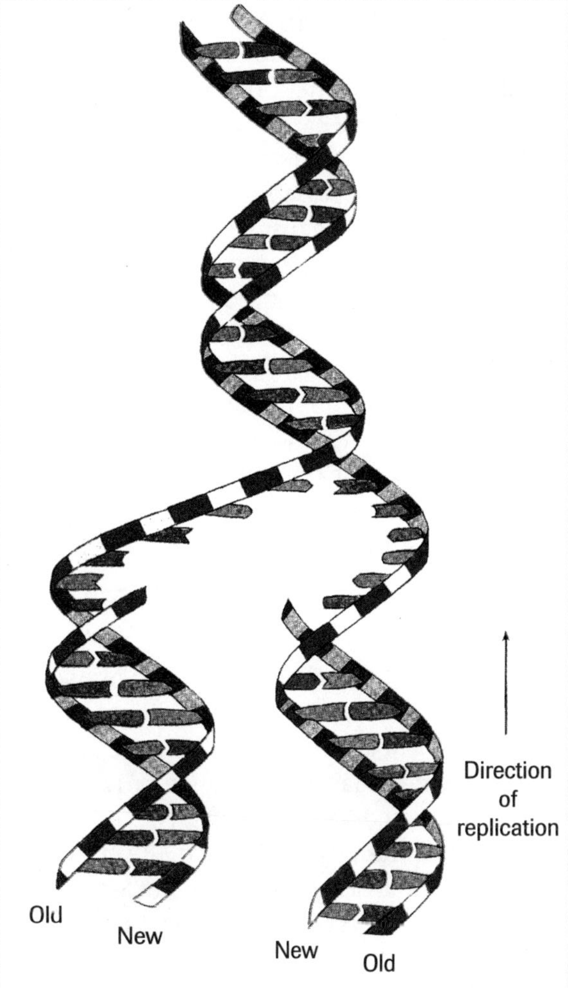

Figure 13.6 Replication of the DNA double helix. The two strands of the helix first unwind, and each old strand serves as a pattern or template for the synthesis of a complementary new strand. Every thymine pairs with an adenine, and every guanine pairs with a cytosine. Replication occurs as the two old strands are unwinding; thus, in the diagram, replication is occurring from the bottom of the figure upward.

As the DNA molecule continues to unwind, more nucleotides are added by complementary base pairing. This continues until, at the end of replication, each original DNA strand has made a complementary DNA strand, resulting in two double helixes. Each DNA molecule, then, is composed of one original strand and one complementary new strand. This is called *semiconservative* replication because half of the original DNA is conserved in each new molecule. DNA replication is a very rapid and extremely accurate process, with about one out of every billion nucleotides being paired incorrectly in a complete molecule. In fact, any damage to a complete DNA molecule from UV light or chemicals, for example, usually can be corrected by different types of repair enzymes.

DNA, RNA, and Protein Synthesis: An Overview

Before one cell becomes two, all the structural and enzymatic proteins that function during interphase must be synthesized. Protein synthesis in all living cells is directed by the genes contained in the double helix of DNA. The information required to synthesize a specific protein is encoded in the genes; the genes, however, do not actually make the protein. Instead, information in the genes, with the assistance of enzymes, is transcribed into a type of ribonucleic acid called messenger RNA, which subsequently carries the coded information from the nucleus to the cytoplasm. Messenger RNA binds to the ribosomes, the cytoplasmic organelles on which protein synthesis actually takes place. Ribosomes are partly composed of another type of RNA called ribosomal RNA, which is also transcribed from DNA in the nucleus of eukaryotic cells. The synthesis of a protein is known as *translation* and is directed by the messenger RNA that is bound to the ribosomes. This step is called translation because the information coded in messenger RNA must be translated into the amino acid sequence of a specific protein. This step requires the assistance of a third type of RNA called transfer RNA, which is also synthesized from nuclear DNA. Thus, the steps in protein synthesis follow the sequence: DNA → RNA → protein. This sequence is the so-called *central dogma* of molecular biology. In summary, the dogma states that the genetic information in DNA is transcribed onto RNA and that this information is then translated into a specific protein. Before describing how the information contained in a sequence of DNA results in the synthesis of a particular protein, we first need to know the nature of this "information."

The Code of Life

We have seen that DNA is composed of two complementary polynucleotide strands, and we have stated that genes, the units of heredity, are composed of DNA. It is apparent, then, that there is a relationship between nucleotides and genes. In the 1960s, researchers determined that the sequence of DNA nucleotides constituted a type of "code" that, when deciphered, consisted of DNA "code words," each of which contained three nucleotides. Thus, the DNA sequence TTAAGACGC, for example, consists of the code words TTA, AGA, and CGC. These three-nucleotide code words are also referred to as *triplets.*

In the synthesis of RNA from DNA, these triplet code words are transcribed in complementary fashion. The DNA triplet TTA, for example, is transcribed as AAU on the RNA strand. Remember that uracil is used in RNA whereas thymine is used in DNA; uracil, therefore, pairs with the adenine of DNA during RNA synthesis.

In any code, the symbols, numbers, or letters, stand for something specific. The dots and dashes of the Morse code designate letters of the alphabet. In the genetic code of DNA, almost all of the triplets designate one of the 20 amino acids found in living organisms. This means that TTA specifies a particular amino acid, AGA specifies a different amino acid, and CGC specifies still another amino acid. As mentioned, each of these DNA triplets is transcribed in complementary fashion in the synthesis of RNA. Thus, AGA is transcribed as UCU on the RNA strand and CGC is transcribed as GCG. On the messenger RNA strand, each triplet is referred to as a **codon.**

As you can see in Fig. 13.7, there are 64 messenger RNA codons, 61 of which code for amino acids. One of these 61 codons also functions as a "start" signal, which indicates the site on messenger RNA where translation is to begin. Each of the other three

The Genetic Code (Triplet Codons of RNA)

First Letter in Codon	Second Letter in Codon: U	C	A	G	Third Letter in Codon
U	UUU UUC UUA UUG	UCU UCC UCA UCG	UAU UAC (UAA) (UAG)	UGU UGC (UGA) UGG	U C A G
C	CUU CUC CUA CUG	CCU CCC CCA CCG	CAU CAC CAA CAG	CGU CGC CGA CGG	U C A G
A	AUU AUC AUA (AUG)	ACU ACC ACA ACG	AAU AAC AAA AAG	AGU AGC AGA AGG	U C A G
G	GUU GUC GUA GUG	GCU GCC GCA GCG	GAU GAC GAA GAG	GGU GGC GGA GGG	U C A G

Figure 13.7 Using nucleotides containing the base uracil (U), cytosine (C), adenine (A), and guanine (G), each of the 64 codons consists of a combination of three of these nucleotides. Sixty-one of the codons specify a particular amino acid. For example, UUU codes for the amino acid *phenylalanine,* and CUU codes for *leucine.* However, three of the codons (UAA, UAG, and UGA) are stop codons, i.e., they signal that protein synthesis is finished. In addition, the codon AUG, which codes for the amino acid *methionine,* is the start codon that initiates protein synthesis on the mRNA strand.

In the chart above, the first letter in a codon is read in the left-hand column, the second letter from the upper, horizontal column, and the third letter is read from the right-hand column.

codons, like a period at the end of a sentence, is a "stop" signal that marks the end of translation. You may be wondering why there are 64 messenger RNA codons when there are only 20 amino acids. This is because all but two of the amino acids are represented by more than one codon. Three amino acids, for example, have six different codons each. Various other amino acids are represented by two, three, or four different codons. This repetitious nature of the genetic code is known as *redundancy* and apparently allows for a margin of error without harming the organism.

The first messenger RNA codon was deciphered in 1961 by Marshall Nirenberg. By linking individual uracil nucleotides, Nirenberg synthesized a messenger RNA strand containing only that one nitrogen base. Thus, the only possible triplet codon on the strand was UUU. When the strand was added to a preparation of amino acids, ribosomes, and other protein-synthesizing components, translation resulted in a polypeptide constructed of only one specific amino acid. Nirenberg demonstrated, then, that UUU codes for that one amino acid. Five years later, through the demanding and meticulous work of many outstanding scientists, the entire genetic code you see in Fig. 13.7 had been cracked.

Recalling that a protein is constructed of a certain sequence of amino acids, the repeating triplets along a DNA strand thus constitute the genetic recipe for specific proteins. The proteins synthesized according to this recipe will genetically determine the type of organism produced and all its physical characteristics. Your DNA determines the color of your eyes, your height, your blood type, the shape of your nose, and every other genetic attribute you have inherited from a long line of ancestors.

Based on our knowledge of the DNA code, we have formulated a comprehensive definition of a **gene.** A gene is a particular sequence of DNA nucleotides that ultimately codes for a specific protein or polypeptide. The combination of genes you inherit is unique (unless you are an identical twin), and this combination is what your cells will translate into specific proteins. When organized, these proteins will form the cells and tissues and organs that genetically are you.

Making Proteins

Transcription

With the above information, we are now ready to discuss the making of a protein as directed by the genetic code of DNA (Fig. 13.8). As the central dogma dictates, RNA is produced using DNA as the template. The synthesis of RNA from DNA is known as **transcription** and involves the participation of certain transcription enzymes. The first major step is the production of messenger RNA, abbreviated mRNA. Essentially, mRNA is transcribed in much the same way as a new DNA strand is formed during replication. As Fig. 13.8 shows, the two DNA strands unwind somewhere in the molecule and *one* of the strands serves as the template on which the mRNA strand is transcribed. Cellular nucleotides that will comprise the new mRNA begin to pair in complementary fashion with each nucleotide along the DNA template. Cytosine (C) nucleotides pair with guanine (G) nucleotides, and adenine (A) nucleotides with uracil (U) nucleotides.

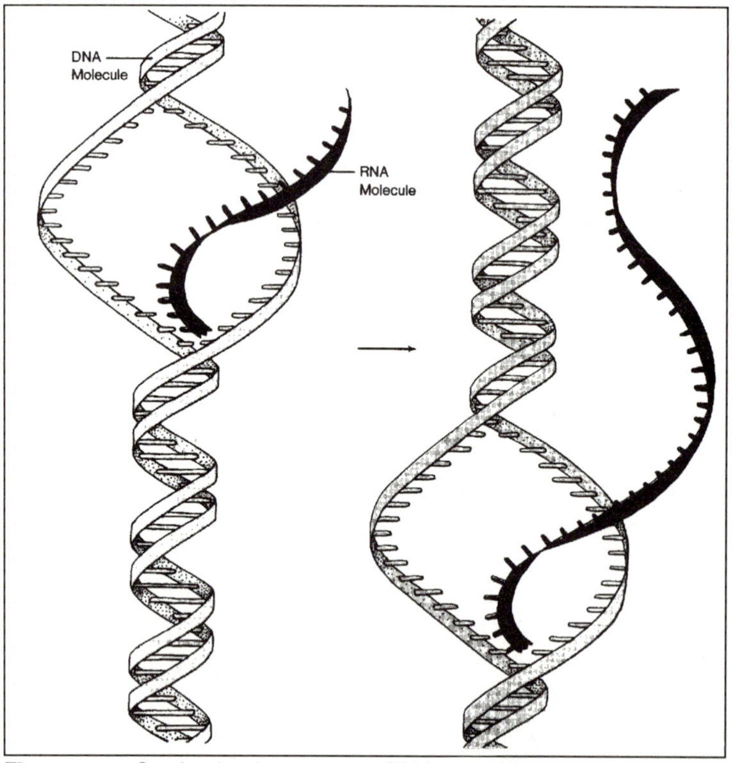

Figure 13.8 Synthesis of messenger RNA from DNA. As the two DNA strands unwind, one strand serves as the template for the synthesis of a complementary strand of mRNA. As synthesis proceeds (from top to bottom in diagram), the DNA strands rewind behind the growing mRNA strand. When synthesis is completed, the mRNA separates from DNA and passes through the nuclear membrane into the cytoplasm.

One by one, the nucleotides of mRNA are transcribed from the DNA template. In this fashion, the genetic information in the code words of DNA is copied as complementary codons onto mRNA. As it forms, the single strand of mRNA peels off the DNA template and the two DNA strands rewind into a double helix. Certain markers on the DNA template instruct enzymes where to start the transcription process and where to stop when the mRNA strand is completed. With the genetic message from DNA coded in mRNA, we now will consider the molecules required to decode the message.

In addition to mRNA, DNA also transcribes other types of RNA that have different cellular functions. One of these is **transfer RNA** (tRNA), a highly folded molecule that will bind to an amino acid and pair it with an mRNA codon that specifies that amino acid (Fig. 13.9). When viewed in two dimensions, tRNA somewhat resembles a cloverleaf; actually, the molecule has many twists in it, and its three-dimensional shape resembles an L. At the open end of the tRNA molecule there is a site for the attachment of a specific amino acid. Many tRNA molecules are transcribed and each one will bind to only one specific amino acid. A specific enzyme carries out the attachment of each amino acid to its particular tRNA molecule.

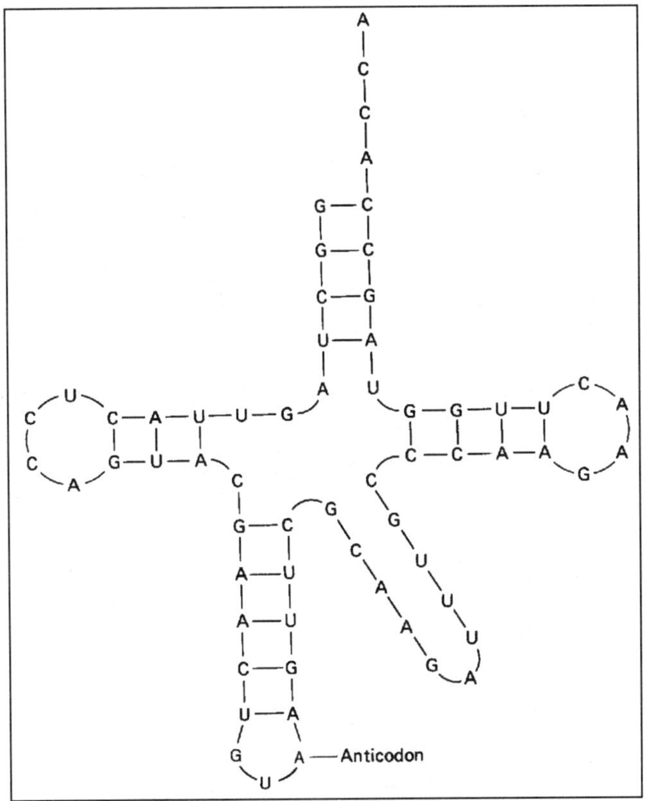

Figure 13.9 Generalized structure of a transfer RNA molecule. The anticodon is complementary to a specific codon of messenger RNA: A = adenine; U = uracil; C= cytosine; G = guanine.

On the loop opposite the amino acid attachment site is a special nucleotide triplet called an **anticodon.** *An anticodon on a given tRNA molecule is complementary to one specific codon on mRNA.* The anticodon UAC, for example, is complementary to the mRNA codon AUG. Anticodon-codon pairing follows the same rules as base pairing in replication or transcription. The function of tRNAs is to bring their attached amino acids to the proper codon on mRNA based on the appropriate codon-anticodon pairing. Thus the triplet code of DNA specifies a particular sequence of mRNA, which, with the help of tRNA molecules, produces a specific sequence of amino acids. This, of course, results in the synthesis of the protein that was coded for originally in the gene.

A third type of RNA required for protein synthesis is **ribosomal RNA** (rRNA). Ribosomal RNA and protein are the constituents of ribosomes, the cellular organelles on which protein synthesis takes place. Because ribosomes have binding sites to which mRNA and tRNA attach, they control the pairing of tRNA anticodons with the correct mRNA codons. Some ribosomes are found attached to the membranes of the endoplasmic reticulum and others lie free in the cytoplasm. In eukaryotic cells, protein synthesis occurs on ribosomes at both sites.

TRANSLATION

With the DNA code transcribed as codons on mRNA, and with the production of tRNA and rRNA, the stage is set to translate the code into a specific protein, as outlined in Fig. 13.10. In eukaryotic cells, the RNAs are transcribed in the nucleus and then pass through the nuclear membrane out into the cytoplasm. Translation is a complex process requiring enzymes at every step, including enzymes that aid in initiating and terminating the process.

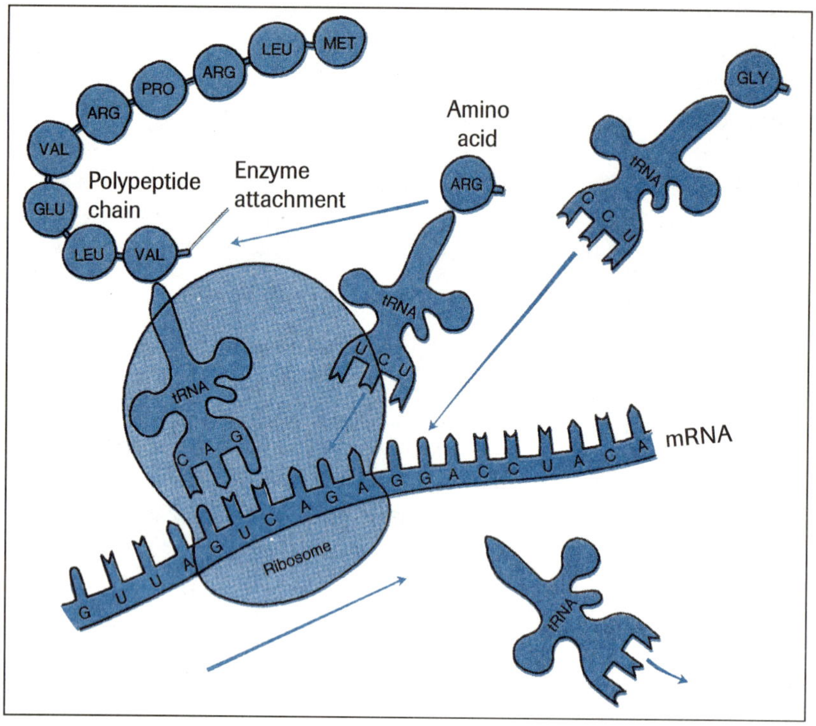

Figure 13.10 Basic scheme of protein synthesis. Within the cell nucleus, one of the strands of DNA synthesizes a complementary strand of messenger RNA. The mRNA passes across the nuclear membrane into the cytoplasm, where it becomes attached to ribosomes. Each type of transfer RNA picks up a specific amino acid and carries it to a ribosome that is moving along the mRNA strand. The anticodon of each tRNA bonds with its complementary codon on the mRNA strand. One by one the amino acids are brought to the mRNA strand and linked by peptide bonds. After its amino acid is linked to the growing polypeptide chain, each tRNA separates from the mRNA strand and moves out into the cytoplasm to await another round of protein synthesis. When synthesis is completed, the polypeptide is released into the cytoplasm of the cell.

In the cytoplasm, a special initiator tRNA binds to a mRNA codon on the ribosome by complementary codon-anticodon pairing. The "start" codon on mRNA is AUG, which means that the anticodon of the initiator tRNA is UAC. Attached to the initiator tRNA is the first amino acid in the protein chain. In essence, this mRNA-tRNA pairing initiates the translation process.

The next stage, called *elongation*, adds amino acids one at a time as each tRNA molecule brings its attached amino acid to the appropriate site on mRNA as specified by its triplet code. Basically, an incoming tRNA binds with mRNA through anticodon-codon pairing, shown in Fig. 13.10. The amino acid carried by tRNA bonds with the first amino acid (the one attached to the initiator tRNA), forming a dipeptide. The first amino acid is removed enzymatically from its tRNA. Messenger RNA slides through the ribosome, shifting the next codon into position. The next incoming tRNA pairs with this codon, and its attached amino acid bonds with the second amino acid in the chain. The growing protein chain now is a tripeptide; that is, it consists of three amino acids. In the meantime, the second amino acid is removed from its tRNA, which is released from the ribosome, leaving the second amino acid as part of the growing peptide chain. This sequence is repeated as each tRNA brings its amino acid to the ribosome, forming a polypeptide. After each new amino acid is bonded to the chain, it is removed from its tRNA, and the tRNA is subsequently released from the ribosome. The tRNAs are reusable and can be reloaded with their respective amino acids in a subsequent round of translation.

As translation continues and mRNA moves through the ribosome, eventually a termination codon of mRNA, such as UAA, reaches the ribosome, signaling the end of translation. A subsequent reaction releases the polypeptide chain from the ribosome, and the finished polypeptide is prepared for use by the cell. This may involve the addition of carbohydrates or lipids to the polypeptide. For example, a complex of protein and polysaccharides is a constituent of human connective tissue. Usually, many copies of a polypeptide can be made simultaneously by several ribosomes attaching to one strand of mRNA and translating the message in assembly-line fashion. Such groups of ribosomes form *a polyribosome,* or polysome for short.

As mentioned, the tRNAs released from the ribosome are reusable and move away from the mRNA strand to await another round of protein synthesis. The mRNA strand may persist for weeks or months, after which it is broken down into its component nucleotides by a specific enzyme. These nucleotides can be recycled to transcribe another RNA molecule in the nucleus.

It needs to be emphasized that the real determinant of protein structure is DNA. The RNAs are only the carriers and interpreters of the genetic message carried on DNA. Therefore, when a cell divides, it is the unique nucleotide sequences in DNA that are passed by the chromosomes to future cells and future generations.

The Universal Genetic Code

Finally, one other aspect of the genetic code of life is most intriguing. In almost every known living organism on Earth, plant or animal, the same DNA code words represent the same amino acids. The only known exceptions include various ciliated protozoans, along with mitochondria and chloroplasts, which have their own DNA. Organisms differ, then, not because each has different code words, but because the various different sequences of their DNA code words result in the synthesis of different proteins. In the nearly universal sameness of the genetic code, there is convincing evidence for the evolutionary relationships thought to exist among all forms of life.

Jumping Genes

For many years, the conventional concept of protein synthesis was that of a rather straightforward, stable mechanism, which, as we have described, involves replicated DNA transcribing a strand of messenger RNA, from which the message is translated into polypeptides. Evidence indicates, however, that the mechanism probably isn't quite that simple or direct. It has been found, for example, that some genes may not be permanently fixed at their positions in chromosomal DNA. It appears, instead, that these genes may move from one site to another on a chromosome, or even leap from one chromosome to another. In some cases, the gene replicates and its copy moves to another location in DNA. The concept of "jumping genes" originated with Barbara McClintock through her breeding experiments on corn in the 1940s. She was studying the effects of gene mutations on the color of corn kernels in an effort to determine the source of their pattern of pigmentation. Some kernels were of one color, others were spotted, and still others lacked color pigment altogether. McClintock proposed that the random insertion of transposed genes in the DNA molecule caused mutations that altered the synthesis of seed pigment. This, she concluded, was responsible for the varied colors of the kernels. For her discovery, McClintock won the 1983 Nobel Prize—over three decades after her original research.

It has been suggested that some jumping genes, or *transposons* as they are called, may have originated from viral genes left in the host DNA by an invading virus. A virus, you will recall, is a segment of DNA or RNA enclosed in a protein coat. The coats of some viruses are surrounded by a membranous envelope derived from the host's membrane. Viruses use the host's cell machinery to replicate their own viral genetic material within the host cell. By so doing, the virus adds its genes to those of the host DNA, which then takes on a new organization. The result is genetic alteration, or *mutation,* of the DNA molecule and thus a change in gene expression. Scientists believe, for example, that some transposons, after jumping tc a site next to a certain gene, may act to turn off the gene; that is, the gene is not transcribed and therefore no protein is produced.

Inactivation of a gene by transposons has been implicated in some human diseases. Turning off a gene involved in the synthesis of a certain enzyme, for instance, could affect functioning of the brain, the heart, or virtually any other vital organ or structure in the body. This does not necessarily mean that the only effects of transposons are harmful. It has been suggested, for example, that by stopping gene transcription, transposons may be effective in regulating cellular development.

Eukaryotic RNA, as well as DNA, is not as stable and inactive as once believed. Messenger RNA, for example, may contain long sequences of meaningless transcribed nucleotides that are snipped out of the RNA molecule by enzymes before translation occurs. The remaining meaningful sequences are spliced together to make the functional mRNA strand that codes for a functional protein. Moreover, there is evidence that RNA in some organisms and viruses may "edit" itself by adding necessary nitrogen bases to its strand or by changing a base. These corrections may be necessary if the genetic information in DNA is not sufficiently clear. In effect, RNA compensates for this lack of clarity in DNA by monitoring and correcting itself until the functional RNA strand is complete.

Mutations: Genetic Errors

Ordinarily, the genetic machinery of the cell functions faithfully and accurately. There are instances, however, when the machinery goes awry as a result of **mutations.** Various environmental factors are known to cause mutations: ultraviolet radiation, x-rays, and an assortment of chemicals, such as pesticide reagents, food additives, and cleaning agents, are well-documented examples. Transposons, as we have noted, can produce mutations, although the mutations arise within the organism rather than from an outside source. Sometimes, for unknown reasons, mutations arise spontaneously. By definition, a mutation is any alteration in the nucleotides of DNA. Such alterations include the *addition* of one or more nucleotides somewhere along a DNA strand, the *loss* of one or more nucleotides, or the *substitution* of one nucleotide for another. From our discussion of protein synthesis and transposons, you can see how any of these mutations could result in a change in the amino acid sequence of a protein by altering the sequence of the DNA triplet codes.

Using a substitution mutation as an example, and a written message as an analogy for a genetic message, the normal phrase "twinkle, twinkle, little star" might become "pwinkle, pwinkle, lipple spar," in which "p" is substituted for "t." Addition or loss of a nucleotide can change the entire "reading" of the genetic message. The sequence [ATG] [CAA] [GGC] on DNA would read [TGC] [AAG] [GC*] if the first nucleotide (A) in the sequence were lost. This one deletion changes all the triplets that follow it. In many cases, such changes may have minor significance, but in other cases, alteration of a protein has deadly consequences.

Sickle-Cell Anemia

One such consequence results from a change in one of the body's most vital proteins: hemoglobin. Hemoglobin is an oxygen-carrying protein in red blood cells, which deliver oxygen to all body cells. Various inherited blood diseases involve mutations that change the amino acids or their sequence in the hemoglobin molecule. Over 250 different abnormal human hemoglobin molecules have been discovered. In some cases, a change in only one amino acid severely interferes with the ability of hemoglobin to combine with oxygen. One disease in which this occurs is *sickle-cell anemia* (Fig. 13.11). This malady is characterized by severe anemia and sickle-shaped red cells that not only have poor oxygen-carrying capacity, but also tend to clog up small blood vessels.

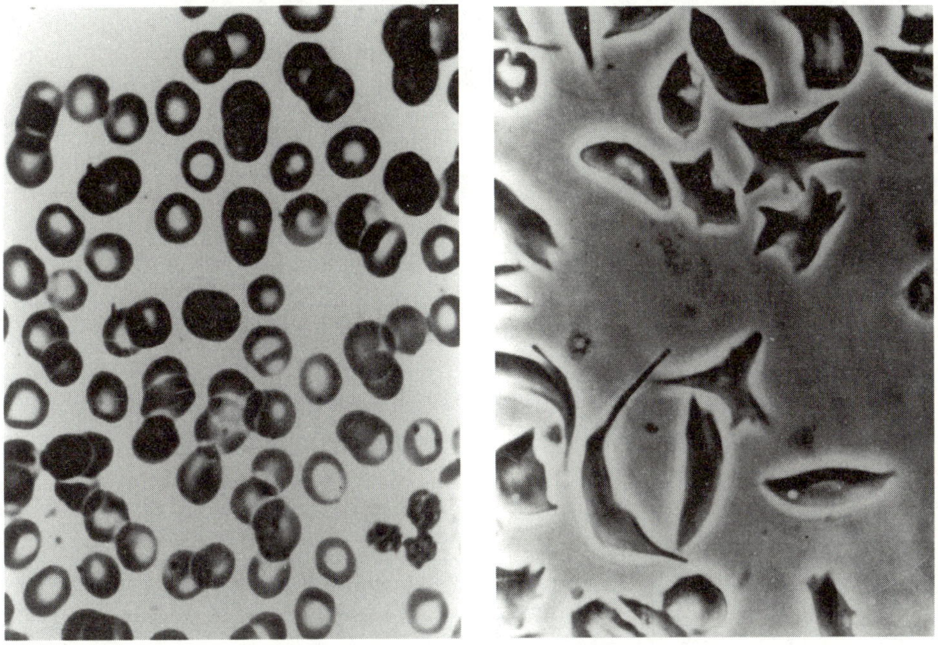

Figure 13.11 Sickle-cell anemia, an inherited blood disease, involves a mutation that changes the amino acids in the hemoglobin molecule.

Sickle-cell hemoglobin is structurally identical to normal hemoglobin, being composed of four polypeptide chains. In two of the chains, however, sickle-cell hemoglobin has one amino acid incorrectly substituted for another. As a result of a mutation that changes one nucleotide in a segment of DNA, there are two incorrect amino acids inserted among the 574 amino acids that make up the hemoglobin molecule. As a consequence of respiratory and circulatory failure, sickle-cell anemia is usually fatal—and all because of one amino acid.

Nuclear Division

Mitosis

It was not until the late 1800s that new staining techniques made possible the observation of chromosomes in living cells. In 1882, the German biologist Walther Flemming described in detail the events of mitosis using the body cells of a salamander. With the pronouncement of the chromosomal theory of heredity in the early 1900s and the discovery of the structure of DNA in 1953, more specific information on mitosis and inheritance began to unfold.

The concept of mitosis is relatively simple. During interphase, the chromosomes are not fully distinct but exist as a long tangled mass of DNA and protein called **chromatin.** Just outside the nucleus of animal cells lie two pairs of tiny microtubular structures called **centrioles.** These structures are absent in most plant cells. Extending

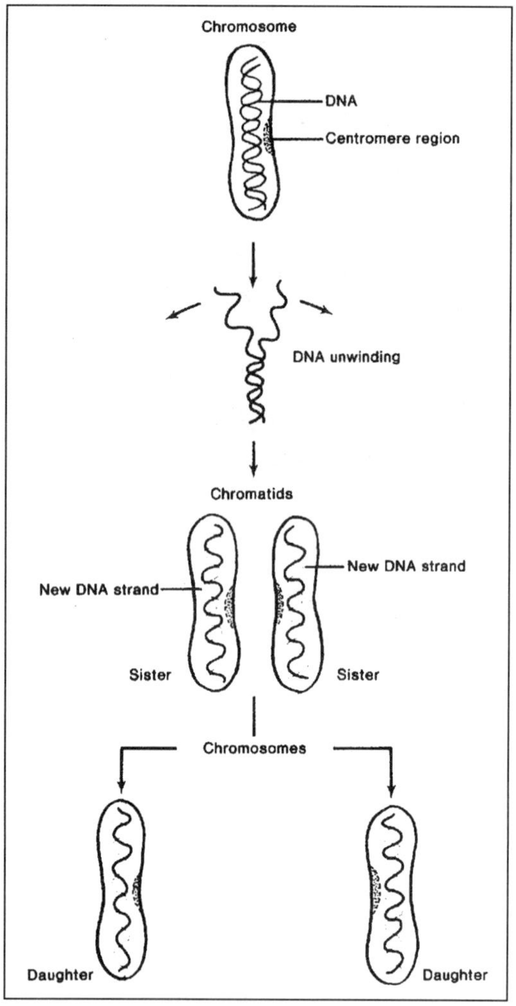

Figure 13.12 A duplicated chromosome consists of two sister chromatids joined at the centromere region. Each chromatid consists of identical helices of DNA. When the sister chromatids separate from each during cell division, they are then called chromosomes.

from the centrioles is another set of microtubules that forms the **mitotic spindle.** While the chromosomes are still diffuse, their DNA is replicated so that each chromosome is composed of two double helixes of DNA joined at a region called the **centromere** (Fig. 13.12). The result, then, is that the total amount of DNA in the cell nucleus is doubled. In human cells, this means that 46 chromosomes are duplicated. The genetic material is now present in two complete copies in such cells. While joined together, however, each duplicate chromosome is referred to as a **chromatid;** these two "sister" chromatids are genetically identical, being composed of the same kind and amount of DNA.

Although mitosis is a continuous, nonstop process, it has been separated into four distinct phases, making it easier to follow the progressive sequence of events (Fig. 13.14). The phases, in order, are *prophase, metaphase, anaphase,* and *telophase.* As stated earlier, the essential function of mitosis is to produce two daughter nuclei in which the chromosomes, and, hence, the amount of DNA, are distributed equally. In most cells, mitosis is followed by division of the cytoplasm (cytokinesis) to form two daughter cells. Cytokinesis results in each daughter cell having a single nucleus and also distributes the organelles and other cellular materials equally to each cell.

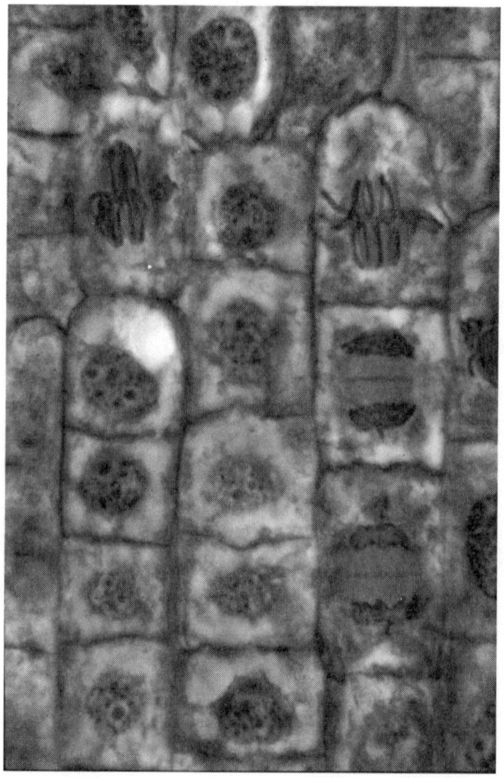

Figure 13.13 Mitosis in the cells of an onion. Identify the stages after looking at Figure 13.14

Prophase

By early prophase of mitosis, which immediately follows interphase, the duplicated chromosomes are fully formed and distinctly visible under the microscope. Each pair of centrioles (if present) outside the nucleus migrates toward opposite ends, or poles, of the cell and the mitotic spindle is formed. The nuclear membrane and nucleoli disperse, leaving the chromosomes suspended in the cytoplasm.

Metaphase

At metaphase, the chromosomes have moved to the center of the cell, forming a *metaphase plate* on which the centromeres are aligned with each other. The centromere region of each duplicated chromosome is attached to mitotic spindle fibers extending from the centrioles, which are now located at opposite poles of the cell. Late in metaphase, the centromere of each chromosome divides, resulting in separate chromatids, each having its own centromere region. After the centromeres divide, each chromatid is called a chromosome.

Anaphase

During anaphase, the now-separated sister chromosomes are moved toward opposite poles of the cell by the spindle fibers. Remember that the DNA content of the chromosomes moving toward one pole is the same as the DNA content moving toward the opposite pole. By late anaphase, each set of chromosomes arrives at its respective pole of the cell.

Telophase

In telophase, nuclear membranes begin to form around each set of chromosomes and the nucleoli reappear. Cytokinesis has begun and, in animal cells, a **cleavage furrow** forms in the cytoplasm between the two nuclei. Shortly, the cell

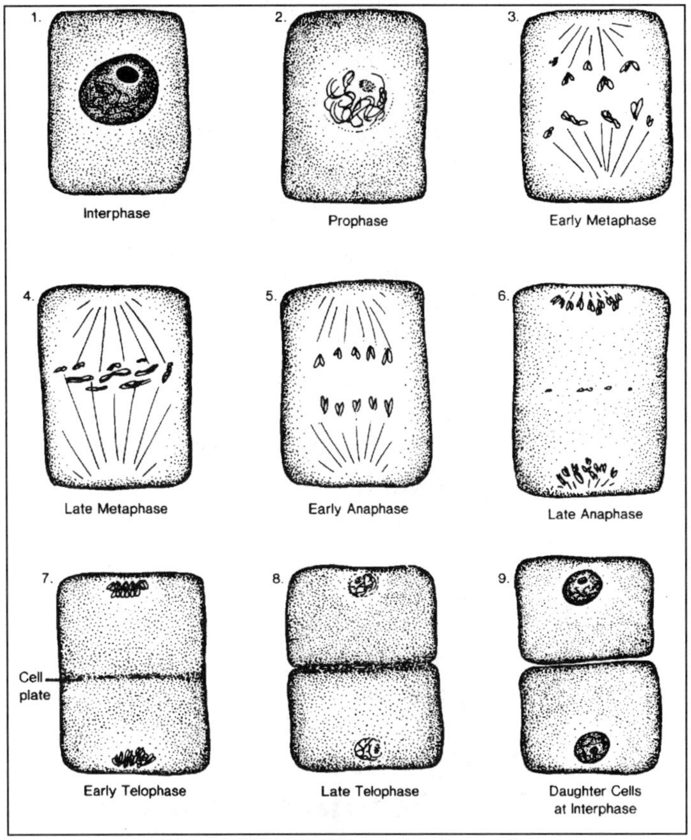

Figure 13.14A Mitosis in plant cells.

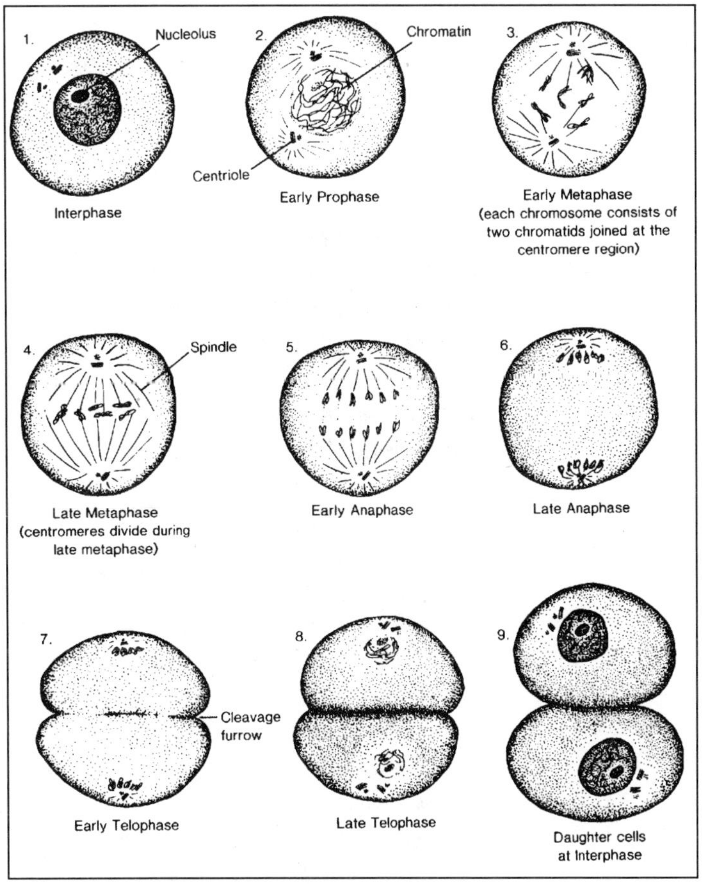

Figure 13.14B Mitosis in animal cells.

pinches in two, forming the genetically identical daughter cells. In plant cells, the new daughter nuclei are separated by a **cell plate** rather than a cleavage furrow. The cell plate arises from membranes of the Golgi apparatus, and eventually the plant cell wall forms between the membranes of the cell plate. Plant cell walls, you will recall, are composed primarily of the polysaccharide cellulose.

As pointed out earlier, some cell types do not undergo cytokinesis after mitosis and thus become multinucleate cells. These include, for example, certain human liver cells, which are binucleate and the multinucleate cells of various algae and fungi. More often, however, nuclear division is followed by cytoplasmic division, or cytokinesis, as you have observed in Fig. 13.14.

One of the most distinctive characteristics of living organisms is their capacity to reproduce others of their own kind. Among some unicellular organisms, such as algae, yeasts, and protozoa, mitotic cell division is essential to the process of reproduction. By various asexual methods to be considered later, these organisms undergo mitotic cell division to produce offspring in basically the same manner as your skin cells produce other skin cells.

Mitosis provides further evidence that life is regulated at the cellular level. In observing mitosis, one can appreciate more fully the fact that the cell is the fundamental unit of life. Without the capacity of living cells to utilize mitosis in making copies of themselves, life would not have progressed much beyond the stage of the primitive prokaryotes.

Meiosis: The Formation of Gametes

Not all the cells formed by sexually reproducing organisms are diploid. In the reproductive tissues, cells arise that are **haploid;** that is, they have one-half the diploid number, or *n* number, of chromosomes. These haploid reproductive cells are called **gametes.** In animals and in some plants, gametes arise by meiotic cell division, or meiosis, which reduces the diploid number by one-half. Some plants also produce haploid *spores* by meiosis.

In humans, all normal gametes have 23 chromosomes, which is one-half the normal diploid number of 46 chromosomes. Gametes of humans and other sexually reproducing organisms are haploid because one gamete from a male and one from a female unite at fertilization to produce a diploid **zygote.** If gametes were diploid instead of haploid, each succeeding generation would have twice as many chromosomes as its parents. In humans, assuming gametes contained 46 chromosomes instead of 23, the second generation would have 92 chromosomes, the third generation would have 184, and the tenth generation would have 23,552 chromosomes!

We said earlier that chromosomes are usually present in the nucleus in pairs. At fertilization, one chromosome of a pair is contributed by the female and one by the male. The two chromosomes constituting each pair are usually the same size and shape and are referred to as **homologous** (Gk. "same proportion") chromosomes. One chromosome of a pair, then, is the *homologue* of the other. Two of the distinguishing events of meiosis are the pairing and separation of homologous chromosomes.

The terminology and most of the steps of meiosis are similar to those of mitosis. There are, however, several crucial differences that have wide-ranging significance, as we will see. Meiosis involves two cellular divisions known as *meiosis I* and *meiosis II,* which result in producing four haploid cells rather than the two diploid cells of mitosis (Fig. 13.15). In addition, the four cells produced from meiosis are usually genetically different from each other and from the parent cell. Meiosis I is separated into *prophase I, metaphase I, anaphase I,* and *telophase I.* Meiosis II consists of *prophase II, metaphase II, anaphase II,* and *telophase II.*

Meiosis I

The interphase preceding meiosis is similar to that of mitosis, involving DNA replication and formation of duplicated chromosomes, each composed of two sister chromatids joined at the centromere.

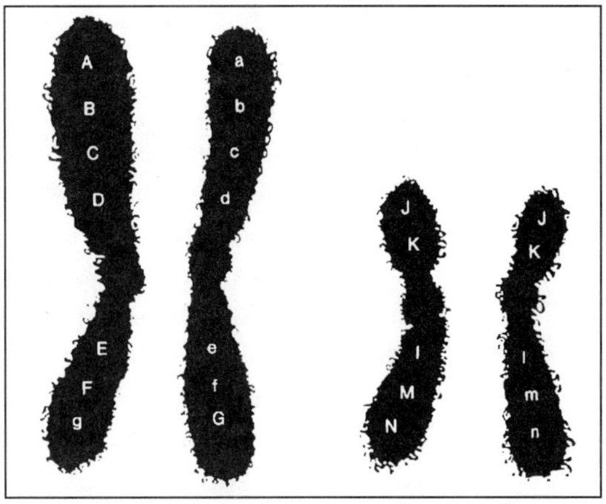

Figure 13.15 Two pairs of homologous chromosomes. Each chromosome pair separates from its partner pair during meiosis I.

Prophase I

In prophase I, the homologous chromosome pairs are aligned together so that the genes on one chromatid are opposite the same genes on the chromatid of the homologous chromosome. In this close proximity, the homologues can break and exchange segments of their chromatids, a phenomenon called **crossing over** (Fig. 13.17). Sister chromatids of a single chromosome also may exchange segments but the genetic consequences are often not detectable in the appearance of the offspring.

Because one chromosome of each homologous pair is donated by the female and the other is donated by the male, the result of crossing over is **genetic recombination;** that is, a mixture of male and female genes on each homologue. New combinations of genes result in increased genetic diversity of the offspring.

Later in prophase I, the centriole pairs, if present, begin to migrate toward the poles of the cell, the spindle forms, and the nuclear membrane and nucleoli break down.

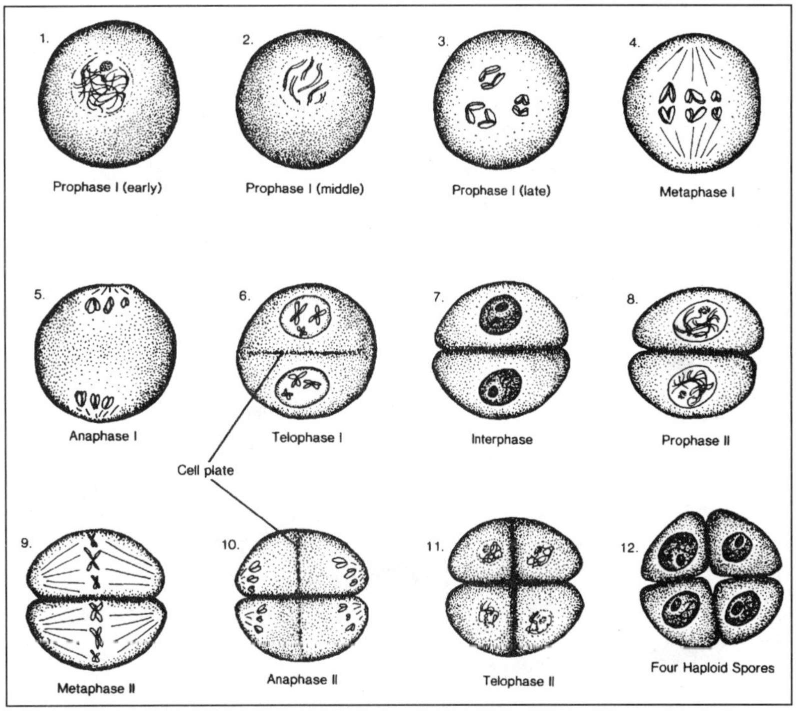

Figure 13.16 Meiosis in plant cells produces four haploid spores.

Metaphase I

At metaphase I, the homologous pairs line up side-by-side perpendicular to the spindle fibers on the metaphase plate; the spindle fibers become attached to the centromeres of each homologue. In metaphase I of meiosis, the centromeres do *not* divide as they do in mitotic metaphase. Each chromosome, therefore, still consists of two chromatids joined at the centromere region.

Anaphase I

During anaphase I, the spindle fibers contract and move each chromosome of a homologous pair toward opposite poles of the cell. Each homologue moving away from its partner still consists of two chromatids. Like genetic recombination, separation of homologous chromosomes has considerable genetic significance. As mentioned, a chromosome from the female parent and a chromosome from the male parent constitute each homologous pair. The manner in which each homologue aligns with its partner on the metaphase plate in metaphase I is essentially random. The female chromosome could line up on one side in one homologous pair, and on the other side in another homologous pair. The same is true for the male chromosome. Thus, depending on how the homologous pairs line up, either of the homologues could move toward one pole of the cell or toward the opposite pole. Moreover, how one pair of homologues lines up does not influence how any other pair lines up. Because of this alignment in metaphase I, the chromosomes contributed by the male and female become assorted randomly at the end of meiosis I and later in the gametes at the end of meiosis II. This phenomenon, known as **independent assortment,** is one of the keys to genetic variability.

Telophase I

At telophase I, each set of homologues has arrived at opposite poles of the cell, and, in some cells, nuclear membranes and nucleoli reform. Cytokinesis gets underway, forming a cleavage furrow in animal cells and a cell plate in plant cells. In most organisms, the cell divides into two daughter cells; in other organisms the cell remains undivided. The spindle fibers disappear and the chromosomes, each still composed of two chromatids, sometimes fade from view under a microscope. If present, the centrioles duplicate in preparation for the second meiotic division. Here at the end of meiosis I, there may be a brief period known as interphase II but there is no replication of DNA as in the first interphase.

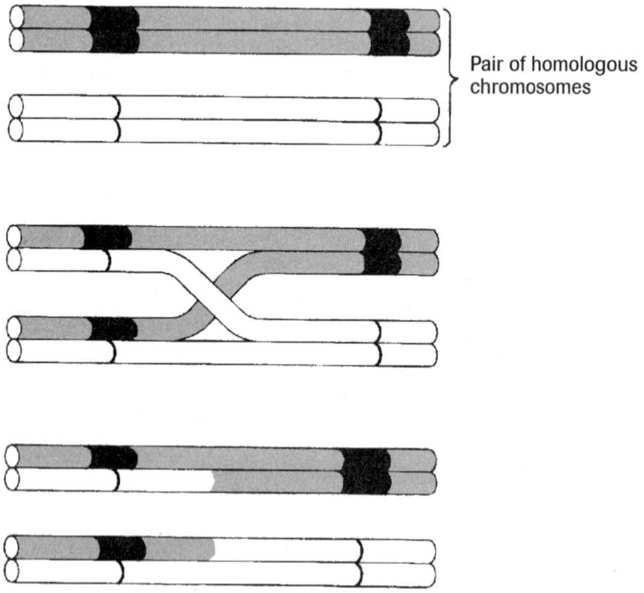

Figure 13.17 An example of crossing over in which there is an exchange of dominant (wide bar) alleles and recessive (narrow bar) alleles. As a result of crossing over, reproductive cells carry a new and different assortment of genes.

Meiosis II

As is evident in Fig. 13.16, meiosis II proceeds in a manner similar to that of mitosis. This is not to say that meiosis II and mitosis are identical; for example, meiosis II is not preceded by chromosomal duplication nor is the genetic composition of the cells formed at the end of meiosis II the same as that of cells formed in mitosis.

In prophase II, spindle fibers form in each of the two cells produced at the end of meiosis I. If present, the nuclear membranes break down, leaving the chromosomes suspended in the cytoplasm.

At metaphase II, the chromosomes in each cell become aligned on the metaphase plate end to end. Spindle fibers from opposite poles of the cells attach to each chromosome at the centromere region. Late in metaphase II, the centromeres divide as they do in mitosis, allowing the two chromatids of a chromosome to exist independently.

During anaphase II, each chromatid (now called a chromosome) moves away from its sister chromatid toward its respective pole of the cell.

In telophase II, the chromosomes arrive at the opposite poles of the cell, the spindle fibers disperse, and a nuclear membrane forms around each of the four sets of chromosomes. Cytokinesis separates the cytoplasm, forming four haploid cells.

Genetic Variability

As a result of crossing over and independent assortment, each of the four haploid cells carries a different assortment of genes. When we consider the hundreds of millions of male gametes produced by many species of plants and animals, the significance of this assortment is astonishing. In terms of its genetic makeup, every gamete is unlike any other gamete. Every gamete produced by the female is also unique.

Union of the male and female gametes at fertilization produces a diploid zygote genetically distinct from any other that ever existed. In the case of human reproduction, any one of 350–400 million sperm released into the female reproductive tract could be the gamete that will unite with the one female gamete. As it turns out, there are about 8 million possible combinations of the 23 pairs of male and female chromosomes assorted in human meiosis. Each gamete, then, contains one of these 8 million combinations.

Meiosis is the paramount source in producing organisms that are genetically variable. Sexual reproduction brings together gametes that produce organisms genetically and physically distinct from their parents and from other members of a population. These genetically variable offspring make adaptation to their environment possible—a circumstance of critical relevance in the evolutionary process. Natural selection acts on these organisms, culling those that do not have survival traits in that environment. The more fit organisms survive to reproductive age and pass their favorable traits to succeeding generations. Because the organisms are variable and there exists a pool of shuffled genes, the chance that some of the organisms will survive if the environment should change is enhanced.

Errors in Cell Division

For reasons not fully understood, meiosis does not always proceed in its normal manner . Members of a homologous pair may not separate from each other at anaphase I or the centromeres may not divide at metaphase II. The failure of chromosomes or chromatids to separate properly is called **nondisjunction** (Fig. 13.18). Nondisjunction in meiosis I is more common and results in both chromosomes of a homologous pair going to one pole of the cell and into a daughter cell. This daughter cell, then, has an extra chromosome and the other daughter cell receives neither of these chromosomes. Following meiosis II, two of the four gametes have an extra chromosome and the other two have one fewer chromosome. In humans, normal gametes have 23 chromosomes. If a gamete having an extra chromosome unites with a normal gamete, the resulting zygote would have 47 chromosomes instead of the normal 46. What are the possible effects of nondisjunction?

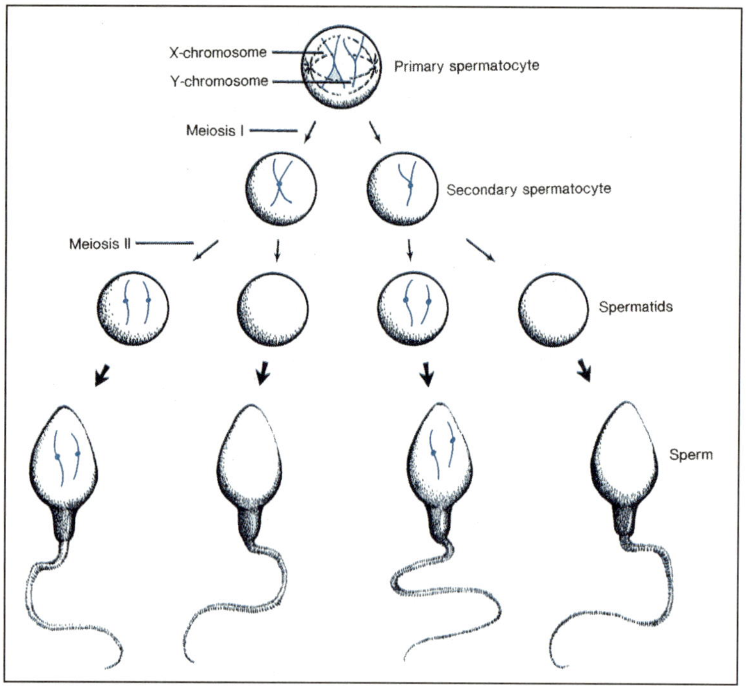

Figure 13.18 Nondisjunction of sex chromosomes in human spermatogenesis (the 22 pairs of autosomes are omitted). The diagram here illustrates nondisjunction in the second meiotic division. The abnormal sperm contain either two X chromosomes, two Y chromosomes or no chromosomes. A normal sperm has either one X chromosome or one Y chromosome.

Down Syndrome

The classic example of extra **autosome** inheritance is *Down syndrome,* which in about 95 percent of the cases results from meiotic nondisjunction (Fig. 13.19). Down syndrome usually involves the failure of the homologues of chromosome pair 21 to separate during meiosis. Consequently, some gametes will have *two* copies of chromosome 21. When one of these gametes unites with a normal gamete that possesses one copy of chromosome 21, the zygote will contain *three* copies of chromosome 21 for a total of 47 chromosomes. Down syndrome is one of the more common human genetic abnormalities, occurring in about 1 of every 700 live births. Characteristics of the condition include mental retardation, short stature, stubby fingers, high susceptibility to infections, and abnormalities of various internal

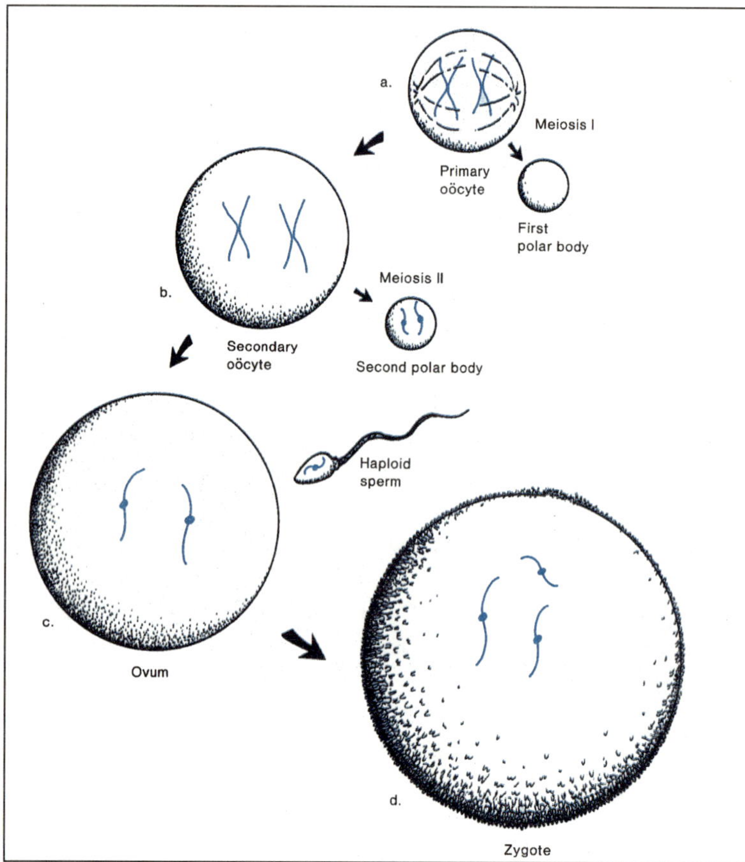

Figure 13.19 Diagram of nondisjunction in a human oöcyte involving one chromosome pair (number 21), as occurs in Down syndrome. (a) Normal primary oöcyte contains two number 21 chromosomes (the other 22 pairs of chromosomes having been omitted for clarity). (b) After meiosis I, failure of the chromosomes to separate from each other (nondisjunction) results in the secondary oöcyte receiving *both* number 21 chromosomes. (c) After meiosis II, the ovum and second polar body each contain *two* haploid number 21 chromosomes. Normally, each should contain only one haploid chromosome 21. (d) At fertilization, a normal sperm with its single number 21 chromosome unites with the ovum, producing a zygote with three number 21 chromosomes. Nondisjunction also may occur in spermatogenesis. (This illustration, for clarity, is highly diagrammatic.)

organs. Life expectancy for individuals with Down syndrome varies from 5 years to 20 years, although the use of antibiotics has increased the lifespan of some Down individuals beyond 20 years.

The risk of giving birth to a child with Down syndrome is about 50 times greater in women aged 45 than in women aged 20. Why this is so has not been determined specifically, but it may be traceable to the age of the oocytes, or egg cells. In humans and other mammals, all the oocytes a female will ever have are present at birth. Evidence suggests that older oocytes have an increased tendency to undergo nondisjunction or to produce structurally altered chromosomes, resulting in production of abnormal gametes. It has been found, however, that in some cases, Down syndrome results from nondisjunction in the male gametes.

Nondisjunction can also occur in mitosis and usually results when sister chromatids fail to separate at anaphase. Mitotic nondisjunction may occur at the first division of the zygote or at any time the embryo is dividing. If nondisjunction occurs at the first mitotic division of the zygote, half the cells will have 47 chromosomes and half will have 45 chromosomes. If mitotic nondisjunction occurs later in embryonic development, three cell lines are established: one with the normal number of 46 chromosomes, one with 47 chromosomes, and one with 45 chromosomes. Survival and the condition of the individual involved depend on which chromosome is affected by nondisjunction.

Patau's Syndrome

Other common examples of extra autosome inheritance include *Patau's syndrome,* which involves an extra chromosome 13. Some of the characteristics of Patau's syndrome include a very small head, extra fingers and toes, seizures, and various internal organ disorders. Most Patau's syndrome fetuses are aborted and of those that are live born, about 90 percent die within a year.

Edward's Syndrome

Edward's syndrome, which is more common in females than in males, involves an extra chromosome 18. Some of the characteristics of Edward's syndrome include a misshapen head, severe mental retardation, stunted growth, and heart defects. Infants born with Edward's syndrome rarely survive beyond one year. Virtually all other fetuses with extra autosomal chromosomes are spontaneously aborted.

Turner's Syndrome

Nondisjunction may also involve the **sex chromosomes;** that is, the X and Y chromosomes. An individual with *Turner's syndrome,* for example, has only one X chromosome; the second X or the Y chromosome is missing. In this case, the affected person has only 45 chromosomes instead of the normal 46. Such an individual is genetically XO and has the external features of a female. Most XO women, however, are sterile because their ovaries fail to develop. The incidence of Turner's syndrome is about 1 in every 2500 live births and is the only condition involving a single missing sex chromosome that is survivable; all other such pregnancies are spontaneously aborted during the first 3 to 6 months of development.

Extra Sex Chromosomes

Examples of extra sex chromosome inheritance include females who have three X chromosomes (XXX) instead of the normal two, or males who have an extra X chromosome (XXY) or an extra Y chromosome (XYY). All of these individuals, then, have 47 chromosomes. The XXX condition results from fertilization of an XX oocyte by an X chromosome-bearing sperm. Some XXX females are sterile and a small percentage have a subnormal IQ.

The XXY condition, known as *Klinefelter's syndrome,* can result from fertilization of an XX oocyte by a Y chromosome-bearing sperm, or from fertilization of an X oocyte by an XY sperm. Individuals with Klinefelter's syndrome have the appearance of males but have enlarged breasts and several other female characteristics, are sterile, and are mildly retarded.

The *XYY-syndrome,* which results from fertilization of an X oocyte by a YY sperm, produces men who are tall, averaging over 6 feet, and below normal in intelligence. First observed in a prison in Scotland, the XYY-syndrome was thought to contribute to antisocial behavior. This is highly speculative, however, and the evidence is inconclusive. Many XYY men exhibit quite normal behavior.

Genetic Screening

Several methods of prenatal (before birth) testing are available to identify human structural and chromosomal birth defects. These methods include ultrasound, fetoscopy, amniocentesis, and chorionic villus biopsy. Such testing can be *noninvasive,* which does not involve penetration of the pregnant uterus, or it can be *invasive,* in which test samples of fluids and/or cells are taken from the uterus or fetus. Today, several hundred prenatal disorders can be screened by these procedures.

Ultrasound

The primary noninvasive test is the use of *ultrasound* (sonar), which produces an image of the fetus that is projected onto a television screen. The fetus then can be examined for any structural abnormalities. Ultrasound is also used to determine the sex of the fetus, progression of normal growth and development, and the volume of fluid surrounding the fetus.

Fetoscopy

An invasive technique called fetoscopy ("viewing the fetus") involves using a thin, light-transmitting tube (fetoscope) that is inserted through a small incision in the abdomen. Once inside the amnion (the thin membrane surrounding the fetus), the fetoscope allows visual examination of the fetus. In addition, fetal blood and tissue samples can be withdrawn by fitting the fetoscope with a needle and syringe. Fetoscopy entails certain risks such as spontaneous abortion and injury to the fetus and, therefore, usually is the procedure of last resort.

Amniocentesis

The invasive technique of withdrawing fluid through a long needle inserted through the abdomen into the amniotic sac is called *amniocentesis* ("puncture of the amnion") (Fig. 13.20). Fluid and fetal cells removed by this procedure are subjected to various tests such as determination of specific protein levels in the fluid and observation of chromosome number and the form and structure of the cells. Amniocentesis is considered a low-risk procedure to both the pregnant woman and the fetus. The primary reason for amniocentesis is fetal chromosome analysis that is recommended for some pregnant women over 34 years of age.

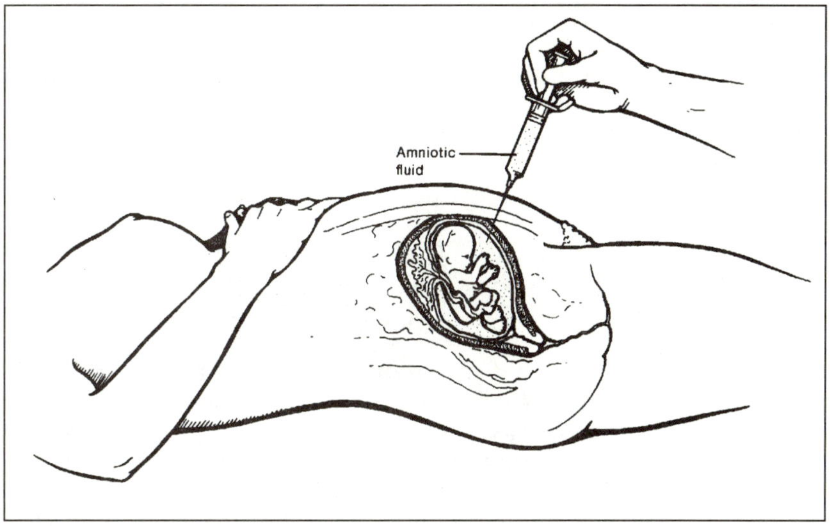

Figure 13.20 Amniocentesis ("puncture of the amnion") involves the withdrawal of amniotic fluid surrounding the developing fetus.

Chorionic Villus Biopsy

Another invasive technique is *chorionic villus biopsy* (CVB), which involves removal of fetal cells from the villi of the outermost fetal membrane, the chorion. Villi are fingerlike projections that extend from the chorion. Villi cells are genetically identical to any other body cell of the fetus. In the CVB procedure, a catheter is passed through the cervix at the top of the vaginal canal and into the chorion area. Using suction, the villi are drawn into the catheter and the villi cells are used for chromosome analysis and other studies. Because of greater risks, including spontaneous abortion, CVB is not performed routinely nor to the same extent as is amniocentesis.

Chapter Summary

1. Bacteria reproduce by a primitive method of cell division called binary fission, which is also employed to some degree by protists. Evolution of eukaryotic cells, with their membrane-bound nuclei, eventually led to mitotic cell division. Mitosis is involved in asexual reproduction and in growth and tissue repair.

    Body cells of eukaryotes contain a specific diploid number of chromosomes. Chromosomes are composed of DNA and proteins and carry the units of heredity called genes. Mitosis refers specifically to nuclear division, which usually is followed by cytoplasmic division, or cytokinesis.

The cell cycle is a sequence of growth and synthesis stages, followed by cell division. Growth stages involve production of organelles and other materials; DNA is replicated during the synthesis stage. DNA is a double stranded helix composed of individual nucleotides.

DNA replication occurs during interphase of the cell cycle and involves the synthesis of new DNA strands using original DNA strands as templates. New DNA strands are synthesized through complementary base pairing. Replication of DNA is semiconservative.

Key Terms:	*binary fission, p. 298*	*nucleotide, p. 302*
	chromosome, p. 299	*replication, p. 305*
	diploid, p. 300	*template, p. 305*

Concept Question: In a living cell, what is the difference between transcription and translation?

2. DNA transcribes its genetic information onto RNA in the nucleus, which translates this information into proteins. The genetic information in DNA is in the form of nucleotide triplets that code for specific amino acids. This DNA code is nearly universal, reflecting the universal relationship among living organisms.

Key Terms:	*triplet, p. 306*	*transcription, p. 308*
	codon, p. 306	*anticodon, p. 309*
	gene, p. 308	*translation, p. 309*

Concept Questions: Assume that in a hypothetical segment of DNA, the sequence of bases is AGGCTA. What would be the sequence of bases in the transcribed mRNA segment? What would be the base sequences of the tRNA anticodons that would bind to this mRNA segment? If the adenine nucleotide at the beginning of the sequence were lost through mutation, how might this affect the synthesis of a polypeptide? What is meant by the "genetic code"?

3. Genes known as transposons do not appear to be permanently fixed at their positions in chromosomal DNA; transposons may have originated from viral genes. Eukaryotic RNA apparently is unstable and may edit itself by adding or changing nitrogen bases.

   Mutations, which are alterations in the nucleotides of DNA, include the addition, loss, or substitution of nucleotides in the DNA molecule. An effect of mutation is seen in sickle cell anemia, in which an incorrect amino acid is substituted for the correct one in two of the polypeptide strands of the hemoglobin molecule.

Key Terms: *jumping genes, p. 311*
mutation, p. 312

Concept Question: Using an example, explain generally how a mutation resulting in the addition of a DNA nucleotide might affect the synthesis and ultimate functioning of a cellular protein.

4. Mitosis occurs after the chromosomal DNA is replicated in interphase; this replication produces two genetically identical chromatids joined at the centromere. Mitosis involves the equal separation of these chromatids to form two genetically identical daughter nuclei. In most cells, cytokinesis follows mitosis, separating the cytoplasm and forming two identical daughter cells.

Key Terms: *chromatin, p. 313*
centromere, p. 314
chromatid, p. 314

Concept Questions: Would mitosis be possible without an interphase? Why or why not? Diagram the chromosomal events of a cell (diploid number = 4) undergoing mitosis. Diagram the same cell undergoing meiosis.

5. Meiotic cell division reduces the diploid number of chromosomes to the haploid number of chromosomes. Meiosis involves two cell divisions called meiosis I and meiosis II. Cytokinesis divides the cytoplasm to form four haploid cells.

   Through crossing over and independent assortment, meiosis contributes to genetic variability. Sexual reproduction adds to this variability by combining gametes that produce genetically unique offspring. Because of their variability, these offspring make adaptation possible, an important factor in the evolutionary process.

   Key Terms: *haploid, p. 316*
   homologous, p. 316
   crossing over, p. 317

   Concept Questions: Explain why human gametes arise by meiosis and not by mitosis. Do you think it is possible for normal mitosis to affect genetic variability? Explain. Why does cell growth and tissue repair occur by mitosis rather than by meiosis?

6. Errors in cell division often involve nondisjunction, which occurs in both body cells and reproductive cells. Extra autosome conditions include Down, Patau's, and Edward's syndromes. A condition resulting from a missing sex chromosome is Turner's syndrome (XO). Extra sex chromosome conditions include XXX, Klinefelter's syndrome (XXY), and the XYY-syndrome.

   Key Terms: *nondisjunction, p. 319*
   autosome, p. 320
   sex chromosome, p. 321

   Concept Question: What is the difference between mutation and nondisjunction? Using examples, explain how each of these affects cells differently.

7. Genetic screening includes noninvasive procedures, such as ultrasound, and invasive procedures that include fetoscopy, amniocentesis, and chorionic villus biopsy.

Chapter Fourteen

Genetics: Determining the Future

A Brief History of Genetics

The Concept of the Gene

Charles Darwin died in 1882 never knowing that the basic genetic laws he needed to complement the principle of natural selection had been published in 1866 by an obscure Augustinian monk named Gregor Mendel (1822–1884). Mendel, through his experiments with garden peas, was the first to determine that hereditary characteristics are carried and passed on to offspring by what he called "factors," known today as **genes.** A gene is a sequence of DNA that encodes the necessary information to synthesize a protein. Moreover, Mendel suggested that these factors occur in pairs in the offspring; that is, one factor is inherited from the male parent and one from the female parent. But Mendel's findings were largely ignored; in fact, it would be another 34 years after publication before Mendel's work was rediscovered.

The modern concept of the gene began to unfold in 1902 when the British physician Archibald E. Garrod demonstrated that the dark urine of some newborn babies resulted from a condition called *alkaptonuria,* which is also often characterized by arthritis and darkening of cartilage. Alkapton is an acid that causes the urine to turn dark. Normal individuals possess a particular protein—an enzyme—that converts alkapton into another, colorless acid. Garrod reasoned that babies with alkaptonuria lacked this enzyme or that the enzyme was not functional. Because the babies were born with alkaptonuria, Garrod proposed that the disease was the result of an **inborn error of metabolism.**

He further proposed that alkaptonuria and other inborn errors of metabolism developed because the affected individuals either lacked one of Mendel's factors or the factor did not function properly. Thus, according to Garrod, one of Mendel's factors is necessary to produce an enzyme. An enzyme actually proven to be responsible for an inborn error of metabolism was not discovered until fifty years after Garrod's proposals. The enzyme that breaks down alkapton was discovered in 1958.

One Gene-One Protein Hypothesis

Largely ignored for more than thirty years, Garrod's work was rediscovered in the 1940s and finally verified in 1951 by the geneticists George W. Beadle and Edward L. Tatum. Using a red bread mold as their experimental organism, Beadle and Tatum induced gene mutations in the DNA of the mold using X-rays. They found that mutated genes could not produce functional enzymes; moreover, they also discovered that a single mutant gene was responsible for the malfunction of a single enzyme. This originated the *one gene-one enzyme hypothesis,* which proposed that each gene specifies the synthesis of one enzyme. Later, however, this hypothesis needed to be modified when it became clear that genes specify the amino acid sequence of any protein—not just the polypeptide sequences that make up enzymes. This concept is now referred to as the **one gene-one polypeptide,** or **one gene-one protein,** hypothesis.

In 1902, a graduate student at Columbia University named Walter S. Sutton combined Mendel's findings with the process of meiotic cell division and proposed that genes—the genetic material—were contained in the chromosomes. This concept came to be known as the *chromosome theory of inheritance.* Established thereafter as an important discipline, the study of heredity was called **genetics,** a term coined in 1905 by an English biologist named William Bateson. Four years later, a Danish biologist named Wilhelm Johannsen referred to Mendel's factors as genes. The gene concept, though modified in later years, was here to stay.

Theories of Inheritance

Attempts to understand heredity date back thousands of years. A 6000 year-old engraved stone discovered near the Persian Gulf has traced various hereditary traits through five generations of horses. In addition, the ancient Greeks believed that blood carried hereditary traits; hence, the source of terms such as "blood relative" and "blood line."

This concept of fluid heredity persisted well into the nineteenth century with Charles Darwin as one of its advocates. Darwin, however, formulated a theory of *pangenesis,* which broke somewhat with the theory that blood or some other body fluid was the only source of hereditary material. Darwin suggested that all parts of the body produce tiny particles called gemmules that were transported in the blood or body fluids to the reproductive organs and into the sex cells. At fertilization, it was assumed that these gemmules could reproduce the same tissues and organs from which the gemmules were originally derived. In other words, a gemmule derived from a parent's liver would develop into a liver in a growing fetus.

It was also during the nineteenth century that the theory of blending inheritance arose. According to this concept, hereditary materials from the male parent and female parent mix, producing offspring that exhibit a blend of parental traits; a process similar to that of mixing red and white paint to produce pink. One of the assertions of this "paint pot" theory was that, once blended, each of the original traits was lost—never again to show up in an individual organism; after all, how is it possible to unmix paint? It was thought that if a brown rabbit and a white rabbit were mated, all the offspring would be tan. There would not be any white rabbits or brown rabbits produced. The theory postulated that, thereafter, all matings between tan rabbits would produce only tan rabbits. Moreover, interbreeding of brown rabbits and white rabbits over many generations would finally produce a population of solely tan rabbits. None of this, however, was supported by scientific evidence. The theory of blending inheritance was incorrect, as we shall see, but it was another attempt to come to grips with the mystery of inheritance—a mystery that awaited the application of science. At this point in the history of genetics, Gregor Mendel and his work enter the picture. We will now take a look at the birth of modern genetics.

Gregor Mendel and the Birth of Modern Genetics

Mendel's Principles

By the age of thirty-two, Gregor Mendel had offered a possible explanation for the variation in individual characteristics evident in different generations of offspring within a population. He also proposed that the factors (genes) responsible for this variation are passed from one generation to the next. Mendel's findings can be summarized as basic genetic principles, or laws, which include *the principle of segregation* and *the principle of independent assortment.*

Mendel's work reflects the cornerstones of scientific inquiry: observing, questioning, hypothesizing, experimenting, and theorizing. He chose a species of garden pea as his experimental organism, with which he performed crosses, or matings, to determine the inheritance of characteristics such as pod color and shape, stem length, and seed color and shape. The results of his research into the basic laws of inheritance provide the framework for this chapter.

Alleles

Mendel determined that there are different forms of a particular factor, each of which is now called an **allele** (Gk. "another form"). Put another way, a single gene may be represented by one of several alternative forms of that gene, each of which is an allele. Genes are inherited in pairs: one allele from one parent and one allele from the other parent. For example, let's say that the allele for tall pea plants is designated as *T* and the allele for dwarf pea plants is represented by *t*. If a given plant inherited the allelic pair *Tt*, then one of the alleles came from one parent and the other allele was inherited from the other parent. A given allele is physically located at a particular position, or **locus,** on a chromosome. We have said that an organism inherits two alleles for a given trait—one from each parent. These two alleles are located at analogous positions, or loci, on two homologous chromosomes; that is, one allele is on one homologous chromosome, or homologue, and the other allele is at the corresponding location on the other chromosome (Fig. 14.1).

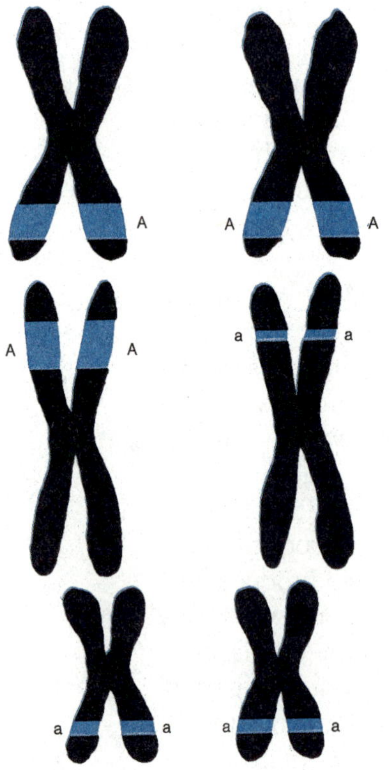

Figure 14.1 Alleles are found at the same location on pairs of homologous chromosomes.

If the two alleles for a trait are the same, such as *TT* or *tt*, the organism is said to be **homozygous** for that trait. If the two alleles are different, as in the case of *Tt*, the organism is said to be **heterozygous** for that trait.

Dominant and Recessive Alleles

Alleles may also be designated as **dominant** or **recessive.** When an allele is dominant, it means that the expression of that allele covers up, or masks, the expression of the other allele—the recessive allele. As an example, the allele for tall garden pea plants *(T)* is dominant to the allele for dwarf plants *(t)*. A plant that is heterozygous for height and possesses the allele pair *Tt* would be tall even though it also carries the dwarf allele. The dominant *T* allele has masked the presence of the dwarf *t* allele and the trait carried by the dominant allele is expressed, or observed in the plant.

As a convention in the field of genetics, the dominant allele is usually designated by a capital letter, such as *T;* the recessive allele is designated by the same letter in lower case, such as *t*. There are three possible combinations of two alleles on a pair of homologous chromosomes; in our example, the combinations would be *TT, Tt,* and *tt*. The allele pair *TT* is designated **homozygous dominant,** *Tt* is **heterozygous,** and *tt* is referred to as **homozygous recessive.**

Interestingly, you may notice that a tall pea plant may either be heterozygous for height alleles *(Tt)* or it could also be homozygous for tall height *(TT)*. For the recessive height trait to be expressed in the pea plant, both alleles must be recessive. A dwarf plant, therefore, could only carry the allele pair *tt*.

Phenotypes and Genotypes

The outward appearance of an organism—its expressed or detectable physical traits—is called its **phenotype.** The phenotype for pea plant height can be either tall or short. The genetic makeup of an organism—the type of alleles it carries—is called its **genotype.** The height genotype of a tall pea plant may be either *TT* or *Tt;* the genotype for the height of a dwarf pea plant must be *tt*.

Although we may know the phenotype of an organism simply by looking at it, we have already noticed that we do not always know its genotype. (Is a tall pea plant genotypically *TT* or *Tt*?) It is apparent, then, that two organisms could have the same phenotype but different genotypes. With dwarf plants or other recessive conditions, the situation is different. If an organism displays a recessive trait, we automatically know its genotype. With dwarf pea plants, the genotype must be *tt*. If a dominant allele were present, the plant would be tall, not dwarf.

The Principle of Segregation

With Mendel's discovery that hereditary traits are carried by pairs of alleles (he called them factors), later work in genetics led to the understanding that pairs of alleles on homologous chromosomes separate from each other during meiosis (Chapter 13) (Fig. 14.2). At the end of meiosis, then, each gamete would contain one allele of a given pair. The separation of a pair of alleles into different gametes at the end of meiosis is known as the **principle of segregation.**

Using the allele pair *Tt* as an example, the *T* allele and the *t* allele separate at meiosis because they are located on separate homologues and eventually pass into different gametes. Therefore, each gamete receives only one of each kind of allele. It is possible for the two alleles to show up together in a new plant

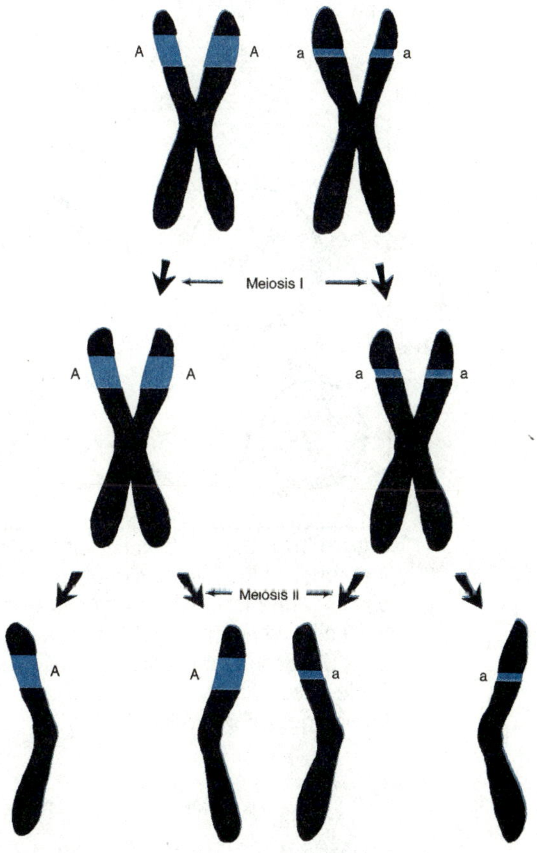

Figure 14.2 Illustration of Mendel's principle of segregation using the allele pair *Aa*. At the end of meiosis, each allele (carried on a chromosome) passes into a different gamete, so that each gamete receives only one of each kind of allele.

when gametes unite at fertilization. This would occur when the female contributed one of the alleles and the male contributed the other.

The Principle of Independent Assortment

In addition to his research into the inheritance of a single trait, Mendel also carried out experiments on his pea plants in which he followed the inheritance of two different traits. The genes for these two traits were located at two different gene loci on two different pairs of homologous chromosomes. These experiments led Mendel to formulate the **principle of independent assortment:** *The alleles for these two different traits segregate independently during meiosis* (Fig. 14.3). Essentially, this means that when one pair of alleles separates in meiosis, it is not affected by how another different pair of alleles on a different pair of homologous chromosomes separates. Let's now take a closer look at the ways in which Mendel could make such determinations.

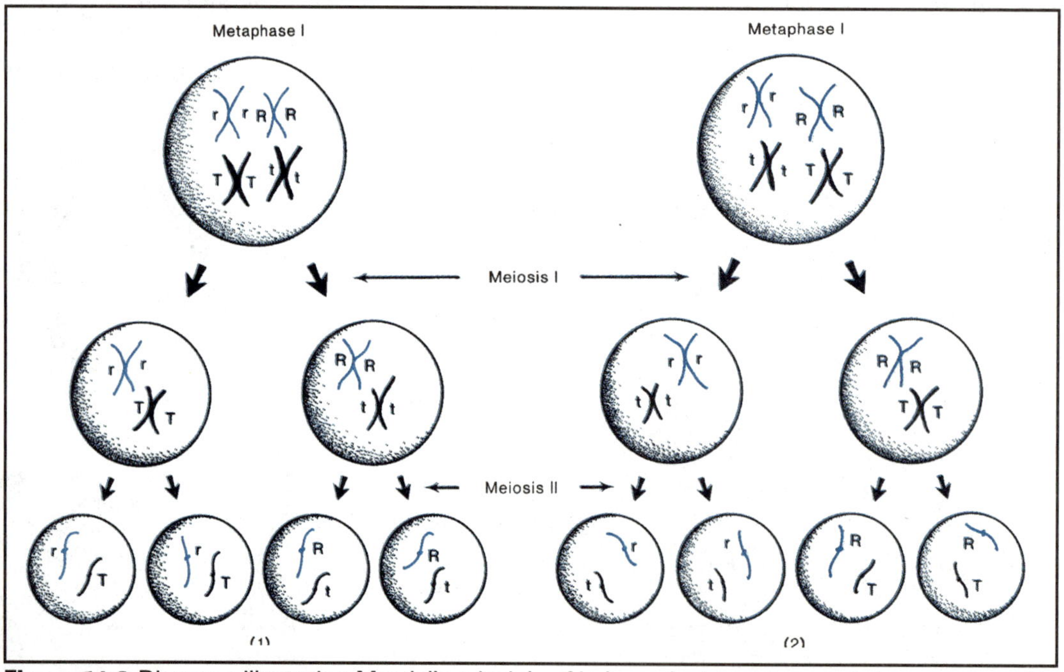

Figure 14.3 Diagrams illustrating Mendel's principle of independent assortment of chromosomes. At metaphase I the two sets of chromosomes may line up as shown in (1) or as shown in (2). These chance chromosomal arrangements result in the production of two different sets of haploid reproductive cells. When the chromosomes with the *R* and *r* alleles separate from each other at anaphase I, their direction of movement is not affected by the movement of the chromosomes with the *T* and *t* alleles; i.e., each pair of homologous chromosomes separates independently of the other pair of homologous chromosomes.

Using the principles of inheritance discovered by Mendel, we can set up various crosses, or matings, between two parental organisms and then determine the genotypes and phenotypes of the offspring expected from these crosses. The two types of crosses that we will consider are ones that involve parents that differ in a single trait or in two traits; these crosses are called *monohybrid* and *dihybrid crosses,* respectively.

Monohybrid Crosses

The first type of cross involves parents that differ in only one trait and is called a **monohybrid, or single-trait, cross.** This type of cross involves one gene locus only. Using our tall and dwarf pea plants as an example, one type of monohybrid cross might involve two homozygotes, such as the mating of parents with the genotypes *TT* and *tt*. The parent generation is designated as the P generation. To analyze any type of experimental genetic cross involving sexually reproducing organisms, the first requirement is to determine the type of gametes each parent can produce during meiosis. In our example, then, the tall *TT* parent can produce gametes containing only a *T* allele; the dwarf *tt* parent can produce only *t* gametes. At fertilization, all the offspring—designated as *the first filial,* or F_1, generation—will have the genotype *Tt;* they will all be heterozygotes. Although they carry the dwarf allele *(t),* the offspring all exhibit a tall phenotype.

Punnett Squares

Suppose that we now use two members of the F_1 generation as parents, and mate a *Tt* plant with another *Tt* plant. What would the members of the second filial, or F_2, generation be like phenotypically and genotypically? Our first task is to determine the types of gametes the parent organisms can produce. In this case, each parent can produce *T* gametes and *t* gametes. In determining all the possible ways in which these gametes could combine to produce offspring, it is often convenient to use a **Punnett square** as shown in Figure 14.4.

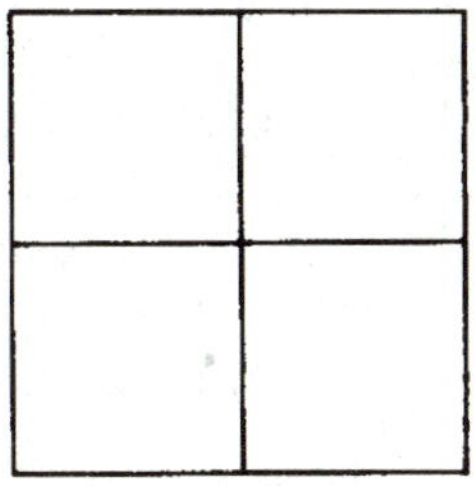

A Punnett square

A Punnett square is a useful device for determining all the possible allele combinations in the offspring resulting from a genetic cross. As you can see in Figure 14.4, the genotypes of the parental gametes are aligned outside the grid and each male gamete is crossed with each female gamete to obtain all of the possible combinations that might occur in such a cross. The dimensions of the Punnett square matrix, then, are determined by the number of types of gametes that can be produced by the parents. The allele combinations from each cross are then recorded in the appropriate square. Each of these crosses, then, represents one of the possible genotypes of the offspring. From these genotypes, the possible phenotypes of the offspring also can be determined.

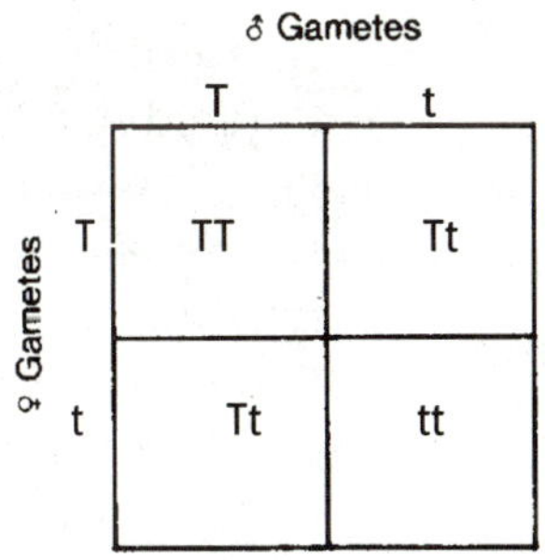

Figure 14.4 A Punnett square is used in crossing parental gametes to determine the possible types of offspring. The example above is a monohybrid cross, since only one genetic trait (height) is involved.

The possible offspring genotypes resulting from a cross between the two heterozygotes are: 1 *TT,* 2 *Tt, 1 tt.* This **genotypic ratio** for the offspring is designated as 1:2:1; that is, one homozygous dominant *(TT),* two heterozygotes *(Tt),* and one homozygous recessive *(tt).* This also could be expressed as 1/4 *TT,* 1/2 *Tt,* and 1/4 *tt.* Phenotypically, three-fourths of the F_2 offspring are tall and one-fourth are dwarf plants. This is referred to as the **phenotypic ratio** and is written 3:1.

Mendel's findings tolled the death knell for the theory of blending inheritance because they demonstrated both that the F_1 generation resulting from a cross between a tall plant and a dwarf plant was not intermediate in height; that is, it was not a "blend" of the parents, and that the dwarf phenotype was not lost since it reappeared in the F_2 offspring.

A Test Cross

We have discovered that if a pea plant displays a dominant trait, such as tall height, the plant may be homozygous *(TT)* for that trait or it may be heterozygous *(Tt)*. Is it possible to determine the actual genotype? In an attempt to answer the same question, Mendel performed an experiment known as a *test cross.* This involves crossing an organism that displays a dominant trait with a homozygous recessive individual to discover whether a phenotypically dominant individual is homozygous or heterozygous for that trait. As we have seen, a tall pea plant may have the genotype *TT* or *Tt;* however, the genotype of a homozygous recessive plant will necessarily be *tt*. Figure 14.5 illustrates the possible results obtained by crossing both possible genotypes of a tall plant *(TT* or *Tt)* with a dwarf *(tt)* .

	T	t
t	Tt	tt
t	Tt	tt

a.

	T	T
t	Tt	Tt
t	Tt	Tt

b.

Figure 14.5 A test cross involves crossing an organism displaying a dominant trait (tall) with a homozygous recessive individual (dwarf). The test cross in (a) produces dwarf offspring, indicating that the organism displaying the dominant trait is heterozygous. In (b) the organism displaying the dominant trait must be homozygous, since no dwarf offspring are produced.

Crossing a *Tt* plant with a *tt* plant produces a phenotypic ratio of two tall to two dwarf plants, or a 1:1 ratio. The fact that the test cross produces *dwarf* plants *(tt)* as well as tall ones *(Tt)* indicates that the tall parent *must* be heterozygous *(Tt)* for this allele. In contrast, test crossing a homozygous dominant *(TT)* plant produces offspring that are *all* tall. In other words, if no dwarf plants are produced from the test cross, the genotype of the tall parent *must* be *TT.*

Dihybrid Crosses

As we mentioned earlier, Mendel pursued his experiments by crossing parents that differed in two traits, which is called a **dihybrid cross.** This type of cross involves two gene loci and these two loci are found on two different pairs of homologous chromosomes. Although there are a few new wrinkles to consider, the same principles apply to dihybrid crosses as they did to monohybrid crosses. Because we have been considering the trait of height in pea plants, let's use these same plants and add a second trait: pod color. In the following examples, green pod color *(G)* will be dominant to yellow pod color *(g)*. Thus, there are tall *(TT* or *Tt)* or dwarf *(tt)* plants that have either green *(GG, Gg)* pods or yellow *(gg)* pods. Since each trait—height and pod color—is determined by a separate pair of alleles that segregate independently during meiosis because they are on different chromosome pairs, the possible combinations for the offspring would be: *GGTT, GgTT, ggTT, GGTt, GgTt, ggTt, GGtt, Ggtt,* and *ggtt.* These allele combinations, of course, represent the genotypes, each of which determines a particular phenotype.

A Test Cross

To illustrate one type of dihybrid cross, let's start as we did with the monohybrid cross, using a homozygous dominant *(GGTT)* and a homozygous recessive *(ggtt)* as parents. At the end of meiosis, the *GGTT* parent will produce gametes containing the alleles *GT,* while the *ggtt* parent produces *gt* gametes. At fertilization, the only possible genotype of the F_1 generation is *GgTt;* that is, the F_1 will be heterozygous for both traits. Phenotypically, the offspring plants will be tall with green pods, as the dominant alleles dictate. Note that in a dihybrid cross, one allele for each trait is present in each gamete.

Using the F_1 *GgTt* offspring as parents, what could we expect in the F_2 generation? Each parent can produce four different kinds of gametes: *GT, gT, Gt,* and *gt.* These can be set up on a Punnett square with the possible F_2 offspring determined as shown in Figure 14.6.

Considering only height, there are 12 tall plants and 4 dwarf plants in the F_2 offspring, which is a phenotypic ratio of 3:1 . This is the same ratio obtained if heterozygous parents are used in a monohybrid cross. Notice, too, that there are 12 F_2 plants with green pods and 4 plants with yellow pods; again, a phenotypic ratio of 3:1. Moreover, the principles of dominance and recessiveness still prevail; there is nothing in this dihybrid cross that conflicts with the principles or results observed in a monohybrid cross.

By referring to Figure 14.6, we can see that the phenotypic ratio of the F_2 offspring is 9:3:3:1. There are 9 offspring expressing the two dominant traits (green and tall), 3 offspring that express one combination of dominant and recessive traits (green and dwarf), 3 offspring that express the other combination (yellow and tall), and only one individual that expresses both recessive traits (yellow and dwarf). This ratio will be the result of any dihybrid cross in which the parents are heterozygous for both traits.

As a final note, this dihybrid cross clearly illustrates how Mendel's experiments led him to propose his principle of independent assortment. When the F_1 generation *(GgTt)* are used as parents, the green and tall alleles and the yellow and dwarf alleles are all combined in one plant. In the F_2 generation, however, these alleles are inherited in different combinations, as if each allele were totally independent of any other allele.

♀ Gametes \ ♂ Gametes	GT	Gt	gT	gt
GT	GGTT	GGTt	GgTT	GgTt
Gt	GGTt	GGtt	GgTt	Ggtt
gT	GgTT	GgTt	ggTT	ggTt
gt	GgTt	Ggtt	ggTt	ggtt

Figure 14.6 A dihybrid cross involving two organisms heterozygous for both traits (GgTt x GgTt). The phenotypic ratio for such a cross is 9:3:3:1.

Variations in Mendelian Genetics

Incomplete Dominance

While Mendel's genetic principles apply to virtually all forms of life, further research has revealed that inheritance in many species is somewhat more complex. Consider, for example, a species of flowers possessing the allele *R* for red petal color and the allele *r* for white petals. Crossing a red flower with a white flower sometimes produces F_1 offspring that are pink (Fig. 14.7). In this instance, the *R* allele is not completely dominant nor is the *r* allele completely recessive. Instead, *both* alleles *partially* express themselves

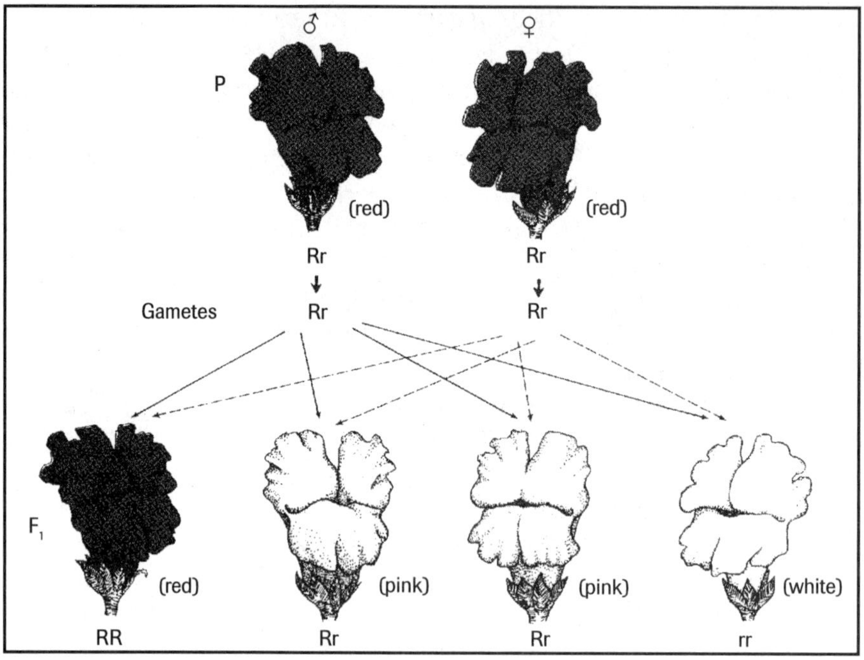

Figure 14.7 Incomplete dominance (R = red, r= white). Although the heterozygous parent flowers are red, the two heterozygous offspring are pink. The R allele in this instance is neither totally dominant, nor is the r allele totally recessive.

in the offspring. This is an example of **incomplete dominance.** Thus, if dominance is incomplete, the *Rr* offspring are pink rather than red. This changes the phenotypic ratio of the offspring.

You may be thinking that incomplete dominance lends some credence to the theory of blending inheritance. That is not the case, however, because crossing the *Rr* offspring can produce both red *(RR)* and white *(rr)* progeny. Thus, the red and white alleles are not inseparable like mixed paint and thus irretrievably lost, but reappear in subsequent generations.

Codominance

In Chapter 13, we referred to the human genetic disease sickle-cell anemia. This disease results from a **point mutation**—the substitution of the base thymine (T) for the base adenine (A)—in the sequence of DNA nucleotides that codes for the blood protein hemoglobin. Consequently, the messenger RNA transcribed from this mutated DNA sequence contains an incorrect codon. During translation, this codon, in turn, specifies an incorrect amino acid that will be inserted into two of the four strands that make up the hemoglobin molecule. Thus, the primary structure, or the precise sequence of amino acids, of each polypeptide strand is slightly altered. The incorrect amino acid substitution produces an abnormal hemoglobin molecule referred to as hemoglobin *S*, or sickle-cell hemoglobin. Normal adult hemoglobin is designated hemoglobin *A*.

As a result of its faulty structure, the hemoglobin *S* molecule becomes contorted into odd shapes and some of the red blood cells deform into a "sickle" shape. The hemoglobin *S* molecules within the red blood cells cannot transport an adequate amount of oxygen to the body cells, and the deformed red cells tend to clump together and clog up small blood vessels. In addition, a normal red blood cell has a life span in the body of about 120 days before being destroyed by various phagocytic cells; sickled red blood cells persist in the body for only 20 days before being destroyed. As a consequence of these factors, individuals with sickle-cell anemia have poor circulation and suffer from severe anemia.

Based on the preceding discussion, let's designate the allele for normal hemoglobin as *A* and the allele for sickle-cell hemoglobin can be represented as *S*. Individuals with the homozygous *AA* genotype have normal hemoglobin and their red blood cells exhibit the biconcave disk shape. In those individuals with the homozygous *SS* genotype, many of the red cells are distorted into a sickle, or crescent, shape. These individuals have the sickle-cell disease and some suffer from such severe anemia that they may die in their early years; others less severely affected seldom live past the age of 20. Individuals with the heterozygous *AS* genotype have *both* normal and sickle-cell hemoglobin. Although possessing the sickle-cell trait, the heterozygote does not have the disease and usually leads a normal life. Thus, neither the *A* allele nor the *S* allele is completely dominant. But because the heterozygote possesses a certain amount of both normal and sickle-cell hemoglobin, the *A* and *S* alleles are said to be **codominant.** In other words, both alleles are expressed and can be detected phenotypically; there is no single, intermediate form of hemoglobin produced.

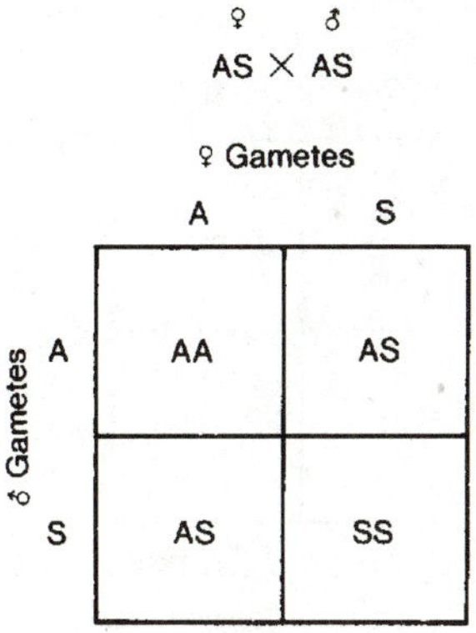

Figure 14.8 Mating between individuals who carry the alleles for both normal hemoglobin *(A)* and sickle-cell hemoglobin *(S)*. The probability that any of their offspring will have sickle-cell anemia (*SS* genotype) is one in four. Since neither the *A* allele nor the *S* allele is dominant, offspring with the *AS* genotpe may lead normal lives.

In a cross between heterozygous individuals who each carry the alleles for both normal hemoglobin *(A)* and sickle-cell hemoglobin *(S)*, the probability that any of their offspring will have sickle-cell anemia (*SS* genotype) is 1 in 4 (Fig. 14.8). About 10 percent or some 3 million African Americans are carriers of the hemoglobin *S* gene. This means that about 1 in 400 infants will carry the *SS* genotype and will develop sickle-cell anemia.

It has been estimated that about 8.5 percent of the African American population of North America carry the sickle-cell trait *(AS)*, whereas the incidence may be as high as 20 percent among the blacks of central and western Africa. Why should the incidence of the sickle-cell trait be so much greater in Africa than in North America? It has been found that individuals heterozygous for the sickle-cell gene have a greater resistance to malaria, a disease common in many parts of Africa, than do individuals who carry two normal hemoglobin alleles. Because the malarial parasite reproduces within red blood cells, the sickled cells protect the heterozygous individual, apparently by preventing reproduction of the parasite. In North America, though, malaria is not a common disease and the heterozygous condition affords no particular adaptive advantage.

There are various types of abnormal hemoglobins present in many other blood diseases as well. As in sickle-cell hemoglobin, other variant hemoglobins usually differ from normal hemoglobin in the substitution of one incorrect amino acid.

Multiple Alleles

The discussion thus far has considered only two possible forms of the alleles responsible for a certain trait; the *T* and *t* alleles in pea plants are examples. In some cases, however, more than two alleles can be associated with the inheritance of a given trait. No organism, however, receives more than two alleles for any given gene, one from each parent.

One of the best-known examples of **multiple alleles** is the human **ABO blood group system.** A similar system is also found in some other mammals. There are three important alleles in this system: A, B, and O—any two of which determine a person's blood type. The possible allele pairs (genotypes) in humans can be: AA, AO, BB, BO, AB, and OO. Both alleles A and B are completely dominant to O, but when found together, A and B exhibit codominance. Therefore, individuals with an AA or AO genotype have type A blood phenotypically; those with a BB or BO genotype have type B blood; those with the AB genotype have type AB blood (since neither allele is dominant); and individuals with the OO genotype have type O blood (Fig. 14.9).

	A	B
O	AO	BO
O	AO	BO

	A	B
A	AA	AB
O	AO	BO

	A	B
B	AB	BB
O	AO	BO

Figure 14.9 Examples of crosses involving the A, B and O blood group antigens.

The alleles A, B, and O indicate the presence or absence of specific markers called **antigens** on the surface of red blood cells. Type A blood is composed of red blood cells that are coated with the A antigen; type B red cells have the B antigen. Red cells of blood type AB express both A and B antigens, and blood type O red cells do not possess, or express, either antigen.

At times, the ABO blood group system proves valuable in paternity suits and in certain criminal cases. Its value, however, lies not in determining who *is* the father or the criminal, but rather in who *is not*. Assume, for example, that a man with type AB blood was involved in a paternity suit that claimed he fathered a child with type O blood. Can the claim be supported genetically? A male with type AB blood cannot possibly contribute the recessive O allele needed to produce a child with type O blood; remember, an individual with type O blood must have two recessive O alleles. Although this proves that a man with blood type AB could *not* have been the father of a child with type O blood, it cannot prove who *is* the father; in fact, any number of men have the appropriate blood type to father a child with type O blood.

As another example, what blood types are possible in the offspring of two parents who are type B and type O (Fig. 14.10)? The possible genotypes of the two parents are BB or BO, and OO. The only possible blood types of the offspring are B and O. Therefore, the children could not have blood types A or AB.

Blood left at the scene of a violent crime can be analyzed as to its type also, but, as in paternity suits, the results can only exclude a particular suspect. Regardless of the blood type found at the scene, there are any number of people with the same type. Blood type alone, then, is not evidence of guilt, but evidence of innocence.

While blood type analysis does not provide evidence of guilt, there is a procedure involving DNA testing that can identify an individual with virtual certainty. In that the DNA of every individual is unique (except for identical twins), a technique providing a *DNA fingerprint* can be

BB × OO

	B	B
O	BO	BO
O	BO	BO

BO × OO

	B	O
O	BO	OO
O	BO	OO

Figure 14.10 Mating between parents with blood types *B* and *O*. Parents with the genotypes *BB* and *OO* produce offspring with blood type *B* only because *B* is the dominant allele. Parents with the *BO* and *OO* genotypes produce offspring with blood types *B* and *O*.

employed to determine the exact identity of an individual. Basically, a DNA fingerprint results from the analysis of certain segments of DNA located between the genes. These segments are quite different for each individual. Through chemical analysis, researchers can detect a pattern, or "fingerprint," of these segments that is unique for each individual. In the case of a criminal act, any small amount of blood or body tissues found at the crime scene can provide a DNA fingerprint, which can be compared to the DNA fingerprint of a suspect.

Pleiotropy

In addition to the obvious effect of producing color or shape or some other phenotypic trait, many genes have related effects that also influence the phenotype. Mendel observed, for example, that the gene responsible for certain colors in flowers also influenced the color of the seeds. The lack of expression of a single gene or a single gene that affects two or more characteristics is known as **pleiotropy** ("much turning"). The single sickle-cell gene, just discussed, may produce up to twenty-five abnormal conditions.

Some human metabolic diseases also exhibit symptoms that result from a single gene having multiple effects. For example, if a gene is defective or absent, the result may be accumulation of an organic molecule that is normally broken down in the body. This occurs in *lysosomal storage diseases,* in which the lysosomal enzymes do not hydrolyze cellular waste molecules or the enzymes are not produced. Large waste molecules, then, tend to accumulate in the lysosomes, which can increase enormously in size. The disruption of metabolism that follows produces crippling, mental retardation, and early death.

Polygenic Inheritance

Phenotypic traits such as height, weight, blood pressure, and shape can be measured and are called **quantitative traits.** Within a population, these traits normally vary between extremes and usually indicate **polygenic inheritance;** that is, each trait is influenced by two or more genes. Quantitative traits, as well as many other traits, are influenced by the environment. Diet, exercise, and living conditions, for example, can influence the expression of genetic traits. Traits resulting from the interaction between genes and the environment are known as **multifactorial traits,** or traits that are determined by many factors. It is thought that most human traits are multifactorial.

To illustrate polygenic inheritance, assume that in a large population of human males the range in height varies between the extremes of 4 feet and 7 feet. Usually, there would be a continuous distribution of height in the population. In other words, there would be individuals of every conceivable height between the extremes, and only a few individuals would be expected to be 4 feet tall or 7 feet tall. The largest proportion of the population normally would have an intermediate phenotype. Each group of individuals of the same height will have the same number of height-determining genes. As a hypothetical example, we will refer to our population in which the tallest males are 7 feet in height and the shortest males are 4 feet in height. We will assume that there are 6 genes responsible for height (the actual number of genes responsible for human height is unknown as yet). In the tallest males, each of these genes may contribute 12 inches of height and in the shortest males, 8 inches of height. Some of the males, however, may have inherited 3 "tall" genes and 3 "short" genes. The height of these individuals, then, would be 5 feet. Similarly, individuals who inherited 4 tall genes and 2 short genes would be 5 feet 4 inches tall, and individuals who inherited 2 tall genes and 4 short genes would be 4 feet 8 inches tall. Thus, all the possible combinations of tall and short genes would result in a population exhibiting a continuous distribution of height.

Gene Linkage and Chromosomes

Linked Genes

It was pointed out in Chapter 13 that chromosomes are assorted randomly in meiosis. This means, of course, that alleles located on different chromosomes are also randomly assorted. Alleles, however, are assorted randomly only if they are on *separate* chromosomes. Some alleles are found on the *same* chromosome and are inherited together, not separately. These alleles, or genes, are said to be **linked.** Therefore, during meiosis, linked genes move together toward one pole of the cell and into a gamete. Obviously, this would change the expected genotypic and phenotypic ratios of the offspring from the ratios expected if all of the genes assorted independently. This, then, is an exception to Mendel's principle of independent assortment.

As an example of linkage, let's again use the *GgTt x GgTt* dihybrid cross in pea plants (Fig. 14.11). You will recall that the principle of independent assortment would dictate formation of four different types of gametes from each parent if *the genes were located on separate chromosomes: GT, Gt, gT,* and *gt*. The expected phenotypic ratio from this cross is 9:3:3:1. Suppose, however, that the *G* and *T* alleles were linked, or *located on the same chromosome,* and that the *g* and *t* alleles were linked on the **homologous chromosome.** What then would be the expected genotypic and phenotypic ratios? Each parent in this instance can produce only *GT* gametes and *gt* gametes. The F_1 genotypes, therefore, would be *GGTT,* 2 *GgTt,* and *ggtt*. Thus, the phenotypic ratio is 3:1 and the genotypic ratio is 1:2:1.

P GgTt x GgTt

Gametes GT gt GT gt

F_1

	GT	gt
Gt	GGTt	Ggtt
gt	GgTt	ggtt

Figure 14.11 Dihybrid cross involving linked genes. The *G* and *T* alleles are located on the same chromosome, and the *g* and *t* alleles are located together on the homologous chromosome. Consequently, the alleles on each chromosome do not assort independently, but move as a single unit.

Crossing Over

Linked genes, however, do not always remain together on the same chromosome. They can be rearranged between homologous chromosomes by the process called *crossing over.* We learned in Chapter 13 that crossing over occurs in prophase I of meiosis and involves the exchange of genetic material between chromatids of homologous chromosomes (Fig. 14.12). These chromosomes intertwine at this stage of meiosis and portions of the two chromosomes may exchange with each other. As the example in Figure 14.12 shows, crossovers involving linked genes can result in an exchange of dominant and recessive alleles between homologous chromosomes. As is the case with any crossover, gametes then carry a different assortment of alleles at the gene loci than they would if crossing over had not occurred.

Crossing over can be of tremendous significance in the evolutionary process. Through the exchange of genetic material, a chromosome might acquire an advantageous allele to go along with another advantageous allele on the same chromosome. In meiosis, these alleles would enter a gamete together and, at fertilization, the offspring receiving this allele pair would have a genetic advantage. In evolutionary terms, this means that these offspring would be favored to reach reproductive age and to pass their advantageous alleles to their offspring.

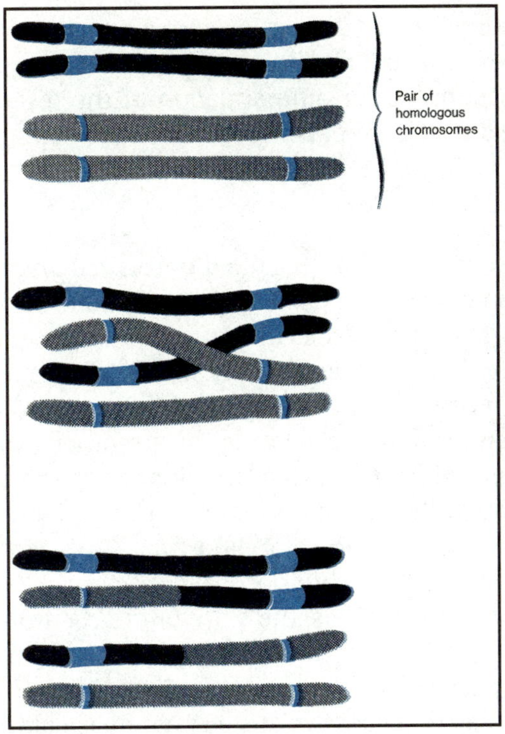

Figure 14.12 An example of crossing over in which there is an exchange of dominant (wide bar) alleles and recessive (narrow bar) alleles. As a result of crossing over, reproductive cells carry a new and different assortment of genes.

In contrast, offspring receiving less advantageous alleles would be less likely to survive and reach reproductive age. Consequently, their alleles would not be passed to future generations, thereby diminishing the frequency of these alleles in the population. Thus, crossing over could change the makeup of a population by producing individuals with different survival potentials. Those individuals with favorable gene combinations have an increased adaptive advantage and will survive in greater numbers. Those individuals with less favorable gene combinations are more likely to be eliminated by natural selection. The result is evolution: the change in the gene pool (total genetic material or kinds of genes) of the population over a period of time.

Sex-Linked Traits

The normal diploid number of 46 chromosomes—or 23 chromosomal pairs—in humans includes 22 pairs of autosomes and one pair of sex chromosomes. Most of the genes responsible for your physical makeup—height, eye color, skin pigmentation, foot size, and so on—are carried on the **autosomes.** But whether an individual is male or female is determined by the single pair of **sex chromosomes.** In humans, the sex chromosomes are designated as X and Y; normal males have the genotype XY and normal females have the genotype XX. In female offspring, one X chromosome is contributed by the mother and one X chromosome is contributed by the father. In male offspring, the mother contributes the X chromosome and the father contributes the Y chromosome. It is the male parent, then, who determines if human offspring will be male or female.

Associated with the X chromosome are certain genetic traits that affect males more frequently than they do females. This is because males possess only one copy of certain alleles that are located on the X, while females possess two copies. A recessive allele on the X chromosome of a male will be expressed because it is the only copy of that allele; however, a recessive allele on an X chromosome of a female will only be expressed if both X chromosomes carry that recessive allele. A genetic trait carried by an allele on the X chromosome is said to be **sex-linked,** or **X-linked.** By comparison, the Y chromosome carries very little genetic information. One gene on the Y chromosome determines production of an antigen involved in immune system responses. A second gene is responsible for differentiation of the testes and thus determines maleness. The Y-linked genes are passed from father to son and, therefore, affect only males.

COLOR VISION

One example of a sex-linked trait is *red-green colorblindness,* a form of defective color vision in which a person experiences difficulty in distinguishing red, yellow, and green. Colorblindness results from the expression of a recessive allele that we can designate as *c,* the dominant allele for normal vision, then, is *C.* Since a female possesses two X chromosomes, she may carry the dominant allele on one X chromosome (depicted as X^C) and the recessive allele on the other (X^c) and still have normal color vision. The reason, of course, is that the dominant allele masks the effect of the recessive allele. Although they have normal vision, women heterozygous for the colorblind condition are *carriers* of the colorblind trait. To be colorblind, a woman must carry recessive alleles on *both X* chromosomes (X^cX^c).

In human males, the story of the X chromosome is a little different. In the case of color vision, the X chromosome of the male may carry the allele for normal color vision or the allele for colorblindness. There is no corresponding allele on the Y chromosome. If he has the recessive allele (X^cY), he is colorblind; there is no heterozygous condition for color vision in males. The lack of a second X chromosome in males that could potentially carry a normal color vision allele to mask the recessive allele makes males more likely to be afflicted with colorblindness than females.

The three possible genotypes associated with color vision in females are X^CX^C (normal vision); X^CX^c (normal vision, carrier of colorblind allele); and X^cX^c (colorblind). In males, there are only two possible genotypes: X^CY (normal vision) and X^cY (colorblind). In determining the possible offspring of matings between these males and females, the same principles apply as for a monohybrid cross. The first step is to determine the types of gametes each parent can produce. Meiosis in the female produces haploid egg cells, each having a single X chromosome. In the male, meiosis produces haploid sperm cells, half of which contain an X chromosome and the other half contain a Y chromosome. With this information, let's consider a cross between a female who is a carrier of the colorblind allele and a male with normal color vision.

The genotype of the female parent in this cross would be X^CX^c and that of the male parent would be X^CY. At meiosis, the female could produce either an X^C egg cell or an X^c egg cell; in the male, one-half of the total number of gametes would be X^C sperm and one-half would be Y sperm. As we have done previously, the gametes are aligned along a Punnett square and then crossed to determine the possible types of offspring as shown in Figure 14.13.

	♂ Gametes	
	X^C	Y
♀ Gametes X^C	X^CX^C	X^CY
X^c	X^CX^c	X^cY

Figure 14.13 A mating between a female who is a carrier of the color blindness trait (X^CX^c), and a male with normal vision (X^CY). In this example, one-half the female offspring are carriers of the color blind allele, and one-half the male offspring are color blind.

All the females from this cross have normal color vision, although one-half of them are carriers of the allele for colorblindness. Only half of the males, however, possess normal color vision. As you can see, the colorblind males inherited the colorblind allele from their mother. Now suppose that the mother had been colorblind instead of a heterozygous carrier. In working this out, you will find that there are still no colorblind daughters produced, although all of them are carriers. Having inherited the dominant color vision allele from their father, the daughters cannot be colorblind. In contrast, all of the sons are colorblind because they all inherited one of their mother's X chromosomes, which carries the colorblind allele.

Colorblindness affects about 1 male in 12, whereas only 1 female in about 250 is affected. For a woman to be colorblind, her father must be colorblind and her mother must be either colorblind or a carrier. Even if her mother is a carrier, a woman whose father is colorblind still has a 50:50 chance of inheriting normal color vision.

♀ Gametes \ ♂ Gametes	X^H	Y
X^H	X^HX^H	X^HY
X^h	X^HX^h	X^hY

Figure 14.14 A mating between a female who is a carrier of the allele for hemophilia (X^HX^h) and a normal male X^HY). One-half of the female offspring will be carriers of the hemophilia allele, and half of the males will be hemophiliacs.

HEMOPHILIA

There are approximately 250 other known examples of X-linked diseases, one of the most serious of which is *hemophilia.* This malady is characterized by the inability of blood to clot. In normal individuals, blood escaping from a wound will clot within minutes, preventing excessive blood loss. Hemophiliacs, however, lack a substance in the blood necessary for normal clotting and may continue to bleed for hours.

As with colorblindness, hemophilia results from the presence of a recessive allele on the X chromosome. The allele for normal blood clotting *(H)* is dominant to the recessive allele for hemophilia *(h).* A cross between a normal male and a female carrier of the disease obeys the same principles of inheritance that we encountered with color vision. Here again, it is rare for females to be afflicted with hemophilia because the father must have the disease and the mother must have it or be a carrier. Hemophilia occurs in only 1 female in about 100 million, but is estimated to affect 1 male in 10,000.

Extranuclear Inheritance

Now that we have strongly emphasized the Mendelian principles of inheritance and the associated concept of nuclear genes, it must be acknowledged that all genes are not located on chromosomes or even in the nucleus. The phenomenon of **extranuclear inheritance** involves genes found in cytoplasmic organelles—principally the mitochondria and chloroplasts. In Chapters 5 and 13, reference was made to the possible origin of eukaryotic cells through the incorporation of prokaryotes into larger cells. Instead of consuming the prokaryote, the host cell eventually may have become dependent on the continued functioning of its guest. The theory proposes that the prokaryote possibly became an efficient energy producer to the extent that the host cell depended on the internalized, smaller prokaryote for energy. Through this symbiotic relationship, the prokaryote became a primitive mitochondrion, and the evolution of eukaryotes was underway.

Similarly, a primitive, photosynthesizing blue-green bacterium may have become a chloroplast after being incorporated into a larger, primitive cell. The point under consideration here is that the prokaryotes incorporated into the larger host cells brought their own genes with them. Today, we know that both mitochondria and chloroplasts possess their own DNA, along with ribosomes and other genetic machinery. Also, the chromosomes of mitochondria and chloroplasts are circular, much like those of prokaryotes. Although mitochondria and chloroplasts are capable of synthesizing some of their own proteins, the great majority of their proteins are synthesized from nuclear DNA.

Mitochondria

In mammals, the inheritance of mitochondria is solely through the female. The egg cells of humans and other mammals contain a large amount of cytoplasm containing mitochondria; the sperm cells, however, do not possess much cytoplasm and their mitochondria are located within the midpiece of the sperm cell, which does not enter the egg cell upon sperm penetration. Thus, a fertilized egg does not receive any mitochondria from the male parent, and human and other mammalian offspring receive their mitochondria solely from their mothers. Mitochondria possess their own complement of specific mitochondrial genes. When mitochondria reproduce, or replicate, they pass their genes directly to new mitochondria. Thus, nuclear genes are not directly involved, and the Mendelian principles of chromosome distribution during meiosis do not apply.

Chloroplasts

In the plant kingdom, evidence for extranuclear inheritance was provided in the early 1900s through research on the four-o'clock plant *(Mirabilis)*. Leaf color in this plant can be green, white, or a patchwork of the two colors. In crossing pollen (male) from a plant of any color with an egg (female) from a green plant, all the offspring leaves were green. In crossing the same plants with a white female, all the offspring were white; and crossing any male with a patchwork female resulted in offspring that were white, green, and patchwork.The phenotype of the male parent made no difference; all the offspring resembled the female. Here again, the explanation lies in the large amount of cytoplasm contributed by the female egg cell.

Chloroplasts, like mitochondria, are cytoplasmic organelles and contain the green pigment chlorophyll. Chloroplasts, then, are passed from the female to the offspring, the leaves of which acquire a green color. If the chloroplasts are defective and cannot produce chlorophyll, the offspring leaves will be white. If they receive both normal and defective chloroplasts from the egg cell cytoplasm, offspring leaves could be green, white, and patchwork. Thus, it is pigment-producing genes in the chloroplasts, not in the cell nucleus, that ultimately determine leaf color in four-o'clock plants.

Genetic Engineering

One of the most exciting and truly revolutionary areas in modern molecular biology is **genetic engineering.** This term refers to the selective alteration of a cell's DNA so that new and different genes from the DNA of another cell type can be inserted directly into this DNA. The cell will then synthesize the proteins coded for by the inserted genes. The gene for human growth hormone, for example, has been inserted into DNA extracted from bacteria. The bacterial DNA then directs the synthesis of this protein hormone. The host DNA and the inserted genes form what is called **recombinant DNA.** The applications of recombinant DNA technology were first realized using bacteria as the host, but now it is possible to insert genes into eukaryotic cells and even into an entire organism. One example involves production of a vaccine against the hepatitis B virus, the infectious agent of a serious liver disease. A gene from the virus is inserted into yeast cells, which synthesize large amounts of the viral protein. The viral protein is then used to prepare the hepatitis B vaccine. To illustrate the basic method of genetic engineering, we will use one of the less complex examples involving bacteria.

Restriction Enzymes

Finding a specific gene on a chromosome full of genes is a formidable task. There are, however, specific bacterial enzymes known as **restriction enzymes** that facilitate this search. Restriction enzymes are used by bacteria as a defense to break down the DNA of foreign invaders, particularly viruses. Discovered in the late 1960s by Werner Arber, Daniel Nathans, and Hamilton Smith, each of these enzymes has the ability to "recognize" a specific sequence of DNA nucleotides and to cut the DNA molecule apart at that precise point in its sequence (Fig. 14.15). Regardless of the source of the DNA, on recognizing a specific nucleotide sequence, a particular restriction enzyme will cut precisely at that point. Using any one of hundreds of restriction enzymes that recognize different sequences of DNA, investigators can produce a variety of DNA fragments of desired lengths. The 1978 Nobel Prize was awarded to Arber, Nathans, and Smith for their discovery and use of restriction enzymes.

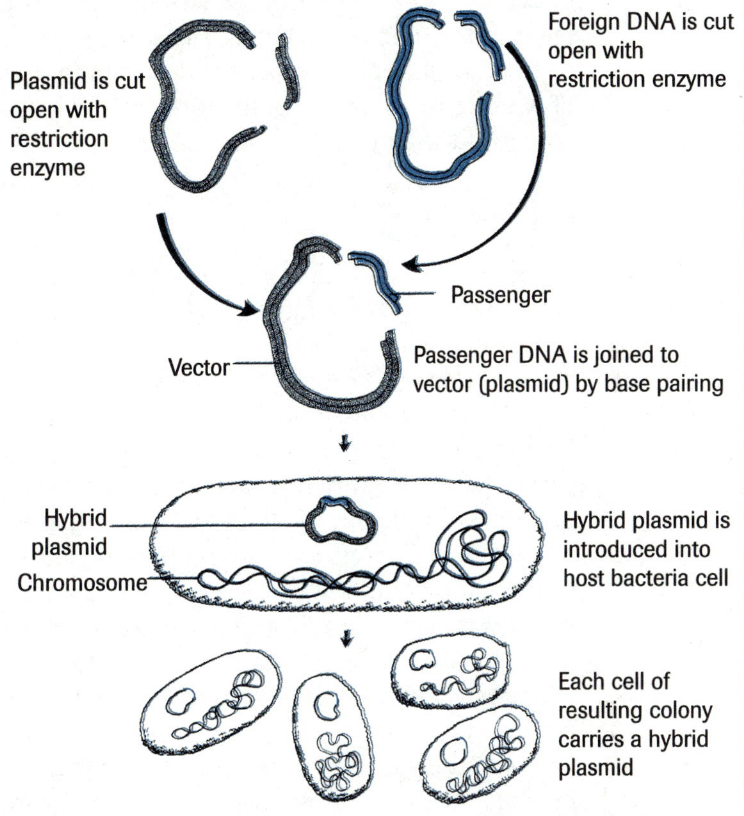

Figure 14.15 Summary diagram of recombinant DNA technique using bacteria. Both the plasmid (isolated from a bacterium) and the foreign DNA (isolated from a different species) are cut open with restriction endonuclease enzymes. A segment of the foreign, or passenger, DNA is mixed with the plasmid (vector), and the two DNA species join by complementary base pairing to form recombinant DNA. The now hybrid plasmid is introduced into a new bacterial host, which, as it divides, produces a clone of hybrid plasmids.

Plasmids

As you are aware, bacteria have a single circular chromosome. Genes are found in a certain sequence along this chromosome, which is composed of double-stranded DNA. In addition to the main chromosome, some bacterial cells contain small circular pieces of DNA called **plasmids.** By experimentally disrupting the bacterial cells, these plasmids can be removed, isolated, and the circular strand cut open and made linear using specific restriction enzymes. When cut, each end of a DNA fragment has a sequence of exposed unpaired bases referred to as "sticky" ends. Next, fragments cut from eukaryotic DNA, as shown in Figure 14.15, are mixed with the now-linear plasmid DNA, and the two types of DNA strands, assisted by an enzyme, join spontaneously at their sticky ends by complementary base pairing. The result is circular "hybrid" plasmids composed of both plasmid DNA and a fragment of eukaryotic DNA, in other words, **recombinant DNA.**

The newly formed hybrid plasmids are then added to a culture of live bacterial cells, which take up the plasmids from the culture solution by transformation. This is simply the process by which the plasmid is inserted into the bacterial cells. When the bacterial cells undergo division, producing thousands to millions of new cells, the enclosed plasmids also replicate along with the cells. It is possible, therefore, to grow numerous colonies of bacterial cells, all containing identical copies of the hybrid plasmids. Because each plasmid contains a copy of a certain eukaryotic gene, the entire process produces many copies of this one gene and is referred to as **gene cloning.**

Bacterial plasmids are not the only method of introducing recombinant DNA into cells. DNA of certain bacterial viruses—bacteriophages—also can be cut by restriction enzymes and foreign DNA inserted between the cuts. This recombinant DNA can be taken up by bacterial cells in the same manner as plasmids are taken up. Within the bacterial cell, replication of the recombinant DNA produces new bacteriophages, each of which carries a copy of the foreign DNA. Here again, a virtually unlimited number of copies of the foreign DNA can be produced.

Gene Cloning

What, you may be asking, is the significance of cloning enormous numbers of the same gene? First, this technique affords researchers the opportunity to study more thoroughly the structural and functional properties of a particular gene. Second, cells transformed by the insertion of foreign genes are capable, in many cases, of synthesizing proteins coded for by the inserted gene. This is the case, for example, with the gene that codes for the protein hormone insulin. Dividing cells in which the insulin gene has been inserted produce numerous copies of this gene, making possible the synthesis of great quantities of insulin for treatment of diabetes. Previously, insulin for human use was extracted from pigs and cattle; however, with the cloning of the insulin gene, the hormone can be obtained more cheaply than animal insulin and in virtually unlimited quantities. Insulin derived from recombinant DNA technology was first approved for human use in 1982. In addition, other human hormones, such as the insulin-regulating hormone somatostatin, thymus gland hormone, and erythropoietin have been produced from gene cloning.

Over the past ten years, a revolutionary technique known as PCR (polymerase chain reaction) has been employed to make numerous copies of a segment of DNA in the laboratory. Using a certain enzyme, a quantity of all four of the DNA nucleotides, and primer RNA molecules, billions of copies of a DNA segment can be synthesized in a few hours. A primer is necessary for the enzyme to initiate DNA synthesis. A solution of DNA containing the segment to be copied is heated to separate the two complementary strands. The enzyme, primers, and nucleotides are then added to the solution. A primer, which is complementary to the end of a DNA segment, binds alongside the end of a segment on each separated strand. The enzyme then adds nucleotides to the primer, using the DNA strands as templates. In this manner, the desired DNA segment is copied. This entire PCR sequence can be repeated as many times as necessary to produce the desired number of copies of the segment.

DNA from organisms thousands or millions of years old or the DNA from very tiny amounts of blood, semen, or skin left at crime scenes can be copied using PCR.

Gene Therapy

Recombinant DNA technology also can be applied in what is called "gene replacement therapy." This involves adding a functional copy of a gene to a cell containing a nonfunctional gene. As an example, a variety of human genetic diseases result from the absence of certain cellular enzymes or from defective enzymes. If the genes coding for the correct enzymes could be inserted into the cells, the diseases could be cured. Gene therapy treatments for cystic fibrosis and certain blood disorders are currently in the experimental stage.

Finally, recombinant DNA technology is also being applied in eukaryotic plant cells. By inserting a given gene into a plant cell, it may be possible to develop better or even new varieties of plants having significant economic importance. One of the most interesting examples related to plant improvement involves the element nitrogen. Nitrogen is a component of proteins and nucleic acids and is one of the nutrients required by all plants. Some species of plants have tiny root nodules containing bacteria that "fix" atmospheric nitrogen (N_2) to produce ammonia (NH_3), a form of nitrogen the plants can use (Chapter 8). Many species of plants, however, such as the major cereal grain crops, cannot support bacteria and must acquire their nitrogen indirectly through industrial fertilizers.

In fixing the nitrogen, the nodule bacteria utilize their *nitrogenase* enzyme system that reduces—adds hydrogens to—nitrogen to form ammonia. Through genetic engineering, it may be possible to insert the nitrogenase enzyme genes into those plants that cannot support the nitrogen-fixing bacteria. Thus, it is possible that the yield of various crops could be vastly improved. And perhaps in the future, advances in genetic engineering may result in the creation of new high-yield varieties of plants that are more nutritious and less expensive to grow.

During the mid-1980s, Stephen Lindow experimented with a species of plant bacteria that secretes a protein around which water freezes to form ice crystals. These bacteria are used by ski resorts to make artificial snow. As inhabitants of citrus and vegetable plants, the bacteria contribute to frost damage of these crops during early winter freezes. By removing the gene that produces the bacterial protein, Lindow developed a strain of bacteria that prevents ice crystal formation on plants. In 1987, these "ice minus" bacteria, as they are called, were sprayed on young strawberry plants to determine if the bacteria could reduce frost damage by replacing the naturally occurring bacterial species. Results from the experiment indicated that the sprayed plants were somewhat more resistant to frost damage than were untreated plants.

Ethical Considerations

With the development of recombinant DNA technology has come the era of the new genetics. It is now possible to identify, excise, insert, and otherwise manipulate genes to the extent that such activities have made headway into agriculture, industry, and medicine. While improved crop productivity, new vaccines and drugs, and the insertion of healthy genes for defective ones offer great promise for the future, these developments are often accompanied to some extent by ethical considerations. What, for example, will be the effects of releasing genetically engineered plants into the biosphere? Could they spread out of control or destroy the natural vegetation of a region? In the case of crop plants, will new strains produce heretofore unknown substances that are harmful if consumed?

The use of genetically altered organisms to clean up chemical wastes is a boon to industry, but how will these organisms interact with other species in the environment and what will be the long term effects? We would also have to consider the possible effects of genetically altered organisms on soil and water contamination.

The visions of treating and curing human diseases with gene therapy are limitless. Currently, there is optimism concerning the use of gene insertion therapy in which a normal gene is inserted into cells having a defective gene. While this technology offers enormous benefits, questions may arise as to the value or necessity of correcting the defect. Should everyone, including violent criminals and other social undesirables, have an equal right to treatment? If the therapy will prolong the life of the very old or otherwise infirm for only a short period of time, is it justified? Who will decide? These are the same ethical questions being asked today concerning medical treatment less dramatic than gene insertion therapy.

In applying gene therapy to the manipulation of human DNA, we encounter the prospect of eugenic engineering, the attempt to improve the human species through selective genetic methods. Here again, we are faced with an array of ethical questions: What is the definition of "improve?" Which genetic traits are most desirable and by whose standards? Should we seek to alter the genetic traits of all human beings or just a select few? What would be the social and cultural implications of breeding a group of distinctly different human beings within a given population? And most importantly, should anyone or any group even attempt to alter the genetic makeup of a human being?

These and other ethical questions, some perhaps not yet imagined, confront the new genetics. While the potential benefits of DNA technology are profound, we must weigh the risks and proceed under the caution flags of reason and common sense.

It is clear that genetics is a complex biological science. We have seen that genes acting alone can determine a particular phenotype or they can be influenced by many environmental factors. A single gene may influence several phenotypic characteristics or more than one gene can affect a single trait. The degree to which a particular gene is expressed can be quite variable, and one gene can interfere with the expression of another gene. These and many other gene interactions add to the complexities of inheritance. All of this complexity, however, can be appreciated in light of the fact that a single human cell contains about six feet of DNA, which works out to a total of some 100 *billion* miles of DNA in one adult human being!

Chapter Summary

1. Early attempts to explain heredity include the concepts of fluid heredity, pangenesis, and blending inheritance. These concepts were later disproved by the plant breeding experiments of Gregor Mendel, who formulated the basic laws of inheritance.

   Alleles are alternative forms of a gene and are located at the same position on homologous chromosomes. Two alleles, one from each parent, are inherited for each trait. Mendel's principle of segregation refers to the separation of allele pairs in meiosis. As a result, each gamete receives only one of each kind of allele. That the separation of one pair of alleles during meiosis is not affected by the separation of any other pair is Mendel's principle of independent assortment.

   Key Terms: *one gene-one polypeptide hypothesis, p. 327*
   allele, p. 328
   homozygous, p. 328
   heterozygous, p. 328
   dominant, p. 328
   recessive, p. 328
   phenotype, p. 329
   genotype, p. 329

Concept Questions: How is it possible for organisms to have the same phenotype but different genotypes? Cite an example of Mendel's principle of segregation. Explain generally how Mendel's discoveries invalidated the theory of blending inheritance.

2. A monohybrid cross involves parents differing in only one genetic trait. From the F_1 offspring, the genotypic and phenotypic ratios can be determined. A dihybrid cross involves parents differing in two genetic traits; following meiosis in each parent, one allele for each trait moves into each gamete.

   A test cross involves mating an organism displaying a dominant trait with a homozygous recessive organism. The results reveal whether the organism with the dominant trait is homozygous or heterozygous for the trait.

Key Terms: *principle of segregation, p. 329*
independent assortment, p. 330
filial (F_l) generation, p. 331
Punnett square, p. 331
genotypic ratio, p. 331
phenotypic ratio, p. 331

Concept Questions: a) Assuming complete dominance, determine the phenotypic and genotypic ratios of the F_1 for the following (*R* = red color; *r* = white color): *Rr* x *rr; RR* x *Rr; Rr* x *Rr*'

b) What would be the phenotypic and genotypic ratios of these crosses if dominance were *incomplete?*

Assuming complete dominance, determine the phenotypic and genotypic ratios of the F_1 for the following (*R* = red color; *r* = white color; T= tall; *t* = short):

a) *RRTTx rrtt*
b) *RRtt x rrTT*
c) *RrTt x RRTT*
d) *RrTt x RrTt*

In corn kernels, smooth texture (S) is dominant to wrinkled texture (s). You observe an ear of corn with only smooth-textured kernels. How could you determine if the smooth texture represents a homozygous or a heterozygous condition?

3. The ABO blood group system is an example of codominance. A and B alleles are dominant except in the AB genotype, in which the alleles display codominance.

Key Terms: *point mutation, p. 334*
ABO system, p. 335
blood antigen, p. 336

Concept Question: A man with blood type A is involved in a paternity suit claiming he is the father of a child with type B blood. Is the claim genetically defensible? Explain.

4. Some genes affect more than one phenotypic trait. The effects of these pleiotropic genes are seen in some human conditions, such as lysosomal storage diseases. Quantitative traits, such as height and weight, usually are determined by polygenic inheritance. Most quantitative traits are influenced by the environment.

Key Terms: *polygenic inheritance, p. 337*
multifactorial traits, p. 337

Concept Question: What harmful conditions do you think might influence multifactorial traits in humans?

5. Alleles may be located on the same chromosome and thus are linked. During meiosis, linked genes move together into a gamete. Linked genes may be rearranged by crossing over, which can provide offspring with additional adaptive traits and a greater survival potential.

   Key Terms: *homologous chromosomes, p. 338*
   crossing over, p. 338

   Concept Question: What is the significance of linked genes in determining the phenotypes of offspring?

6. Sex-linked, or X-linked, traits in humans are determined by alleles on the X chromosome. Color blindness and hemophilia, both caused by recessive alleles, are two of many sex-linked conditions. Males are affected more frequently by recessive sex-linked traits, which are inherited from the mother.

   Key Terms: *autosome, p. 339*
   sex chromosome, p. 339

   Concept Question: One of the nonlethal forms of muscular dystrophy results from a sex-linked recessive gene. A man who has the disease marries a woman who is a carrier of the dystrophy gene.

   a. What percentage of their sons could be expected to have muscular dystrophy ?
   b. What percentage of their daughters could be expected to be carriers of the dystrophy gene?
   c. What percentage of all their children could be expected to be normal?

7. Genetic engineering, or recombinant DNA technology, involves insertion of a foreign gene into a host cell that, as it divides, produces numerous copies of the gene. This process, known as gene cloning, allows certain genes to produce large quantities of a protein, such as insulin or growth hormone. Genes are cut from DNA using restriction enzymes; when inserted, the foreign gene and the host DNA form recombinant DNA. Polymerase chain reaction (PCR) can be used to make numerous copies of a DNA segment.

   Key Terms: *recombinant DNA, p. 342*
   restriction enzyme, p. 342
   plasmid, p. 343
   clone, p. 344

   Concept Questions: Eugenics is the discipline concerned with improving human hereditary characteristics. Positive eugenics is concerned with promoting reproduction by individuals having desirable or superior traits; negative eugenics is concerned with preventing reproduction by individuals having undesirable or inferior traits. What are some of the possible ethical issues at stake in each of these positions? Do you consider these issues to be valid? Explain. How might DNA technology affect both positive eugenics and negative eugenics?

   Explain how recombinant DNA technology might be applied toward curing the disease sickle-cell anemia.

Life Arising from Life

Chapter Fifteen

Why Sex?

On the face of it, the heading for this section would seem to call for the most obvious response; namely, if the author thinks he needs to explain the question, he may also spend a goodly portion of his time pursuing other equally futile pastimes, such as rummaging for gold pots at rainbow terminals. While this sentiment is duly acknowledged, it so happens that the functional significance of *sexual* and *asexual reproduction* in the evolutionary scheme is an exceedingly thorny issue. We do know that sexually reproducing organisms are more widespread over Earth than asexually reproducing species, but why this should be so is an unanswered question. Not all plants and animals engage in sex, yet for some it is indispensable.

Asexual reproduction is highly efficient, and many asexually reproducing organisms, such as potatoes, are widely distributed as well. Some organisms are totally asexual, which is a form of reproduction that requires only one parent (Fig. 15.1). By splitting in two or by growing a bud somewhere on its body or even fragmenting into several pieces, the asexually reproducing organism gives rise to a lineage of offspring similar to itself.

Asexual reproduction has its advantages; an individual need not find a mate and is able to produce numerous offspring in a relatively short period of time. Because the parents of asexually reproduced offspring have proved to be adaptive, asexual reproduction is thought to be advantageous in a stable or unchanging environment. But asexual reproduction also has its disadvantages. Many asexually reproducing organisms are haploid and thus have only one copy of a given gene. In these organisms, a harmful recessive gene is always expressed since there is no dominant allele to mask its effects. Furthermore, mutation is the only source of variation in asexually reproducing organisms and it might require many generations before an advantageous mutation appeared.

Sexual reproduction, in contrast, produces offspring that are variable but similar to either parent. This variability may be of adaptive value in some situations, particularly in environments that are changing rapidly. But various asexually reproducing organisms also exhibit enormous variation and are known to adapt to changing environments as well. Still, there assuredly is *some* significant evolutionary distinction between sexual and asexual reproduction. One hypothesis suggests that the distinction may involve cell mortality. Because of errors in DNA replication and environmental influences, the DNA of body cells becomes damaged over time. Mitotic cell division, the process by which some organisms reproduce asexually, apparently offers limited opportunity for these cells to repair this type of damage, and the cells continue to age and degenerate.

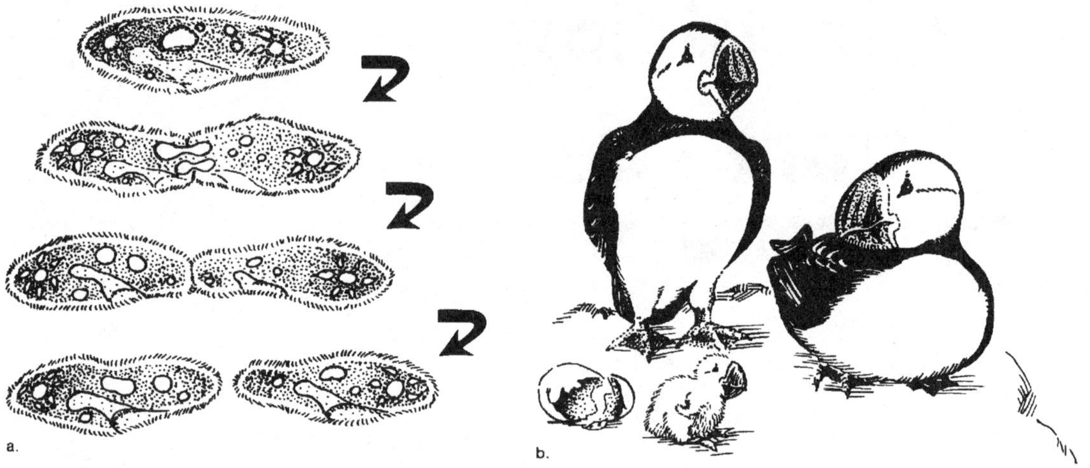

Figure 15.1 Reproduction results in the production of offspring that are similar copies of the parent or parents. (a) Asexual fission in *Paramecium* produces offspring genetically identical to the parent. (b) Sexual reproduction in puffin birds produces a single chick having a combination of traits derived from both parents.

Meiosis, in which chromosomes are paired and recombined, may provide a means of DNA repair different from that of mitosis. This means that the DNA of a reproductive cell, further assisted by the DNA contributed by another reproductive cell at fertilization, can be repaired more frequently and thereby delay the aging process. The diploid condition resulting from the union of gametes confers some protection against harmful recessive genes. Also, two copies of each gene provide a safety factor if one copy goes bad.

In passing from generation to generation, reproductive cells represent a virtually immortal cell line. In addition, sexual reproduction provides variability in natural populations, a factor underlying our basic understanding of the evolutionary process. Furthermore, sexual reproduction often results in production of large numbers of gametes, a factor that increases the odds of fertilization and the continuation of the species.

While both sexual and asexual reproduction have persisted for millions of years, the relative significance of each in the evolutionary scheme remains unclear. Reproduction itself, however, is one of the essentials of life, and our intent in this chapter is to examine some of the basic mechanisms by which life is perpetuated. Bacteria undergo the simplest and most primitive form of reproduction in the evolutionary scheme, followed by the protists, such as algae, in which reproduction involving mitosis and meiosis is first evident. Thought to have arisen from protistan ancestors, fungi are unique organisms that employ spore formation in their reproductive processes. Plants are also considered to have evolved from protists, specifically some type of green algae. Reproduction in the nonvascular plants, such as most mosses, reflects their aquatic origin in that water is essential for the transfer of sperm from the male to the female. The earliest vascular plants lacked seeds and are represented today by the ferns and a few other groups, which also depend on water for the transfer of sperm. The most highly evolved seed plants are the angiosperms, which are vascular plants that possess flowers. Flowers bear the reproductive structures from which the male and female sex cells arise.

Invertebrate evolution has resulted in extremely varied reproductive methods, although the more highly evolved invertebrates do not necessarily possess the most complex reproductive systems. Among the vertebrates, the reproductive systems exhibit a basic similarity.

We will use these diverse groups of organisms to illustrate the variations in reproductive methods that have evolved on Earth over the past 3.5 billion years.

Bacterial Reproduction

Binary Fission

The prokaryotic bacteria and blue-green bacteria reproduce asexually by a type of cell division called **binary fission,** or "splitting in two" (Fig. 15.2). Although binary fission apportions the genetic material equally to each cell, neither mitosis nor meiosis occurs in prokaryotes.

In preparation for fission, the single bacterial chromosome replicates in the cytoplasm, and each chromosome then attaches to the plasma membrane. The bacterial cell begins to elongate, and the plasma membrane eventually pinches inward between the two chromosomes. A cell wall then forms between the chromosomes and the newly formed daughter cells may separate from each other or remain attached, depending on the bacterial species.

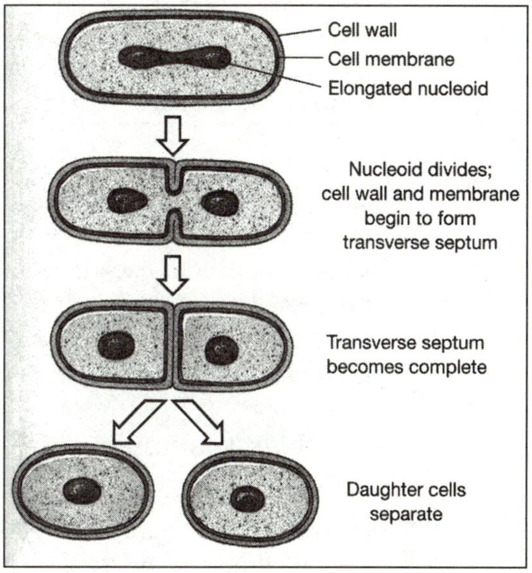

Figure 15.2 (a) Binary fission in a bacterium. Reprinted from *Microbiology: Principles and Applications,* Third Edition, by Jacquelyn G. Black, 1996, Prentice-Hall, Inc.

Conjugation

Some bacteria carry out a version of sexual reproduction known as **conjugation,** which involves the transfer of genetic material from one bacterial cell to another (Fig. 15.3). As a requirement for this type of conjugation, the donor cell must contain a specific type of plasmid, which, you will recall, is a small, circular DNA molecule lying free in the cytoplasm. The recipient cell does not contain this type of plasmid. The bacterial cell that is donating DNA forms a tiny extension that attaches to the recipient bacterial cell. A cytoplasmic tube forms between the two cells for the transfer of DNA. One strand of the double-stranded DNA plasmid is passed through the conjugation tube to the recipient cell. Both cells then make a second, complementary strand of DNA, which yields two cells, each with a complete copy of this plasmid. In receiving the plasmid DNA, the recipient cell is converted to a donor cell. If it subsequently undergoes conjugation, this new donor cell may now transfer a copy of its plasmid DNA to another recipient cell.

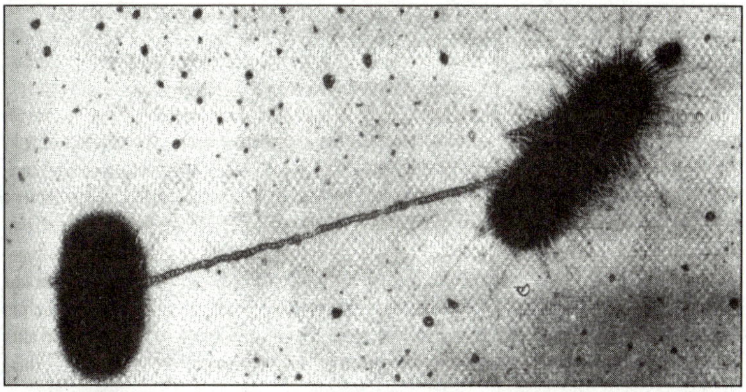

Figure 15.3 Conjugation: bacterial "mating." Conjugation—the transfer of genetic material—occurs through a special large, hollow sex pilus, which is shown here connecting a pair of Escherichia coli. One bacterium (seen here on the top right) acts as a donor, transferring DNA to the recipient. In this photo, the donor bacterium is bristling with nonsex pili that probably help it attach to surfaces. Reprinted from *Biology: Life on Earth,* Fourth Edition, by Gerald and Teresa Audesirk, 1996, Prentice-Hall, Inc.

Occasionally, one of these conjugation plasmids becomes incorporated into the main bacterial chromosome. During conjugation, this plasmid passes through the conjugation tube into the recipient cell, dragging a portion of the attached chromosomal DNA along with it. The entire bacterial chromosome does not often make it all the way through the tube before conjugation is disrupted in some way. Thus, the recipient cell receives only a segment of the donor's chromosome. This segment can align with a homologous segment on the recipient's chromosome, and the two segments can exchange DNA through crossing over (see Chapter 14). This genetic recombination produces a recipient cell with a new genotype. In a changing environment, this new assortment of genes may provide an adaptive advantage.

Reproduction in Algae, Fungi, and Plants

Algal Reproduction

Algae are photosynthetic protists that lack the supporting and conducting tissues xylem and phloem, and which usually lack true roots, stems, and leaves. Algal species are quite diverse and include unicellular, multicellular, and colonial forms. Reproduction in algae occurs both asexually and sexually, with many species utilizing both reproductive processes at different points in their life cycles. To illustrate algal reproduction, we will use two common species—*Chlamydomonas,* a unicellular alga, and the filamentous, multicellular alga *Spirogyra.*

Chlamydomonas

This microscopic, unicellular green alga possesses a single chloroplast and two anterior flagella. *Chlamydomonas* is a **haploid** cell and reproduction may be asexual or sexual, although the asexual method is more common (Fig. 15.4). The sexual phase is initiated when two cells of opposite mating types (designated + and –) act as gametes and fuse to form a diploid zygote. After a period of dormancy, this zygote undergoes meiosis to produce haploid algal cells, which may develop into new *Chlamydomonas* cells of opposite mating types. During the asexual phase, however, these haploid cells undergo cell division to form new + and – haploid cells that are the same as those produced in the sexual

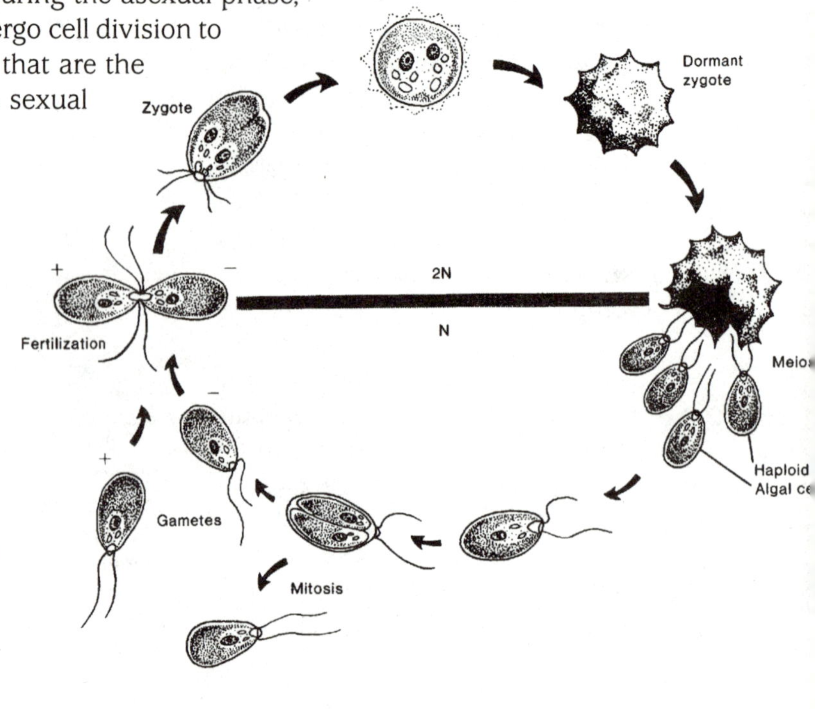

Figure 15.4 Sexual and asexual reproduction in the alga *Chlamydomonas.* At the beginning of the sexual (2n) phase, opposite mating types (designated + and –), which are haploid cells, act as gametes and fuse to form a diploid (2n) zygote. After a period of dormancy, the zygote undergoes meiosis to produce haploid (n) algal cells. During the asexual (n) phase, these cells undergo mitosis to form new + and – haploid cells.

phase. Thus, the only **diploid** stage in the life cycle of *Chlamydomonas* is the zygote.

Spirogyra

This filamentous green alga, commonly known as "pond scum," consists of rectangular cells arranged like boxcars placed end-to-end. One of the unusual and distinguishing features of *Spirogyra* is its ribbonlike chloroplast that forms a spiral along the length of each cell in the filament. Also unusual for green algae is its method of sexual reproduction; it is a process known as **conjugation,** which is suggestive of the same process in bacteria (Fig. 15.5). Two filaments, each haploid, align themselves side-by-side, and hollow conjugation tubes form between opposite cells of each filament. The entire contents of one cell, the *supplying cell,* then pass through the conjugation tube and fuse with the contents of the *receiving cell* in the other filament. The supplying cell thus provides a motile gamete that moves into the receiving cell and fuses with the nonmotile gamete. Fusion of the contents of the two cells produces a diploid zygote. Later, each zygote is released from its receiving cell and undergoes meiosis to produce new haploid *Spirogyra* filaments. Here again, the only diploid stage in the life cycle is the zygote. Asexual reproduction in *Spirogyra* occurs by fragmentation, in which each broken piece grows into a new filament.

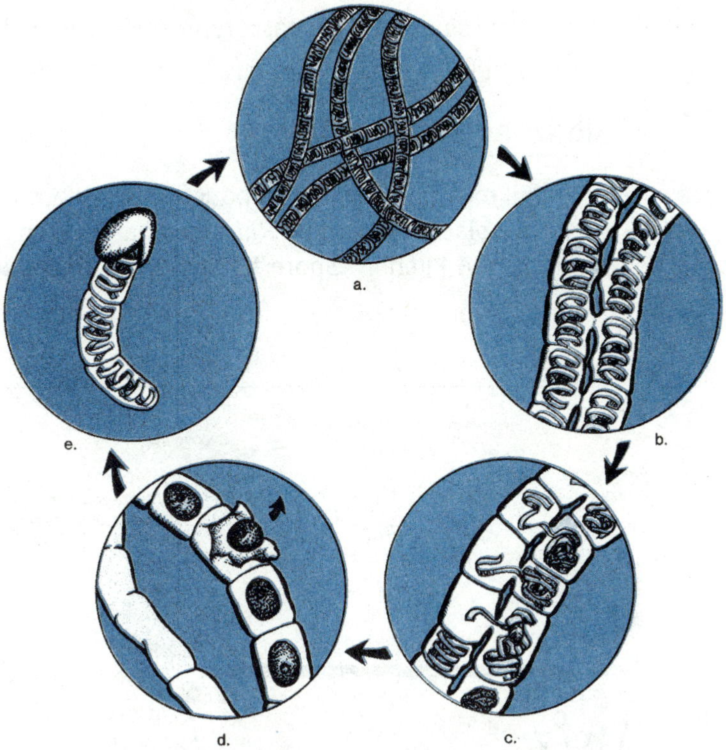

Figure 15.5 Reproduction in the multicellular green alga *Spirogyra.* (a) Filaments of rectangular *Spirogyra* cells. (b) Two haploid filaments showing their spiral-shaped chloroplasts. Conjugation tubes are forming between opposing cells of the filaments. (c) The contents of the supplying cells pass through the conjugation tubes to the receiving cells. (d) Fusion of the cellular contents in each receiving cell forms a diploid zygote, one of which is shown being released from the receiving cell. (e) The zygote (or zygospore) undergoes meiosis to produce new haploid filaments.

Fragmentation and other forms of asexual reproduction, such as budding, occur more frequently when the environment is stable. The offspring of asexually reproducing organisms exhibit virtually no genetic variability, which is an advantage when environmental conditions are relatively constant. Sexual reproduction, in contrast, is more common when environmental conditions are unfavorable because the genetically variable offspring have a greater likelihood of adapting to those conditions.

Fungal Reproduction

Almost all fungi reproduce by means of **spores,** which are cells capable of germinating into an adult organism without uniting with another cell. A single mature fungus of some species can release up to a billion spores. Some fungi also reproduce asexually by fragmentation in which the parent breaks up into parts that regenerate into new organisms. The single-celled yeasts reproduce asexually by budding and also reproduce sexually by spore formation.

RHIZOPUS

We will examine fungal reproduction by spore formation using a common representative: the black bread mold *Rhizopus* that is often found growing on bread or cake left too long in the kitchen. Spore formation in *Rhizopus* occurs both asexually and sexually (Fig. 15.6).

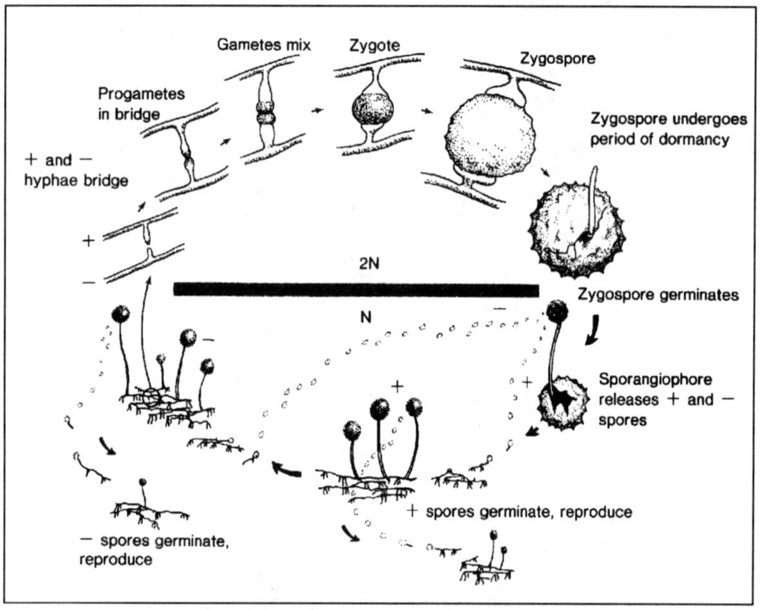

Figure 15.6 Life cycle of the common black bread mold *Rhizopus*. The upper (2n) diagram is the gamete-producing sexual cycle. The lower (n) diagram is the spore-producing asexual cycle.

Asexual Reproduction

Rhizopus and other fungi, with the exception of yeasts, are composed of tiny tubular strands called *hyphae* (singular, hypha), which arise from germinating spores. When landing on a food source, *Rhizopus* spores germinate to form horizontal hyphae that branch out over the food. These hyphae are anchored by tiny rootlike structures that penetrate into the food source. In asexual reproduction, the horizontal hyphae give rise to upright hyphae that have globelike spore cases at their tips. Hundreds of haploid spores develop within each spore case, which bursts when mature, releasing the spores. If conditions are favorable and the spores land on a suitable nutrient source, they germinate to form new hyphae.

Sexual Reproduction

At times, however, environmental conditions, such as temperature and moisture, may be unfavorable. Under these conditions, some species of *Rhizopus* may reproduce sexually. Sexual reproduction in *Rhizopus* involves hyphae of opposite mating types, designated + and –, and is somewhat suggestive of conjugation in the green alga *Spirogyra*.

When + and - hyphae align closely, each hypha forms a lateral extension containing several haploid nuclei. Tips of the extensions make contact with each other and the wall between the two tips breaks down. Nuclei from each hypha pair up and become encased in a tough spore case. The spore case is highly resistant to drought, temperature extremes, and other harsh environmental conditions. When conditions again become favorable, the paired nuclei fuse to form a diploid nucleus that immediately undergoes meiosis to produce haploid spores. Soon thereafter, the spores are released and may germinate to form new hyphae once settled on a suitable food source.

Nonvascular Plant Reproduction

Mosses: Alternation of Generations

There are several groups of nonvascular land plants, the most numerous of which are the mosses. Mosses and their allies do not possess true roots, stems, or leaves. The plant body, however, is differentiated into leaflike and stemlike parts, along with rootlike structures called *rhizoids*. While some mosses have conducting tissues similar to xylem and phloem in parts of the plant, most mosses lack true vascular tissues and thus cannot retain water for long periods of time nor can water be efficiently transported up the plant body. Consequently, water must be absorbed directly from the surrounding air or from a nearby source of moisture.

Mosses, although fairly primitive plants as evidenced by their inability to retain and transport water, provide insight into some of the reproductive features that characterize many of the more highly evolved forms of both plant and animal life. First, there are often separate, distinguishable sexes in the mosses, each producing unique sex cells, or gametes. Second, fertilization is internal, within the body of the female. Third, development of the zygote occurs for a time within the female and is dependent on the female for nourishment.

Mosses undergo a reproductive cycle known as **alternation of generations,** in which a sexual stage, represented by the **gametophyte** ("gamete plant"), is followed by an asexual stage, the **sporophyte** ("spore plant") (Fig. 15.7). In mosses and similar plants, it is usually the gametophyte that is conspicuous; that is, it is the stage in the life cycle that you ordinarily notice. The sporophyte in these plants is usually quite small and often escapes detection.

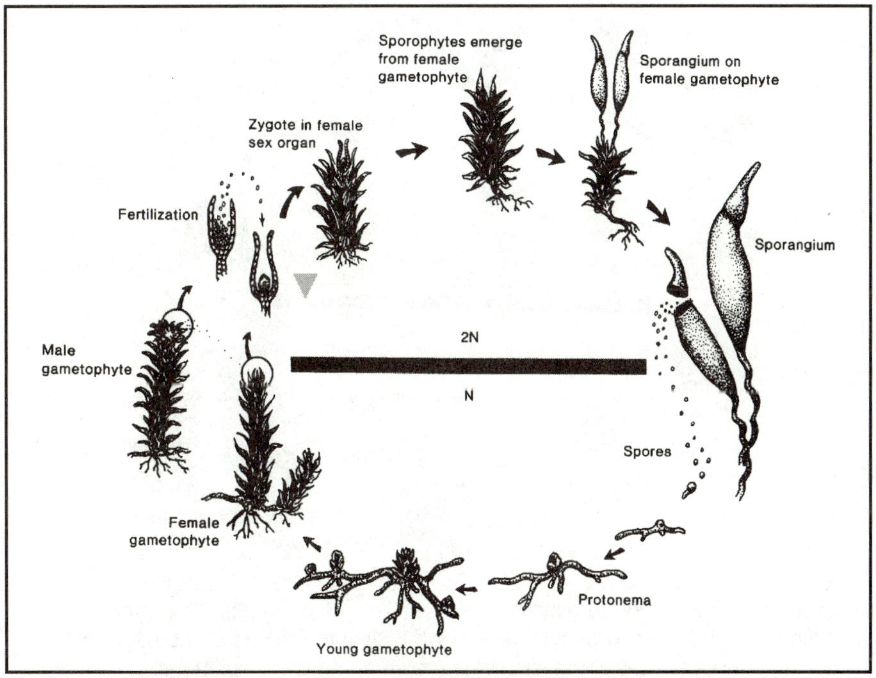

Figure 15.7 Alternation of generations in a moss. The gametophyte generation is haploid (N) and the sporophyte generation is diploid (2n).

We have mentioned that there are separate sexes in most mosses, a haploid male gametophyte and a haploid female gametophyte. These are the common mosses you notice growing on a rock or a tree. The male plant produces haploid sperm by mitosis within reproductive structures called **antheridia** ("little flowers"); in the female, haploid eggs develop within **archegonia** ("primitive parents"), which are found at the tip of the plant. During the fertilization process, the flagellated sperm require water, ordinarily in the form of rain or dew, to swim to the female plants. In lacking vascular tissue, the mosses cannot retain water and must rely on an external source of moisture for the transfer of sperm. Union of sperm and eggs occurs within the archegonia; the resulting diploid zygotes, which will develop within the female plants, are the first cells of the sporophyte generation.

Each zygote divides by mitosis to form an elongated stalk with a capsule at its tip. Throughout development, this sporophyte, both stalk and capsule, remains attached to and nourished by the female gametophyte. Within the capsule, haploid spores are produced by meiosis and are released when the capsule matures. Spores are dispersed by the wind and, if they fall into suitable soil, each may germinate into a tiny branched filament called a **protonema** ("first thread"). These microscopic structures very much resemble filamentous green algae. Each protonema divides by mitosis to give rise to new haploid male and female gametophytes, completing the cycle. This new gametophyte generation will again be followed by a sporophyte generation, perpetuating the species.

Vascular Plant Reproduction

Vascular plants are so named because of the presence of the conducting tissues xylem and phloem. Both xylem and phloem are distributed throughout the roots, stems, and leaves. Vascular plants are immensely diverse, but all undergo life cycles involving alternation of generations. We will use two groups of vascular plants to illustrate contrasting reproductive methods among the more advanced land plants: the seedless ferns and the seed-bearing angiosperms.

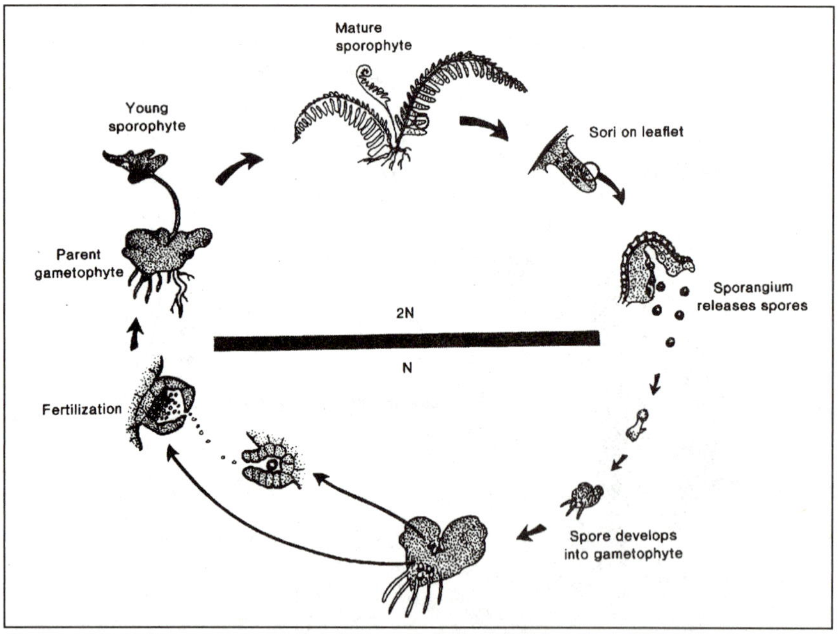

Figure 15.8 Alternation of generations in a fern. The gametophyte generation is haploid (n) and the sporophyte generation is diploid (2n), as in the mosses. In ferns, however, the sporophyte generation is conspicuous.

Ferns

Although possessing a well-developed vascular system and true roots, stems, and leaves, the ferns, like the mosses, are only partially adapted to land. The flagellated sperm require water to swim to the egg, and the gametophyte, which is very much reduced in ferns, does not contain vascular tissue and must absorb water from the soil. In comparing alternation of generations in ferns and mosses, we can note a few similarities and some significant differences (Fig. 15.8). For example, the fern plant found hanging in a florist shop is the diploid sporophyte generation. In ferns, then, it is the sporophyte generation that is conspicuous and not the gametophyte, as in the mosses.

On the underside of fern leaves, haploid spores produced by meiosis are contained within clusters of spore cases, each cluster of which is called a **sorus** ("a heap", plural, sori). When a spore case breaks open, the haploid spores are released and dispersed by the wind. A germinating spore gives rise to a photosynthetic heart-shaped gametophyte called the **prothallus** ("first shoot"). Male and female reproductive structures, the antheridia and archegonia, respectively, develop on the underside of the haploid prothallus. Within the antheridia and archegonia, haploid gametes (sperm and eggs) are produced by mitosis. If there is adequate water when the mature flagellated sperm are released, they swim to the archegonium and one will fertilize the egg cell. The resulting diploid zygote grows out of the prothallus and becomes the new diploid sporophyte. When the fern sporophyte develops sori on its leaves, the cycle is completed.

In the ferns and other vascular plants, the sporophyte is conspicuous and the gametophyte, although multicellular, is quite small. In these plants, the sporophyte is not totally dependent on the gametophyte for its existence. As we saw in the mosses, the sporophyte grows out of and is nourished by the female gametophyte. Vascular plants, in contrast, develop independent sporophytes that must absorb and transport water and minerals, photosynthesize, and otherwise exist independently.

Angiosperms

The most highly evolved and biologically successful vascular plants are the angiosperms, or flowering plants. The flower of angiosperms functions in reproduction and, depending on the species, may bear only the male reproductive structure, the **stamen,** only the female reproductive structure, the **carpel,** or both (Fig. 15.9).

The **anther** at the tip of the male stamen consists of pollen sacs containing cells that give rise to haploid pollen grains, the male gametophytes. Pol-

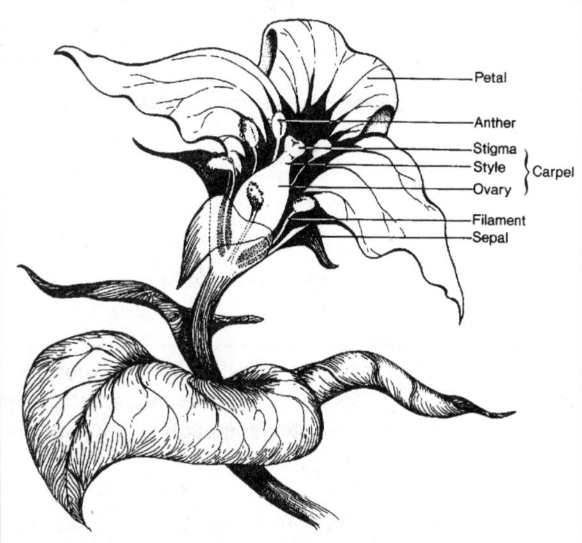

Figure 15.9 External flower structure. The female reproductive structure is the *carpel,* which consists of the stigma, style, and ovary. The male structure, the *stamen,* is composed of the anther and its elongate stalk, the filament.

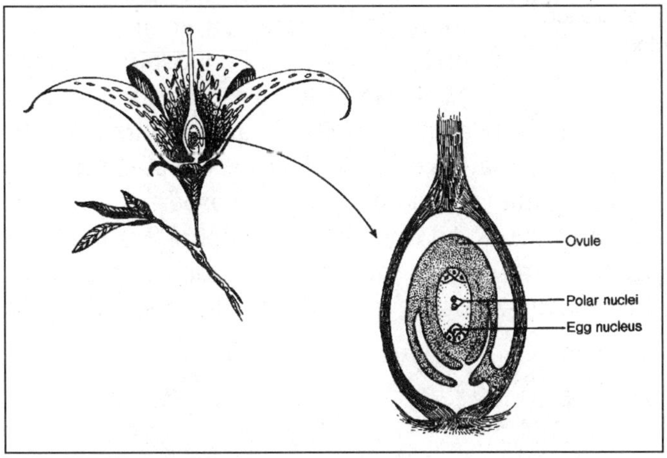

Figure 15.10 Female gametophyte of an angiosperm. The gametophyte, or embryo sac, consists of eight haploid nuclei. The egg nucleus will fuse with a sperm cell to form the diploid zygote. The two polar nuclei fuse with another sperm cell to form the triploid endosperm.

len grains, in turn, produce haploid sperm cells. The base of the female carpel is swollen, forming the **ovary,** which contains one or more **ovules.** The female gametophyte develops within the ovule (Fig. 15.10). Cell division in the ovule produces seven haploid cells, three of these seven are arranged at one end of the ovule and three at the other end. One of these six cells becomes the functional egg. The seventh cell, situated in the center of the ovule, contains two haploid nuclei, referred to as **polar nuclei.** This entire structure, consisting of seven cells with eight haploid nuclei, is the female gametophyte, also known as the **embryo sac.**

When mature, the pollen sacs rupture, releasing the haploid pollen grains, which are blown about by the wind, transported on the feet or bodies of insects, or carried in the fur of animals. As pollen settles on a flower, the grains adhere to the sticky **stigma** at the tip of the carpel. The pollen grains begin germinating to form pollen tubes, one of which eventually grows downward into the carpel until it enters the ovary and penetrates an ovule. Two sperm cells are released into the embryo sac, and one sperm fertilizes the egg, forming the diploid zygote. The zygote, then, is the first cell of the sporophyte generation.

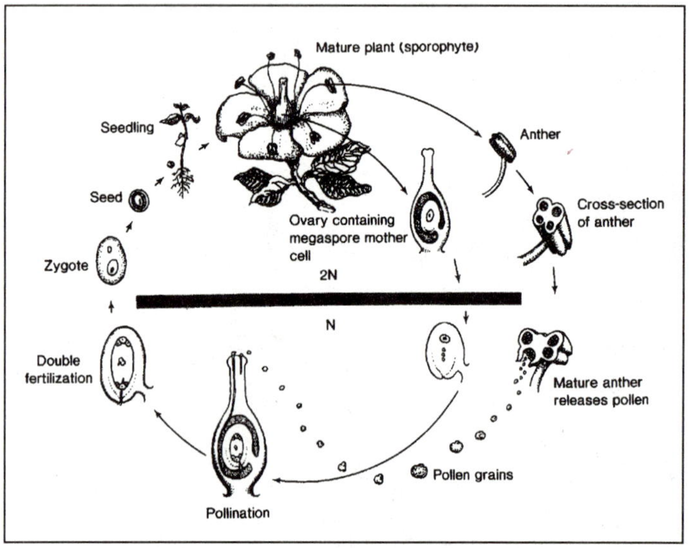

Figure 15.11 As in the ferns, the diploid sporophyte generation of angiosperms is conspicuous, while the haploid gametophyte generation is much reduced.

The other sperm cell unites with the two polar nuclei, which had previously fused, to form a triploid nucleus. This nucleus undergoes several mitotic divisions to produce a triploid tissue known as **endosperm.** During embryonic development, the surrounding endosperm serves as a source of nourishment. In some plants, the embryo depletes all the endosperm before the seed matures, in other plants, some endosperm persists for use by the embryo during germination. The kernels, or grains, of plants such as wheat and rice, for example, are composed predominantly of endosperm tissue. In consuming these and other grains, humans and many other animals are nourished by triploid endosperm tissue.

This reproductive process involving two sperm cells is called **double fertilization** and, along with the accompanying formation of triploid tissue, is a phenomenon unknown in any other group of living organisms. The ovule of angiosperms develops into a seed with the nutritive tissue and young sporophyte (developing embryo) enclosed within a seed coat. In angiosperms, the seed is entirely surrounded by the ovary. When the seed germinates and the resulting sporophyte develops flowers, a new life cycle can begin anew (Fig. 15.11).

Reproduction in Animals

INVERTEBRATE REPRODUCTION

About 95 percent of all known animal species are invertebrates. Thus it would be impossible to present even a small fraction of the varied reproductive methods of this diverse group of animals. That being the case, we will cite a couple of quite different reproductive methods and then comment generally on a few other methods that illustrate the adaptive measures employed by some invertebrates in reproducing their kind.

Cnidaria: Aurelia

Our first example involves a member of the Cnidaria, which include hydras, jellyfish, sea anemones, and corals. Some cnidarians, including the hydras, sea anemones, and corals, exist only in the cylindrical **polyp** form. Others, such as *Aurelia,* go through both polyp and **medusa,** or "jellyfish," stages in their life cycles (Fig. 15.12).

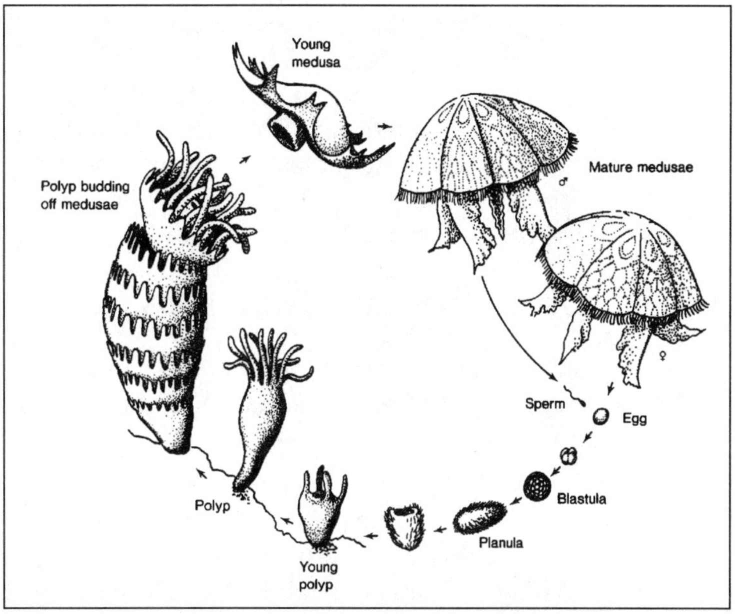

Figure 15.12 Life cycle of the cnidarian, *Aurelia.* The medusae (jellyfish) give rise to the polyp sexually, while the polyp buds asexually to produce the medusae.

The diploid medusae of *Aurelia* are of separate sexes and produce haploid eggs and sperm by meiosis. The medusa, then, represents the sexual stage. When released into the water, an egg and sperm unite to form the diploid zygote, which develops into a ciliated larva. After swimming about for a time, the larva settles down and grows into a polyp. The diploid polyp soon undergoes mitosis to produce new medusa buds asexually. The buds are released from the polyp and develop into sexually mature medusae. T polyp, therefore, is the asexual stage in the life cycle. The l cycles of *Aurelia* and other similar species differ from alternation of generations in plants, since both the polyp and medusa are diploid.

Rotifers

A group of unusual little animals, which look much like protozoans, are the *rotifers* ("wheel bearers") (Fig. 15.13). These microscopic, aquatic animals acquired their name from an anterior ring of motile cilia, which gives the appearance of a revolving wheel. One of the bizarre and fascinating features of rotifers is that each member of a given species has exactly the same number of body cells, usually a few hundred. Furthermore, beyond a certain point in the development of rotifers, mitotic cell division is no longer possible, and the animals cannot grow or replace missing or broken body parts.

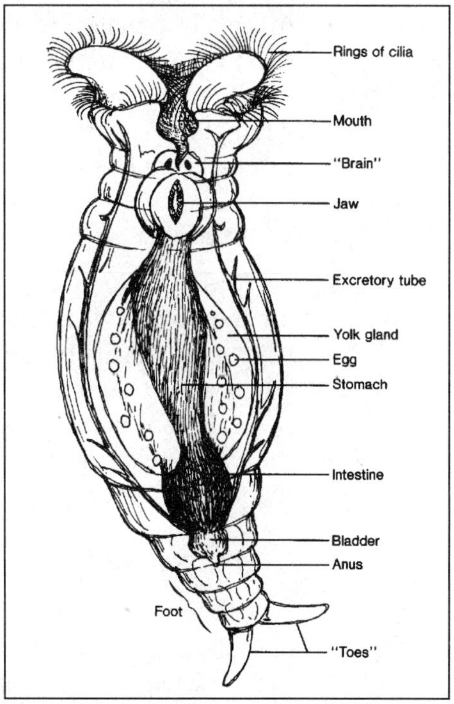

Figure 15.13 Although looking like a type of protozoan, the microscopic rotifer actually is a roundworm. The complexity of the organism is apparent from the drawing.

As in some species of worms, insects, and lizards, most species of rotifers have no males. Eggs laid by the females develop by **parthenogenesis** ("virgin origin"); that is, the eggs develop without being fertilized by sperm. In the few species in which they are found, males live only about 3 days, only long enough to reproduce, since they lack a mouth, an anus, and most other digestive structures and thus are unable to feed.

The explanation for the reproductive oddities of rotifers involves the production of two different types of eggs by the females. One type arises by meiosis and is haploid; the other type arises by mitosis and is diploid. If the haploid eggs are not fertilized, they develop into males; diploid eggs always develop into females. Both of these egg types develop parthenogenetically. If the haploid eggs are fertilized, however, they also develop into females. In species of cold-climate rotifers, the male is important because only the fertilized haploid eggs can survive the winter months. These eggs hatch into females in the spring. The short-lived males, however, are no longer around to observe the fruits of their labor.

Other Reproductive Methods

Invertebrates have evolved elaborate methods of reproducing, often engaging in exotic rituals, dances, and other displays to entice a mate.

Spiders

In some species of spiders, for example, the male repeatedly circles the female and at times raises up on his back legs, allowing the female to crawl beneath him. When mating is completed, the female proceeds to devour the entire body of the male.

Slipper Shells

Among the mollusks, certain snails known as slipper shells often exist stacked one on top of the other. The snails are arranged so that the penis of the upper individual can reach the female opening of the individual below. The young are always males, but there is subsequent degeneration of the reproductive system and a male then develops into a female or back into a male. Males remain as males as long as they are attached to a female; if separated from the female, they develop into females. Apparently, this sex reversal is one-way; females do not develop into males but remain females.

Snails

The reproductive process of another species of snails is equally unusual. These snails are **hermaphroditic;** that is, each organism possesses both male and female reproductive structures. Hermaphroditism is widespread in the animal world. The female opening in each snail contains tiny carbonate spicules, or darts, each honed to a very fine point. Before mating, a pair of snails situate themselves about an inch apart and commence firing darts at each other. After the battle, the participants, equally scarred, intertwine, copulate, and both lay eggs.

The Squids

Our final example of invertebrate reproduction involves the mating of another group of mollusks, the squids. In the male squid, sperm are stored in a sac that opens into the mantle cavity surrounding the head. Squids have two long tentacles and eight shorter arms. The right or left fourth arm in male squids is modified to form a sperm-transferring organ. In mating, the male grabs the female and they swim about face-to-face. Using his modified arm, the male removes sperm packets from the sac in the mantle cavity and inserts the packets into the mantle cavity of the female. The sperm packets erupt, liberating the sperm. The female then discharges eggs from the oviducts; as the eggs are released, they are strung together by a sticky gelatinous secretion and held by the arms of the female. After fertilization, the eggs, which may number from 1000 to 5000, are attached to an underwater rock or some other object. Soon after mating, the adult squids die.

Vertebrate Reproduction

The reproductive systems of vertebrates—agnathans, fish, amphibians, reptiles, birds, and mammals—are quite similar, although some variations do exist. The major differences are seen in the structure of various male copulatory organs and in the shape of the uteri among females. Some oddities among the vertebrates include species of lizards and snakes in which no males are found, and species of fish and frogs in which individuals can change from one sex to the other. In some vertebrates, the temperature at which the eggs are incubated determines the sex of the offspring. In some species, such as crocodiles, higher temperatures produce females and lower temperatures produce males; but in lizards and some other vertebrates, the reverse is true. Some lizards and fish even reproduce by parthenogenesis.

Egg-Laying Mammals

One group of vertebrates, the *monotremes,* includes the only egg-laying mammals. Although prominent millions of years ago, the only present-day monotremes are the duckbill platypus and the spiny anteater, both found mainly in Australia. After hatching from the egg, the young monotreme is nourished by milk secreted by the mammary glands of the female.

Pouched Mammals

Another group, the pouched mammals, or *marsupials,* includes kangaroos, koala bears, and wombats, all from of Australia, and the opossum, the only marsupial representative in the Americas. When born, the young marsupial is in a very immature stage of development and must crawl up the abdomen into the mother's pouch, where development is completed.

Placental Mammals

Our emphasis in this section will center on only one type of mammalian reproduction, specifically human reproduction. The reproductive structures of other mammals, however, are quite similar to those of humans. Humans are *placental* mammals, a group in which the young develop within the uterus for a relatively long period of time and often are well developed at birth. The **placenta,** an organ unique to mammals, is formed from the lining of the uterus and the outermost membrane surrounding the developing fetus.

Human Reproduction

The Male Reproductive System

The primary reproductive function of the human male is sperm production and the transfer of sperm to the female (Fig. 15.14). Sperm are produced in the male gonads, or **testes,** which are enclosed within a sac called the **scrotum** located at the base of the penis. About a month before birth, the testes of a developing male fetus normally descend from the lower abdominal cavity into the scrotum. Failure of the testes to de-

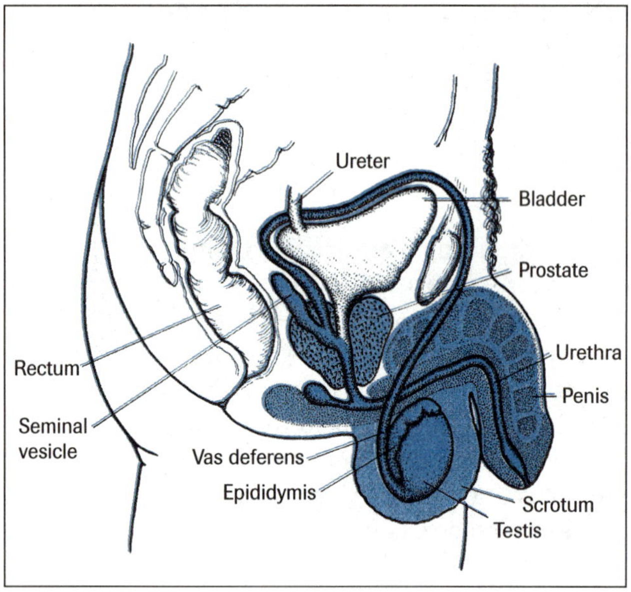

Figure 15.14 Reproductive structures of the human male.

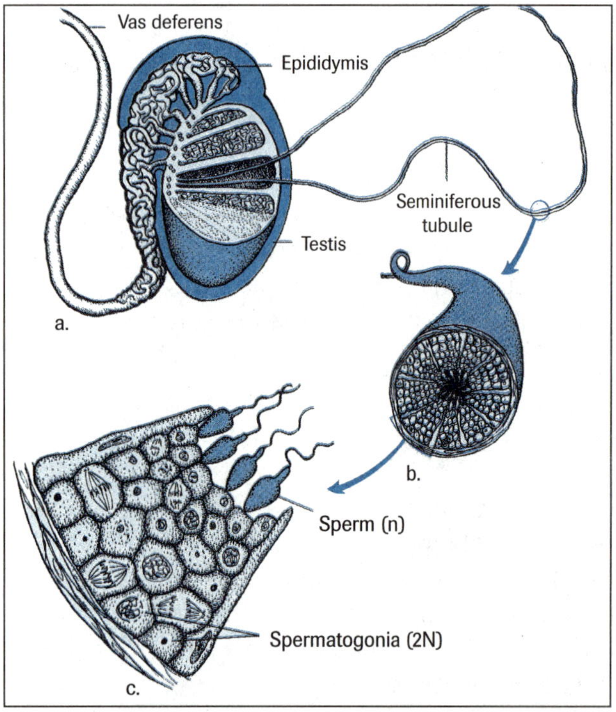

Figure 15.15 (a) Cutaway diagram of a testis, showing a single uncoiled seminiferous tubule. (b) Cross section of a seminiferous tubule, showing the reproductive tissue in which spermatogenesis occurs. (c) Enlarged section of reproductive tissue. Diploid spermatogonia undergo meiosis to produce the haploid sperm. (Sperm drawn much larger than normal in relation to other cells.)

scend results from inadequate secretion of the male sex hormone **testosterone** from the fetal testes. Body temperature is slightly higher than the temperature within the scrotum and, as a consequence, cells within the undescended testes degenerate and sperm cannot be produced.

Within each testis, about 800 elongated, tightly coiled **seminiferous tubules** produce haploid sperm by meiotic cell division (see Chapter 13) (Fig. 15.15). The process of sperm production is called **spermatogenesis,** which results in the production of up to 500 million sperm cells per day. Sperm arise from germ cells, or **spermatogonia,** that divide by mitosis to form **primary spermatocytes.** Some of the spermatogonia divide by mitosis to produce additional spermatogonia. It is the primary spermatocytes that undergo meiosis to produce sperm.

Mature human sperm cells contain 23 chromosomes; that is, 22 autosomes, or body chromosomes, and either an X or a Y sex chromosome. The seminiferous tubules within each testis also contain large nurse cells that nourish the sperm. Clusters of interstitial cells surround the tubules and secrete the male sex hormone testosterone.

From the seminiferous tubules, sperm move through tiny ducts into the **epididymis,** an enlarged duct in which sperm mature for several days. Leading from each epididymis is an elongated sperm duct, or **vas deferens,** which joins a sac called the **seminal vesicle.** The seminal vesicles secrete a fructose-rich solution that nourishes sperm. The end of the vas deferens and a short duct from a seminal vesicle join to form the **ejaculatory duct,** which empties into the urethra. The urethra extends from the urinary bladder to the end of the penis. In the male, then, both urine and sperm are transported in the urethra.

At the base of the bladder is the **prostate gland,** which releases a thin, milky secretion into the urethra. The secretion is alkaline, which, along with secretions of the seminal vesicles, helps neutralize the acid secretions of the urethra and the female reproductive tract. Two other small glands, one on either side of the urethra, are the **bulbourethral glands,** which secrete a lubricating mucus into the urethra. Sperm and secretions of all the reproductive glands form a fluid called **semen.**

A normal ejaculate usually consists of about 2–5 ml of semen, which contains 75 to 400 million sperm cells. A sperm cell count below 20 million per milliliter of semen may result in sterility. Each sperm cell consists of a head and a flagellum, which has a middle piece and a tail. The head, which contains the chromosomes, also contains enzymes necessary for penetration of the sperm cell into the female egg cell. The middle flagellar piece is rich in mitochondria, which produce the ATP that powers movement of the tail as the sperm cell travels through the female reproductive tract.

Regulation of Male Reproductive Hormones

It was pointed out in Chapter 11 that the hypothalamus secretes hormones that regulate the secretion of hormones from the pituitary gland. The anterior pituitary hormones that influence the function of the male testes are **follicle stimulating hormone (FSH)** and **interstitial cell-stimulating hormone (ICSH).** Secretion of these hormones is regulated by **gonadotropin-releasing hormone (GnRH)** from the hypothalamus.

Follicle-stimulating hormone is named for its function in the female; but, in the male, FSH stimulates germ cells in the seminiferous tubules to undergo spermatogenesis. In addition, FSH also stimulates the testes to release **inhibin,** a hormone that exhibits negative feedback by inhibiting FSH secretion. ICSH also acts on the testes but specifically stimulates interstitial cells in the testes to secrete testosterone. Testosterone stimulates the accessory sex organs and is responsible for development and maintenance of the male secondary sexual characteristics, such as pitch of the voice, beard growth, and muscle development.

Testosterone, along with inhibin, regulates both spermatogenesis and the correct level of testosterone in the blood through negative feedback. If the level of testosterone in the blood rises to levels that are above normal, neurons in the hypothalamus reduce the output of GnRH, and the anterior pituitary decreases its secretion of FSH and ICSH. Testosterone levels in the blood remain fairly constant in the male from puberty until about 40 years of age, after which the levels slowly decline.

Among the numerous functions of testosterone is its promotion of protein synthesis and muscle growth. Some weight lifters, body builders, and other athletes use a synthetic form of testosterone, an "anabolic steroid," to increase muscle mass. Continued use of large doses of testosterone produces side effects that are potentially dangerous. Chief among these effects are damage to the heart, liver, and kidneys, and degeneration of the testes. Through feedback inhibition of the hypothalamus and pituitary gland, high levels of testosterone in the blood can suppress the normal production of testosterone and GnRH. It has been reported that the suppression of GnRH secretion may be permanent. Women taking testosterone may develop irregular menstrual periods, experience hair loss, and alter their secondary sexual characteristics.

The Female Reproductive System

Reproductive functions of the human female include production of egg cells, reception of sperm from the male, preparation of the uterus for pregnancy, and giving birth (Fig. 15.16). Production of the egg cells occurs by meiotic cell division within the two **ovaries,** which are suspended in the pelvic cavity by ligaments (Fig. 15.17). Production of the egg cells, or **oocytes,** is called **oogenesis,** which specifically refers to the production of *secondary* oocytes, as explained shortly. Unlike the male, human females are born with all

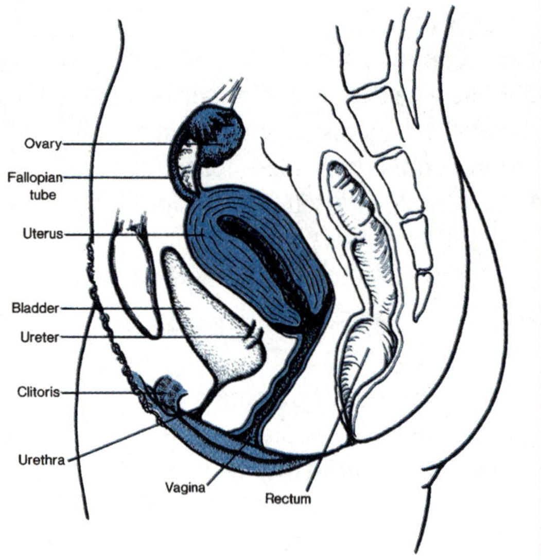

Figure 15.16 Reproductive structures of the human female.

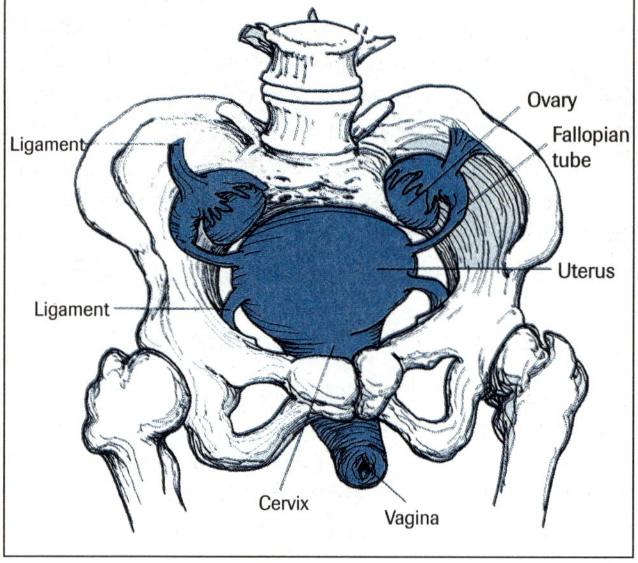

Figure 15.17 Location of the female reproductive structures in the pelvis. The ligaments anchor the structures to the surrounding bones.

Figure 15.18 Diagram of oögenesis as it occurs in female humans and other higher animals. In humans, however, meiosis II is not initiated until the sperm penetrates the secondary oöcyte.

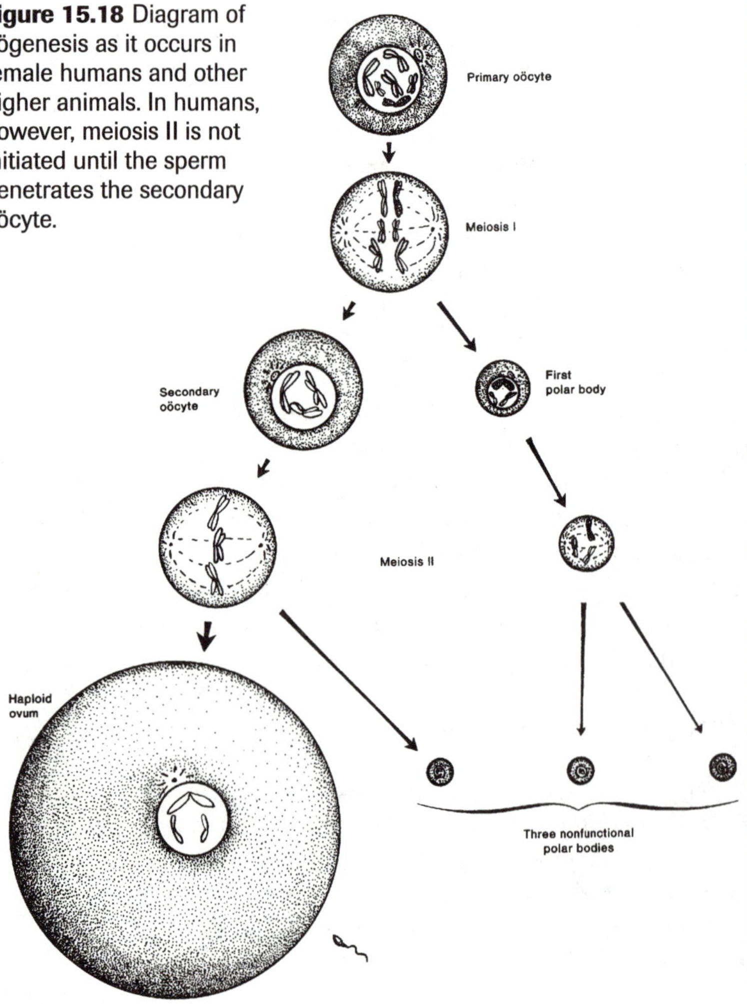

the reproductive cells they will ever possess; in other words, females do not produce new egg cells during their lifetime.

During prenatal development, the ovaries contain millions of **oogonia,** the diploid reproductive cells from which oocytes are produced (Fig. 15.18). Before birth, the oogonia have already begun meiotic division, but only to the stage of prophase I. At this stage, the reproductive cells are called **primary oocytes.** By puberty, at age 12 to 14, each ovary contains about 200,000 of these primary oocytes. Within the ovary, a primary oocyte is enclosed within a layer of cells, forming a **primary follicle.** With the onset of the menstrual cycle, hormonal changes stimulate increased development of the primary follicle, which acquires additional cell layers and becomes a **secondary follicle.** Continued enlargement and development of the secondary follicle results in a fully mature **Graafian follicle.**

As follicle growth proceeds, the primary oocyte enlarges as

it accumulates yolk from surrounding follicle cells. The primary oocyte then continues dividing to complete meiosis I; this produces a secondary oocyte and a tiny first **polar body.** Cell division during oogenesis is uneven, with most of the cytoplasm going to the oocytes and very little to the polar bodies. The secondary oocyte then begins the second meiotic division, which proceeds only to metaphase II and then stops. The first polar body may or may not divide, which is of little significance since all the polar bodies are nonfunctional and eventually degenerate.

Regulation of Female Reproductive Hormones

Oogenesis and development of the follicles and oocytes are controlled by fluctuating levels of various reproductive hormones (Fig. 15.19). Accompanying these activities in the ovaries are changes in the **uterus,** the muscular organ in which the human embryo and fetus develop. The *embryo* refers to the developing human from the second to the eighth week of development; from the ninth week of development until birth, the developing human is called *a fetus.* The entire series of events in the ovaries and uterus generally are referred to as the **menstrual cycle.** Typically, the cycle is about 28 days long, but may vary within a normal range of 21 to 35 days.

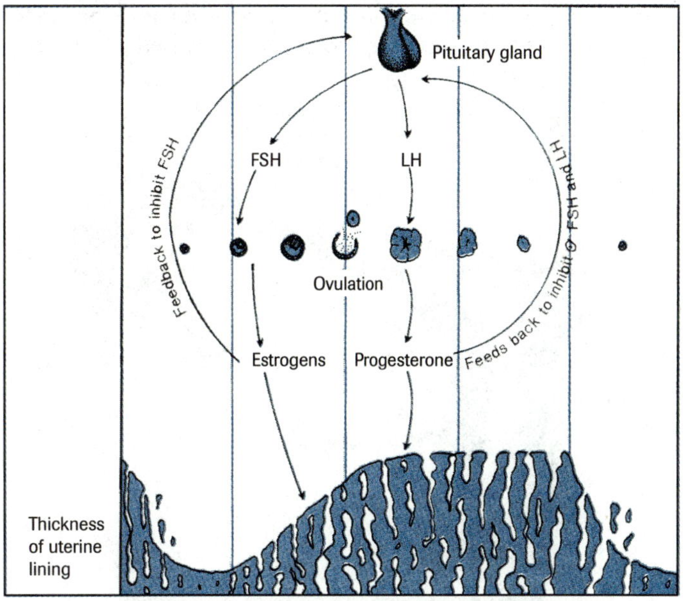

Figure 15.19 The interactions between the pituitary and sex hormones in regulating the human menstrual cycle (see text for details).

At puberty, the hypothalamus increases its output of GnRH, which stimulates the release of FSH and **luteinizing hormone (LH)** from the anterior pituitary gland. Luteinizing hormone is an alternative name for interstitial cell-stimulating hormone (ICSH) of the male. FSH, as its name attests, stimulates development of the primary follicles, although normally only 1 of perhaps 25 maturing follicles becomes the Graafian follicle during a single cycle. All the other follicles fail to develop and soon degenerate. As the primary follicle matures, it secretes **estrogen,** one of the female sex hormones. Estrogen functions in development and maintenance of the reproductive organs and the secondary sexual characteristics, such as breast development, distribution of body hair, and voice changes. Estrogen also inhibits GnRH secretion from the hypothalamus and FSH from the anterior pituitary gland, but triggers increased LH secretion.

Ovulation

On about day 14 of a 28-day cycle, there is a surge of high levels of LH released from the pituitary gland, and the follicle ruptures, releasing the oocyte (Fig. 15.20). LH is assisted in this action by a surge of FSH several hours later. LH also stimulates the primary oocyte to complete meiosis I to form the secondary oocyte. Release of the secondary oocyte from the follicle is called **ovulation.** With release of the oocyte, the follicle begins to reduce its output of estrogen.

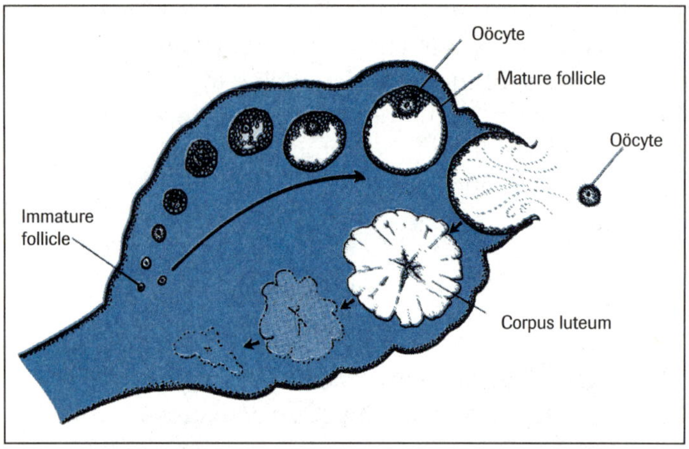

Figure 15.20 Diagram of follicle development, ovulation, and corpus luteum formation in the human ovary. As described in the text, each of these processes is regulated by hormonal activity.

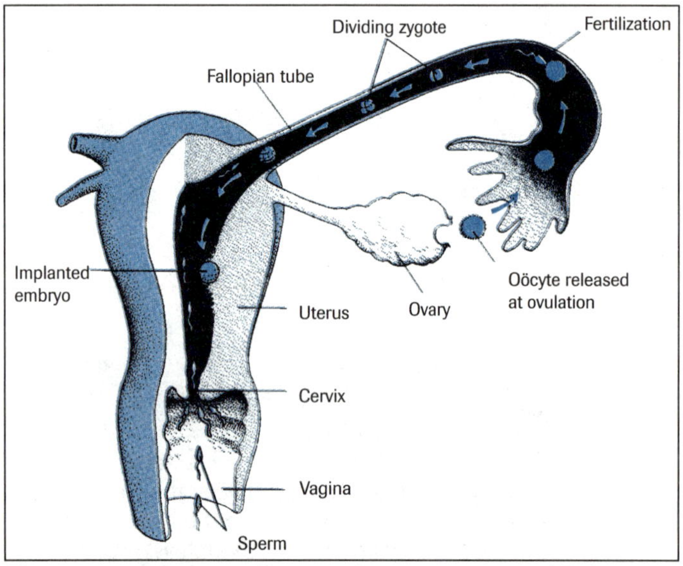

Figure 15.21 Cutaway diagram of vagina, uterus and Fallopian tube, showing the events leading to conception and implantation of the embryo. Following ovulation, the oöcyte is drawn into the upper region of the Fallopian tube where fertilization occurs. The diploid zygote undergoes cleavage as it moves down the Fallopian tube toward the uterus. After several days, the now multicellular embryo arrives at the uterus where it soon implants in the thick uterine lining.

Following ovulation, the empty Graafian follicle is transformed into the **corpus luteum** ("yellow body"), which begins to secrete the second female sex hormone **progesterone,** along with smaller amounts of estrogen. Progesterone and estrogen secretion exert a negative feedback effect on the hypothalamus, inhibiting GnRH and thus decreasing secretion of FSH and LH. The primary function of progesterone is to prepare the uterine lining for reception of the embryo if fertilization occurs.

Progesterone and, to a lesser extent, estrogen prepare the uterus by stimulating cell division and enlargement in the tissue lining the uterus. The uterine lining thickens and develops a rich network of blood capillaries that will bring nutrients and oxygen to the developing embryo. If pregnancy does not occur, the corpus luteum gradually shrivels and usually disappears after several months. As the corpus luteum disintegrates, it can no longer secrete progesterone and estrogen, which decline to very low levels in the blood. This removes the inhibitory effect of these hormones on the hypothalamus, and GnRH can initiate a new cycle.

Following ovulation, the secondary oocyte is normally drawn into a **uterine tube,** or *Fallopian tube,* one of which extends from each ovary to the uterus. If fertilization does not occur, the uterine lining begins to disintegrate because of the low levels of sex hormones in the blood. Cells sloughed off the uterine lining along with mucus and blood form the menstrual fluid, which is expelled from the uterus into the vaginal canal.

Fertilization

Sperm ejaculated into the vagina pass through the cervix and into the uterus and uterine tubes (Fig. 15.21). Propulsion of the sperm through the female tract results from movement of the sperm flagella and muscular contractions of the uterus and uterine tubes. Of the millions of sperm released into the vagina, only about 10,000 survive past the uterus, and barely a hundred or so actually reach the oocyte. Sperm can remain functional in the female reproductive tract for up to 72 hours, and the oocyte is functional for about 24 hours after ovulation. Thus, for conception, sexual intercourse must occur between 3 days before ovulation and one day after ovulation.

Fertilization, which normally occurs in the upper portion of a uterine tube, requires the presence of at least a hundred or so sperm cells. Although only one sperm cell will fertilize the secondary oocyte, the additional sperm are necessary to provide a sufficient amount of enzyme to break through the layer of follicle cells surrounding the oocyte. Once there is an opening in the cell layer, a number of sperm cells may slip through, but only one fuses with the oocyte membrane and penetrates to the interior of the oocyte.

Entrance of a sperm cell into the oocyte results in two major events. First, reactions within the oocyte occur to prevent penetration of the cell by any additional sperm—the presence of more than one sperm within the oocyte produces a nonfunctional zygote—and second, the oocyte completes meiosis II. Meiotic cell division produces the tiny second polar body and the haploid ovum. The nucleus of the ovum, called the **female pronucleus,** fuses with the **male pronucleus,** or head of the sperm cell, to form the diploid zygote. With formation of the zygote, the first cell of a new human being, fertilization is completed.

About a week after ovulation, the uterine lining, under the influence of estrogen and progesterone, is prepared to receive the embryo. As long as adequate levels of these hormones are maintained in the blood, the uterine lining remains intact. Maintenance of estrogen and progesterone secretion is the function of the corpus luteum, but as noted, this structure begins to disintegrate following ovulation. If fertilization occurs, however, the placenta begins to secrete the hormone **human chorionic gonadotropin (HCG),** which prevents the corpus luteum from disintegrating. Consequently, estrogen and progesterone levels rise in the blood and are maintained by HCG for about twelve weeks, although the level of HCG declines to a lower level throughout the remainder of pregnancy. This is of minimal importance because by the end of the twelfth week of pregnancy, the placenta has developed and begins to secrete its own estrogen and progesterone. It is hormonal secretion by the placenta, then, that maintains the uterine lining for the final six months of pregnancy. No longer needed, the corpus luteum degenerates.

We mentioned that fertilization normally occurs high in a uterine tube; as it develops, the embryo moves down the tube to the uterus, where it implants. Instances arise, however, in which the embryo becomes implanted elsewhere, resulting in an **ectopic pregnancy.** Ectopic means "out of place." As an example, the embryo may implant in the tissues of the ovary, in the abdominal cavity or, more commonly, in a uterine tube. A uterine tube is only about 0.6 cm (one-quarter inch) in diameter; as the embryo grows, the tube usually ruptures, causing massive hemorrhaging. The embryo and the uterine tube must then be removed surgically. Ectopic pregnancies rarely come to term because the embryo cannot receive adequate nutrients that normally would be supplied by the placenta.

Menopause

At the age of 40 to 50, menstrual cycles become increasingly irregular and eventually stop. Cessation of ovulation and the menstrual cycles is called **menopause.** Few follicles remain in the ovaries of menopausal women and those follicles that remain become altered as the ovaries age. By age 50, there are no primary follicles remaining in the ovaries; in instances of premature menopause, this can occur before the age of 40. The result is that less progesterone and estrogen are produced. As the levels of these sex hormones decline, there is a reduction in the size of the breasts and uterus, and an increase in the incidence of osteoporosis. Osteoporosis is the abnormal loss of bone density and thinning of the skeleton because of the lack of minerals, such as calcium and phosphorus. As a consequence, some bones become brittle and more susceptible to fractures. Physical and emotional changes, such as fatigue, irritability, depression, and "hot flashes" often precede menopause. Although not recommended in all cases, administering decreasing amounts of estrogen over a period of time often is an effective treatment for these symptoms.

Contraception and Abortion

Contraception refers to any method that will prevent pregnancy. The most effective contraceptive measures, aside from abstinence, include physical and chemical agents and surgical procedures. Reports on the effectiveness of some of these physical and chemical agents vary between their effectiveness when "used properly" and their "actual" effectiveness. Spermicidal agents, for example, are foams or creams that kill sperm cells. Inserted into the vagina before sexual intercourse, these agents are reportedly about 97 percent effective, but their actual effectiveness may be as low as 80 percent.

Methods of Contraception

Birth Control Pills

The various methods of birth control include birth control pills, diaphragms, intrauterine devices (IUDs), condoms, and surgical procedures.

Birth control pills are one of the most effective methods of contraception, being about 98 to 99 percent effective. Birth control pills usually contain estrogen and progesterone, which apparently act by preventing ovulation through inhibition of GnRH and LH release. For a small percentage of women, there is some risk involved in taking birth control pills, particularly those containing high levels of female sex hormones. The major risks are abnormal menstrual bleeding, cardiovascular problems, and uterine and breast cancer, particularly in women over 35 years of age.

Diaphragms

The diaphragm, a dome-shaped device that is inserted into the vagina, acts as a barrier to prevent sperm from entering the uterus. The diaphragm can be used along with a spermicide. The use of a diaphragm, with or without a spermicide, is less effective—only 80–85 percent effective—than using a condom.

The cervical cap and vaginal sponge are other devices that are also inserted into the vagina. Both are usually used in conjunction with a spermicide. Cervical caps are about 92 percent effective, while the effectiveness of vaginal sponges used alone varies from 78 to 97 percent.

Intrauterine Devices

Another physical agent used by women is the intrauterine device, or IUD. The mechanism by which an IUD works is not certain, but it prevents implantation of the embryo in the uterine lining. IUDs are approximately 97 to 99 percent effective. In recent years, some IUDs have come under fire for producing potentially dangerous side effects, such as perforation of the uterus; consequently, many IUDs are no longer sold or in use.

Vaginal rings and skin inserts containing sex hormones are recent female contraceptive devices, although their use is limited, primarily because of cost. The rings placed in the vagina and tiny rods inserted under the skin release the hormones at a steady rate.

Condoms

The only effective male contraceptive presently in widespread use is the condom, a membranous sheath that fits over the penis. The condom can be used in association with a spermicide, in which case its actual effectiveness is about 95–96 percent. Without a spermicide, the condom is about 90 percent effective. A male contraceptive being tested is an extract of cottonseed oil that reduces sperm count. The extract, named *Gossypol,* is absorbed through the skin; when treatment is discontinued, the normal sperm count is restored within a year or so.

Vasectomy and Tubal Ligation

Surgical procedures to prevent pregnancy include **vasectomy** in the male (Fig. 15.22A) and **tubal ligation** in the female (Fig. 15.22B). Vasectomy, which can be done with local anesthesia in a physician's office, involves cutting and tying the sperm ducts. This prevents sperm from entering the ejaculated fluid, most of which is secreted by the prostate and other accessory glands. After leaving the testes, sperm cells can travel no farther than an epididymis, where they eventually are destroyed by phagocytic cells. Vasectomy is virtually 100 percent effective and does not affect testosterone secretion or male sexual performance.

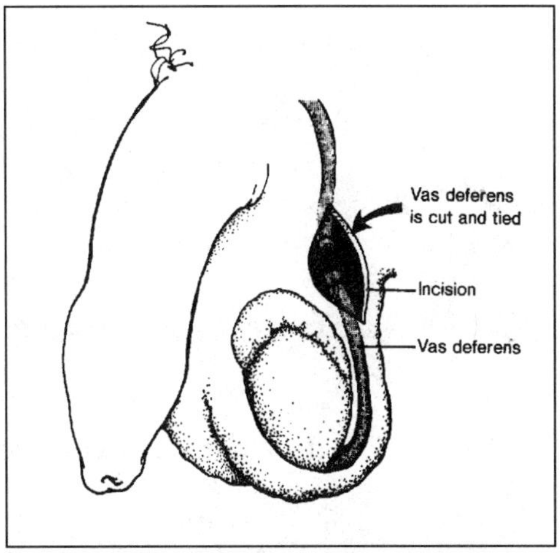

Figure 15.22A Vasectomy. A small incision is made in the scrotum, exposing the vas deferens, or sperm duct. The vas deferens is cut and either end is tied. This prevents sperm from reaching the urethra.

Tubal ligation involves cutting and tying or clamping the uterine tubes through an incision in the abdomen. This procedure, which is also 100 percent effective in preventing pregnancy, blocks the uterine tubes, preventing the union of sperm and oocyte. Tubal ligation does not alter the normal secretion of sex hormones and has no adverse effect on femininity.

Abortion

Some pregnancies, rather than being prevented, are terminated by abortion. The term is from the Latin abortus, "to miscarry." Because of physiological or developmental problems, some abortions, or miscarriages, occur spontaneously. Typically, these occur early in pregnancy. Therapeutic abortions are carried out when the life or health of the mother is at stake. At times, an embryo may implant very low in the uterus near the cervix, a condition known as **placenta previa.** As pregnancy proceeds, part of the growing placenta may extend across the opening in the cervix and, as the uterus stretches, this extension may tear

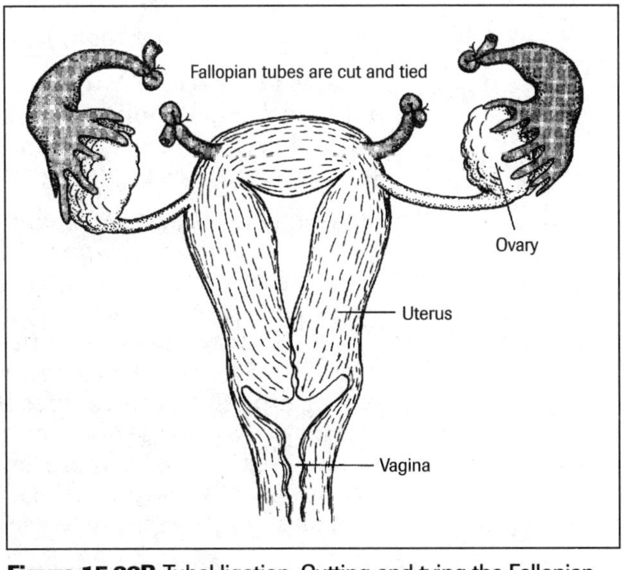

Figure 15.22B Tubal ligation. Cutting and tying the Fallopian tubes prevents sperm from reaching the oöcyte.

and cause hemorrhaging. This can threaten the life of the mother and often results in a miscarriage.

Abortions performed as a means of birth control are called induced abortions. One type of induced abortion involves dilation of the cervix and drawing the embryo from the uterus using a suction device. Another method is called dilation and curettage (D & C), in which the cervix is dilated and the lining of the uterus is scraped to remove the embryo and surrounding tissues. Both suction and D & C are performed within the first 12 weeks of pregnancy; after this time, the placenta is attached firmly to the wall of the uterus, and an attempted abortion often results in extensive hemorrhaging. A drug known as RU-486, which is available to women in France, is used to induce abortions during the first few weeks of pregnancy. RU-486 acts to block the effects of progesterone on the uterus, resulting in the sloughing of the uterine lining.

After the fifteenth week of pregnancy, abortions may be induced by injecting salt solutions into the amniotic cavity, the fluid-filled chamber that houses the developing fetus. Within an hour or two after injection, the fetus dies and is subsequently expelled along with the placenta.

The death rate for women undergoing legal abortions during the first 12 weeks of pregnancy is about 3 per 100,000 women; for abortions undergone after 12 weeks, the number rises to about 25 per 100,000 women.

Chapter Summary

1. Reproductive processes may be asexual or sexual; both types produce offspring that are environmentally adaptable. The evolutionary significance of asexual and sexual reproduction may involve cell mortality, genetic variability, and the expression of harmful genes; meiosis may offer more opportunities for DNA repair than does mitosis. Bacterial reproduction may occur by binary fission or conjugation.

   Key Terms: *asexual reproduction, p. 349*
   sexual reproduction, p. 349
   binary fission, p. 351
   conjugation, p. 351

   Concept Question: After reproducing sexually for many generations, a hypothetical population of organisms suddenly loses this capacity and converts to asexual reproduction. What are the basic reproductive advantages and consequences resulting from this conversion?

2. Algae may reproduce asexually, sexually, or by both methods. Sexual reproduction in *Chlamydomonas* involves fusion of haploid cells of opposite mating types to form a diploid zygote. Meiotic division of the zygote produces new haploid cells of opposite mating types. Asexual reproduction in *Chlamydomonas* involves mitotic division of a single cell to form new haploid cells. The alga *Spirogyra* reproduces sexually by conjugation. The resulting diploid zygote divides by meiosis to produce new haploid *Spirogyra* filaments. Asexual reproduction in *Spirogyra* is by fragmentation. Almost all fungi reproduce by spores, which, in some species, may be formed asexually or sexually. Various fungal species reproduce asexually by fragmentation, and yeasts reproduce asexually by budding.

   Key Terms: *haploid, p. 352*
   diploid, p. 352
   supplying cell p. 353
   receiving cell, p. 353
   spore, p. 353

Concept Question: What are the differences and similarities in sexual reproduction in *Chlamydomonas* and *Spirogyra*?

3. Nonvascular land plants such as mosses exhibit alternation of generations, which includes a haploid gametophyte generation and a diploid sporophyte generation. Reproduction in vascular plants, such as ferns and angiosperms, also involves alternation of generations.

Key Terms: *gametophyte, p. 355*
sporophyte, p. 355
antheridium, p. 355
archegonium, p. 355

Concept Question: How do you think the reproductive processes of mosses and ferns have influenced their ability to adapt to a given environment?

4. In angiosperms, the flower bears the male and female reproductive structures. At fertilization, a pollen grain, the male gametophyte, forms a pollen tube that extends down to the ovary at the base of the carpel and into an embryo sac, or female gametophyte. Two sperm cells are released and one sperm fertilizes the egg, forming the new diploid sporophyte; the other sperm unites with the two fused polar nuclei to form the triploid endosperm, which nourishes the developing embryo.

Key Terms: *stamen, p. 357*
carpel, p. 357
anther, p. 357
ovary, p. 358
endosperm, p. 358
double fertilization, p. 358

Concept Question: In terms of evolutionary adaptation, what do you think is the significance of double fertilization in angiosperms?

5. Cnidarians display polyp and/or free-swimming medusa forms in their life cycles. The medusa stage of the jellyfish *Aurelia* is diploid and produces male and female gametes by meiosis. Fusion of an egg and sperm produces a diploid larva that grows into a polyp. The polyp produces new medusae asexually by mitosis. Rotifers are aquatic roundworms, most species of which have no males. Unfertilized eggs develop by parthenogenesis. Female rotifers produce haploid eggs by meiosis and produce diploid eggs by mitosis. Unfertilized haploid eggs develop into males; fertilized haploid eggs develop into females. Diploid eggs always produce females.

Key Terms: *polyp, p. 359*
medusa, p. 359
parthenogenesis, p. 359
hermaphroditic, p. 360

Concept Question: Select an invertebrate group and suggest how its method of reproduction reflects an environmental adaptation.

6. In placental mammals, such as humans, the young develop for an extended period of time in the uterus. The placenta is an organ formed from the lining of the uterus and the outermost membrane surrounding the developing fetus. Human males produce haploid sperm in the seminiferous tubules of the testes. The ovaries of human females contain oogonia that produce haploid oocytes. Fertilization of the oocyte requires a number of sperm for penetrating the follicle cells surrounding the oocyte. Only one sperm penetrates the oocyte, which then completes meiosis to form the haploid ovum and second polar body. The male and female pronuclei fuse within the oocyte to form the diploid zygote.

Key Terms: *seminiferous tubules, p. 362*
spermatogonia, p. 362
follicle-stimulating hormone, p. 363
interstitial cell-stimulating hormone, p. 363
oogonia, p. 363
Graafian follicle, p. 364
luteinizing hormone, p. 365

Concept Question: The reproductive tissues of virtually all sexually reproducing organisms contain an enormous number of potential reproductive cells, such as sperm, oocytes, and spores. What do you think is the significance of producing such a large number of these cells? How does this phenomenon relate to the adaptability of the offspring of these organisms?

7. The uterine lining is maintained by high levels of progesterone and estrogen secreted by the corpus luteum. If fertilization does not occur, the corpus luteum degenerates. With disintegration of the corpus luteum, progesterone and estrogen levels decline, reducing their inhibitory effect of GnRH. An ovulated oocyte is usually drawn into a uterine tube, where fertilization normally occurs. If fertilization does not occur, the low levels of sex hormones in the blood permit the uterine lining to be sloughed off in the menstrual flow. If fertilization does occur, the placenta secretes human chorionic gonadotropin (HCG), which maintains the corpus luteum for about 12 weeks. Thereafter, the uterine lining is maintained by progesterone and estrogen secreted by the placenta.

Key Terms: *corpus luteum, p. 366*
human chorionic gonadotropin, p. 367

Concept Questions: Explain basically what initiates the human menstrual cycle and what terminates it on a regular basis. What happens to a secondary oocyte in humans if fertilization does not occur? What happens to it if fertilization does occur? How are each of these related to the structural changes of the uterus?

8. Female birth control agents include spermicides, birth control pills, the diaphragm, IUDs, cervical cap, vaginal sponge, vaginal rings, and skin inserts. Male contraceptives include condoms and an experimental extract of cottonseed oil. Surgical birth control procedures are vasectomy in the male and tubal ligation in the female. Vasectomy involves cutting and tying the sperm ducts, and tubal ligation involves cutting and tying the uterine tubes. Pregnancy may be terminated by abortion, which can be spontaneous, therapeutic, or induced.

Key Terms: *contraception, p. 367*
abortion, p. 369

Part Eight

Life Feature 6: Growth and Development

Chapter Sixteen

Life Emerging

Biological Rhythms

Throughout the kingdoms of life there are numerous instances of physiological and behavioral activities that occur on a fairly regular schedule (Fig. 16.1). Some nocturnal animals leave their lairs and forage for food only once in a 24-hour period. The skin of fiddler crabs is darkened during the day but becomes lighter at night. This cycle continues for a time even if the crabs are kept in total darkness. In some plants, sleep movements—the lowering of leaves in the evening and the raising of leaves in the morning—occur on a day-night schedule.

Humans, too, often function on a regular schedule. Many people awaken at the same hour every morning regardless of their desire not to do so. Human body temperature is usually at its lowest point in the early morning and rises in the late afternoon. Hormone levels, enzymatic activity, and mental abilities also fluctuate during a 24-hour day.

These and many other activities of organisms occur on a fairly regular 24-hour cycle and are called **circadian** (Latin "about a day") **rhythms.** Scientists believe that such cyclical activity is controlled by an internal *biological clock.* Believed to originate within an organism, such a clock is some type of timing mechanism that controls the biological rhythms of many activities.

In addition to daily cycles, there are also annual cycles. You may be familiar with the yearly hibernation periods of squirrels and bears and with the seasonal migration of birds. Males and females of many animal species pair and mate only during the spring. In addition, there are cycles that occur on a weekly or monthly basis, such as reproductive cycles in various female animals. All of these cycles, like circadian rhythms, are believed to be controlled by internal biological clocks.

Although cyclical activities are influenced mainly by the nervous system or hormones or both, the basic mechanism underlying biological rhythms remains unidentified. In some plants, however, there is strong evidence that the cell nucleus may be the site of regulation of these rhythms. In other plants, the mechanism for this regulation may be in the cytoplasm or associated with a cytoplasmic organelle, such as a chloroplast. There is also evidence that certain plants may have biological clocks in both the nucleus and cytoplasm, and that the nuclear regulation of biological rhythms can control the clock in the cytoplasm.

While it has been suggested that biological rhythms are actually nothing more than responses to external stimuli, such as gravity, cosmic radiation, light, darkness, and temperature, this does not appear to be the complete explanation. In the absence of these stimuli, biological rhythms continue for a certain period of time. Biological clocks, however, are undoubtedly responsive to external stimuli. In fact, the clocks may be set or reset by environmental conditions. Biological rhythms are so pervasive throughout the living world that it has been suggested that they be considered one of the basic life features.

Figure 16.1 Some biological rhythms. Much of life is rhythmic in its manifestations. The rhythms are usually adaptations to natural daily, lunar, tidal, or annual cycles. (a) Wildflowers in bloom (Alaskan irises, *Iris setosa,* in the Barren Islands, Alaska). (b) Migrating wildebeests in Africa. (c) A herd of grazing bison in Yellowstone National Park. Many animals feed only at certain times. (d) Two bull elk contending for mastery of a harem of does before mating. Like migratory behavior, reproductive behavior is noticeably cyclic in most species. (e) A Maine hardwood forest in autumn. One of nature's more impressive displays of color is seen for a few weeks each year before the leaves begin to fall. (f) Like this arctic fox, many animals undergo seasonal changes in coat color. As with the arctic fox, such changes usually serve to camouflage the animal.

This chapter emphasizes the life feature of growth and development by examining the fate of the fertilized egg cell in selected organisms. Under the influence of various chemicals or hormones, this single cell, usually a zygote, undergoes division to form a multicellular embryo, a process known as *embryogenesis.* This process occurs in all multicellular organisms, although in parthenogenesis the egg cell is not fertilized. The cells that arise from the zygote increase in number and later develop into different, specialized cell types that will eventually form the tissues of the organism. As development proceeds, these tissues give rise to the organs and other structures of the body. One group of organisms, the sponges, do not develop discrete tissues during embryogenesis; a few other groups, such as some cnidarians and flatworms, although possessing tissues, do not form organs. Nonetheless, early embryogenesis in all multicellular organisms follows a similar pattern of cell division and cell specialization.

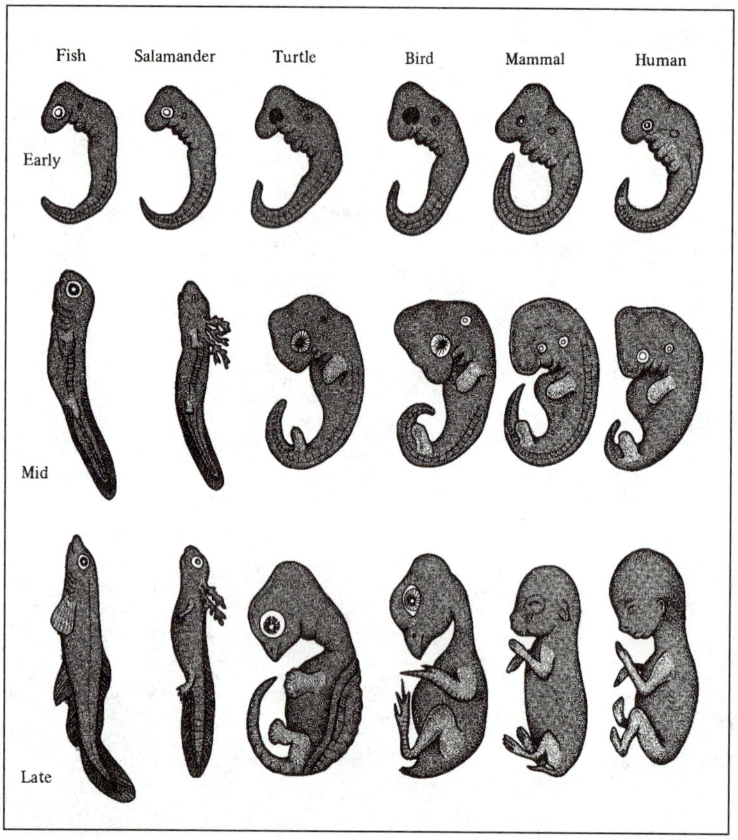

Figure 16.2 Embryonic stages of diverse vertebrates. Note similarity in the external features of early embryos of all vertebrates.

Embryology, the study of developing embryos, has proven invaluable to our understanding of the theory of biological evolution. Early in the nineteenth century, the German biologist Karl Ernst von Baer observed that the younger embryos of different species resemble each other more closely than do their older embryos. In the vertebrates, this means that the young embryos of agnathans, fish, amphibians, reptiles, birds, and mammals are quite similar. In fact, it is virtually impossible to distinguish these early embryos by observation (Fig. 16.2). Later in development, however, the embryos of each vertebrate group become increasingly different. These observations have led to the inference that similar embryonic development indicates a close evolutionary relationship.

To illustrate the similarities and differences in embryogenesis among different species, we will consider development in three major groups of organisms: angiosperms, echinoderms, and mammals. First however, we will comment on some of the factors that influence development in living organisms.

Plant Cycles

Biological rhythms are quite apparent in the growth and development of plants, particularly those rhythms that occur on a seasonal basis. Seeds germinate at a certain time of the year, leaves fall in autumn, and many flowers unfold in the spring. The major stimulus responsible for seasonal rhythms in plants is the relative lengths of daylight and darkness. The response to a change in the lengths of daylight and darkness is called **photoperiodism.**

Photoperiodism and Phytochrome

Part of the explanation for photoperiodism involves a plant pigment called **phytochrome** (meaning "plant color"), which promotes or inhibits development of various plant structures. Phytochrome occurs in two forms, P_r and P_{fr}, that can be interconverted by light (Fig. 16.3). P_r, the inactive form of phytochrome, absorbs red wavelengths of light, which convert it to P_{fr}, the active form in most plants. In turn, P_{fr} absorbs far-red wavelengths; that is, wavelengths of light outside the red spectrum and barely visible to the human eye. The far-red wavelengths convert P_{fr} back to the inactive form, P_r. Both red and far-red wavelengths are present in daylight, which results in both forms of phytochrome being present in a plant during the day. In darkness, however, P_{fr} reverts to P_r.

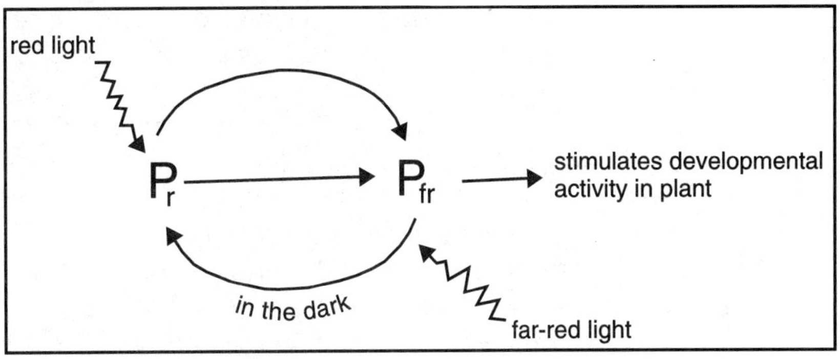

Figure 16.3 The inactive form of phytochrome (P_r) is converted to the active form (P_{fr}) by red light, far-red light or darkness causes P_{fr} to revert to P_r and developmental activity is shut down.

Thus, the developmental process is turned on when P_r absorbs red wavelengths and is converted to its active form (P_{fr}). In darkness, the developmental process is usually shut down as P_{fr} is converted to the inactive form (P_r). Light is an important factor in both promoting and inhibiting seed germination in different species of plants. Lettuce and tomato seeds, for example, will not germinate in darkness but require a brief exposure to red light before they will germinate. If the seeds are immediately exposed to far-red light after being exposed to red light, however, they will not germinate. Exposure to far-red light, then, cancels the effect of red light. But if the exposure to far-red light is followed by another few minutes of exposure to red light, germination proceeds.

Phytochrome is known to alter the permeability of membranes within the plant cell and to activate substances such as enzymes and hormones that regulate cell function. Briefly exposing wheat seeds to red light, for example, apparently alters the permeability of organelle membranes, allowing a growth-regulating hormone to pass through the membranes into the cytoplasm. In darkness, however, the hormone is confined within the organelles and does not pass through the membranes.

Photoperiodism and Flowering

Angiosperms, or flowering plants, are grouped into *long-day* plants, *short-day* plants, and *day-neutral* plants. Long-day plants, such as garden beets and spinach, usually flower in late spring or early summer when the period of daylight is *longer* than a certain minimum number of hours. Some plants, for example, will not flower if the length of daylight is less than 12 to 14 hours. Depending on the plant, the critical photoperiod for flowering may be more than or less than 12 hours. Flowering in short-day plants, such as soybeans and poinsettias, requires a length of daylight shorter than some critical period; these plants, then, usually flower in late summer or early fall when the days are shorter. Day-neutral plants are unaffected by photoperiodism and apparently will flower whenever they reach a certain stage of maturity. Various crops, such as corn, rice, tomatoes, and peas, are examples of day-neutral plants.

Interestingly, it turns out that the controlling stimulus for flowering is actually the length of uninterrupted *darkness* to which the plants are exposed rather than the length of daylight. Thus, it is more accurate to speak of short-night (long-day) and long-night (short-day) plants. The terms *long-day* and *short-day,* however, have persisted in the vocabulary of plant biology; nonetheless, for the remainder of this discussion we will use the terms short-night and long-night in an effort to minimize confusion.

A long-night plant that requires at least 8 hours of continuous darkness to flower is the cocklebur. But if this dark period is interrupted by only a few minutes of *red* light, the plant fails to flower. An interruption by far-red light, however, does not inhibit flowering. This is because far-red light serves to inactivate phytochrome; the phytochrome, however, is already in the inactive state due to the darkness. Thus there is no change in flowering. If the cocklebur is exposed to red light, the phytochrome is converted to its active form, which inhibits flowering in long-night plants. Thus, phytochrome, in association with a biological clock, in some manner determines whether or not flowering will occur. Apparently, phytochrome informs the clock of the presence of light or darkness, and the clock measures the length of darkness.

In addition, experiments have shown that night-length measurements occur specifically in mature plant leaves, and that some "flowering substance"—presumably a hormonelike chemical—is transmitted from the leaves to the bud where flowering occurs. This is evident by the fact that exposure of only one cocklebur leaf to long nights, while exposing the rest of the plant to short nights, results in flowering. Moreover, a single cocklebur leaf exposed to long nights can be removed and grafted onto a host plant exposed only to short nights, and the host plant will flower. It is apparent, then, that the flowering substance is transmitted from the single leaf to the entire host plant. Also, the leaves exposed to short nights apparently do not produce any substance that influences flowering. Despite years of research, the specific identity of a flowering substance in leaves is unknown.

Phytochrome is involved not only in seed germination and flowering, but in other developmental responses in plants as well, such as growth of seedlings, synthesis of chlorophyll, pigment formation, and leaf growth. Therefore, the effects of phytochrome, along with those of photoperiodism and biological clocks, play an important role in the overall development of many plants.

Embryogenesis

The recipe for life, including every enzyme, hormone, cell, tissue, and organ, is contained within microscopic coils of DNA. This recipe, if followed without serious error, gives rise to any and all life forms known. It is characteristic of life that, in whatever form, it always begins with a single cell. In multicellular organisms, this original unit of life, the zygote, undergoes a series of developmental events known as **embryogenesis.** Basically, embryogenesis involves the interaction of three processes: **growth, cellular differentiation, and morphogenesis** (Fig. 16.4). The information required to direct these processes is largely contained within the zygote itself. The human zygote, for example, receives most of this information in the form of 46 chromosomes, 23 from the father and 23 from the mother.

Growth

Growth is a complex phenomenon, but basically refers to an increase in the number and size of cells. In the more highly evolved organisms, the single-celled zygote grows by cell division to produce a multicellular **embryo.** The embryonic stage begins at fertilization of the egg cell and ends at a certain point in development that is characteristic of each species. An increase in cell size is seen, for example, in the elongation of root cells in growing plants.

Cellular Differentiation

Cellular differentiation refers to the formation of different types of cells as the embryo develops. For example, some embryonic cells are destined to become muscle cells or epidermal cells or leaf cells. In human beings, more than 200 different cell types develop, each type having a specific task to perform during the life of the organism.

Morphogenesis

Morphogenesis means "development of form" and pertains to the movement and arrangement of cells and tissues that eventually give shape to the body structures and organs. The location to which an embryonic cell moves and resides during development often determines its ultimate fate. Indeed, the arrangement of cells in relation to other types of cells is a critical factor in directing normal embryonic development.

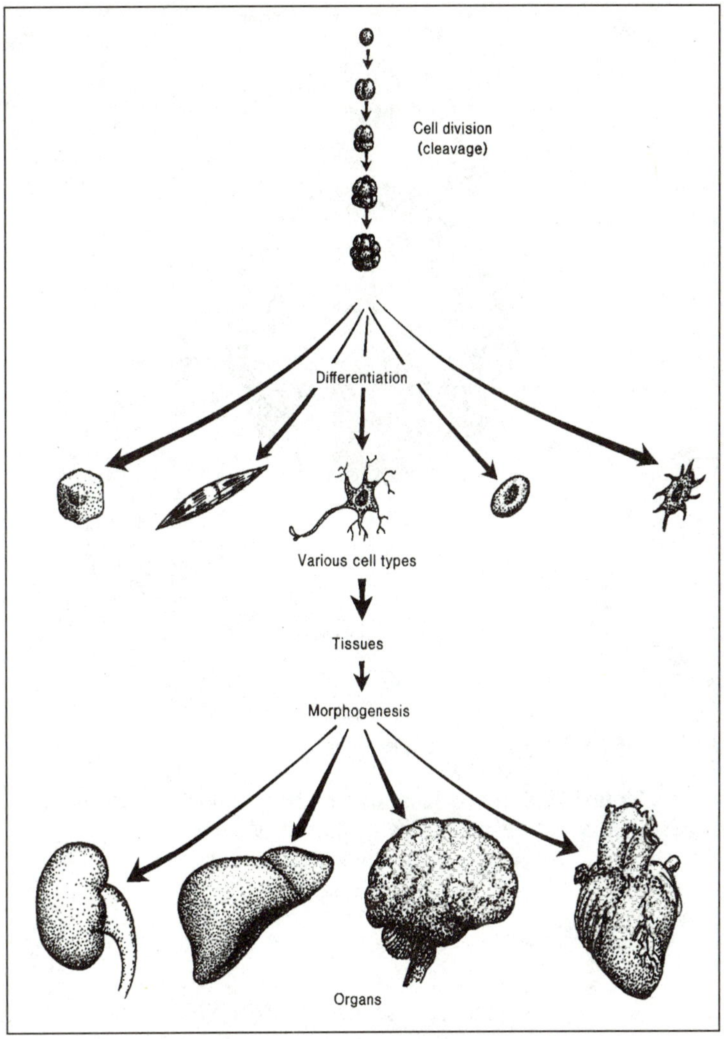

Figure 16.4 The process of development involves growth, cellular differentiation, and morphogenesis, illustrated here as it occurs in some vertebrates.

Control of Development

As you can see in Fig. 16.4, growth and development proceed from division of a single cell, which ultimately gives rise to a variety of different cells, tissues, and organs. This single cell, or zygote, therefore, is said to be **totipotent** ("wholly or totally able"). How, you might ask, is it possible for one cell to serve as the source of all the structural parts of an organism? What factors might influence embryonic cells arising from the zygote to "change," or differentiate? Although some of the factors are known, the exact mechanisms by which they operate to influence differentiation are not fully understood. Some clues, however, do exist.

Cleavage of the zygote occurs by mitosis and subsequent cell division, a process that results in each daughter cell having the same set of chromosomes as the parent

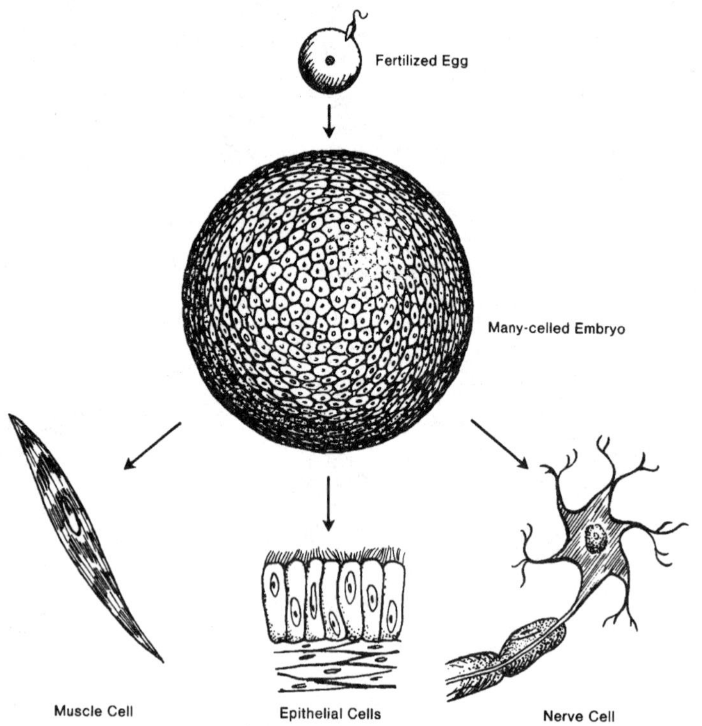

Figure 16.5 The process of development includes *cellular differentiation.* Many different types of cells may arise from a multicellular embryo.

cell. Throughout childhood, adolescence, and adulthood in humans, mitotic cell division continues to provide the cells necessary for body growth and tissue repair. Each of these cells is a descendant of the original single-celled zygote; it follows, then, that all body cells, that is, all the cells except gametes, of human beings are genetically identical. Further, all body cells of each species of multicellular organisms are genetically identical as well. It is obvious, however, that human skin cells do not look or function like muscle cells and red blood cells do not resemble nerve cells. Nonetheless, each of these cell types arises from the single cell of the zygote through a process of differentiation (Fig. 16.5).

Can it be demonstrated that these and other body cell types really *are* genetically identical? The answer to this question is important because it eliminates one of two possibilities underlying the basic mechanism of differentiation. First, if the answer is no, body cells *do not* contain identical genetic material, it would suggest that each different cell type is distinct because it contains different genetic material, and it would further suggest that some of the original genetic material in the zygote was lost from these various cells at some point during the developmental process. Second, if the answer is yes, all body cells *do* contain the same DNA, we are led to conclude that only part of the original entire genetic material is being expressed in each different cell type; that is, differentiation of cells depends on different genes being expressed in each particular cell type.

The Totipotency of Cells

Classic experiments designed to answer this important question were carried out in the late 1950s by Robert W. Briggs, Thomas J. King, and John G. Gurdon. The nucleus of an unfertilized frog egg was destroyed by ultraviolet radiation and replaced with a nucleus removed from an intestinal cell of a young tadpole (Fig. 16.6). It was found that the egg containing the transplanted nucleus could develop into a fully grown frog. Thus, the nucleus of the specialized, differentiated intestinal cell, like the nucleus of the original frog egg, did indeed contain all the genetic instructions required to form a new adult organism.

Still further evidence was provided by Frederick C. Steward using phloem cells removed from a carrot (Fig. 16.7). The phloem cells were placed in nutrient and growth substances and grew into a complete carrot plant. Thus, a phloem cell nucleus must contain all of the genetic information necessary to direct development of the entire plant. The phloem cells, then, are totipotent. Today, nuclear transplant and cell isolation experiments are performed routinely on a number of species.

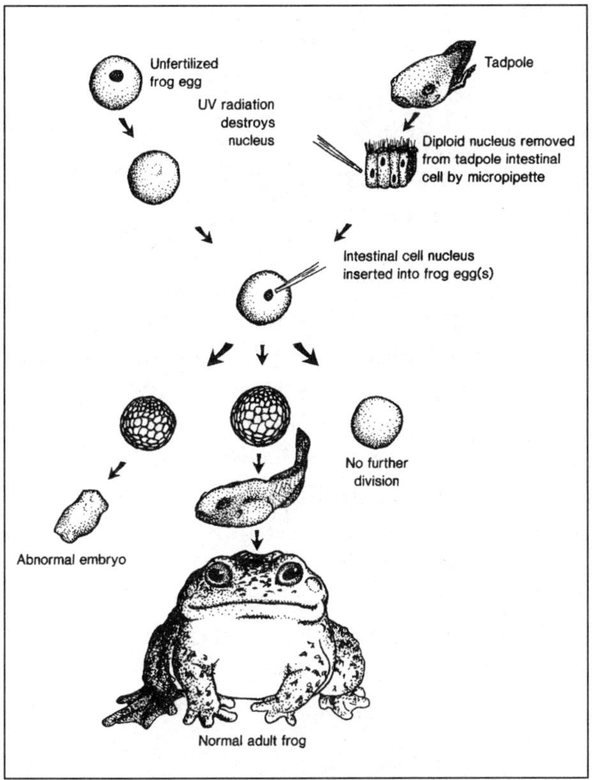

Figure 16.6 Nuclear transplant experiments have provided evidence that the nuclei of most types of body cells contain all the genetic information required to produce a complete adult organism. In the experiment illustrated here, the frog eggs into which the intestinal cell nuclei were inserted sometimes produced an abnormal embryo or failed to divide at all; at other times, however, one of the eggs developed into an adult frog.

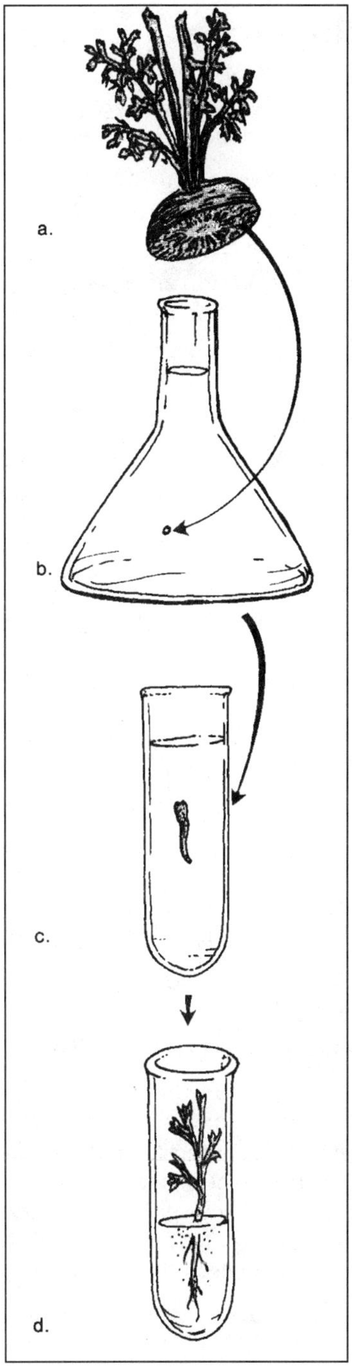

Figure 16.7 F. C. Steward experiment. (a) Individual phloem cells were isolated from a carrot and (b) placed in a liquid growth medium. (c) the cells developed roots and, when placed in a solid agar growth medium, (d) developed into mature carrot plants. This provided evidence that a single cell of an organism is totipotent.

From these and similar experimental findings, it is apparent that all body cells of a multicellular organism are indeed genetically identical. Therefore, as stated earlier, we are led to conclude that cellular differentiation is a consequence of the expression of different genes in different cell types. Differential gene expression, then, gives rise to the production of a particular set of proteins needed to carry out the specialized function of a given cell type. But what influences gene expression in a given embryonic cell? Why are some genes expressed and some not expressed? This, in somewhat different form, was the original question. On the basis of the foregoing information, we will try to shed some light on a few of these influences.

Regulation of Gene Expression

The immediate environment of a cell nucleus is the surrounding cytoplasm. There is evidence to suggest that the cytoplasm contains stimulatory and inhibitory substances that affect the activity of certain genes. One of the experiments that demonstrates cytoplasmic regulation of development involves the study of cleavage, or cell division, in frog embryos (Fig. 16.8). After fertilization, a small, half-moon-shaped mass of material called the *gray crescent* appears in the cytoplasm of the frog egg cell. As cleavage of the zygote begins, the first mitotic division normally divides the gray crescent equally between the two daughter cells. If these two cells are separated from each other experimentally, both cells will develop normally to form two individual tadpoles. By experimentally altering the first mitotic division, however, it is possible to produce one daughter cell that contains all of the gray crescent material, with the other daughter cell receiving none. When these daughter cells are separated, only the cell containing the gray crescent material develops into a normal tadpole; the other daughter cell will not develop normally but divides for a time to form an unorganized mass of undifferentiated cells.

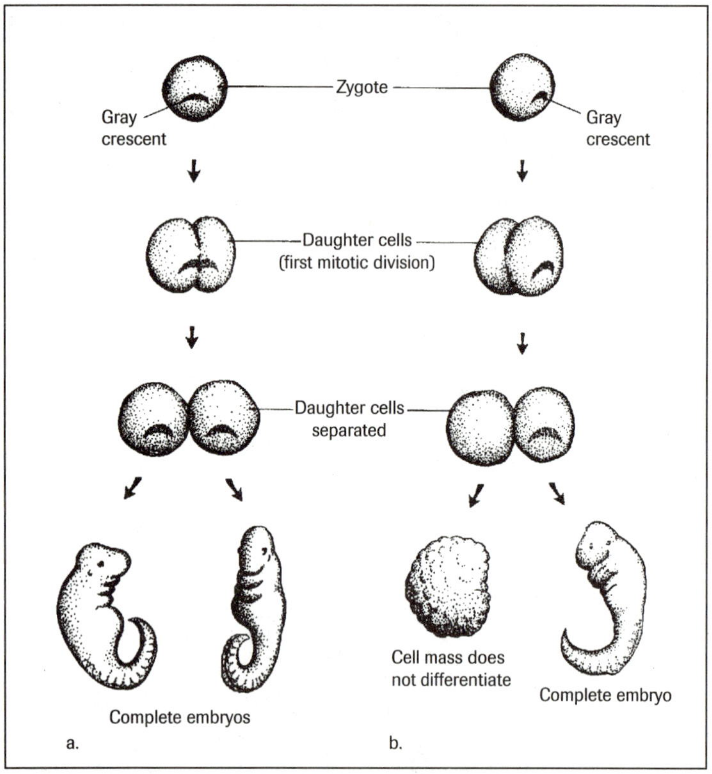

Figure 16.8 Influence of cytoplasmic constituents on development. If the first mitotic division divides the gray crescent material equally between (a) the two daughter cells, a complete frog embryo will develop from each cell. However, if (b) the first mitotic division is altered, only the cell receiving the gray crescent material will produce a complete embryo.

From this experiment, it is apparent that equal distribution of cytoplasmic materials at the first cell division is required for the production of two cells having the same developmental potential; that is, two cells that possess the same genetic information needed to form two normal embryos. Later in cleavage, however, a stage is reached at which embryonic cells separated from each other will not form complete embryos, but are destined to become only a specific part of the adult organism. In other words, as development proceeds, embryonic cells become increasingly differentiated or specialized. In effect, their fate is sealed.

Why should these cells that are produced at a later stage during development have such limited potential while the two daughter cells formed at the first cleavage both retain full developmental potential? As we just discovered, part of the answer involves distribution of the cytoplasm. Although the first cleavage of the frog zygote divides the cytoplasm equally between the two daughter cells, later cleavages divide the cytoplasm unequally in each of the various cells. Accordingly, the nucleus of each of these dividing cells is exposed to larger or smaller amounts of cytoplasm and, hence, to different cytoplasmic materials, such as enzymes or hormones. In some manner, the particular cytoplasmic materials affect gene expression, which controls cellular differentiation and thereby directs the future course of embryonic development.

Transcription Regulation

Differentiation ultimately depends on cellular changes at the molecular level; that is, at the level of the DNA molecules. Cellular differentiation comes about, then, by activation or repression of certain genes that control the synthesis of specific proteins in a given cell or group of cells. Obviously, if every body cell synthesized the same proteins, all the cells would be identical. This would mean that the DNA in each cell would be transcribing the same messenger ribonucleic acid (mRNA), which, in turn, would direct the synthesis of the same proteins, and ultimately yield identical cells performing identical tasks (see Chapter 13). Thus, if cells are to differentiate, only specific segments of DNA in a given cell can be active in protein synthesis. Other DNA segments would be active in a different cell type. It would be expected, then, that each kind of differentiating cell would transcribe different sequences of mRNA.

Chromosomes of various organisms can be stained to show mRNA being transcribed. Those parts of the chromosome that are not actively transcribing mRNA do not exhibit the same staining pattern. The location of segments of active mRNA transcription on the chromosomes differs in different types of tissues. The result is the synthesis of proteins that are specific for each type of tissue. Therefore, even though differentiating cells have a complete complement identical DNA, at given times they produce different kinds of mRNA.

Induction

Not only do the nucleus and cytoplasm influence the course of embryonic development, but entire tissues also play a role. The type of tissue and its position or location in the developing embryo can influence differentiation of other tissues in the embryo. The phenomenon of one tissue influencing development of another is known as **induction,** a discovery made by Hans Spemann, who received the 1935 Nobel Prize for his work.

Briefly, Spemann's experiments demonstrated the inductive effect of mesoderm on ectoderm, two of the embryonic tissues that, along with endoderm, form the three primary germ layers. Mesoderm, by virtue of its position next to ectoderm in an amphibian embryo, induces ectoderm to form nerve tissue. Ectoderm that is isolated away from the influence of mesoderm fails to develop, forming only a blob of undifferentiated cells.

As is almost always the case in science, an important discovery opens the floodgates to many new questions. The first and most obvious question related to the inducing effect of mesoderm is: How is this induction carried out? In terms of the scientific method, this question is legitimate because it is testable and relates to an objective observation. To say that induction results from physical contact between tissues does not offer much information unless we can explain why this close proximity is important. As a hypothesis, we might even ask if the tissues must be in physical contact at all. At this point, it is necessary to provide experimental evidence that supports or refutes the hypothesis.

In an attempt to obtain this evidence, researchers separated the ectoderm and mesoderm layers by a slight amount and inserted a porous filter between them, preventing the tissues from touching each other. They discovered that induction still took place. If a nonporous material is placed between the tissues, however, induction is inhibited. Thus, the experimental evidence indicated that some type of communication between ectoderm and mesoderm was necessary for induction to occur. Apparently, then, some chemical substance diffuses from one tissue to the other to influence the induction process. More specifically, chemical substances probably influence induction by activating or repressing certain genes which ultimately results in the production of certain proteins and not others. It should be noted that some chemicals may inhibit tissue development, while others stimulate the process. Inhibition of development is important; once a certain tissue or organ fully develops in the body, it would be pointless and even harmful to continue development of that same cell type. We might wind up with a chest full of hearts or a couple of livers.

Although it is evident that chemical inducers are active in development, the specific identity of any particular inducer is unknown. Suggestions have included strands of RNA, various proteins, hormones, and inorganic ions, but, for now, the substance or substances remain a mystery.

We have seen that embryonic development is controlled in part by the interactions of the nucleus, the cytoplasm, and even entire tissues. Through the particular distribution of cytoplasmic materials and the process of induction, the nuclei of embryonic cells are exposed to varied environments and substances, which are thought to affect expression of certain genes. The result is that cells differentiate or become specialized. Each of these cells, however, retains the same genetic makeup as the original zygote from which it arose.

This is by no means the entire story of control of embryonic development. There are, for example, physical environmental factors, such as light, temperature, and gravity that affect embryonic cells. An increase in temperature, for instance, speeds up the metabolic rate of cells. If only some of the embryonic cells are so affected, their rate of development could be thrown out of phase with that of the other cells. The result might be cessation of development of the individual or abnormalities in the mature organism.

Figure 16.9 The effect of temperature on coloration in the Siamese cat. The dark fur results from the presence of melanin, a pigment produced at lower body termperatures; lighter fur results when higher body temperatures reduce the production of melanin. The enzyme involved in the production of melanin is active only at lower body temperatures.

The physical factors that affect development, however, are not necessarily harmful. A case in point is the effect of temperature on coloration in the Siamese cat. These animals are albinos with blue eyes, but they have dark areas on their tails, faces, feet, and ears (Fig. 16.9). These dark areas contain more of the dark-brown pigment melanin than does the remainder of the body. The production of melanin is controlled by a heat-sensitive enzyme that is formed during development. This particular enzyme is less active at higher temperatures, a condition that reduces the production of melanin. Most of the cat's coat is warm enough to inactivate the enzyme, which results in a lighter colored fur. On the body extremities, however, the temperature is lower than the temperature over the rest of the body. At this lower temperature, the enzyme actively produces melanin, resulting in the characteristic dark patches on the fur.

While there is a great deal known about organismal development and its control, there is much that is not known. New research techniques will undoubtedly be forthcoming as scientists probe the intricacies of gene structure and function. And if answers continue to be elusive, we may well find that we simply were not asking the right questions.

In the sections that follow, we will examine growth and development as it occurs in echinoderms and in two of the most highly evolved groups of organisms: angiosperms and mammals. In that embryonic development is similar among the placental mammals, we will emphasize the process as it occurs in human beings.

Embryogenesis in Angiosperms

In the preceding chapter, we left the fertilized angiosperm egg in the embryo sac, along with the typically triploid endosperm nucleus. The endosperm nucleus divides by mitosis to form the nutritive endosperm, and the zygote undergoes mitotic divisions to form the plant embryo.

The entire female gametophyte is surrounded by a seed coat derived from outer tissues of the ovule. The embryo, endosperm, and protective seed coat form the **seed.** As the seed develops, the surrounding ovary wall thickens and forms the fruit with the seed inside.

The first mitotic division of the zygote establishes the *polarity* of the embryo; that is, an upper part, which is the shoot or stem, and a lower part, which is the root. Subsequent divisions produce a mass of cells that soon undergoes differentiation to form the various embryonic tissues. These tissues will give rise to adult tissues such as the epidermis, xylem, and phloem. This is followed by morphogenesis, in which the structures of the plant begin to take shape.

Shoot and Root Development

The embryonic seed leaves begin to develop on the shoot, and elongate as growth of the embryo proceeds. Seed leaves store or absorb food used by the embryo as it develops. Two of the most important tissues of the developing embryo are the apical meristems of the shoot and root. An **apical meristem** is a growing area at the tip of the shoot and root. Virtually all of the cells of the young and adult plant are derived from the apical meristems.

In addition, woody plants have **lateral meristems,** which are strips of cells extending along the lengths of shoots and roots. Division of these cells produces lateral growth, increasing the thickness of the shoots and roots. Once these fundamental tissues and structures are established, and if supplied with adequate water and oxygen at the right temperature, the seed is prepared to germinate.

The first structure to emerge during seed germination is the **radicle,** or embryonic root, which will eventually develop into a highly-branched system as a result of cell division of the root apical meristem (Fig. 16.10).

The shoot, along with its seed leaves, then emerges from the soil into the air. Their stored food consumed by the developing embryo, the seed leaves begin to wither and eventually fall off the young seedling. In some species, however, the seed leaves remain in the soil; in other species, they are photosynthetic. At this stage, the young angiosperm is an independent organism capable of photosynthesizing and existing on its own.

THE YOUNG SEEDLING

Having emerged from the soil, the apical meristem of the shoot begins to produce the leaves and branching stems of the plant. These develop from new apical meristems produced along the length of the shoot. Finally, one or more of the meristems differentiates to form a flower, the reproductive and distinguishing structure of the angiosperms. Thus, a new mature angiosperm is established and a new life cycle may begin once more. With its unique adaptations for solving the problems of worldwide survival, the young seedling stands with all other angiosperms as one of the most biologically successful of all living plants.

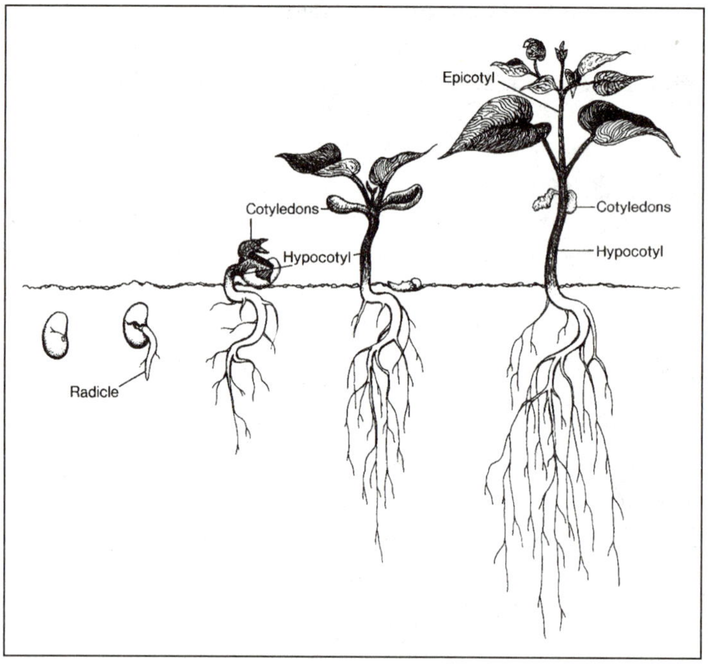

Figure 16.10 Bean germination. The first embryonic structure to break through the seed coat is the radicle, or primary root. Later, the hypocotyl pushes through the soil surface, pulling the cotyledons (seed leaves) along with it. As the epicotyl lengthens and the large foliage leaves form, the cotyledons progressively wither until they are lost. The hypocotyl and epicotyl are the regions below and above the seed leaves.

Embryogenesis in Echinoderms

The echinoderms include the starfish, brittle stars, sand dollars, sea urchins, sea cucumbers, sea lilies, and feather stars, all of which are marine animals. Most echinoderms possess both male and female reproductive structures but copulation does not occur. Eggs are usually released and fertilized in the surrounding sea water. Division of the zygote and early embryonic development follow a similar pattern in all echinoderms (Fig. 16.11). We will describe echinoderm embryogenesis using a species of starfish as a representative.

After fertilization in open waters, the resulting starfish zygote undergoes a series of cell divisions called **cleavage.** Cell division continues until, by the second day, a spherical, hollow ball of cells called the **blastula** is formed. The blastula possesses cilia and is capable of swimming through the water. The wall of the blastula begins to form an inpocketing, or invagination, that produces a two-layered sac called the **gastrula** (*Gk. gaster* stomach). This inpocketing is analogous to pushing in on a tennis ball. The invagination in the wall of the blastula continues to deepen and will eventually form the primitive gut of the embryo. Later, the primitive gut will become the digestive tract, and the opening into the gut will form the anus. The primitive gut deepens until it touches the inner wall on the opposite side of the blastula; this junction will give rise to the mouth of the embryo.

With formation of the gastrula, the starfish embryo possesses an outer layer of cells, the **ectoderm,** and an inner layer of cells, the **endoderm,** which line the primitive gut. Cells derived from the primitive gut later form the **mesoderm,** which is laid down between the ectoderm and endoderm. These three cells layers of the gastrula are the primary *germ layers,* which will eventually give rise to all the tissues and organs of the adult starfish. Development of a three-layered embryo is characteristic of early embryogenesis in most animals.

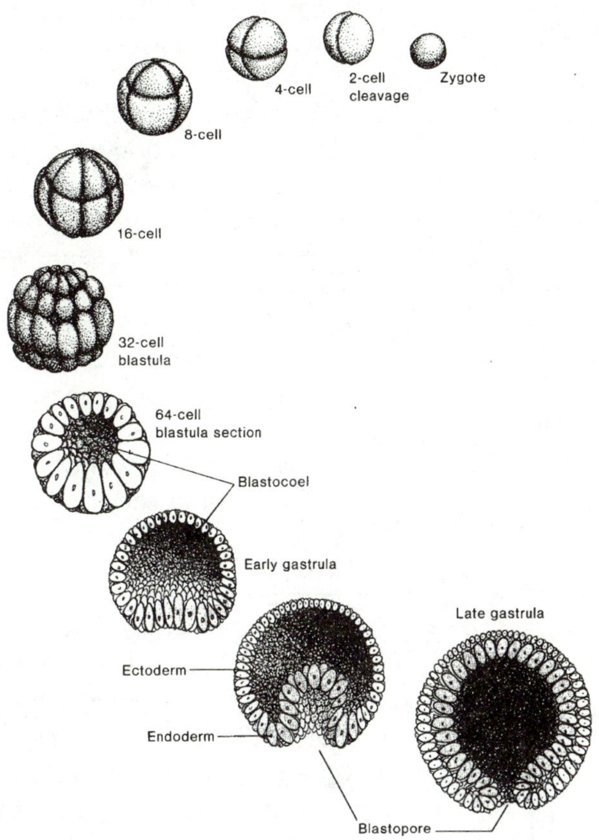

Figure 16.11 Cleavage and gastrula formation as it occurs in many invertebrates. Cleavage of the diploid zygote proceeds by mitotic cell division. As the embryo grows, there is increasing cell specialization, beginning with the formation of endoderm in the early blastula stage.

Following the gastrula stage, the embryo first develops into a ciliated larva and then into a larva with armlike projections on its body. The larva then settles on the ocean bottom where it develops into a young starfish barely 1 mm in diameter.

Human Embryogenesis

In the human zygote, the genetic information to produce a fully formed adult is all contained within the DNA of 46 chromosomes. This profound statement was not, and could not have been, expressed until well into the twentieth century. It was only realized in the nineteenth century that both parents contribute to the inheritance of their offspring. Following the first observation of sperm cells by the Dutch lens maker Anton van Leeuwenhoek in the seventeenth century, hypotheses related to human development began to flourish.

One proposal—known as the *preformation theory*—suggested that the sperm and eggs contained preformed miniature body parts that simply enlarged during development. Extending this concept another step, it was proposed that sex cells contained entire miniature embryos and the embryo's sex cells contained miniature embryos, and so on like boxes within boxes (Fig. 16.12).

Some microscopists even believed they *saw* a preformed human embryo within a sperm cell, others thought they saw the embryo within an egg cell. The former group, known as *spermists,* concluded that the female merely serves to incubate the embryo and that the male determines all inheritable traits. The latter group, the *ovists,* proposed that the mother determines all inherited characteristics and that sperm only stimulate growth of the egg cell.

In the latter half of the eighteenth century, observations of developing chick embryos revealed that embryonic tissues and organs arise gradually from the zygote. This progressive development of form is called **epigenesis** and eventually replaced preformation as the explanation for embryogenesis. The fundamental concepts of modern embryology are based on the scientific evidence provided by the original theory of epigenesis.

In comparing chick embryonic development with that of humans, we find that development in both species proceeds in much the same way. This, as stated earlier, indicates a close evolutionary relationship. The chick, however, requires only 3 weeks to complete development, whereas human development requires 9 months. Development of the human embryo from conception until birth is separated for convenience into three trimesters, each three months in length. In the following sections, we will consider the major developmental events that occur in each of these trimesters.

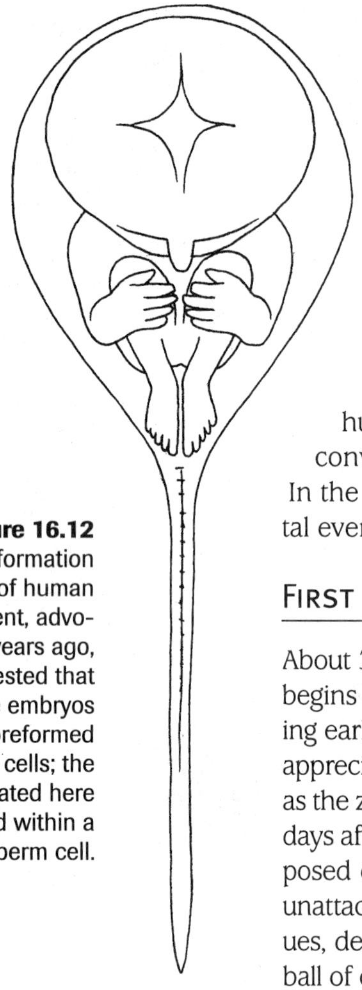

Figure 16.12 The preformation theory of human development, advocated 200 years ago, suggested that miniature embryos existed preformed within sex cells; the embryo illustrated here is depicted within a sperm cell.

First Trimester

About 30 hours after fertilization in a uterine tube, the human zygote begins a series of rapid cell divisions called *cleavage* (Fig. 16.13). During early cleavage, the zygote divides into several cells, but there is no appreciable increase in the total mass of cytoplasm. Divisions continue as the zygote moves down the uterine tube toward the uterus. About 3 days after fertilization, the cell mass enters the uterus and is now composed of a solid ball of cells called the **morula.** The morula remains unattached in the uterus for two or three days and, as cleavage continues, develops into a fluid-filled, hollow ball of cells called a **blastocyst.**

The outer layer of the blastocyst wall will form the **chorion,** one of the four extraembryonic ("outside the embryo") membranes and a component of the placenta. Within the hollow blastocyst, an aggregation of cells forms at a location along the inner wall. This group of cells is the **inner cell mass,** which will give rise to the human embryo. Approximately 5 to 7 days after fertilization, the blastocyst becomes attached to the uterine wall and begins to burrow into the uterine lining, a process called *implantation.* Even before implantation, the placenta secretes the hormone human chorionic gonadotropin (HCG), which maintains the corpus luteum. The corpus luteum, you will recall, is the structure formed from the ovarian follicle after ovulation. Progesterone and estrogen secreted by the corpus luteum maintain the uterine lining, preventing menstruation and loss of the embryo.

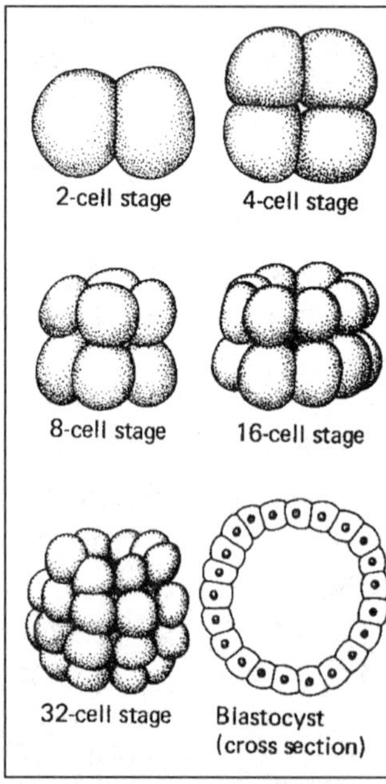

Figure 16.13 Cleavage of the zygote consists of a series of mitotic divisions that give rise to the hollow blastocyst. This generalized sequence is characteristic of many animals.

Germ Layer Formation

By the time the blastocyst begins to implant, a small space forms between the inner cell mass and the blastocyst wall. This space increases in size to form the fluid-filled **amniotic cavity.** The inner cell mass then flattens out and differentiates to form the **embryonic disk,** which is composed of two cell layers. The upper layer is ectoderm and the lower layer is endoderm. A short time later, a third layer, the mesoderm, is laid down between the ectoderm and endoderm. These three layers are the primary germ layers and will give rise to all the tissues of the adult human (Fig. 16.14).

Structures ultimately derived from ectoderm include the brain and spinal cord, epidermis of the skin, teeth, and the lens and cornea of the eye. Endoderm gives rise to the lining of various ducts and tubules, the anterior pituitary gland, the thyroid gland, and the thymus gland. Mesodermal structures include muscles, bones, the circulatory system, blood, and dermis of the skin. With formation of the germ layers, the first two-week stage of development is completed.

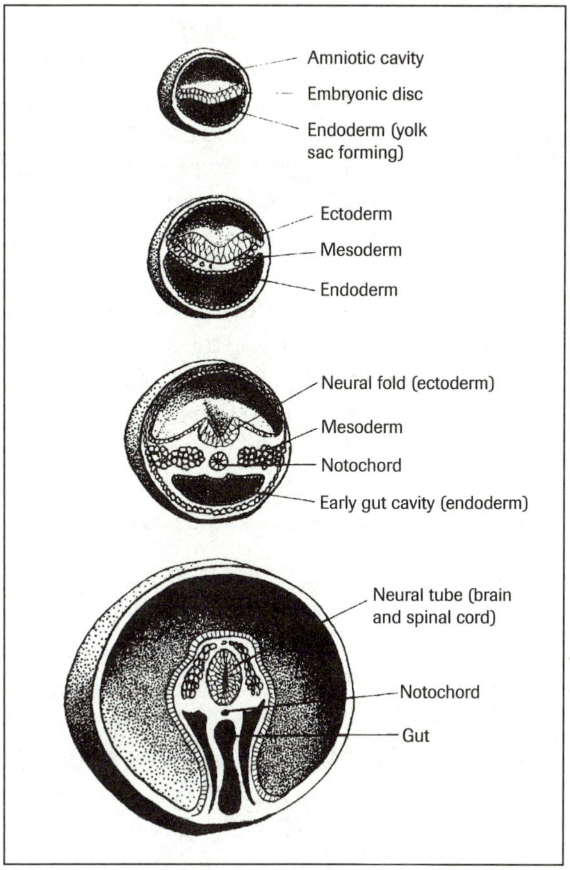

Figure 16.14 Development of embryonic tissues and associated structures in the human blastocyst.

Morphogenesis and Formation of the Placenta

During the following six-week period of embryonic development, morphogenesis results in formation of specific body organs derived from the germ layers, and the extraembryonic membranes and placenta are formed (Fig. 16.15). The critical difference between this period and the first two weeks of development is that the developing embryo now requires specific environmental, nutritional, and organizational conditions to survive.

As morphogenesis proceeds within the embryo, the first of the membranes surrounding the embryo to form is the **yolk sac,** which appears as a pouch suspended below the embryonic disk. The human yolk sac actually contains no yolk, but serves as a site of blood cell formation in the early embryo. Later in development, the yolk sac degenerates, leaving its stalk attached to the embryonic disk.

Cells of the embryonic disk also spread upward at this time, forming a lining around the inner surface of the amniotic cavity. This lining forms the **amnion,** a membrane that immediately surrounds the embryo and will also surround the fetus from the ninth week of development until birth. The amniotic cavity is filled with *amniotic fluid,* which cushions and protects the developing infant. Near the base of the yolk sac, a small outpocketing forms the third extraembryonic membrane, the **allantois.** Part of the allantois later gives rise to the urinary bladder and also to the umbilical arteries and vein.

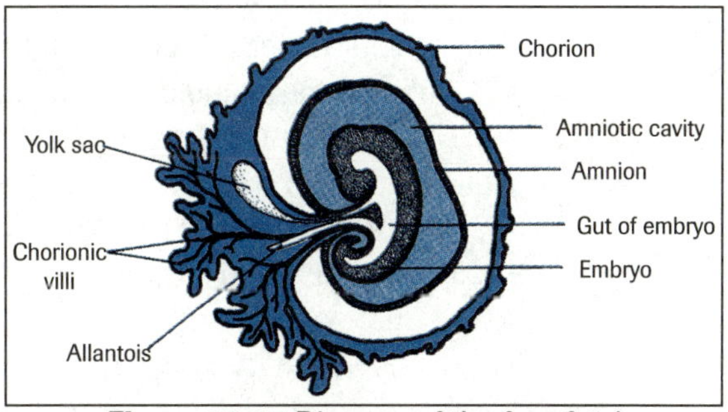

Figure 16.15 Diagram of the four fetal membranes–chorion, amnion, allantois, and yolk sac–and their relationship to the human embryo.

The fourth extraembryonic membrane is the **chorion,** which arises from the outermost layer of the blastocyst. As mentioned, the chorion becomes part of the placenta. The placenta, the structure by which the embryo is attached to the uterine wall, is formed from the chorion and uterine lining. Blood vessels develop in the chorion, with embryonic circulation beginning in the third week as the embryonic heart begins to beat. Formation of chorionic blood vessels signals the beginning of development of the placenta. Extending from the inner wall of the chorion are tiny finger-like projections called *chorionic villi,* which penetrate into the uterine lining; the villi are in close proximity to blood vessels in the lining. There is established, then, a close association between the circulations of the mother and the embryo. Exchange of nutrients, respiratory gases, and wastes occurs by diffusion between the embryonic and maternal blood vessels. In addition, the baby receives certain antibodies from the maternal bloodstream. This passive immunity against foreign antigens such as bacterial infection lasts for about two months following birth; thereafter, the child's immune system has developed to respond and function on its own.

By the fourth week of development, the chorion has expanded greatly within the uterine lining, enclosing the amnion with the embryo inside. The yolk sac remains suspended below the embryo. A portion of the allantois, along with its blood vessels, forms the body stalk, which attaches the embryo to the chorion. The body stalk fuses with the yolk stalk, forming the umbilical stalk, which attaches the embryo to the placenta. The umbilical stalk develops into the *umbilical cord,* which carries deoxygenated blood to the placenta through paired umbilical arteries and returns oxygenated blood to the embryo through a single umbilical vein.

Embryonic Development

During embryonic development, a fibrous, rodlike structure called the **notochord** grows along the dorsal length of the embryo. The notochord is present in all chordate embryos, but in adult humans and other vertebrates, it is eventually replaced by the vertebral column. Remnants of the notochord in humans remain as part of the intervertebral discs; the pads of cartilage between the spinal vertebrae. Ectoderm overlying the notochord curves dorsally to form the **neural tube,** which will give rise to the central nervous system, which includes the brain and spinal cord. In addition, rudiments of the skeletal and muscular systems are developing from embryonic mesoderm.

By the end of the first month, the head of the embryo is quite large, and the eyes, ears, and nose have begun to form. The digestive system, including the liver and pancreas, is also developing, while the tiny heart is actively pumping blood. Present, too, are small *limb buds,* which will develop into arms and legs. Small slits and "gill pouches" resembling those of a fish embryo are evident along either side of the head and neck. In fish and some amphibians, the slits develop into gills, but in humans and other land vertebrates, they normally close before birth.

Tissues around the pouches, however, develop into such structures as the eardrum and Eustachian tube.

During the second month, morphogenesis continues as the embryo acquires somewhat more distinctive human characteristics. Organ development continues and there is increasing prominence of facial features, limb buds, and the musculoskeletal system. Internal sex organs, either ovaries or testes, may be distinguished, but the sex of the embryo cannot yet be determined from external embryonic structures. As in other vertebrate embryos, a small tail develops, but it regresses and finally disappears by the end of the second month. There have been rare cases, however, in which the tail was still present at birth. Fortunately, this condition can be corrected surgically.

Human Fetal Development

The term **fetus** (Latin "offspring") is applied to the human organism from the ninth week of development until birth. By the end of the third month, the fetus possesses the rudiments of all the major organs. Facial expressions, movements of the arms and legs, and reflexes, including the sucking reflex, become quite evident during the third month.

In addition, the sex of the fetus can be determined externally, breathing movements are evident, and there is degeneration of the notochord. At the end of the third month, the fetus is slightly less than 76 mm (3 in.) long from head to rump and weighs about 14 g (0.5 oz.).

Second Trimester

From the fourth month to the end of the fifth month, the fetus more than doubles in length and increases in weight to about 227 g (0.5 lb.). During the sixth month, it will attain a length of about 305 mm (12 in.) and will weigh about 680 g (1.5 lb.). Its heart, which can be heard with a stethoscope, beats at a rate of about 140 times per minute. The average heart rate in an adult human is 70 to 80 beats per minute. The body cells of a human fetus metabolize at about twice the rate of the cells of adult humans. The more rapid heart rate of the fetus insures that its cells will be supplied with the oxygen required to maintain this high metabolic rate. During the fifth month, the mother can feel movements of the fetus as it shifts positions within the amniotic cavity. By the sixth month of development, the fetus has grown hair on its head, and its skin is red and wrinkled.

Third Trimester

The final three months of prenatal life are marked by rapid growth of the body and final development of the organ systems. The brain and entire nervous system rapidly become more complex as an array of new nerve cells and tracts develops. The bones of the skull are soft and flexible, allowing for continued growth of the brain.

By the end of the eighth month, the fetus is about 457 mm (18 in.) long and weighs approximately 1135 g (2.5 lb.). The baby also acquires a deposit of fat tissue, causing the skin to appear less wrinkled.

At the end of the final month of pregnancy, human fetuses average over 500 mm (about 20 in.) in length and weigh about 3175 g (7 lb.) (Fig. 16.16). Usually, the time required for complete development of a human baby—measured from conception until birth—is 266 days. Many normal babies, however, are born 2 or 3 weeks before or after this time period. Human fetuses born during the seventh or eighth month of development have a good chance of survival; but a fetus born before the twenty-sixth week often does not survive, primarily because of inadequate development of the respiratory system.

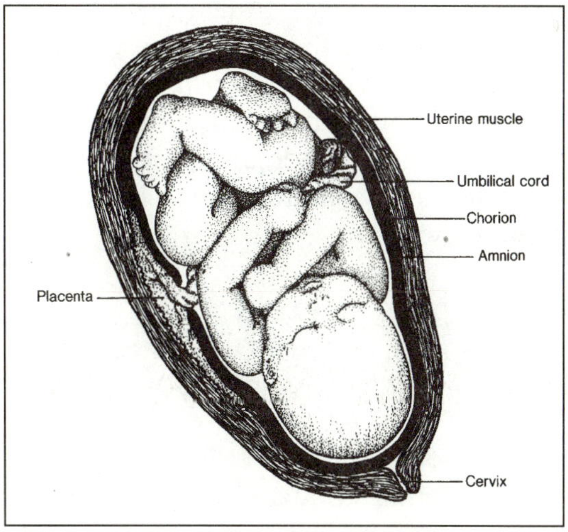

Figure 16.16 Position of the human fetus in the uterus shortly before birth.

Labor and Birth

The act of childbirth, known as parturition (Latin "to be in labor"), typically begins with strong contractions of the uterus. These contractions are considered to result from stretching of the uterus and cervix, caused by movements of the fetus and by hormonal factors. It is probable that stretching of the cervix by the head of the fetus initiates muscular reflexes in the wall of the uterus through positive feedback. Resulting contractions of the uterine muscle help push the baby into the birth canal.

We mentioned that the uterus secretes large amounts of estrogen and progesterone throughout most of pregnancy. The effect of progesterone is to inhibit contractions of the uterus, whereas estrogen increases the intensity of contractions. It is thought that increased uterine contractions late in pregnancy result in part from a proportionately greater secretion of estrogen than progesterone. In addition, stretching of the uterus and cervix causes reflex stimulation of the posterior pituitary gland to release the hormone **oxytocin.** This hormone also acts to increase the rate and force of uterine contractions.

The uterus itself contributes to its own contractions late in pregnancy by secreting a group of fatty acids known as **prostaglandins.** These substances are secreted by virtually all body tissues and, in addition to stimulating smooth muscle contraction, have many other effects, such as inducing fever and inflammation, dilation of blood vessels, and enhancing blood clotting.

The exceptionally strong contractions of the uterus, known as *labor,* occur about once every 30 minutes as the birth process begins. These contractions become progressively stronger and at shorter time intervals as the process continues, and soon occur every minute or so. The cervix progressively dilates during the first stage of labor and fully expands to 10 cm, usually within 24 hours. Women who have given birth previously, however, may achieve maximum dilation of the cervix in much less time. Once this occurs, the amnion may rupture, releasing its fluid, and the head of the fetus enters the birth canal. Continual uterine contractions help force the fetus through the birth canal until it emerges, usually head first.

After birth of the infant, the placenta and extraembryonic membranes, collectively known as the *afterbirth,* are expelled through the vagina. For a short time, the uterus continues to contract mildly, which constricts local blood vessels, limiting the loss of blood. Thereafter, the uterus begins to revert to its normal pre-pregnancy condition and, within 4 to 5 weeks, usually returns to the size it had been before pregnancy.

Multiple Births

The birth of twins, triplets, quadruplets, or any multiple birth usually occurs for two basic reasons (Fig. 16.17). First, the female may ovulate more than one oocyte, each of which is fertilized by a separate sperm. Fraternal twins, for example, arise from two separate oocytes fertilized by two separate sperm. Triplets, quadruplets, and others could arise in the same manner. These offspring are no more alike genetically than any other brother or sister.

Second, following division of the zygote, embryonic cells may separate from each other or the inner cell mass may divide, producing two or more identical embryos. In either case, the embryos arise from the same fertilized oocyte and genetically are virtually identical. These are the ways in which identical twins, triplets, and others may be produced. In some cases, multiple births may result from a combination of multiple ovulations and the separation of cells or division of the inner cell mass.

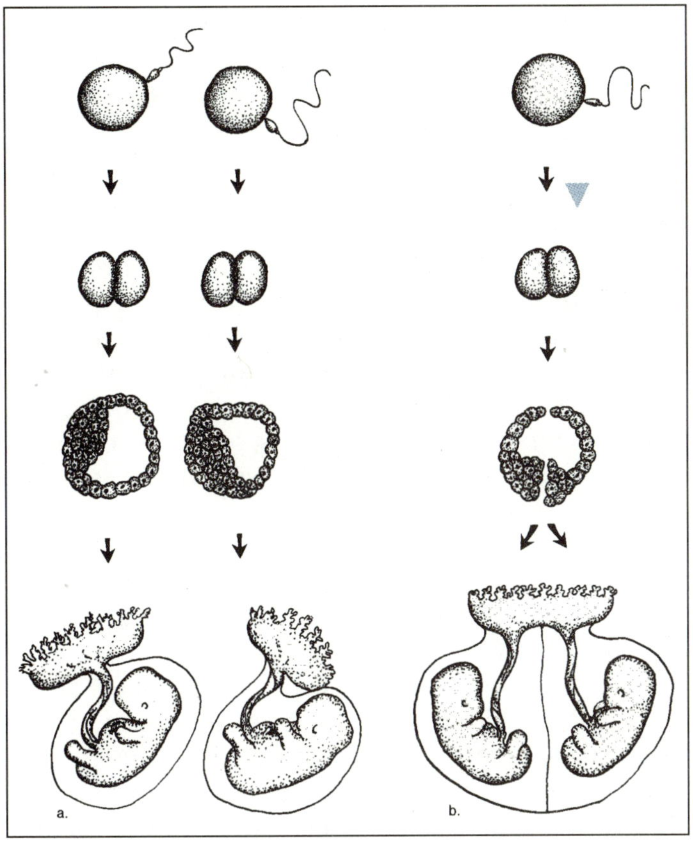

Figure 16.17 (a) Fraternal twins develop from two separate eggs, each fertilized by a different sperm. Each zygote develops into a separate embryo with its own placenta. Fraternal twins are no more alike than other brothers or sisters. (b) Identical twins develop from a single zygote that eventually separates into two embryos. Both embryos share the same placenta and fetal sac. Having arisen from the same fertilized egg, identical twins are genetically the same.

Infertility: Causes and Treatment

Causes of Infertility

Infertility, the inability to have children, may result from male factors or from female factors or both. Infertile males usually do not produce a sufficient number of motile sperm or the sperm ducts cannot transport sperm, while infertile females do not produce functional oocytes, or they experience malfunction of the uterine tubes or cannot maintain a developing embryo in the uterus.

A low sperm count can result from several factors, such as inflammation of the sperm-producing tubules and prolonged use of various drugs. Even if an adequate number of sperm are produced, constriction of the sperm ducts caused by infection or injury may block the transport of sperm.

The failure to produce functional oocytes is often associated with inadequate secretion of follicle-stimulating hormone and luteinizing hormone from the pituitary gland. Even if functional oocytes are produced, inflammation or collapse of the uterine tubes may block the passage of sperm. If fertilization does occur, an inflamed or collapsed uterine tube may prevent the embryo from moving into the uterus.

In some cases, a displaced uterus can interfere with conception. If the ligaments that hold the uterus in place become stretched or weakened, the uterus may descend and bulge into the vaginal canal. Finally, if fertilization can occur and the uterine tubes are functional, the uterus may not be able to initiate a pregnancy because of a low level of progesterone secretion by the corpus luteum. Normally, progesterone from the corpus luteum prepares the uterus for pregnancy and maintains the uterine lining during early pregnancy.

Treatment of Infertility

Male infertility that results from a low sperm count can often be corrected by collecting several ejaculates to increase the number of sperm. This collection of semen is then introduced into the female reproductive tract, a procedure known as *artificial insemination.*

For women unable to ovulate, the so-called *fertility drugs* can increase oocyte production by stimulating the release of follicle-stimulating hormone from the anterior pituitary gland. The high level of FSH in the blood stimulates maturation of more follicles. Because it is difficult to determine the degree of follicle stimulation necessary, treatment with fertility drugs sometimes produces several oocytes, resulting in multiple births. Also, providing synthetic progesterone sometimes is effective in maintaining the uterine lining if pregnancy should occur.

In cases where the uterine tubes restrict the passage of sperm or cannot transport the embryo, two procedures known as *in vitro* (Latin "in glass") *fertilization* and *embryo transfer* have made pregnancy possible for many women for nearly 20 years. In these procedures, oocytes are removed from the ovaries and incubated in a dish. After six hours at body temperature, the oocytes are fertilized with sperm. The zygotes are allowed to divide for 24 to 48 hours and the resulting embryos are transferred to the uterus. Although only one embryo usually completes development in the uterus, the transfer of several embryos to the uterus has occasionally resulted in multiple births.

A most intriguing pattern that is presently unexplainable with regard to natural multiple births is that the ratio of twins to single births in the United States is about 1:89; triplets occur at a ratio of $1{:}89^2$; quadruplets at a ratio of $1{:}89^3$; and so on.

Aging and Death

Development has been discussed primarily from an embryonic standpoint so far, but development is actually a lifetime process. Throughout an organism's existence, it continues to change until it dies. An integral part of that change is the gradual degeneration of body cells and tissues that cannot be replaced. The biological process of degeneration, known as aging, eventually renders the entire organism physiologically unable to survive. While a great deal is known about the physiological changes that accompany aging, the specific *causes* of aging are less apparent.

Genetics and Aging

Through experiments using cell cultures, it has been shown that some human embryonic cells can undergo only about fifty or so cell divisions before becoming nonfunctional; in other words, one cell can produce about fifty generations before cessation of cell division. This finding suggests that cells may be genetically "programmed" to divide only a finite number of times during the life of the organism.

It is known, for example, that the number of functional nerve cells in the body declines steadily after age 30. The progressive loss of cell vitality suggests that there is nothing we can do to extend our allotted life span. This is apparently where genetics may come in. There is no conclusive evidence to date that exercise or vitamins or consuming sunflower seeds will prolong life. But long-lived parents and grandparents may be a determining factor in living a long life yourself. There are exceptions, of course, but there is increasing evidence that there is a strong genetic component to aging. We really don't know exactly what this means, but the author can offer a small ray of hope by recommending that you choose your parents wisely.

Another hypothesis assumes that aging of cells results from gene mutations that alter the genetic information coded in DNA. Such changes might result in the synthesis of defective enzymes, which could upset the normal metabolism of various cells, or they may alter cells in such a manner that lysosomes release increased amounts of tissue-destroying enzymes.

After skeletal muscle cells die, they are not replaced by new muscle cells. Instead, the dead cells are replaced by a tough connective tissue. With fewer functional cells, a working muscle may become overtaxed, a factor that may hasten aging of the remaining muscle cells. Similarly, the amount of connective tissue in other organs and structures increases with time. Blood vessels infiltrated with excessive connective tissue become hardened and less elastic, thereby contributing to arteriosclerosis and hypertension, two common diseases of aging.

AGING: CAUSES OR CONSEQUENCES?

Other factors, such as loss of natural immunity, mutations, and hormonal imbalance, also have been suggested as contributing factors in the aging process. With a loss of immunity, the body becomes more susceptible to infection and disease, and the body may even be attacked by its own antibodies. Mutations, which can occur throughout life, affect the functioning of body cells. A decline in the level of hormones alters the chemistry of the body, leading to malfunction of the endocrine organs and an upset of homeostasis.

In considering the few factors presented here, perhaps the fundamental question is: Are these factors the *cause* of aging or simply the *result?* or both? The truth is we do not know. The science of the physiology of aging is called *gerontology* and is a relatively new discipline among the biological sciences. The questions it seeks to answer are formidable indeed: What is aging? Why do organisms age? Is aging inevitable? For now, the answers remain hidden from our view, locked somewhere within a living cell.

Chapter Summary

1. Some biological rhythms occur on a 24-hour cycle; others occur weekly, monthly, or annually. Rhythms are thought to be controlled by an internal biological clock and have been observed to continue for a time in the absence of external stimuli. Photoperiodism involves the plant pigment phytochrome, which occurs in two forms: P_r and P_{fr}. In most plants, P_{fr} is the active form that affects developmental processes. The controlling stimulus for flowering in plants is the length of darkness to which they are exposed. Phytochrome determines in part whether or not flowering will occur.

   Key Terms: *circadian rhythm, p. 374*
   photoperiodism, p. 377
   phytochrome, p. 377

   Concept Questions: What is the general significance of biological rhythms in living organisms? What might be one way in which a particular rhythm could be upset? What is the general role of phytochrome in the development of flowering plants?

2. The zygote of a multicellular organism undergoes a series of developmental events called embryogenesis, which include growth, cellular differentiation, and morphogenesis. All human somatic cells arise by mitosis and cell division of the zygote; these cells, therefore, are genetically identical. There is evidence that the cytoplasm of embryonic cells contains stimulatory and inhibitory substances that affect gene activity. As embryonic development proceeds, cells become increasingly differentiated. This is due in part to unequal division of the cytoplasm; different amounts of cytoplasm and different cytoplasmic materials in some manner affect gene expression, which controls cellular differentiation. Entire tissues affect embryonic development through induction. Induction is thought to be controlled by chemical substances that diffuse from one tissue to another. These chemicals stimulate or inhibit development by activating or repressing certain genes. At the molecular level, cellular differentiation is controlled through the synthesis of mRNA at specific sites on the DNA molecules of a particular cell. The result is the synthesis of proteins specific for each type of tissue.

   Key Terms: *cellular differentiation, p. 378*
   morphogenesis, p. 378
   embryo, p. 379
   totipotent, p. 379
   induction, p. 383

   Concept Questions: Describe an experiment indicating that all body cells are genetically identical. What is the experimental evidence that the cellular cytoplasm influences differentiation? What is induction and what is its significance? Cite an example of this process. Why is it said that cellular differentiation is ultimately controlled at the molecular level?

3. The angiosperm zygote and endosperm nucleus divide by mitosis within the female gametophyte. The embryo, endosperm, and seed coat, which surrounds the female gametophyte, form the seed. During germination of the angiosperm seed, the radicle, or embryonic root, emerges first, followed by the shoot and sometimes the seed leaves. In addition to producing new roots, leaves, and branches, apical meristems also give rise to the flower.

   Key Terms: *seed, p. 385*
   apical meristem, p. 385
   lateral meristem, p. 385
   radicle, p. 385

   Concept Question: What is the basic relationship between an apical meristem and cellular differentiation in plants?

4. Starfish embryogenesis involves cleavage and formation of a blastula and a gastrula. The gastrula develops the three primary germ layers, which give rise to all the tissues and organs of the starfish. After a brief larval stage, the young starfish develops into an adult.

   Early attempts to explain human embryonic development included the preformation theory, which proposed that miniature human embryos were contained within the sex cells. Epigenesis, the theory of progressive development of form, eventually replaced the preformation theory.

   Key Terms: *cleavage, p. 386*
   blastula, p. 386
   gastrula, p. 386
   ectoderm, p. 386
   endoderm, p. 386
   mesoderm, p. 386
   epigenesis, p. 387

Concept Questions: What is the fundamental difference between the preformation theory of human development and epigenesis? What is the role of the scientific method in determining the validity of these two concepts? Compare the general scheme of embryogenesis in starfish and humans.

5. Human development begins with cleavage of the zygote to form the morula, which hollows out to form the blastocyst. As the blastocyst is implanting in the uterine lining, the amniotic cavity is formed and the inner cell mass develops into the embryonic disk. The disk originally is composed of ectoderm and endoderm; mesoderm is laid down later. Ectoderm, endoderm, and mesoderm are the germ layers from which all body tissues are derived.

   The extraembryonic membranes formed during the embryonic period are the yolk sac, amnion, allantois, and chorion. The placenta, composed of the chorion and uterine lining, serves to exchange nutrients, respiratory gases, and wastes between the blood of the mother and blood of the developing infant.

   The first month of embryonic development is characterized by the appearance of the notochord and neural tube. During the third month, the fetus possesses the rudiments of all the major organs. The final three months of development are characterized by rapid body growth and final development of the organ systems. Human development is measured from conception until birth and normally requires 266 days.

   Key Terms: *blastocyst, p. 388*; *inner cell mass, p. 388*; *amnion, p. 389*; *chorion, p. 389*; *notochord, p. 390*; *fetus, p. 391*

   Concept Questions: What are the general roles of the four extraembryonic membranes in human development? During human development, circulating maternal and fetal blood do not mix. Why do you think this is important?

6. Childbirth results from a combination of factors that induce contractions of the uterus and cervix. Movement of the fetus and increased secretion of estrogen and oxytocin, along with prostaglandins, set up strong uterine contractions that help push the baby into the birth canal.

   Multiple births may result from separation of cells after the zygote divides, fertilization of multiple oocytes, or from division of the inner cell mass. Human infertility results from both male and female factors. Some infertility problems can be treated through artificial insemination, *in vitro* fertilization, and embryo transfer.

   Key Terms: *oxytocin, p. 392*; *prostaglandins, p. 392*; *infertility, p. 394*

   Concept Question: A woman who cannot bear children has her embryos, obtained through *in vitro* fertilization, inserted into the uterus of a substitute, or surrogate, mother. Following a successful pregnancy and birth, the surrogate mother claims that the child belongs to her. Whose rights do you think are paramount in this case—the embryo donor or the husband who is the sperm donor or the surrogate mother? Who would have the right to sue for custody of the child after a divorce? What do you see as some of the other ethical and legal problems accompanying this claim?

7. Aging is an integral part of the development of living organisms. Suggestions as to the causes of aging include limited cell division, a loss of natural immunity, mutations, and hormonal imbalance. Whether these are actually the causes of aging or the results is unknown.

Part Nine

Life Feature 7: Evolutionary Adaptation

Chapter Seventeen

Evolution, Darwinism, and Man

Evolution

The Meaning of Evolution

There is no single concept more essential to a unified view of the living world than that of **evolution,** a term derived from the Latin *evolvere,* literally meaning "to unroll or unfold." The term *evolution* can be used in a cultural sense, as when we speak of the evolution of the automobile, hair styles, clothing, or ideas. In biology, our concern is *biological* evolution: the "unrolling or unfolding" of life. Inherent in both the cultural and biological concepts of evolution is the idea of *change,* an "unrolling" or modification of form or kind. The high-powered automobile of today represents a vast modification of the Model T Ford of the early 1900s. Similarly, all of the more recent forms of life are modifications of earlier forms. However, we certainly cannot relate cultural evolution to biological evolution in terms of mechanisms or results. As used here, cultural evolution proceeds by human design or intent, often with a predetermined goal in mind. Biological evolution, however, involves mechanisms that operate without known design or intent, and which have no specific goal. But the key difference between cultural and biological evolution is that cultural evolution is not hereditary.

In previous chapters, it was pointed out that life can be characterized in part by its unity and its diversity. This might seem like an unlikely characterization unless unity and diversity could be reconciled through some unifying concept. This unifying concept is biological evolution. Specifically, one of the cornerstones of evolutionary theory is that all organisms have descended from a common form of life. The unity of life, therefore, can be understood in terms of a common ancestry. Diversity, on the other hand, has resulted primarily from the capacity of organisms or species to adapt to Earth's many environments.

Through the process of chemical evolution when Earth was very young, interactions between various forms of nonliving matter gave rise to primitive cells, the original forms of life. At this point, biological evolution had begun. There was established, then, an evolutionary bond between life and Earth itself, a bond that has persisted down through the ages.

History of Evolutionary Theory

Evolution is actually an ancient historical concept, contemplated thousands of years ago by the early Egyptians and Chinese. Ancient Egyptian science was permeated with an array of mythological gods, however, and some Chinese beliefs concerning nature were grounded in superstition and magic. Rigorous scientific inquiry with regard to evolution, therefore, was lacking in both of these early civilizations. Later, during the Greek era, philosophers such as Anaximander (611–547 B.C.) attempted to provide naturalistic rather than mythological explanations for natural phenomena. Anaximander believed that animals originated in a moist environment and that humans evolved by a process similar to that of other organisms. These beliefs were eventually supported by scientific evidence some 2500 years later.

Still, many of the Greek ideas relating to evolution were as ill-founded as those of the early Egyptians and Chinese. The great Greek philosopher Aristotle (384–322 B.C.), for example, believed that organisms could not change in form over time; that is, they did not evolve. Further, he thought that life could be spontaneously generated from nonliving matter. This view, which persisted into the seventeenth century, held that maggots could be produced naturally from rotting meat, or mice could arise from dirty rags or decaying leaves.

The Greek philosopher Empedocles (495–435 B.C.), however, believed that higher forms of life evolve; that is, they do not arise by intentional design, and he believed that the best-adapted organisms survive. In these instances, Empedocles closely foreshadowed the modern view of biological evolution.

For hundreds of years after the Greek era, evolution was openly discussed, but it was eventually buried under the intellectual stagnation of the Middle Ages that began with the fall of Rome late in the fourth century. Even in the creative atmosphere of the Renaissance, spanning the fourteenth through the sixteenth centuries, evolution and, in fact, biology in general received little attention. The revival of interest in evolution was not to begin until the eighteenth century.

Throughout history, opposition to the concept of evolution has been based largely on religious grounds. Such opposition is founded to a significant extent on belief in the "fixity of species" concept, which holds that each species was created and fixed in its present form and thus is incapable of changing over time. As we mentioned, even Aristotle held this view.

An accompanying belief that was accepted for many centuries held that Earth came into existence instantly through one sudden, explosive act of creation. Moreover, early Christian theologians calculated that Earth was about 6000 years old. Beginning in the eighteenth century, however, the science of geology, which explores the history and structure of Earth, began to provide evidence that Earth has a much longer history.

Uniformitarianism

An eighteenth century Scottish physician named James Hutton (1726–1797) proposed that natural processes, over long periods of time, gradually brought about geologic changes in Earth's surface. These processes, he reasoned, were in effect in the present as they had been in the distant past. This concept, later expanded by the Scottish geologist Charles Lyell (1797–1875), came to be known as **uniformitarianism.** In other words, geologic events from Earth's ancient past can be explained in terms of existing geologic processes; in fact, the current geologic changes and processes taking place on Earth have occurred in the same manner throughout geologic time. Lyell asserted that, with time, there is a uniform balance between geologic processes that erode Earth's surface and those that build it up. For the most part, this concept of an ancient Earth undergoing continuous, gradual change over time was ignored or considered antireligious. But it challenged the accepted opinions concerning the age of Earth. For Lyell, geological processes seemed to require not thousands of years, but millions of years to alter the face of Earth.

Fossil Discoveries

During the latter part of the eighteenth century, support for the antiquity of Earth began to be provided by fossil discoveries, which indicated that the history of life was one of diversity and change and, quite importantly, that it accompanied the developmental history of Earth. A **fossil** is either the remains of an organism that once existed or an imprint of an organism in soil or rocks (Fig. 17.1). Much of the original fossil evidence was uncovered by the English geologist William Smith (1769–1839). In his study of Earth's surface, Smith found that each different rock layer contained distinctive groups of fossils. Older, deeper rock layers, laid down early in Earth's history, contain fossilized organisms older than those in the upper, younger rock layers. From one stratum to the next, different species are apparent. Moreover, the older the stratum, the less the fossilized organisms exhibit a similarity to modern organisms. Smith's observations attested to the extreme age and formative processes of Earth and to the relationship between

Figure 17.1A Fossil remains of a fern preserved in shale.

Figure 17.1B Fossils of extinct coiled mollusks.

Figure 17.1C Trilobite fossil preserved on the surface of rock. Trilobites were ancient arthropods apparently related to spiders and crabs.

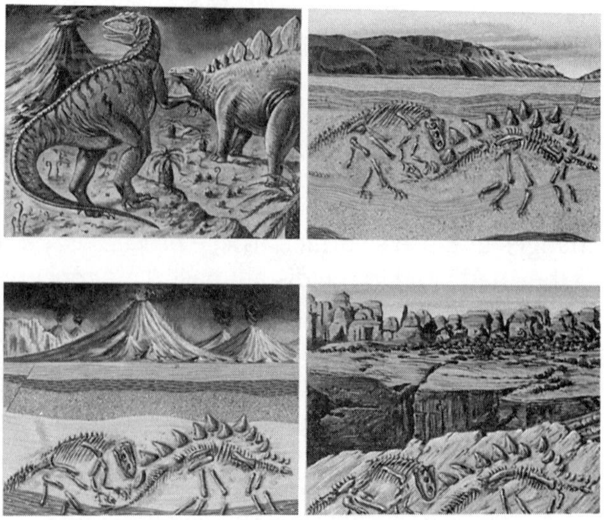

Figure 17.1D Illustrations depicting the death and fossilization of two dinosaurs–the predatory *Allosaurus* and *Stegosaurus,* an herbivore.

these factors and the history of life. Other geologists confirmed Smith's findings, and the view of the world began to change.

In its initial stages, then, evolutionary theory received its greatest impetus from the science of geology. Geology revealed not only the ancient and developmental history of Earth, but also provided evidence for the mutability, or changeability, of living forms. Both of these precepts were later to be found indispensable to an integrated theory of biological evolution.

Although the geological revelations of the eighteenth century were impressive, there were many scientists of the time who were adamantly opposed to both uniformitarianism and any theory of evolution. One of the staunchest and most influential opponents was the French zoologist Georges Cuvier (1769–1832), who virtually founded the science of studying fossils, called *paleontology.* Although an expert on fossil animals, Cuvier strongly advocated the fixity of species concept and belittled the idea of an extremely ancient Earth. In his interpretation of the fossil record, Cuvier proposed that, following the creation of all species, there was a series of sudden climatic and geologic upheavals, one after the other, that wiped out great numbers of each species. Survivors of each catastrophe repopulated the world, only to have many of their numbers eliminated by the next upheaval. Again, the world was repopulated by the survivors, and this cycle continued.

To account for the different kinds of fossil species observed in younger and older rocks, he proposed that a devastated area was repopulated by other species that had immigrated to that area. Since Cuvier proposed that all species were created simultaneously, he believed it should be possible to trace the fossils of all exterminated species back to the original group of organisms present at the time of creation. Fossil evidence to support this sequence has never been found. Nonetheless, Cuvier's theory, known as **catastrophism,** dominated the thinking of many scientists throughout the early part of the nineteenth century.

Lamarck's Theory of Evolution

Another Frenchman, Jean Baptiste Lamarck (1744–1829), disagreed with the theory of catastrophism and proposed that all species are descended from preexisting species. In 1809, Lamarck published his own theory of evolution that included two concepts generally accepted during his day. The first was the principle of "use and disuse." Basically, Lamarck believed that, because of environmental conditions and internal drives or "desires," an organism develops certain "needs," which are met by alterations of its existing structures or organs or by development of new ones. Continual use of these altered or new structures increases their size and functional capacity, while disuse causes them to become smaller, less functional, weaker, and even lost. These variations were thought to be heritable by the offspring. This idea constituted the second Lamarckian concept: the **inheritance of acquired characteristics.** For example, according to Lamarck's theory, an organism that lives mostly in the dark, such as a mole, has tiny, degenerate eyes because they are so infrequently used. Moles apparently inherited this characteristic from ancestors whose eyes had progressively degenerated through lack of use.

Lamarck and Giraffes

Another classic example of Lamarckianism explains how the modern giraffe acquired its long neck (Fig. 17.2). When food was scarce, the short-necked ancestors of giraffes needed to stretch their necks to forage on leaves growing high in a tree. This caused their necks to increase slightly in length; the longer neck was then inherited by the offspring. These offspring, in turn, stretched their necks to obtain food. Lamarckian theory proposed that this cycle continued for generations until all descendant giraffes had much longer necks than their short-necked ancestors. Thus, through the use of a their necks, giraffes acquired longer necks that could be inherited by the offspring. We can summarize Lamarck's theory as follows:

Figure 17.2 Lamarck's concept of inheritance of acquired characteristics. Driven by some inner need, the short-necked ancestral giraffe kept stretching its neck until it became progressively longer. According to Lamarck's theory, the acquired longer neck could be inherited by subsequent generations.

(1) Environmental conditions cause an organism to have certain needs.

(2) Through changes in existing body structures or development of new ones, these needs are satisfied.

(3) The use of a body structure increases its size and efficiency, while disuse of a structure causes it to degenerate or disappear.

(4) The changes in body structure acquired through use or disuse can be inherited by the next generation.

Within the context of Lamarck's theory, it may appear that any acquired characteristic could be inheritable. For example, the theory seems to support the position that a 97-lb weakling who later acquires a "Mr. Universe" physique through weight lifting could pass his acquired muscles to his offspring, or that the accidental loss of a finger would result in all succeeding generations being fingerless.

Lamarck, however, would have disagreed with these examples because he was careful to distinguish between the types of traits that could and could not be inherited. He did not believe that an acquired physique or loss of a finger or any other trait acquired by direct intervention of the environment could influence the evolutionary process. He did not believe that environmental effects themselves, such as weight lifting, brought about heritable changes in organisms. Lamarck's crucial point was that the use or disuse of an organ or structure, in conjunction with the basic needs of an organism, was a central factor in evolution. In other words through the use or disuse of a bodily structure an organism is responding to its environment, and this altered structure, in meeting the needs of that organism, can be passed to its offspring. As these needs are met, the organism becomes better adapted to its environment.

DNA and Heredity

Today, we are virtually certain that inherited characteristics result primarily from information coded in the DNA of chromosomes, which are passed from parent to offspring. Therefore, variations observed in future generations result mainly from genetic factors. What all organisms inherit, then, is a particular set of genes—the genotype—and not necessarily any particular set of physical characteristics—the phenotype. Herein lies the scientific problem with Lamarckism. For offspring to inherit a characteristic acquired solely because of its use or disuse, the DNA for that characteristic would have had to pass first into the gametes (egg cells or sperm) of the parents. This occurrence has not been supported by scientific evidence. The concepts of "fluid heredity" (see Chapter 14) and inheritance of acquired characteristics were generally accepted in Lamarck's day; nothing was known of chromosomes, genes, and DNA.

Actually, Lamarck's theory of evolution was of little significance to the scientific community of the early 1800s. In that era, creationism ruled; species were fixed and evolution was irrelevant. The irrepressible Cuvier, the chief foe of evolution, lambasted Lamarck, who subsequently endured lifelong hardship, went blind, and died a pauper. History, however, has vindicated Lamarck to a small extent. Although his theory of inheritance of acquired characteristics was incorrect, he challenged the fixity of species concept and advanced the idea that evolution is central to an explanation of the diversity of life.

Charles Darwin and Evolutionary Theory

The most notable figure ever associated with evolutionary theory was the English naturalist Charles Robert Darwin (1809–1882) (Fig. 17.3). Darwin was born in Shrewsbury, England, and, at an early age, displayed a consuming interest in the natural world. At the insistence of his physician father, Darwin entered medical school at the University of Edinburgh when he was 16 years old, but he found medicine boring and unappealing. He left Edinburgh and entered Cambridge University to study for the ministry. After receiving his degree in 1831, Darwin learned that a naturalist's position was available on the H.M.S. *Beagle,* a scientific research ship that was about to embark on a 5-year voyage around the world. Darwin signed on at the age of 22, beginning a journey that would take him to South America and the Galapagos Islands, where he spent much of his time on shore collecting thousands of plant and animal specimens.

Figure 17.3 A mid-nineteenth century painting of Charles Robert Darwin.

Aboard the *Beagle,* Darwin read a newly published book entitled *Principles of Geology,* that was written by the geologist Charles Lyell. Lyell did not agree with the current theory of catastrophism; rather he believed that Earth had undergone a process of long, continuous change through the effects of natural causes. Darwin seized on the idea, reasoning that organisms also might have undergone a similar change. Influenced by Lyell's ideas, Darwin began to doubt the prevailing opinion of an unchanging Earth that was only a few thousand years old. As he began to consider that life on Earth may have evolved over time, Darwin realized that a few thousand years was probably too brief a span for any major evolutionary change to occur. Armed with Lyell's theory, however, Darwin could proceed with the conviction that Earth was much older and therefore there was, indeed, time enough.

After his return to England in 1836, Darwin continued to accumulate data on his theory of evolution, an endeavor he was to pursue for more than 20 years. In 1858, Darwin received a letter from a fellow naturalist Alfred Russell Wallace (1823–1913) who had conceived an evolutionary theory virtually identical to that of Darwin. Dismayed, Darwin wrote to his friend Charles Lyell, "I never saw a more striking coincidence . . . he could not have made a better short abstract." To avoid any potential controversy, Lyell and an associate presented the papers of both Darwin and Wallace at a meeting of the Linnean Society in London on July 1, 1858. This proved to be an acceptable resolution, and Wallace, by his own admission, felt that Darwin deserved most of the credit.

In the following year, Darwin worked feverishly to complete his manuscript. He presented his work publicly in his book *On the Origin of Species by Means of Natural Selection, or the Preservation of Favored Races in the Struggle for Life.* Published on November 24, 1859, the book immediately created a storm of controversy; there were impassioned confrontations, not only between scientists and the public but also among the scientists themselves. Fundamentalists, the creationists, and advocates of a literal interpretation of biblical writings, were outraged; catastrophists scorned the book; and churches attacked the book with vengeance. Darwin was even accused of asserting that humans had descended from apes, a proposition that does not even appear in his work. Darwinian evolution had provoked an uproar that would not subside for decades.

Natural Selection

Given that there was adequate time during Earth's history for organisms to evolve, what natural process might affect their evolution? For Darwin, that process was **natural selection.** Although the concept did not originate with Darwin, he maintained that natural selection was the essential force behind biological evolution. In arriving at this conclusion, Darwin considered the following:

(1) There are many more organisms born in most populations than will survive or reach reproductive age.

(2) Over long periods of time, however, the size of most natural populations tends to remain fairly constant. This occurs mainly because there are limited environmental resources, and, because of a struggle for existence, many individuals die before reaching reproductive age.

(3) Individual organisms in a given population have variable characteristics; that is, they are not all exactly the same in their anatomical, physiological, and behavioral makeup.

(4) Because of certain characteristics, some organisms within a population have a better chance of surviving and reproducing than others; that is, some organisms are better adapted to their environment than others.

(5) Favorable characteristics tend to be transmitted to the offspring in greater proportion than less favorable characteristics because more individuals with favorable characteristics reach reproductive age than do individuals not as well adapted.

Over time, Darwin proposed, the cumulative effects of these factors could produce new species of organisms. Basically, natural selection refers to different degrees of success in reproducing different genotypes, which results from the effects of the environment on heritable phenotypes. You will recall that **fitness** in evolutionary terms refers to the relative ability of organisms to produce offspring. The phrase "survival of the fittest" refers to reproductive ability and not merely to survival, no matter how long an organism may live.

What, then, determines who is fit and who is not? In other words, what determines the selection of some organisms and not others? The answer, as just mentioned, is the *environment:* all the forces that affect an organism, such as climate, availability of food, disease, and enemies.

Depending on the variable physical characteristics of organisms, environmental conditions tend to favor the perpetuation of some of these characteristics; that is, they are selected for, and tend to eliminate the others. It is important to point out that natural selection does not *produce* variations in organisms; rather, it acts on *existing* variations, usually culling those heritable variations that are of minimal adaptive value in a given environment.

Often, however, this selection is not necessarily adaptive. If a climatic or geologic upheaval altered a certain habitat, the surviving organisms may not be the ones best suited to their changed environment. From this, it is apparent that an adaptive characteristic in one environment may be of no advantage or may even be a disadvantage in a different environment.

Actually, natural selection can operate within a variable population in three different ways: stabilizing selection, directional selection, and disruptive selection.

Stabilizing Selection

First, there is **stabilizing selection,** which tends to favor the intermediate phenotype and exclude extreme phenotypes. In other words, average individuals possess the greatest fitness. This is seen, for example, in human birth weights. Babies weighing about 3–4 kg (6.5–9 lbs.) have higher survival rates than do very small or very large babies. Stabilizing selection is also evident in the number of offspring produced by a given species. If there were six puppies normally produced in a litter, more of them, on the average, would tend to survive than would puppies from larger or smaller litters.

Directional Selection

Second, **directional selection** tends to eliminate individuals at one extreme in the population or those that deviate from the average, while favoring individuals at the other extreme. As a result of directional selection, a given population progressively includes more of the extreme phenotypes. This usually occurs when the environment is changing in a certain direction, such as a change in climate, or if an expanding population in a stable environment is subjected to new environmental conditions as it moves into a different area. Directional selection also comes into play in the predator-prey relationship. In the evolution of animals that chase their prey, there is a premium on speed. Slow runners or animals of average speed tend to have lower survival rates because they cannot catch their prey. Directional selection, then, favors the faster animals, which feed regularly and thus tend to survive in greater numbers. At the same time, directional selection also acts on the prey to increase its potential for survival.

Disruptive Selection

Third, **disruptive selection** favors individuals at *both* extremes, while tending to eliminate the intermediate phenotypes. This has the effect of separating a population of individuals into two subpopulations, each with the potential for becoming quite different. The effects of disruptive selection can be seen when a stable environment is disturbed in some manner. The disturbance may create new habitats and opportunities for new ecological niches. This type of selection would most probably be involved in the development of new species. For example, members of a plant population that become separated by an expanse of water or even a drainage ditch may experience new environmental influences, such as a change in soil composition or exposure to the wind. Those plants that possess adaptive characteristics for coping with the new environmental extremes would tend to survive, while plants with an intermediate phenotype would not. In time, the two subpopulations may become so unlike that each could be classified as a different species.

Darwin and Giraffes

Natural selection can also be illustrated using the example of the evolution of long necks in giraffes (Fig. 17.4). According to Darwinian theory, an ancestral population of giraffes probably exhibited variations in neck length. Some necks were a little longer than others and some were a little shorter. Once the leaves and fruit on the lower levels of a tree had been consumed, only the longer-necked individuals could reach food at the higher levels. Consequently, it was more likely that a greater number of the longer-necked individuals would survive. These organisms would tend to reach reproductive age and perpetuate themselves in greater numbers than would the shorter-necked individuals. Through repetition of this activity over many generations, the descendant giraffe population would have longer necks than would the ancestral population. This is an instance of directional selection, which favors organisms at one extreme in a population.

Figure 17.4 The Darwinian concept of evolution by natural selection. In a given population of giraffes there is variation in neck length. Natural selection would tend to favor the longer neck individuals because they are able to feed on leaves higher up and thus would tend to survive in greater numbers. This favored characteristic would be passed to proportionally more offspring. After many generations, a population still may exhibit variations in neck length, but the population overall shows an increase in neck length.

Within several years after publication of *The Origin of Species,* many biologists came to accept the reality of naturalistic evolution, although few agreed with Darwin's theory of natural selection. *Naturalistic evolution* is a vitalistic philosophy that attempts to explain natural processes by relying on an indefinable force or drive that is thought to influence or determine these processes. While environmental effects on organismal variation was an essential key in his theory, Darwin could not answer with convincing evidence a most crucial question: What is the source of the variations observed in natural populations? It was to be decades before the answer emerged, long after Darwin died in 1882.

The Progression of Evolutionary Theory

In 1866, seven years after the publication of Darwin's *The Origin of Species,* the genetic discoveries of Gregor Mendel (see Chapter 14) appeared in the *Proceedings of the Natural History Society* in the town of Brunn, Austria. Darwin was apparently unaware of Mendel's work; in fact, even those who were familiar with it were not particularly impressed. For another 34 years, Mendel and his basic laws of heredity were ignored.

Chromosomal Theory of Heredity

During the last quarter of the nineteenth century, though, advancements in cell biology, such as new staining techniques and studies of the cell nucleus, led to the discovery of chromosomes and the processes of mitosis and meiosis. By the turn of the century, Mendel's work was "rediscovered" and subsequently incorporated into the *chromosomal theory of heredity;* that is, hereditary traits are determined primarily by genes, which are the constituents of chromosomes. The Dutch botanist Hugo de Vries (1848–1935), one of the discoverers of Mendel's work, contributed the important concept of *mutations* to the chromosomal theory. In fact, an increasing number of scientists were delving into the mysteries of heredity, confirming Mendel's principles and establishing new ones. In 1908, G.H. Hardy and Wilhelm Weinberg applied hereditary principles to natural populations, establishing the science of *population genetics.* In his study of the fruit fly *Drosophila* in 1910, Thomas Hunt Morgan (1866–1945) introduced the concept of sex-linked traits and provided further evidence that genes are located on chromosomes. The novel science of genetics was on its way.

Although genetics began to make tremendous strides in the early twentieth century, the study of evolution was largely ignored until 1935. By the early 1940s, however, the fields of anatomy, paleontology, population genetics, biogeography, embryology, and others had made important contributions to evolutionary theory. Some brief examples will illustrate the significance of these contributions.

Homologous Structures

Anatomical similarities that exist among different species indicate descent from a common ancestor (Fig. 17.5). For example, the forelimb bones of humans, dogs, walruses, bats, and all other mammals share a common construction. While an arm or a wing or a flipper has its unique function, each is similar to the others anatomically. Evidence indicates that mammals evolved from an ancestor with five-toed feet, a single upper arm bone, and two smaller forearm bones. This same general

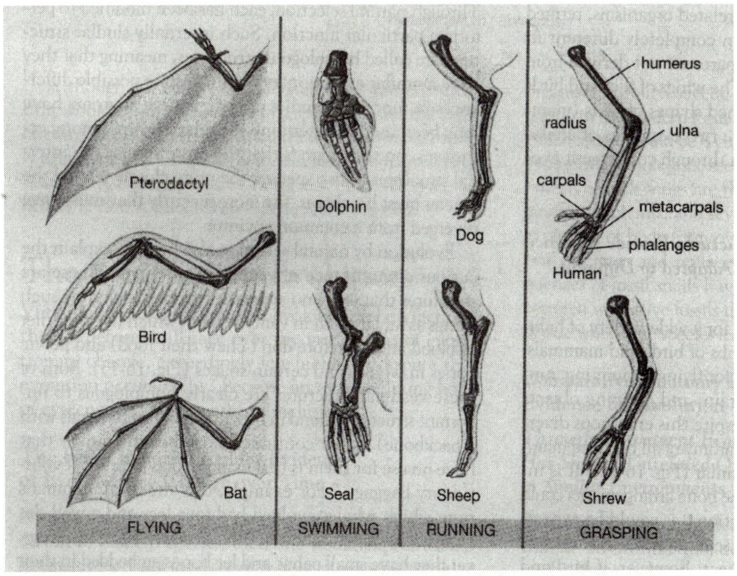

Figure 17.5 The forelimbs of these animals are homologous, an indication of descent from a common ancestor. Reprinted from *Biology: Life on Earth,* Fourth Edition, by Gerald and Teresa Audesirk, 1996, Prentice-Hall, Inc.

plan is also found in all modern mammals. Similarity in construction resulting from a common ancestry is called **homology.** We refer, therefore, to the wing of a bat and the flipper of a walrus as *homologous* structures.

Vestigial Structures

Some of the most convincing anatomical evidence for evolution is seen in **vestigial structures.** These are remnants, or leftovers, of ancestral structures that no longer are functional in modern organisms. About a hundred vestigial structures, such as the appendix and caudal (tail) vertebrae, are found in modern humans. Horses, rabbits, and some other mammals have a structure called the *cecum* in which bacterial digestion occurs. The cecum is homologous to the human appendix, which has no significant function in modern humans. It is possible, however, that the appendix may have had a digestive function or some other function in our distant ancestors.

Whales have remnants of pelvic bones and some snakes have vestigial hindlimbs. Both of these characteristics suggest an evolutionary relationship to a four-legged ancestor.

The presence of caudal vertebrae in humans strongly suggests that our ancestors had tails. Although it is rare, some human babies today are born with short, well-developed tails.

Developmental Similarities

Evidence for evolution is also provided by **comparative embryology.** In observing the embryos of vertebrates, for instance, it is virtually impossible in the early stages to determine which embryo is that of a fish, which is human, and so on. Early embryogenesis in closely related organisms is quite similar. Moreover, it is sometimes possible to determine homologous structures among the embryos of related species. Here again, the similarities in embryonic development among different species indicate descent from a common ancestor.

The Fossil Record

We commented earlier on the significance of fossil evidence to evolutionary theory. It is usually the inorganic, or hard, structures, such as bones and teeth, that survive in the fossil record. Soft body parts are normally decomposed by bacteria and fungi. Fossil remains tell us not only about past life, but also about historical changes in Earth. Discovering fossilized sea organisms or shells at the top of terrestrial mountains indicates that the mountains once were under water. Fossils are also preserved as molds, in which only the imprint of an organism is left. Leaves of ancient plants or feathers of extinct birds are often imprinted in rocks. Insects that are 25 million years old have been found preserved intact in amber, the dried resin of ancient cone- bearing trees. Finally, there are footprints of many animals, including great dinosaurs and early humans, that have been preserved for thousands and millions of years.

Genetic Studies

Thomas Henry Huxley (1825–1895), a staunch defender of Darwinian evolution, proposed in 1863 that there was a close relationship between humans and the chimpanzee and gorilla. Darwin later supported this position. Huxley and other scientists of the nineteenth century, however, had very little experimental evidence to support their view of human-ape similarities. Today, experimental techniques, such as DNA hybridization, protein sequencing, and immunological analysis, provide data that tend to support Huxley's position.

DNA hybridization studies, which involve comparing the DNA of closely related species, indicate that human DNA is most similar to that of chimpanzees, with the gorilla running a close second. In contrast, the DNA of orangutans and monkeys is much less similar to human DNA, indicating a more distant evolutionary relationship between humans and these two primates. DNA hybridization involves separating the DNA strands of the double helix and using one of the strands to determine if it will bond with a single DNA strand acquired in the same manner from another species. The single DNA strands of humans and chimpanzees, for example, bond to form a hybrid double-stranded DNA. This hybridization indicates that the DNA sequences in humans and chimpanzees are similar. Single DNA strands from organisms too distantly related will not bond to form the hybrid DNA. This is because unrelated organisms have quite different DNA sequences.

Comparing differences in the sequence of amino acids of proteins reveals that humans and chimpanzees differ very little in the amino acid composition and sequence of their proteins. The differences are slightly greater between humans and gorillas and much greater between humans and orangutans and monkeys.

Immunological analyses reveal about the same pattern of relationships as the other tests, except that the gorilla is a little closer immunologically to humans than are chimpanzees. Immunological analyses compare the reactions of proteins of one species with antibodies from another species. Antibodies derived from humans, for instance, are allowed to react with proteins from gorillas, orangutans, and monkeys. The extent of the reaction between the human antibodies and these proteins is a measure of the evolutionary relationship between humans and the other primates. The greater the reaction, the closer the relationship. The reaction of gorilla proteins to human antibodies is slightly greater than that of chimpanzee proteins.

Other data from molecular genetics studies have suggested a close genetic relationship between humans and chimpanzees, a relationship closer than either shares with the gorilla. More recently, however, these genetic data have been challenged, and the argument has been presented that there is an equally close genetic relationship shared by humans, chimpanzees, and gorillas.

Synthetic Theory of Evolution

In the decades following the 1940s, the structure and function of DNA were revealed and the genetic code, found to be similar in almost all organisms, was deciphered. The code further attested to the fundamental unity of almost all forms of life. Further, new dating techniques confirmed the antiquity of Earth and its first living inhabitants. Through input from these sources, attention was again focused on the study of evolution. This time, however, evolutionary theory was more integrated and comprehensive, using Darwinism coupled with genetics and the other sciences. This modern evolutionary theory is known as the **synthetic theory of evolution,** or *neoDarwinism.* It is important to note that not all scientists today agree with some of the claims of the synthetic theory, a point on which we will comment after we consider the basis of the theory.

We have mentioned that Charles Darwin was unable to account for variation within a population because the principles of inheritance were unknown to him. He was convinced that evolution—the process of change in populations from generation to generation—resulted primarily from the action of natural selection on heritable variations within a population. But he did not know the *source* of the variation nor its relationship to the reproductive process. Darwin knew nothing about chromosomes, DNA, genes, or mutations.

Today, we know that the cells of living organisms contain some type of chromosomal material, genes are the units of heredity that compose this material, genes are actually specific sequences of nucleotides in DNA that produce RNA, and a mutation is a change in the genetic makeup of a cell. Thus, we know that heritable variation among members of a sexually reproducing population results primarily from genetic factors. With this information, we can define evolution in a very general sense as the process of *genetic* change in populations from generation to generation. Still, the essential question remains: What is the source of genetic variation? The answer is that there are two sources: mutation and recombination.

Mutation

Basically, a mutation is a heritable change in a DNA molecule. A **point mutation** of a gene is a change in one nucleotide of DNA. Point mutations may involve substitution of one nucleotide for another nucleotide in DNA, or the insertion or deletion of a single DNA nucleotide. Most point mutations are thought to be relatively harmless unless, of course, they result in the synthesis of abnormal proteins. Insertions and deletions usually result in **frameshift mutations,** which shift the reading frame of a nucleotide sequence. Frameshift mutations usually involve the insertion or deletion of two or more nucleotides in a gene.

Point Mutations

Substitutions

A substitution replaces one nucleotide on one DNA strand and its paired nucleotide on the complementary strand with another complementary pair of nucleotides. The messenger RNA transcribed from this altered DNA sequence would consist of codons different from those transcribed from the original DNA. Suppose, for example, that the DNA triplet GAG mutated to GCG. The mRNA codon transcribed from GAG would be CUC, but the GCG mutation is transcribed into RNA to read CGC. During protein synthesis, CUC may code for one amino acid while CGC codes for another. This mutation, then would change one amino acid that makes up the protein.

This type of substitution ordinarily has little or no effect on the functioning of the affected protein, but occasionally such a mutation improves a protein to the benefit of the organism. In other cases, however, a substitution mutation is harmful, producing an abnormal protein that causes malfunction or disease in an organism. A classic example of a disease caused by an abnormal protein is sickle-cell anemia. In two of the four polypeptide strands of the sickle-cell hemoglobin, one of the amino acids is different from the amino acid found in normal hemoglobin. The result caused by this single amino acid change is a distorted hemoglobin molecule incapable of transporting adequate oxygen to the body cells.

A.	Sequence of **triplets** in segment of DNA:	AAA — GAC — GGG — TTC
		↓ ↓ ↓ ↓
	Sequence of codons in transcribed messenger RNA:	UUU — CUG — CCC — AAG
		↓ ↓ ↓ ↓
	Sequence of amino acids in polypeptide:	Phenylalanine — Leucine — Proline — Lysine
B.	DNA:	AGA — GAC — GGG — TTC
		↓ ↓ ↓ ↓
	mRNA:	UCU — CUG — CCC — AAG
		↓ ↓ ↓ ↓
	Polypeptide:	Serine — Leucine — Proline — Lysine
C.	DNA:	AAA — CGA — CGG — GTT—C
	mRNA:	UUU — GCU — GCC — CAA .
		↓ ↓ ↓ ↓
	Polypeptide:	Phenylalanine — Alanine — Alanine — Glycine

Figure 17.6 Examples of gene mutations. (a) A segment of DNA, showing four triplets that transcribe complementary codons in a segment of messenger RNA. The mRNA codons are translated into a sequence of amino acids (polypeptide). (b) Substitution mutation. The same segment of DNA, except that guanine (G) in the first DNA triplet has been substituted for adenine (A). The transcribed RNA triplet UCU codes for the amino acid serine, which is different from the amino acid phenylalanine coded by the mRNA triple UUU. Thus, the polypeptide has been altered and may affect the organism. (c) An addition mutation. Cytosine (C) has been added, or inserted, between the first two triplets in the original DNA segment. This changes not only the second triplet, but those to the right of it as well. Consequently, the polypeptide is quite different from the one in (a).

FRAMESHIFT MUTATIONS

Insertions and deletions involve adding nucleotides somewhere along a DNA strand or losing nucleotides from the strand (Fig. 17.6). These mutations alter the nucleotide sequence in DNA, which then transcribes messenger RNA codons that are different from those of normal DNA. For example, if a segment of DNA consisted of the nucleotide sequence GTA – TGC – AAA, the corresponding triplet codons of the transcribed mRNA would be CAU – ACG – UUU. Suppose that two nucleotides, one containing adenine (A) and one containing cytosine (C), were inserted between the first two nucleotides in the DNA segment given. The DNA nucleotide sequence then would be GACATATGCAAA, and the transcribed codons would be CUG – AUA – CGU – UU*. The insertion of these nucleotides has thrown off the normal triplet reading frame of the DNA sequences, resulting in a different mRNA sequence. This type of mutation is known as a **frameshift mutation** because it literally shifts the reading frame of the DNA sequence. The amino acids coded for by these codons are thus different from the amino acids coded for by the original sequence of codons. Thus, the two proteins containing different amino acid sequences would be dissimilar.

If nucleotides were deleted, or lost, from the DNA strand, the results are essentially the same, with a shift in the reading frame of the DNA. In almost all cases, insertions and deletions result in the synthesis of nonfunctional or dysfunctional proteins.

Rarely, inheritance of a genetic mutation may be of adaptive value to an organism, particularly in a changing environment where the organism is subjected to new conditions. Inheritance of such a mutation in a DNA sequence is important in the evolutionary process if its effect brings about some change in the offspring. In other words, a mutation must alter the *phenotype,* or physical appearance of an organism, before it can be determined if the mutation is of adaptive value in a given environment. If the mutation results in an advantageous variation, natural selection would dictate perpetuation of the variation in succeeding generations. If the variation were a disadvantage to the organism, making it less able to adapt to its environment, the organism may not survive long enough to reproduce, and its DNA would be eliminated from the population.

Chromosomal Alterations

Chromosomal mutations, or alterations, involve changes in the normal structure or number of chromosomes (Fig. 17.7). Various physical and chemical influences, such as X-rays, ultraviolet light, pesticides, and mustard gas, may cause a chromosome to break, resulting in the loss of one or more DNA segments. A segment may subsequently reattach to the same chromosome in an inverted position. Alternatively, a segment may reattach to its homologous chromosome, producing a duplication of that segment on a single chromosome, or it may attach to a nonhomologous chromosome. This transfer of a chromosomal segment to another, nonhomologous chromosome is called a *translocation*. More commonly, translocations are reciprocal, with two nonhomologous chromosomes exchanging segments. In any of these instances, either the number or the arrangement of the genes on the chromosome is adversely affected. As with mutations in a single gene, chromosomal alterations ultimately affect the synthesis of cellular proteins.

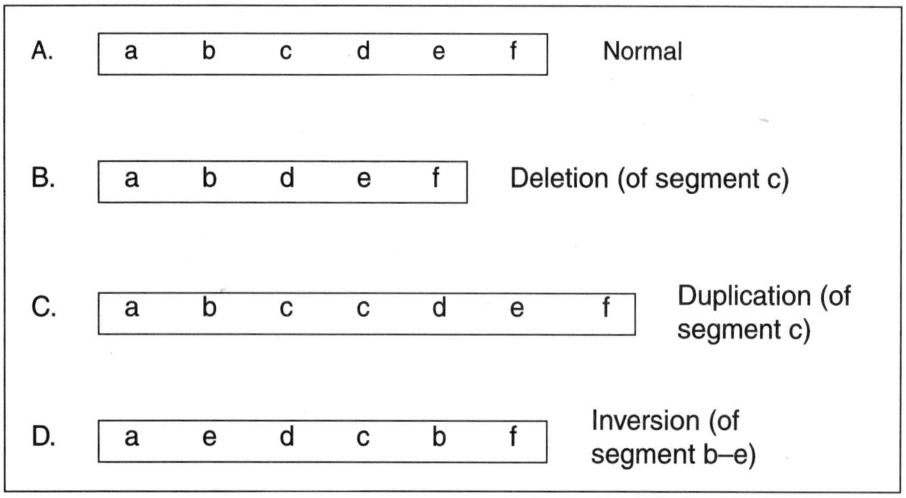

Figure 17.7 (a) Representation of a normal chromosome, showing segments a through f. (b) Because of a break in the chromosome, segment c has been deleted. (c) An aberration in the chromosome has resulted in a duplication of segment c. (d) Following a break in the chromosome, segment b–e has rejoined the chromosome in an inverted position.

Nondisjunction

Changes in normal chromosome numbers often occur because of **nondisjunction,** as discussed in Chapter 13. Failure of homologous chromosomes to move apart in meiosis I or sister chromatids to separate in meiosis II results in one gamete receiving two of the same chromosomes, while the other gamete receives neither of the two chromosomes. Union of either of these abnormal gametes with a normal gamete produces a zygote with an abnormal chromosome number. One of the most common examples of an abnormal chromosome number (47 rather than 46) in humans is Down syndrome. This malady is characterized in part by poor muscle tone, a flattened face, thick, short hands and feet, heart problems, and an IQ in the range of 25–70.

Extra Chromosomes

Some organisms, particularly plants, may have one to several additional sets of chromosomes, a condition known as **polyploidy** ("many fold"). You will recall that n designates a haploid number of chromosomes, and that $2n$ designates a diploid number of chromosomes. An organism that has one complete, additional haploid set of chromosomes is $3n$, or triploid; a $4n$, or tetraploid, organism has two additional haploid sets of chromosomes. Extra chromosome sets usually arise by accident during cell division.

Polyploidy is a major mechanism of evolution in some plants. For instance, if a tetraploid plant arose from a diploid parent population, it could interbreed with other tetraploids that may arise, but it cannot interbreed successfully with plants of the parent population. This is because the gametes of the tetraploid are $2n$, whereas those of the diploid plants in the population are n, and the chromosomes of each plant either do not pair or they mispair. Each gamete, then, cannot receive the same number of chromosomes. When the gametes are united at fertilization, the resulting $3n$ hybrid offspring are sterile because their reproductive tissues cannot undergo normal meiosis. Some of these plants, however, are able to reproduce asexually. This is also true of other polyploids with uneven numbers of chromosome sets, such as pentaploids ($5n$). Thus, the tetraploids are reproductively isolated; that is, they cannot interbreed successfully with the parent population and represent a new species.

Polyploids with uneven numbers of chromosome sets can also produce new species. Through accidents in cell division, particularly nondisjunction, the chromosome number in the reproductive tissue of sterile hybrids may double. These plants are then capable of making gametes because they now have an even number of chromosomes, and each chromosome has a homologue with which it can pair during meiosis. When gametes from these plants unite, the resulting offspring are fertile and can interbreed. If an offspring has a chromosome number different from that of its parents, it is reproductively isolated from the parent population and thus represents a new species.

Recombination

Recombination refers to the methods by which genes are brought together in new combinations with other genes during sexual reproduction. In meiosis, *crossing* over, which is the exchange of genetic material between male and female homologues, results in new combinations of alleles on each chromosome. In addition, when a pair of homologous chromosomes moves away from each other at anaphase I of meiosis, their movement is independent of any other pair of homologues. At anaphase II, separation of the sister chromatids distributes the alleles randomly into separate gametes (see Chapter 13). Meiosis, then, results in a shuffling of the alleles, producing a variety of allele combinations in the gametes. Finally, alleles from parent organisms are recombined when the two gametes, the sperm and ovum, unite. The resulting zygote contains a unique assortment of alleles; that is, a unique genotype that will be expressed as a new phenotype. As we have said, natural selection will act on this phenotype, determining its adaptability to a given environment.

Genetic variability is the key to evolutionary advancement. In the final analysis, it is genetic recombination that contributes most significantly to the variation essential to evolutionary change. Genetically, every population—a group of one species living in a certain geographical area—is characterized by its **gene pool,** the total of all the genes of all the organisms in that population. The concept of a population is central to the study of evolution because it is the population that evolves and not the individual organism. A population may vary in its genetic makeup from one generation to the next, but an individual organism normally has the same genetic makeup all its life. Mutations and recombination are the sources of variation within the gene pool of a natural population and ultimately provide the phenotypic variability on which natural selection acts.

Speciation

The development of new species from preexisting ones is known as **speciation.** The evolution of new species involves genetic changes within a population. We have seen that evolutionary change comes about when there is an alteration in the gene pool of a population. Such a change in the gene pool over successive generations is called **microevolution,** the mechanisms of which can lead to the formation of new species. Two of those mechanisms are mutations and natural selection. Two additional mechanisms are **genetic drift,** which refers to random changes in the gene pool of a population from generation to generation, and **gene flow,** the change in a gene pool that results from organisms or seeds or pollen entering or leaving a population (Fig. 17.8).

Genetic Drift

Random fluctuations in the gene pool of a population occur because the size of a population and its gene frequencies, that is, the proportions of the different alleles, are determined by inheritance from a previous generation. If the next generation is smaller in size, it is more likely that its gene frequencies will differ from the previous generation simply by chance. In small populations, genetic drift is significant because only a few organisms may possess a particular genotype. If these genotypes are lost by some chance geologic or climatic event, the gene pool of subsequent generations would be altered significantly. The loss of a particular genotype is much less likely to occur by chance in large populations.

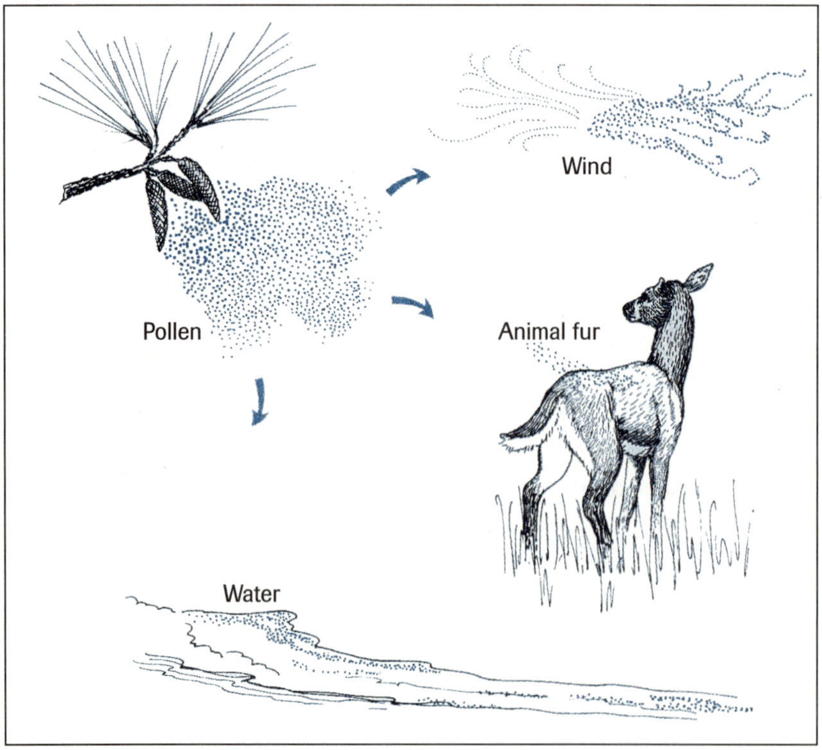

Figure 17.8 An example of gene flow. Pollen grains, containing the male gametes, may be distributed to other pine forests by wind or water, or in the fur of wide-ranging animals.

Genetic drift is evident among a splinter group of Pennsylvanian Amish, in which there is an abnormally high incidence of a rare form of dwarfism caused by an autosomal recessive allele. The Amish generally intermarry and rarely allow outsiders into their population. On occasion, some families leave the general Amish population to found new, smaller communities. One founding male possessed the allele for dwarfism, which eventually increased in frequency in the new community far beyond its frequency in the general Amish population. Thus a single individual was responsible for a change in the gene frequencies between two related populations. This example illustrates why genetic drift has been referred to as evolution by chance.

Gene Flow

Gene flow results in the introduction of new genes into a population as individuals or reproductive cells enter that population, or in the loss of genes from the population as organisms emigrate. Gene flow may be either a limiting force or a creative force in evolution. Migrating organisms that interbreed with members of their new population may introduce different genes into the population. Because of gene flow, the two populations may, over time, become so similar genetically that they will not evolve into separate species. In contrast, the introduction of new genes into a population may create adaptively superior genotypes, which may enhance the evolutionary potential of the population.

Geographic Isolation

We noted in Chapter 2 that, over 200 million years ago, all of the present-day continents were united in one enormous land mass known as Pangaea. In time, Pangaea split into two separate masses, both of which later broke up into the individual continents of today. This phenomenon of continental drift was the major geographic factor in isolating great numbers of species. The fossil record indicates that many species were distributed worldwide during the time the continents were united. After breaking up, the continents became separated by ocean barriers that geographically isolated untold numbers of plant and animal populations. Such isolation can lead to the development of new species. This is evident today in that each continent, experiencing millions of years of isolation, has evolved its own unique species of plant and animal life. One of the most familiar examples are the marsupial mammals that once were widespread over North America and Europe. For tens of millions of years, however, except for the opossums of the Americas, marsupials have been found almost exclusively in Australia.

Various natural barriers, such as mountains, deserts, and bodies of water, can physically separate a species into two or more isolated populations (Fig. 17.9). Such geographically isolated populations are referred to as **allopatric** (Greek *allos,* "other," and Latin *patria,* "fatherland"). If the change in environmental conditions is sufficiently drastic, a population begins to experience new environmental pressures; some of the members may become adapted to the new conditions, while others may not. Speciation may occur, particularly in a small population, if a gene pool becomes sufficiently variable to produce phenotypes different from those of another isolated population.

Reproductive Isolation and Physical Characteristics

The major criterion that defines a biological **species** is *reproductive isolation.* This means that a species is defined primarily in terms of its members' inability to interbreed with members of any other species. Members of the same species usually interbreed with each other and produce fertile offspring. A second criterion that defines a species is that members of the same species have physical characteristics more like each other than like the members of any other species. Another definition of a species emphasizes reproductive adaptations that heighten successful mating among members of the same population. In other words, recognition between mating individuals, and not reproductive isolation, is the major criterion in defining a species. Whatever the case, the development of an all-inclusive definition of a species has proved to be rather difficult. For one thing, reproductive isolation cannot be applied to asexually reproducing organisms; for another, is it not possible to state, without testing, that crossbreeding positively cannot occur between some members of two sexually reproducing species. Despite these and a few additional considerations, the species concept is generally workable within the overall framework of biological evolution.

Figure 17.9 Geographical isolation. A classic example of isolating mechanisms affecting speciation involves the Kaibab squirrel, an inhabitant of the north rim of the Grand Canyon, and the Abert squirrel, found on the south rim. Although closely related, they do not interbreed because they are separated by the Grand Canyon. Consequently, the two populations have evolved differently. The Kaibab squirrel has a dark underside and an all-white tail, while the Abert squirrel is lighter in color with a white underside.

The factors responsible for reproductive isolation may operate to prevent mating between members of two species or, if copulation has taken place, to prevent the production of offspring. Among animals, the prevention of mating may result from factors such as incompatibility of the reproductive structures of members of two species, differences in periods of sexual activity, different habitats, or different behavioral patterns, such as courtship displays that are not recognized as such by either species. In any of these instances, members of one species are eliminated as possible mates for members of another species, and gene flow between the populations is prevented.

If copulation between members of different species has taken place, sperm from one species may not survive in the female reproductive tract of the other species. Alternatively, if fertilization should occur, the zygote may fail to develop. Even if the hybrid offspring should survive, it may be poorly adapted to the environment or it may be sterile. The mule, which is a hybrid offspring of a female horse and a male donkey, is an example of hybrid sterility. These isolating factors, then, like those that prevent mating, inhibit gene flow between the two different populations.

In our earlier discussion of mutations, we saw that polyploidy is a major mechanism of evolution in some plants, and that some polyploid organisms constitute a new species. When species arise within a single population, it is known as **sympatric** (Greek *syn,* "together, " and Latin *patria,* "fatherland") speciation. In this case, species become reproductively isolated without being geographically isolated. Sympatric speciation does not involve genes entering or leaving a population. As discussed, a polyploid is a new species because, as a result of its chromosome number, it is reproductively incompatible with members of its parent species.

Punctuated Equilibrium or Gradual Change?

The Darwinist view of the speciation process is that it occurs over time through small cumulative changes that result in the formation of new species. There is some contention, however, as to the rate at which speciation occurs. If we diagrammed the development of species from ancestral forms, the diagram is considered by some biologists to appear as a branching tree with each species gradually changing step-by-step over long periods of time. There is, then, according to this view, a gradual divergence from ancestral forms to succeeding species.

A competing view, arguably also Darwinian, is known as **punctuated equilibrium** and contends that changes in form during speciation occur rapidly in spurts, and that this is followed by a long period of very little physical or genetic change in a species. In other words, species are in a state of general equilibrium (little or no change) throughout most of their existence until this equilibrium is interrupted (punctuated) by periods of rapid speciation. This view was proposed by the American paleontologists Niles Eldredge and Stephen Jay Gould in 1972. In comparing the gradual and punctuational views, the apparent bone of contention is the definition of "rapid." In dealing with geological time, we are usually referring to many thousands of years.

One of the problems common to both points of view centers around the lack of transitional forms in the fossil record. A transitional form is an organism intermediate between one species and its evolutionary descendants; a form having some anatomical characteristics like the older species and some characteristics like the new ones. The fossil record, however, rarely shows gradual transitional forms for any one group of organisms. Indeed, new species often appear rather suddenly in the fossil record. This would appear to support the position of punctuationists, who contend that speciation occurs too rapidly for the development of various intermediate forms. In contrast, those who advocate that evolution proceeds gradually through a series of changes in a population argue that transitional forms existed but that very few have been preserved in the fossil record. Alternatively, the changes in these forms may be internal and thus are not detectable from examining only the external anatomy left in the fossil record.

Regardless of the opinions held on evolutionary theory, it is agreed by most biologists that natural selection—the mechanism by which organisms become adapted to their environments—is the central concept in biological evolution. Punctuationists, however, emphasize what is known as *species selection,* in which changes that occur during development of a species are the major factors in determining evolutionary trends. The emphasis, then, is on a species rather than on individual organisms. It has been argued, however, that species selection by itself cannot account for the small, virtually imperceptible changes necessary for the evolution of complexity in the living world. Furthermore, it has also been proposed that species selection and, indeed, the entire theory of punctuated equilibrium is but a minor ripple in the sea of neo-Darwinian theory. The debate, no doubt, will continue.

Humans from the Past

It would be hard to imagine that anyone, at some point in his or her life, has not wondered where we, the human species, really came from. Unfortunately, the fossil record of human evolution is sketchy, and investigators admit to considerable disagreement regarding its interpretation. Nonetheless, it is all but certain that humanlike species have inhabited Earth for at least 5 million years and possibly much longer. Compared with other organisms, humans are quite recent arrivals on the biological scene.

Human beings belong to a group of mammals known as *primates (Greek* "first" or "highest"). Our closest primate relatives are the **anthropoid** (humanlike) apes: chimpanzees, gorillas, orangutans, and gibbons. All primates have several distinguishing features: hands or feet adapted for grasping; three-dimensional vision; a shortened nose; upright posture; wide, flattened nails; and a highly developed brain.

Twenty-Five Million Years Ago

Anthropoid apes became generally widespread over a large part of Earth some 25 million years ago. One large group of these animals has been named *Dryopithecus,* meaning "oak ape," which is a reference to the trunks of oak trees found in the same deposit where the fossil apes were discovered. Actually, *dryopithecines,* as their group is called, include not only *Dryopithecus,* which inhabited Europe, but also other similar types that were found in Africa, Asia, and other parts of Europe. Apparently, dryopithecines were about the size of a present-day chimpanzee, although they may have been lighter in build. Some of them had faces slightly more flattened than those of modern apes, and their teeth were partially humanlike. Although dryopithecines roamed Earth for about 20 million years, there is no definitive fossil evidence bridging the gap between these early apes and later forms. Although their own fossil evidence is fragmentary, consisting mostly of jaws and teeth, it has been suggested that the dryopthecine line is the forerunner from which modern apes and humans evolved. Recent fossil evidence from Africa and fossils discovered earlier in China and Pakistan, however, make this relationship unclear.

Four Million Years Ago

The next significant find, a group of fossils at least 3.5 to 4 million years old, was recovered from various sites in eastern and southern Africa (Fig. 17.10). The original specimen, discovered in 1924, was described by the British anthropologist Raymond A. Dart as the skull of a juvenile female. He named the group to which the skull belonged *Australopithecus africanus* ("southern ape of Africa"). Later, similar fossils were found and, although these australopithecines had brains only about one-third the size of modern humans and possessed jutting, apelike jaws, they had humanlike hands and teeth and walked erect.

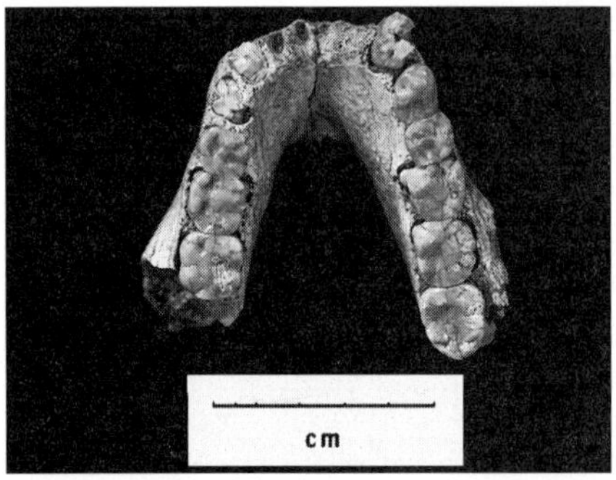

Figure 17.10 The lower jaw bone and dentition thought to be those of an Australopithecine, perhaps 3 million years old.

In 1974, an australopithecine skeleton, about 40 percent complete, was discovered in Ethiopia by Donald Johansen and his coworkers. The skeletal remains, memorialized with the name "Lucy," were of a slightly built female about 1 m (3 ft.) tall, who lived approximately 3 million years ago. Because of some anatomical variances from A. *africanus,* Lucy and others of her kind have been designated *Australopithecus afarensis* (from the Afar region of Ethiopia). This group is considered the truly human stock from which later humans evolved. While

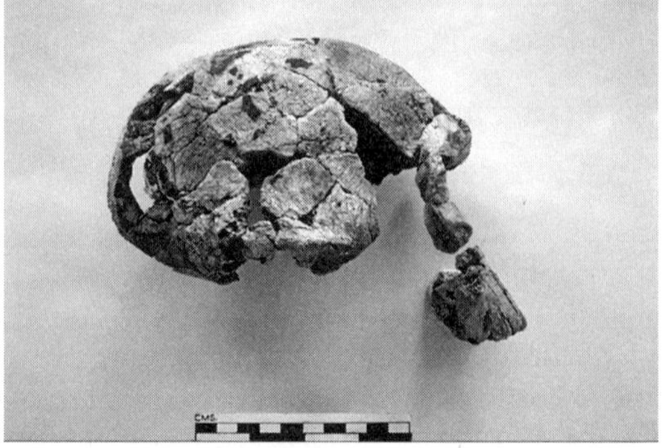
Figure 17.11 A partially reconstructed skull of *Homo habilis* approximately 1.8 million years old.

the matter is in no way settled, some anthropologists consider A. *africanus* a descendant of A. *afarensis.*

In 1961, Louis Leakey recovered a group of fossils from Olduvai Gorge in Tanzania (Fig. 17.11). Leakey called the remains *Homo habilis* ("skillful man"), which some scientists consider a variant of *Australopithecus.* Other investigators argue that, with its larger brain, *Homo habilis* is a more advanced form and warrants inclusion in the genus *Homo,* the genus of modern humans. This would place *Homo habilis* in the direct line leading to modern humans. Apparently, *Homo habilis* and the later australopithecines coexisted for nearly a million years. As a group, the australopithecines persisted for a span of about 2 million years.

From 500,000 Years Ago to Today

In the early 1890s, Eugene Dubois, a Dutch physician, traveled to the island of Java, where he uncovered a group of humanlike fossils some 500,000 years old. Dubois referred to the fossils as *pithecanthropines* ("ape men"), and designated them specifically as *Pithecanthropus erectus* ("upright ape man"). The name was selected because Dubois originally thought he had found the "missing link" between humans and apes. Subsequently, *Pithecanthropus* was renamed *Homo erectus* ("upright man") (Fig. 17.12).

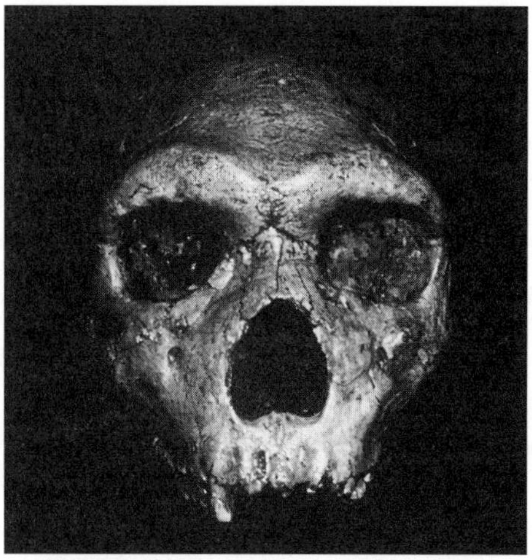

Figure 17.12 A skull from Petralona in northern Greece. It may be the remains of a *Homo erectus* specimen.

The brain size of the pithecanthropines was about double that of the australopithecines, and they were quite heavily built and muscular. They apparently lived in caves and had fashioned a variety of stone tools. Charred bones found in their caves attest to their use of fire, and they had probably fashioned clothing for protection against the cold. Apparently, the pithecanthropines evolved from the African australopithecines and later spread to Europe and parts of Asia. The most recent species of australopithecines, *A. robustus,* lived at the same time as the early pithecanthropines. Although the two groups coexisted for a time, presumably the pithecanthropines eventually replaced the australopithecines in Africa.

Some investigators think that the pithecanthropines *(Homo erectus)* gave rise to and later coexisted with the Neanderthals, who first appear in the fossil record about 130,000 years ago. They have been placed in the genus *Homo* and once were given the name *Homo neanderthalensis.* A characteristic separating Neanderthals from their forebears was an increase in brain size. Neanderthals had brains equal to or larger in size than those of modern humans. A skull discovered in 1921 in the African country of what is now Zambia is thought to be an intermediate between the skulls of pithecanthropines and Neanderthals.

The average brain size of a pithecanthropine was 1000 cm^3 (cubic centimeters), and that of a modern male is 1450 cm^3. The Zambian skull indicated a brain size of about 1250 cm^3. Similar skulls were later found in Greece, China, and other locations. All of these skulls suggest a gradual change from the pithecanthropine stage to the Neanderthal form.

It appears that Neanderthals were a little over 5 feet tall, with thick, jutting brows, and teeth quite similar to those of modern humans. Although widely distributed throughout Europe, parts of Asia, and the Middle East, the Neanderthals suddenly disappeared from the fossil record about 40,000 years ago. The cause of their disappearance remains a mystery.

After the extinction of the Neanderthals, populations closely resembling modern humans arose in Europe and other parts of the Old World. These people—the Cro-Magnons—were among the first representatives of modern *Homo sapiens.* Cro-Magnons are well known for their elaborate paintings on the walls of caves; some of these paintings are 10,000 to 25,000 years old.

The exact course of human evolution following the Neanderthal and Cro-Magnon stages is unclear. The earliest evidence of *Homo sapiens* consists of remains from Africa dated about 120,000 years ago. Some investigators believe Neanderthals were the direct ancestors of modern humans, who are designated *Homo sapiens,* while other investigators contend that, because both Neanderthals and early *Homo sapiens* coexisted in Europe, Neanderthals probably were not the ancestors of modern *Homo sapiens.* Whatever the case, *Homo sapiens* survived to populate Earth, when all others failed. For some 40,000 years, *Homo sapiens* has been the only species of humans on Earth.

The Mitochondrial DNA Debate

One of the most intriguing probes into human evolution does not involve skull fragments or shin bones, but rather a type of DNA found in mitochondria, the tiny, energy-producing organelles found in the cytoplasm of eukaryotic cells (see Chapter 5). While nuclear DNA is inherited from both parents, mitochondrial DNA (mtDNA) is inherited solely from one's mother. This is because the female ovum contains almost all of the cytoplasm with its attendant organelles that is contributed to the zygote.

In 1987, Rebecca Cann and other biochemists reported their analyses of the mtDNA of 147 women from Africa, Europe, Asia, Australia, and New Guinea. It was found that there were very few differences in the mtDNA among the women, indicating that the mtDNA they all shared was recently inherited. It was also discovered that the mtDNA of women of African descent exhibited the most differences from the other samples of mtDNA. This was interpreted as a result of a greater accumulation of mutations as compared to the mtDNA of the other women. This was so, according to the researchers, because the mtDNA of the African descendants was the oldest form of all the mtDNAs analyzed; in other words, the most ancient DNA would have more time to undergo mutation and would exhibit more variation than would more recent mtDNA forms. It was suggested, therefore, that the oldest mtDNA is found in Africa and, hence, the origin of all modern humans can be traced to Africa.

It was also proposed that all humans are descendants of a single, original "mitochondrial Eve," who is thought to have resided somewhere in Africa about 200,000 years ago. Subsequent research by other investigators, though, has questioned Cann's findings. There is disagreement, for example, over the time scale for the origin of modern humans. Some researchers, relying on fossil evidence, date the origin of modern humans back only about 120,000 years rather than 200,000 years. Still others have concluded that modern humans arose in at least three different regions of the world simultaneously and that there is no African "Eve."

More recent reports contend that there is no evidence for a definite geographic origin for a common mitochondrial ancestor nor is there convincing support for the simultaneous evolution of humans in several regions of the world. A current theory holds that modern humans evolved from *Homo erectus* in Africa and expanded into Europe and Asia about 100,000 years ago.

Thus, the answer as to where we, the human species, came from and exactly when we came to be continues to be elusive. For now, the comprehensive story of human evolution remains, as it has for decades, a challenging puzzle awaiting a conclusive solution.

We began this chapter with the premise that evolution is the essential concept that unifies our view of the living world. Throughout this book, we have attempted to justify this premise by presenting seven life features that underlie the unity and diversity of life. From the life feature of organization to the life feature of growth and development, evolution weaves its unifying thread, revealing the fundamental relationships that exist among all forms of life. We have seen the organizational similarity of living organisms revealed at the subcellular and cellular levels. We have followed the common pathway most living organisms employ in their use and conversion of energy. Through specialized cells or chemicals, all life forms sense and respond to their environment. We have observed a variety of control mechanisms that regulate the activities of living organisms, normally maintaining these activities within homeostatic limits. The perpetuation of life is carried out by reproductive processes, followed by growth and development sequences, that are fundamentally similar in almost all forms of life.

Superimposed on these unifying relationships is the immense diversity of life, a consequence of the innate capacity of living organisms to adapt to specific, and sometimes vastly dissimilar, environments. In the diversity of life, we are witness to the creative force of biological evolution. It is through the evolutionary process that diversity arises among living organisms, which are all united by the seven life features. Underlying this statement are centuries of arduous work that culminated in discoveries and scientific theories; each of these, to quote Aristotle, "adds a little to our knowledge of nature and from all the facts assembled there arises a certain grandeur."

Chapter Summary

1. Inherent in the concept of biological evolution is change through time. Evolution, the concept that reconciles the unity and diversity of life, was first contemplated by the ancient Egyptians and Chinese and later by the Greeks. Following the Greek era, interest in evolutionary theory stagnated, but interest was revived in the eighteenth century. Historically, opposition to evolution was based on the "fixity of species" concept. This view began to change in the eighteenth century when geologists contributed the concept of uniformitarianism and provided fossil evidence attesting to the extreme age of Earth and of life.

   Key Terms: *uniformitarianism, p. 401*
   fossil, p. 402
   catastrophism, p. 403
   inheritance of acquired characteristics, p. 403

   Concept Questions: Describe a specific theory that supports the "fixity of species" concept. What contributions by early geologists challenged this concept? Explain generally why Lamarckian evolutionary theory is scientifically unacceptable.

2. Charles Darwin developed a theory of biological evolution based on the principle of natural selection as the primary force in the evolutionary process. Natural selection does not produce variation in organisms, but acts more favorably toward some variations than toward others. Natural selec-

tion operates within a population in three ways: through stabilizing selection, directional selection, or disruptive selection.

Key Terms: *natural selection, p. 406*
fitness, p. 406
stabilizing selection, p. 407
directional selection, p. 407
disruptive selection, p. 407

Concept Question: Cite an example of natural selection. What is the general significance of each of the three types of natural selection?

3. During the early twentieth century, advancements in cell biology, along with the discovery of Mendel's basic laws of genetics, resulted in the chromosomal theory of heredity. Some of the evidence for evolution includes anatomical and embryological similarities among different species, vestigial structures, and fossils. In addition, DNA hybridization, protein sequencing, immunological analysis, and genetics provide evidence for the close relationship between humans and chimpanzees.

Key Terms: *homology, p. 410*
vestigial structure, p. 410
comparative embryology, p. 410

Concept Questions: Explain how the contributions of anatomy, embryology, and molecular genetics support the modern theory of biological evolution. What is the general significance of mutation and recombination to evolutionary theory?

4. The synthetic theory of evolution couples Darwinian theory with genetics and other disciplines. Evolution is then defined as the genetic change in a population over time. The genetic sources of variation within a population are mutation and recombination. A mutation is a change in the genetic makeup of a cell. Recombination is the major factor contributing to variation within a population. Changes in chromosome number are seen in nondisjunction, which can result in polyploidy. Polyploidy is a major evolutionary mechanism in some plants and can result in reproductive isolation within a population.

Key Terms: *synthetic theory of evolution, p. 411*
point mutation, p. 412
frameshift mutation, p. 413
nondisjunction, p. 414
polyploidy, p. 415
recombination, p. 415
gene pool, p. 415

Concept Question: Why are the basic laws of genetics crucial to understanding Darwinian evolutionary theory?

5. A population is characterized genetically by its gene pool. This concept is important because it is the population that evolves and not the individual organism. Speciation involves genetic changes in the gene pool of a population. The mechanisms of microevolution leading to formation of new species include natural selection, mutations, genetic drift, and gene flow. Geographic isolation can also contribute to speciation. A biological species is defined primarily in terms of reproductive isolation. A species is also defined in terms of recognition between mating partners. As a result of natural barriers, a species may become geographically isolated into two or more populations; such populations are said to be allopatric. Sympatric speciation occurs within a population; in this case, a species is reproductively isolated but not geographically isolated. According to two competing views,

speciation may occur gradually over time through small cumulative changes or speciation may occur more rapidly in spurts. The gradualist theory maintains that natural selection is the major concept in biological evolution; punctuated equilibrium favors the concept of species selection.

Key Terms: *microevolution, p. 416*   *allopatric, p. 417*
genetic drift, p. 416   *species, p. 417*
gene flow, p. 416   *sympatric, p. 418*

Concept Question: What is reproductive isolation and how can it come about? What is the significance of reproductive isolation in the evolutionary process?

6. Humans and anthropoid apes are classified as primates, all of which share certain common characteristics. The primates most closely related to humans are chimpanzees, gorillas, orangutans, and gibbons. Dryopithecines and other anthropoid apes became widespread over Earth about 25 million years ago. A group of australopithecine fossils about 4 million years old were discovered in Africa in 1924 that possessed both apelike and humanlike characteristics. An australopithecine skeleton discovered in 1974 was named *Australopithecus afarensis.* Modern humans may have evolved from *A. afarensis* stock. Some investigators consider *Homo habilis* a variant of *Australopithecus,* while others place *H. habilis,* with its enlarged brain, in the direct line leading to modern humans.

Pithecanthropines were a group of humanlike fossils about 500,000 years old. Thought to be the "missing link" between apes and humans, *Pithecanthropus erectus* was renamed *Homo erectus.* Apparently, the pithecanthropines evolved from and later replaced the australopithecines. Pithecanthropines may have given rise to the Neanderthals 130,000 years ago. Neanderthals mysteriously disappeared from the fossil record about 40,000 years ago. Neanderthals were replaced by the Cro-Magnons, some of the first representatives of modern *Homo sapiens.* The earliest evidence of *Homo sapiens is* from African remains dated 120,000 years old.

Original mitochondrial DNA studies suggested that the oldest mtDNA was found in African women and, hence, that the origin of modern humans can be traced to Africa; specifically to a 200,000 year-old African "Eve." Other findings deny this origin and claim that modern humans arose in at least three different world regions simultaneously. A more recent theory holds that modern humans arose from *Homo erectus* in Africa and expanded into Europe and Asia about 100,000 years ago.

Key Terms: *anthropoid, p. 419*
australopithecine, p. 420
pithecanthropine, p. 421
Neanderthal, p. 421
Homo sapiens, p. 421

Concept Question: Do you think natural selection applies to the evolution of humans? Explain.

Glossary

acetyl coenzyme A A compound formed from a two-carbon fragment (acetyl group) of pyruvic acid, which is bonded to coenzyme A; it functions as a means of entry for compounds into the Krebs cycle.

acid Any substance that increases the hydrogen ion concentration of a solution.

actin A thin protein filament found in skeletal muscle; it acts with myosin to bring about muscle contraction and relaxation.

active artificial immunity A type of immunity acquired when an antigen is introduced into the body by vaccination; the body responds by activating its T cells and by producing antibodies.

active natural immunity The type of immunity acquired by natural exposure to a disease-causing antigen; the body responds by activating its T cells and by producing antibodies.

active site The specific area of an enzyme that interacts with its substrate during an enzymatic reaction.

active transport The energy-requiring transport of a solute across a cell membrane by carrier molecules from a region of low solute concentration to a region of higher solute concentration.

adaptation (1) Any structural, functional, or behavioral characteristic that increases an organism's ability to survive in a given environment, (2) The genetically determined ability of an organism to adjust to environmental demands.

adrenal cortex The outer tissue layer of the human adrenal glands; it secretes several hormones, including aldosterone and cortisol.

adrenal medulla The inner tissue layer of human adrenal glands; it secretes the hormones epinephrine and norepinephrine.

adrenocorticotropic hormone (ACTH) An anterior pituitary hormone that stimulates release of cortisol from the adrenal cortex.

aerobic Requiring oxygen for respiration and/or growth.

aldosterone An adrenal cortex hormone that regulates the concentration of sodium and potassium ions in the blood.

allantois One of the four vertebrate extraembryonic membranes; formed from an outpocketing of the yolk sac; it and its blood vessels form the body stalk in humans.

allele An alternative form of a gene.

allergy A heightened sensitivity to a substance (antigen) that in normal amounts is harmless to most people.

allotropic Referring to a geographically isolated population.

alpha cells Cells in the pancreas that secrete the hormone glucagon.

alternation of generations A plant reproductive cycle in which a haploid (n) generation is followed by a diploid ($2n$) generation, which produces the gametophyte, completing the cycle.

alveoli Microscopic, thin-walled sacs, clusters of which are found at the terminals of respiratory bronchioles; alveoli are the sites of respiratory gas exchange in the lungs.

amniocentesis A technique involving puncture of the amnion to examine fetal cells suspended in the amniotic fluid.

amnion One of the four vertebrate extraembryonic membranes; it lines the amniotic cavity and immediately surrounds the developing embryo and fetus.

amniotic cavity A fluid-filled cavity that forms between the inner cell mass and the blastocyst wall; it houses the developing embryo and fetus.

amniotic egg A fluid filled egg that, having a drought-resistant shell, allows some vertebrates to reproduce on land rather than in water.

anabolism The building up, or synthesis, phase of metabolism in which larger molecules are built from smaller molecules.

anaerobic Not requiring oxygen for respiration and/or growth.

angiosperm A plant possessing flowers; the seeds of angiosperms are enclosed in an ovary.

angiotensin A blood plasma protein hormone that constricts blood vessels, causing an increase in blood pressure.

anther In angiosperms, a group of spore sacs in which the male pollen grains are produced, and which is located at the tip of the filament.

antibody A protein produced in response to the presence of an antigen; antibodies are secreted by plasma cells and react with the antigen that caused their production.

anticodon Located on a loop of transfer RNA, a three-nucleotide sequence that is complementary to a specific codon of messenger RNA.

antidiuretic hormone (ADH) A posterior pituitary hormone that increases the permeability of the kidney collecting ducts to water.

antigen Any substance capable of stimulating the production of antibodies.

aorta The large artery arising from the left ventricle of the vertebrate heart; oxygenated blood leaves the heart via the aorta to be distributed throughout the body.

apical meristem Embryonic plant tissues found at the tips of developing shoots and roots; meristems supply cells for virtually all parts of the young and adult plant.

archaebacteria A group of ancient prokaryotes that usually inhabit extreme environments.

artery A blood vessel that conducts blood away from the heart.

atom The smallest unit of an element that retains the properties of that element.

atomic number The number of protons in the atomic nucleus of an atom.

ATP (adenosine triphosphate) A nucleotide consisting of adenine, ribose, and three phosphate groups (PO_4); it is the major form of stored cellular energy.

atrium A upper chamber of the heart through which blood passes before entering a ventricle.

autoimmunity Development of immunity against one's own body antigens.

autonomic nervous system (ANS) The division of the peripheral nervous system that controls the internal involuntary activities of the body; consists of the sympathetic and parasympathetic divisions.

autosome Any body chromosome other than a sex chromosome.

autotroph An organism that can synthesize organic nutrients from inorganic compounds using solar energy or inorganic molecules.

axon The neuron process, or fiber, that conducts the nerve impulse away from the cell body of a neuron.

bacteriophage A virus that infects bacteria.

base Any substance that lowers the hydrogen ion concentration of a solution.

B cell Activated in bone marrow, a type of lymphocyte that differentiates into an antibody-producing plasma cell.

beta cells Cells in the pancreas that secrete the hormone insulin.

bile A secretion produced in the liver and stored in the gall bladder; bile emulsifies lipids.

binary fission A type of reproductive cell division in prokaryotes in which each new cell receives a copy of the parent chromosome.

biodegradable Capable of being broken down and incorporated into the natural cycles of matter.

blastocyst In human pre-embryonic development, the hollow ball structure surrounded by the chorion and containing the inner cell mass; it arises from the morula.

blood A liquid tissue consisting of plasma and formed elements.

body stalk Formed from the allantois, a structure that attaches an embryo to the chorion; it later fuses with the yolk stalk to form the umbilical cord.

Bowman's capsule The anterior, double-layered portion of a nephron that surrounds the glomerulus; involved in producing the plasma filtrate.

bronchiole A small respiratory tubule in the lungs; formed from branching of a bronchus.

bronchus One of the two large respiratory tubes leading into the lungs; formed from branching of the trachea.

buffer A substance capable of neutralizing a solution by bonding with hydrogen ions or hydroxide ions.

calorie The amount of heat energy required to raise the temperature of one gram of water 1 degree C.

capillaries Tiny, thin-walled blood vessels through which various substances are exchanged between the blood and body cells; Also, thin-walled vessels in the lymphatic system that carry lymph.

carbaminohemoglobin The compound formed when carbon dioxide complexes with hemoglobin in red blood cells.

carbohydrate An organic compound consisting primarily of carbon, hydrogen, and oxygen atoms organized to form monosaccharides, disaccharides, and polysaccharides.

cardiac/coronary veins Veins in the heart that return deoxygenated blood from heart muscle to the right atrium.

cardiac sphincter A muscular constriction at the junction of the esophagus and the upper stomach.

carnivore An organism that feeds on flesh only.

carpel The female reproductive structure of flowering plants; consists of the stigma, style, and ovary.

carrying capacity The greatest number of organisms a given environment can support for an extended period of time.

catabolism The energy-releasing metabolic process whereby larger, more complex molecules are broken down into smaller, simpler ones.

cell The structural and functional unit of life; the smallest biological unit possessing life.

cell body The expanded region of a neuron that contains the nucleus and other cytoplasmic organelles.

cell cycle A sequence of stages through which a developing and dividing cell passes; includes G_1, S, G_2, and cell division.

cell plate Formed during cytokinesis, a membranous structure formed between a dividing plant cell; the new plant cell wall forms between the cell plate.

cell theory All organisms are composed of one or more cells, which are the structural and functional units of life.

cellular differentiation The developmental process whereby cells become different and specialized for specific functions; results from different genes being expressed in each cell type.

cellular respiration The cellular catabolic process involving glycolysis, the Krebs cycle, and the electron transport system.

cellulose A polysaccharide constituent of plant cell walls; it is constructed of chains of bonded glucose molecules.

cell wall The tough, outer wall surrounding the plasma membrane of plant cells; it consists primarily of cellulose.

central nervous system (CNS) The division of the vertebrate nervous system consisting of the brain and spinal cord.

centrioles Paired microtubular organelles that lie outside the nucleus of animal cells; they apparently are associated with spindle function.

centromere The region at which two sister chromatids are joined together.

cephalization Development and specialization of the anterior, or head, end of the nervous system.

cerebellum A section of the hind brain involved in regulating muscle coordination and balance.

cerebrospinal fluid (CSF) Derived from blood plasma, a fluid found in the brain and spinal cord.

cerebrum The largest structure of the vertebrate brain; composed of right and left hemispheres, it is the center for thought, memories, special sense interpretation, control of motor activities, and other functions.

chemical evolution A sequence of interactions between simple chemicals leading to formation of complex chemical compounds.

chemoreceptor A specialized nerve cell capable of detecting chemical stimuli.

chitin A polysaccharide found in the body coverings of insects and crustaceans, and in fungal cells; it consists of chains of bonded glucose molecules.

chlorophyll The green light-trapping pigment located within chloroplasts.

chloroplast The chlorophyll-containing organelle in which the reactions of photosynthesis occur.

chorion One of the four extraembryonic membranes of vertebrates; it is the outermost membrane and forms part of the mammalian placenta.

chorionic villi Extending from the inner wall of the chorion, tiny fingerlike projections that penetrate the uterine lining; they establish a close association between the blood circulations of the mother and embryo.

choroid The tough, vascularized middle layer of the eyeball.

chromatid One of the two strands of a duplicated chromosome.

chromatin The diffuse mass of DNA and protein that forms the chromosomes in eukaryotic cells.

chromosomes Found in the nuclei of eukaryotic cells, thin, threadlike bodies consisting of DNA and protein.

chyme The mixture of partially-digested food and digestive secretions in the stomach and duodenum.

cilia Tiny hairlike structures that function in cell movement and in filtering small particles.

circadian rhythm An activity of organisms that occurs on a 24-hour cycle.

cleavage A series of mitotic divisions of the zygote; it results in formation of a hollow ball of cells.

cleavage furrow An indentation on either side of a dividing animal cell; it eventually separates the cytoplasm, forming two daughter cells.

climax community The final and most stable community stage in ecological succession.

cochlea The fluid-filled organ of hearing in the inner ear; it contains the organ of Corti.

codominant Pertaining to a pair of heterozygous alleles that express both of their traits in the phenotype.

codon Copied from DNA, a three-nucleotide sequence on messenger RNA that specifies a particular amino acid or that initiates or terminates protein synthesis.

coevolution Simultaneous adaptation by interacting populations.

cohesion The clinging together of molecules of the same substance.

collecting duct The tubular structure that receives urine from the distal tubules of nephrons; it empties urine into the ureters.

commensalism A symbiotic relationship between two different species in which one member benefits and the other member is unaffected.

community Different species that live in the same geographical area.

compound Any substance consisting of two or more different elements.

cones Photoreceptor cells in the retina responsible for color vision.

conjugation A method of sexual reproduction among lower organisms in which genetic material is passed between mating organisms.

connective tissue Tissues that function in support, protection, binding together of body structures, and production of blood cells; includes loose connective tissue, fat tissue, tendons, ligaments, bone, cartilage, and blood.

continental drift The movement of continental land masses as a result of shifting crustal plates.

coronary arteries The two major arteries that supply oxygenated blood to heart muscle.

corpus luteum A structure in the ovary that forms after the oocyte is discharged from the follicle and that secretes estrogens and progesterone.

corticotropin releasing hormone (CRH) A hypothalamic hormone that controls release of ACTH from the anterior pituitary gland.

cortisol An adrenal cortex hormone released in response to stress; it promotes release of glucose, fatty acids, and amino acids from most body cells and promotes uptake of amino acids by the liver.

covalent bond An attractive force between atoms resulting from the sharing of their outer energy level electrons.

cross bridges Tiny protuberances on the thick myosin filaments of skeletal muscle; they pull the thin actin filaments during muscle contraction.

crossing over Reciprocal exchange of genetic material between chromatids of homologous chromosomes in prophase I of meiosis.

crust The outermost layer of Earth, about 80 kilometers thick.

cytokinesis Cytoplasmic division that forms separate cells following mitosis or meiosis.

cytoplasm In living cells, the semifluid contents containing the organelles and bounded by the plasma membrane.

deactivating enzyme An enzyme that breaks down a neurotransmitter at a chemical synapse.

decomposer Bacteria or fungi that break down dead organic matter.

dehydration synthesis A metabolic reaction in which a large molecule is formed from smaller molecules by the removal of a water molecule.

dendrite The neuron process that conducts the nerve impulse toward the cell body of a neuron.

density-dependent factor Any population-regulating factor that has a greater influence as the number of organisms living in a particular area increases; includes competition, predation, and parasitism.

density-independent factor Any population-regulating factor that has the same influence on a population regardless of its size.

deoxyribonucleic acid (DNA) The double-stranded nucleic acid that carries the genetic information for determining cell protein structure.

deoxyribose A 5-carbon sugar constituting part of the structure of a DNA molecule.

diaphragm The large, flat sheet of muscle separating the thoracic and abdominal cavities.

diastolic pressure The minimum pressure in a heart chamber when it relaxes.

diffusion The spontaneous movement of a substance from where that substance is in higher concentration to where it is in lower concentration, until equilibrium is reached.

dihybrid cross A mating between parents differing in two genetic traits.

diploid number The double, or $2n$, number of chromosomes in a cell; one set of chromosomes is inherited from each parent.

disaccharide A carbohydrate consisting of two bonded simple sugar molecules, or monosaccharides.

distal tubule The terminal end of a nephron tubule; it empties urine into a collecting duct.

dominant Pertaining to an allele that covers up the phenotypic effects of a recessive allele.

duodenum The upper 25 cm of the small intestine.

ecological succession A sequence of usually gradual changes in an ecosystem, often culminating in a climax community.

ecology The biological science concerned with the interactions living organisms have with each other and with their nonliving environment.

ecosystem A community (different populations) plus its nonliving environment.

ectoderm One of the three germ tissue layers of the human embryo; it originally forms the outermost layer of embryonic cells.

effector organ Usually a muscle or gland that responds to a nerve impulse.

electron A negatively-charged atomic particle having a charge of –1.

electronegativity Pertaining to an atom that tends to attract electrons to itself

electron transport chain A cellular system of carrier molecules that transports electrons from higher to lower energy levels.

element A substance that cannot be broken down into a simpler substance by ordinary chemical means; a basic form of matter.

embryo Term applied to the developing human from the beginning of the third week until the eighth week of development; also refers to many developing plants and animals.

embryogenesis A series of events undergone by a zygote in its development into an embryo; includes growth, cellular differentiation, and morphogenesis.

embryonic disk Formed from the inner cell mass, the structure that gives rise to the human embryo.

emulsification The breaking up of large lipid droplets into small lipid droplets by the action of bile.

endocytosis Engulfment of substances by a cell membrane; includes phagocytosis and pinocytosis.

endoderm One of the three germ tissue layers of the human embryo; it originally forms the innermost layer of embryonic cells.

endoplasmic reticulum In eukaryotic cells, a membranous network within the cell cytoplasm; some ER is studded with ribosomes and some is not.

endorphin One of the brain neuropeptides that acts to suppress pain signals.

energy The capacity to do work; that is, to bring about some action or effect. Potential energy is the energy stored in a substance; kinetic energy is the energy of motion, the energy that does work.

energy efficiency The ratio of useful energy to the total energy of a system.

energy flow The entry of solar energy to Earth and the subsequent passing of this energy in converted form from one organism to another.

enkephalin One of the brain neuropeptides that acts to suppress pain signals.

environmental resistance The total of the factors that limits growth of a population; includes density-dependent and density-independent factors.

enzyme An organic molecular catalyst (protein or RNA) that increases the rate of a chemical reaction.

epigenesis The progressive development of form of an embryo.

epinephrine A hormone secreted by the adrenal medulla in response to sympathetic nervous system stimulation; it usually is released during periods of stress.

epithelial tissue Tissue that forms the outer covering of plant and animal structures, and that lines the inner walls of organs and tubular structures in animals.

erythropoietin A hormone that stimulates red blood cell production by bone marrow tissue.

estrogens The female sex hormones secreted by the ovaries.

eubacteria A group of prokaryotes that includes the majority of bacteria.

eukaryotic Pertaining to cells having a membrane-bound nucleus and a number of cytoplasmic organelles; refers to all organisms other than bacteria.

eustachian tube The passageway connecting the middle ear to the pharynx.

eutrophication The process by which a body of water or an entire ecosystem becomes enriched through the addition of organic nutrients; may occur naturally or through the effects of pollution.

evolution Changes in the gene pool of a population from one generation to another as a result of natural selection acting on heritable variation.

exocytosis The cellular process by which substances are expelled through the cell membrane to the outside of the cell; essentially the reverse of endocytosis.

extranuclear inheritance Acquisition of genetic traits from cytoplasmic organelles rather than from nuclear genes.

facilitated diffusion Transport of a solute across a cell membrane—assisted by carrier molecules—from regions of higher solute concentration to regions of lower concentration.

fallopian tube In females, the muscular tube extending from an ovary to the uterus.

farsightedness The ability to see distant objects, while near objects appear blurry; results when light rays focus behind the retina.

fast-twitch fibers Skeletal muscle fibers that contract rapidly and fatigue quickly.

fat A lipid that is usually solid at room temperature.

fatty acid An organic acid that comprises part of the structure of neutral fats, oils, waxes, and phospholipids.

fermentation The chemical breakdown of an organic molecule in the absence of oxygen.

fetus Term applied to the developing human fetus from the ninth week of development until birth.

filtrate The portion of blood plasma and its constituents that passes through the glomerular walls into Bowman's capsule of a nephron.

first law of thermodynamics The law of conservation of energy: energy cannot be created or destroyed.

fitness The relative ability of an organism to survive and produce offspring.

flagellum An elongated, whiplike locomotor organelle found of the bodies of some bacteria, protists, and on sperm cells.

follicle-stimulating hormone (FSH) An anterior pituitary hormone that stimulates maturation of the ovarian follicle and primary oocyte.

food chain A sequence of organisms, each serving a food for the next group of organisms in the chain.

food web A certain number of interconnected food chains.

formed elements Collectively, the red and white blood cells and platelets found in blood.

fruit The mature plant ovary, along with associated structures; the fruit contains the seeds.

gamete A haploid reproductive, or sex, cell.

gametogenesis The cell division process resulting in the formation of gametes.

gametophyte In plants with an alternation of generation cycle, the haploid stage that produces gametes.

ganglion A group of nerve cell bodies lying outside the central nervous system.

gastrula A stage in embryonic development in which the blastula becomes a three layered embryo composed of ectoderm, mesoderm, and endoderm.

gene A certain sequence of nucleotides along a strand of DNA that codes for a specific protein or polypeptide; the unit of heredity.

gene cloning Production of copies of a gene from a single gene.

gene flow The change in the gene pool resulting from organisms, seeds, or spores entering or leaving a population.

gene pool The total of all of the genes of all the organisms in a population.

genetic drift Random changes in the gene pool of a population from generation to generation.

genetic engineering Insertion of a foreign gene into a host DNA.

genetic recombination A unique arrangement of genes resulting from crossing over, independent assortment, and sexual reproduction.

genetics The biological science concerned with the principles governing inheritance.

genotype The genetic makeup of an organism; the type of genes it has.

genotypic ratio The relative proportions of different offspring genotypes resulting from a genetic cross.

germ layers The three embryonic tissues—ectoderm, mesoderm, and endoderm—from which all adult animal tissues are derived.

glomerulus A mass of capillaries surrounded by Bowman's capsule of a nephron; it filters blood plasma and blood constituents into Bowman's capsule.

glucagon A hormone released from the alpha cells of the pancreas; it raises the blood sugar level, preventing hypoglycemia.

glycerol A sugar alcohol consisting of a 3-carbon backbone with a hydroxyl (OH) group bonded to each carbon atom; it is the basic structural compound of neutral fats, oils, and phospholipids.

glycogen A polysaccharide that functions as the storage form of glucose in humans and other animals; it consists of bonded chains of glucose molecules.

glycolysis Occurring in the cell cytoplasm, a series of reactions that split glucose into two molecules of pyruvic acid.

Golgi apparatus In eukaryotic cells, an organelle consisting of a stack of flattened membranous sacs; it functions in packaging, distributing, and secreting cellular products.

granum A stack of thylakoid membranes within a chloroplast.

gravitropism The response of an organism to gravity.

growth hormone An anterior pituitary hormone that stimulates body cells to grow and multiply.

growth regulator A plant chemical that influences various aspects of plant growth and development.

guard cells Paired photosynthetic cells that border a stomate; they function in regulating the size of the stomate.

habitat The environment in which an organism lives and interacts.

haploid number The single, or *n*, number of chromosomes in a cell; one-half the diploid number.

heat of vaporization The amount of heat energy required to convert a liquid to a gas.

hemoglobin The red, oxygen-transporting protein pigment in red blood cells.

heterotroph An organism that must acquire its nutrients by feeding on other organisms or some other outside source.

heterozygous Pertaining to paired alleles that are different for a given trait, as *Tt*.

homeostasis The ability of an organism to maintain an internal physiological stability, as in temperature regulation; also may refer to communities or ecosystems that are resistant to change.

homologous Pertaining to chromosome pairs that are of the same size and proportion; one chromosome is contributed by the female and one by the male.

homology Similarity of structure resulting from a common ancestry.

homozygous Pertaining to paired alleles that are the same for a given trait, as *TT*.

hormone A chemical messenger released from endocrine structures and transported to a target organ, where the hormone exerts a specific biological effect.

human chorionic gonadotropin (HCG) A hormone released from the placenta early in pregnancy; it stimulates the corpus luteum to continue its secretion of progesterone.

hydrogen bond A weak attractive force between a hydrogen atom and another atom with a negative charge.

hydrolysis A metabolic reaction in which a large molecule is split into smaller molecules by the addition of a water molecule.

hyperglycemia An abnormally elevated concentration of glucose in the blood.

hypertonic Pertaining to a solution that has a higher solute concentration than that of another solution.

hypoglycemia An abnormally low concentration of glucose in the blood.

hypothalamus A mass of neurons located in the forebrain above the pituitary gland.

independent assortment The principle that chromosomes, and hence, genes, are distributed randomly in meiosis; the members of one pair of homologous chromosomes do not influence the way the members of any other pair are distributed in the gametes.

induction The process whereby one group of embryonic cells influences development of another group.

inferior vena cava The great vein that returns all deoxygenated blood from vessels below the heart to the right atrium of the heart.

inner cell mass Located within the human blastocyst, a group of cells that becomes the embryonic disk from which the human embryo arises.

inorganic Nonliving; not containing the element carbon.

insulin A hormone released from the beta cells of the pancreas; it lowers the blood glucose level, preventing hyperglycemia.

interneuron Found only in the central nervous system, a nerve cell that relays nerve impulses between sensory and motor neurons.

interphase The period of the cell cycle in which DNA replication and production of organelles and other constituents occurs; includes G_1, S, and G_2 stages.

interspecific Occurring between or among different species.

interstitial cell A cell of the testes that secretes the male sex hormone testosterone.

interstitial cell-stimulating hormone (ICSH) A protein pituitary gland hormone that stimulates the interstitial cells of the testes to secrete testosterone.

intraspecific Occurring within a single species.

invertebrate An animal that lacks a spinal column.

ion An electrically-charged (either positive or negative) atom or molecule.

ionic bond An attractive force between ions of opposite electrical charges.

iris The colored portion of the eye that regulates the size of the pupil.

islets of Langerhans A group of hormone-secreting cells in the pancreas; the cells include alpha and beta cells.

isotonic Pertaining to a solution that has the same concentration of solute as that of another solution.

jejunum The middle section of the small intestine between the duodenum and ileum; about 1 m in length.

karyokinesis Division of a cell nucleus.

kinetic energy The energy of motion; the energy that does work.

kingdom The broadest and most inclusive category of biological classification.

Krebs cycle Occurring in mitochondria, a cyclic series of cellular respiration reactions that break down glucose into carbon dioxide, hydrogen ions, and electrons.

lacteal Located in the core of an intestinal villus, a lymphatic vessel involved in absorption of digested lipids.

lactic acid An acid, $C_3H_6O_3$, normally produced in animal cells as a result of the anaerobic metabolism of glucose.

larynx The voice box or "Adam's apple", located at the entrance to the trachea.

law of conservation of matter Matter cannot be created or destroyed; thus, the quantity of matter in the universe is constant.

memory cell A T or B lymphocyte that has been exposed to an antigen and therefore able to react against the antigen on second and subsequent exposures. Memory cells remain in lymphoid tissues as a long-lasting reserve of immune cells.

meninges The three layers of tissues covering the brain and spinal cord.

mesoderm One of the three germ layers of the human embryo; it forms the middle layer of cells.

messenger RNA (mRNA) Transcribed from DNA, a ribonucleic acid that attaches to ribosomes and carries the genetic message for the amino acid sequence of a specific protein.

metabolism All the chemical processes occurring in living organisms; includes anabolism and catabolism.

microfilament In eukaryotic cells, a thin, elongated organelle that functions in cell support and movement.

microtubule In eukaryotic cells, a small, hollow cylinderlike organelle that functions in cell support and movement.

minerals Nutrient ions required in varying amounts by living organisms.

mitochondrion In eukaryotic cells, a small organelle that functions as the major site of ATP synthesis.

mitosis Division of a cell nucleus to form two genetically-identical nuclei.

mitotic spindle An organization of microtubules that controls chromosome movement in mitosis and meiosis.

mixed nerve A nerve containing both sensory and motor components.

molecule Two or more chemically bonded atoms, the smallest unit of a compound that retains the properties of that compound.

molting The shedding of the outer covering, or exoskeleton, of an animal.

monohybrid cross A mating between parents differing in one genetic trait.

monosaccharide A simple sugar, some of which serve as constructive units of disaccharides and polysaccharides.

morphogenesis The development of organ and body shape or form during embryogenesis.

morula In embryogenesis, the stage at which the embryo is a compact, sixteen-cell spherical mass.

motor neuron A nerve cell that transmits a nerve impulse from the central nervous system to an effector organ.

motor unit A motor neuron and the skeletal muscle fibers it supplies.

multiple allelism The condition in which more than two alleles are responsible for a given genetic trait.

mutation A heritable change in the nucleotide sequence of DNA.

mutualism A symbiotic relationship between two species in which both members benefit.

myoglobin The red, oxygen-transporting protein pigment found in muscle tissue.

myosin A thick protein filament found in skeletal muscle; it acts with actin to bring about muscle contraction and relaxation.

NADP A cellular coenzyme that functions as a hydrogen ion and electron acceptor in photosynthesis; NAD and FAD serve the same function in cellular respiration.

natural selection The process whereby individuals more successfully adapted to their specific environment tend to produce a greater number of surviving offspring than less successfully adapted members of a population; differential success in reproduction within a population.

nearsightedness The ability to see near objects, while distant objects appear blurry; results when light rays focus in front of the retina.

negative feedback A body mechanism that maintains homeostasis by shutting off or decreasing the effect of a process that otherwise tends to magnify in effect.

neo-Darwinism Darwinian evolutionary theory coupled with the laws of genetics.

nephron The microscopic functional unit of the kidneys; it forms urine through filtration, reabsorption, and secretion.

nerve A bundle of nerve fibers located outside the central nervous system.

nervous tissue A tissue composed of neurons, which are responsible for the sensation—response mechanisms of an animal.

neural tube Derived from ectoderm overlying the notochord, an embryonic tubular structure that gives rise to the central nervous system.

neuromuscular junction The synaptic connection between a motor neuron and a muscle fiber.

neuron A nerve cell, the structural unit of the nervous system; it consists of a cell body, dendrite(s), and an axon.

neurotransmitter Released by the tips of axons, a chemical that relays the nerve impulse across a synapse by stimulating an adjacent neuron or effector.

niche The total role an organism or species plays in nature.

nondisjunction The failure of homologous chromosome pairs or sister chromatids to separate during meiosis.

nonspecific immunity The form of immunity involving internal and external factors that protect the body against invading substances without having to "know" the specific identity of the substances; the factors include skin, enzymes, chemicals, and phagocytes.

nonvascular Pertaining to a plant that lacks the internal conducting tissues xylem and phloem.

norepinephrine A hormone released from the adrenal medulla in response to sympathetic nervous system stimulation; it exerts effects similar to those of epinephrine.

notochord A fibrous tissue rod extending along the dorsal length of all chordate embryos; in humans and other vertebrates, it is replaced by the vertebral column.

nucleotide The basic building block of nucleic acids; it consists of a nitrogen base, a 5-carbon sugar, and a phosphate group.

nutrient cycle A natural sequence in which matter is converted in form and repeatedly circulated through the living and nonliving worlds.

oil A lipid that is usually liquid at room temperature.

olfactory cell Responsible for the sense of smell, a sensory receptor that responds to chemicals in the air; located in the nasal cavities.

omnivore An organism that feeds on both plant and animal matter; that is, an organism that is both an herbivore and a carnivore.

oogenesis The production of eggs or oocytes by the female reproductive structure.

organ A group of different tissues organized into a body structure—such as a heart—to perform a particular function.

organelle Any one of the discrete cytoplasmic bodies that performs a specific function in eukaryotic cells.

organic compound As applied to living organisms, a carbon-containing substance consisting of two or more atoms.

organ of Corti The functional unit of hearing located in the cochlea.

osmosis The spontaneous movement of water across a selectively permeable membrane from where there is proportionately more water to where there is less water, until equilibrium is reached.

otoliths Small grains of calcium carbonate found in the utricle and saccule of the inner ear.

ovulation The release of an oocyte from an ovary.

oxidation The loss of electrons or hydrogen ions from an element or compound.

oxytocin A posterior pituitary hormone that causes contraction of the muscular wall of the pregnant uterus during labor, and also stimulates release of milk from the breasts.

pancreatic amylase A carbohydrate-digesting enzyme secreted by the pancreas.

pancreatic lipase A lipid-digesting enzyme secreted by the pancreas.

Pangaea The ancient giant land mass the eventually split into Laurasia and Gondwanaland; from these two land masses all the continents were derived.

parasitism A symbiotic relationship between two species in which one—the host—is harmed and the parasite benefits.

parasympathetic division One of the two divisions of the autonomic nervous system; it acts in opposition to the sympathetic division.

parturition A term for childbirth.

passive immunity A type of immunity acquired by the transfer of antibodies from mother to child or by injection of an antiserum containing antibodies prepared from the blood of an animal.

pepsin A protein-digesting enzyme secreted by cells in the stomach.

peripheral nervous system (PNS) The spinal, cranial, and autonomic nerves, all of which extend from the central nervous system.

peristalsis The rhythmic contraction and relaxation of the smooth muscle wall of the digestive tract.

PGAL A three-carbon sugar generated in the light-independent reactions of photosynthesis; it is used in the synthesis of various sugars and other organic compounds.

pH The relative acidity or alkalinity of a solution.

phagocytosis The engulfment of large particulate matter by a cell membrane; particles are taken to the interior of the cell and digested or destroyed.

pharyngeal slits In developing chordates, small openings leading from the pharynx to the outside; the slits persist in adult invertebrate chordates.

pharynx The muscular section of the digestive tract below the mouth cavity; the throat.

phenotype The outward appearance of an organism.

phenotypic ratio The relative proportions of different offspring phenotypes resulting from a genetic cross.

pheromone Secreted by many animal species, a chemical that functions as a communication signal between members of the same species.

phloem The internal vascular tissue that transports organic nutrients throughout a plant.

photoperiodism In plants, a response to a change in the length of daylight and darkness.

photosystem A specific chlorophyll molecule and its primary electron acceptor.

phototropism The turning of a plant or plant structure toward a light source.

phytochrome A plant pigment that promotes development of various plant structures.

phytoplankton Microscopic algae and protists that float on the surface of oceans and lakes.

pinocytosis The uptake of water droplets by tiny vesicles in the plasma membrane.

pituitary gland An endocrine organ located in the forebrain beneath the hypothalamus; it secretes at least nine hormones, seven from its posterior lobe and two form its posterior lobe.

placenta In mammals, an organ comprised of the lining of the uterus and the outer membrane (chorion) of the embryo; it functions in the exchange of nutrients, respiratory, gases, and wastes between the blood of the mother and the blood of the embryo.

plankton Blue-green bacteria, protists, and tiny animals that float on the surface of oceans.

plasma The liquid portion of blood.

plasma cell An antibody-producing cell derived from the differentiation of B lymphocytes.

plasma membrane The boundary membrane that surrounds the cytoplasm of cells; it consists of a phospholipid bilayer in which proteins are embedded.

plasmid A small circular piece of cytoplasmic DNA found in bacteria.

plasmolysis In plant cells, the collapse of the plasma membrane and cytoplasmic contents as a result of water loss.

plastid In plant cells, a membranous organelle that stores pigments and various nutrients; includes leucoplasts, chromoplasts, and chloroplasts.

platelet One of the formed elements of blood; tiny cell fragments involved in the clotting of blood; also called a thrombocyte.

platelet plug An aggregate of platelets that stops bleeding from small wounds.

polyploidy The condition of possessing one to several additional sets of chromosomes.

polysaccharide A carbohydrate consisting of a chain of individual simple sugars, or monosaccharides, bonded together.

population A group of organisms of the same species living together in a particular geographic area.

potential energy Energy stored in a substance or object.

predation The act of preying on another organism.

primary consumer An herbivore that feeds on autotrophs, or producers.

primary electron acceptor Located in the thylakoid membranes, a molecule that traps and holds electrons boosted from chlorophyll molecules.

primary structure The sequence of amino acids in a protein molecule.

principle of independent assortment When one pair of alleles separates in meiosis, they do not affect how another, different pair of alleles separates.

principle of segregation Separation of a pair of alleles into different gametes at the end of meiosis.

producer An autotroph; an organism capable of synthesizing its own food using the energy of sunlight or energy from inorganic molecules; the original link in a food chain.

product The substance (s) formed when an enzyme acts on its substrate.

progesterone A steroid hormone produced by the corpus luteum and placenta; its major function is to maintain the lining of the uterus in preparation for or maintenance of pregnancy.

prokaryotic Pertaining to cells that lack a distinct membrane-bound nucleus and all eukaryotic organelles except ribosomes; only bacteria are prokaryotic.

prolactin An anterior pituitary hormone that stimulates mammary gland development and milk production in humans and other mammals.

property Any particular characteristic, quality, or attribute of a substance.

prostaglandins A group of cellular fatty acids with a wide variety of functions, including stimulation of contractions of the pregnant uterus.

protease One of several protein-digesting enzymes secreted by the pancreas.

proton A positively-charged atomic particle found in the nucleus of an atom; it has a charge of +1.

proximal tubule The portion of the kidney nephron between Bowman's capsule and the loop of Henle.

pulmonary artery The large artery that carries deoxygenated blood from the right ventricle of the heart to the lungs.

pulmonary veins Blood vessels that carry oxygenated blood from the lungs to the left atrium of the heart.

Punnett square A checkerboard grid used to cross gametes to determine offspring types.

pupil The circular opening in the iris through which light enters the eye.

pyloric sphincter A muscular constriction at the junction between the lower stomach and duodenum.

pyruvic acid A 3-carbon acid derived from the splitting of a glucose molecule in glycolysis.

quaternary consumer A carnivore that feeds on the tertiary consumers.

radicle The embryonic root of a plant.

recessive Pertaining to an allele that is not expressed in the phenotype when paired with a dominant allele.

recombinant DNA A hybrid DNA resulting from combining a foreign DNA with a host DNA.

recombination A method by which genes are brought together in new combinations with other genes. In eukaryotes, recombination may occur through crossing over or at fertilization.

red blood cell One of the formed elements of blood; it functions mainly in the transport of oxygen.

reduction The gain of electrons or hydrogen ions by an element or compound.

reflex An automatic, unlearned response to a stimulus performed without thought or prior knowledge.

reflex arc The functional unit of the nervous system; involves a receptor, sensory neuron, usually an interneuron, motor neuron, and an effector organ.

renin An enzyme released by the kidneys when arterial blood pressure falls; it is involved in formation of the hormone angiotensin.

respiratory center A mass of neurons in the medulla oblongata, it is sensitive to the level of carbon dioxide in the blood, and regulates the respiratory rate by sending impulses along nerves to the respiratory muscles.

restriction enzyme An enzyme that recognizes certain base sequences on DNA and cuts the DNA molecule apart at that point.

retina The inner layer of the retina; it contains the rods and cones.

retrovirus An RNA virus that employs a reverse transcriptase enzyme in transcribing DNA from RNA.

ribosomal RNA (rRNA) Transcribed by DNA, a ribonucleic acid that, along with protein, constitutes a ribosome.

ribosomes In all living cells, tiny spherical organelles that function as the sites of protein synthesis.

RNA transcriptase A viral enzyme that allows an RNA template to transcribe new RNA.

root pressure The pushing force exerted by osmosis of water into the root cells of plants.

RuBP (ribulose bisphosphate) The 5-carbon sugar to which carbon dioxide bonds on entering the chloroplast in the light-independent reactions of photosynthesis; it is regenerated in the plant through the bonding of PGAL molecules.

saccule One of the organs of equilibrium in the inner ear; it detects linear acceleration and deceleration.

salivary amylase The starch-digesting enzyme secreted by salivary glands in the mouth.

sclera The tough, outermost covering of the eyeball.

secondary consumer A carnivore that feeds directly of herbivores.

secondary structure The spiral or pleated shape of a protein molecule.

second law of thermodynamics In an energy conversion, some of the energy escapes in the form of nonusable, nonrecoverable heat; thus, the amount of useless energy in the universe continues to increase, while the amount of useful energy declines.

seed In angiosperms, or flowering plants, the plant embryo, endosperm, and protective seed coat.

semicircular canals The three tubelike organs of equilibrium in the inner ear; they detect angular motion.

seminiferous tubules Within the testes, the tiny tubules in which sperm are produced by meiosis.

sensory neuron A nerve cell that conducts a nerve impulse away from a sensory receptor toward the central nervous system.

sex chromosome In humans and some other animals, the X and Y chromosomes that determine the sex of an organism.

sex-linked In humans, pertaining to a genetic trait carried by alleles on the X chromosome; also X linked.

sinoatrial node (SA node) A small mass of specialized tissue located high in the right atrium of the heart; it initiates the electrical impulse responsible for the normal heartbeat; also called the pacemaker.

slow-twitch fibers Skeletal muscle fibers that contract slowly and do not fatigue readily.

sodium/potassium pump Located in cell membranes, an ion carrier system that transports sodium ions out of cells and transports potassium ions into cells.

society A population of individuals that interact to mutual advantage; characterized by interbreeding, division of labor, and somewhat rigid organization.

solution A liquid and the substance, or solute, dissolved in it.

speciation The development of new species from preexisting species.

species Organisms that are alike structurally and functionally, and that normally interbreed to produce fertile offspring.

specific heat The amount of heat energy that must be gained or lost to change the temperature of 1 gm of a substance 1 degree C.

specific immunity The type of immunity resulting from the action of T cells and B cells, it involves previous exposure to an antigen, resulting in recognition and memory of the antigen.

spermatogenesis The production of sperm by the male reproductive structure.

spinal reflex A reflex involving the muscles supplied by the 31 pairs of spinal nerves.

spindle An arrangement of microtubules involved in the movement of chromosomes during cell division.

sporangium In some plants, a structure in which spores are produced.

spore An asexual reproductive cell capable of germinating into an adult structure or organism.

sporophyte In plant alternation of generations, the diploid ($2n$) spore-producing generation.

stamen The male reproductive structure of flowering plants; it consists of the filament and anther.

starch A polysaccharide stored in plants and found in a variety of foods; it consists of chains of bonded glucose molecules.

steroid Any of a group of lipid compounds structured of four interlocking carbon rings; includes cholesterol, testosterone, and others.

stimulus Any environmental factor capable of causing a response by an organism.

stomata Tiny openings on the surface of leaves through which oxygen and carbon dioxide diffuse out of and into a plant.

stretch reflex A spinal reflex, such as a knee jerk, involving a stretch receptor located in a muscle.

stroma The fluid-filled space surrounding the thylakoids in a chloroplast, and the site of the light-independent reactions of photosynthesis.

substrate The substance on which an enzyme acts.

superior vena cava The great vein that returns all blood from above the heart to the right atrium of the heart.

symbiosis A relationship in which two organisms live in close association with each other.

sympathetic division One of the two divisions of the autonomic nervous system; it acts in opposition to the parasympathetic division.

sympatric Pertaining to speciation within a single population.

synapse The neural junction between an axon and another cell; the nerve impulse is conducted across the synapse by a neurotransmitter.

system A group of different organs working together to perform a particular function, such as digestion.

systolic pressure The maximum pressure in a heart chamber during contraction.

taste bud The sensory receptors for the sense of taste; found mostly on the tongue.

taxis An orientational response toward or away from a stimulus.

taxonomy The biological science concerned with naming, describing, and classifying organisms.

T cell A lymphocyte of the immune system that, after activation in the thymus gland, moves into lymphoid tissue; some T cells directly attack foreign antigens.

template A pattern from which something is copied, as DNA is the template from which RNA is transcribed in living organisms.

tendon A band of tough connective tissue that binds muscle to bone.

Ten Percent Rule In that approximately 90 percent of the energy at each level in a food chain escapes as heat, only ten percent of the energy at each level is available to the next level.

territoriality Defense of a section of an animal's home range against intruders.

tertiary consumer A carnivore that feeds directly on secondary consumers.

tertiary structure The contorted, folded shape of a protein molecule resulting from various bondings and attractive forces.

test cross A mating between an organism displaying a dominant trait and an organism homozygous for that trait.

testes The male sperm-producing and testosterone-producing organs.

testosterone The sex hormone of male vertebrates.

theory A hypothesis that is supported by experimental evidence.

thylakoid Found within chloroplasts, a flattened membranous sac, a stack of which forms a granum; thylakoids are the sites of the light-dependent reactions of photosynthesis.

thyroid-stimulating hormone (TSH) An anterior pituitary hormone that stimulates the release of thyroxine and other hormones from the thyroid gland.

thyroid-stimulating hormone releasing factor (TRF) Released from the hypothalamus, a hormone that stimulates the pituitary gland to release thyroid-stimulating hormone (TSH).

thyrotropin-releasing hormone (TRH) A hypothalamic hormone that stimulates the pituitary gland to secrete thyroid-stimulating hormone (TSH).

thyroxine The major hormone secreted by the thyroid gland; it increases the rate of body metabolism and is necessary for normal growth and maturation of tissues.

tissue A group of the same type of cells organized to perform a particular function.

totipotent Referring to a cell capable of giving rise to all parts of an organism.

trachea The respiratory tube leading from the pharynx to the bronchi; the windpipe.

transcription The copying of a complementary strand of RNA from a DNA template.

transfer RNA (tRNA) Transcribed from DNA, a ribonucleic acid that transports amino acids from the cytoplasm to their appropriate codon sites on the messenger RNA-ribosome complex.

translation The synthesis of a polypeptide as directed by the sequence of codons on a strand of messenger RNA.

transpiration The evaporation of water from the surface of plant leaves through the stomates.

tropism The bending of a plant body toward or away from a stimulus.

turgor movement A response in plants resulting from an increase or decrease in turgor pressure within the cells.

turgor pressure In plant cells, the internal fluid pressure created by osmosis.

umbilical cord A cordlike structure that connects the human fetus to the placenta; it contains the umbilical arteries and vein.

umbilical stalk The structure formed from the fusion of the body stalk and yolk stalk; it develops into the umbilical cord.

ureter A muscular tube that conveys urine from the kidney to the urinary bladder; one ureter extends from each kidney.

urethra The muscular tube that conveys urine from the urinary bladder to the outside of the body.

urinary bladder The muscular sac that receives and stores urine coming in from the ureters.

uterus In female mammals, the thick, muscular organ in which the young are carried and nourished; in other animals it may serve to store eggs.

utricle One of the organs of equilibrium in the inner ear; it detects the pull of gravity.

vacuole A membrane-bound space within the cytoplasms of many cells; it functions in the storage of ions, crystals, water, and other substances.

valence The bonding, or combining, power of an atom, as determined by the number of electrons in its outermost energy level.

vascular Pertaining to a plant that contains the internal conducting tissues xylem and phloem.

vein A blood vessel that conducts blood toward the heart.

ventricle A lower chamber of the heart that pumps blood to the lungs or out to the remainder of the body.

vertebrate An animal having a spinal column.

villus A microscopic, fingerlike projection that extends from the lining of the small intestine.

visceral reflex A reflex that causes smooth muscle and cardiac muscle to contract.

water vascular system In echinoderms, a system of internal vessels that branch into tiny tube feet; functions in locomotion, feeding, and gas exchange.

white blood cell One of the formed elements of blood; it functions in defense of the body against infection and disease. Also called a leukocyte.

withdrawal reflex A reflex resulting in the movement of a body structure away from the stimulus; involves a sensory receptor, sensory neuron, usually an interneuron, motor neuron, and an effector organ.

xylem The internal tissue forming the woody part of plants; it functions in transporting water and as physical support of the plant.

yolk sac One of the four extraembryonic membranes of vertebrates; a membranous pouch suspended below the human embryonic disc. It serves as a site of blood cell formation in the early embryo, but later degenerates.

yolk stalk The basal portion of the yolk sac that fuses with the body stalk to form the umbilical cord.

zooplankton Tiny crustaceans, fish eggs, and invertebrate larvae that float on the surface of oceans and lakes.

zygote A fertilized egg; the cell formed by the union of male and female gametes (fertilization).

Common Word Roots Used in Biology

Root	*Word Example*
a (Gk.) not, without	*acellular,* without, or not having, cells
ab (L.) away, off	*aboral,* on the side away from the mouth
aer (Gk.) air, atmosphere	*aerobic,* requiring or living in the presence of oxygen
agglutin (L.) glued together	*agglutinate,* to clump or stick together
alg (L.) seaweed; also (Gk.) pain	*algae,* photosynthetic aquatic plants lacking true leaves, stems, or roots; *neuralgia,* pain in a nerve
allelo (Gk.) one another, parallel	*allele,* alternative form of a gene
alve (L.) a cavity or socket	*alveolus,* microscopic air sac in a lung
amoeb (Gk.) change	*amoeba,* microscopic protozoan that changes shape as it moves
amphi (Gk.) double, on both sides	*amphibian,* a vertebrate that lives both in water and on land
an (Gk.) not, without	*anaerobic,* not requiring or living in the absence of oxygen
anti (Gk.) against, opposite	*antibody,* a protein molecule that binds to a foreign antigen
archeo (Gk.) ancient	*archaeology,* the study of history from the remains of ancient human cultures
arthr (Gk.) a joint	*arthropod,* an invertebrate having jointed appendages
aut (Gk.) self	*autotroph,* an organism capable of synthesizing its own food
bi (L.) two, double	*bilateral,* pertaining to or having two sides
bio (Gk.) life	*biology,* the scientific study of living things
blast (Gk.) a bud	*blastula,* a hollow ball of cells formed during embryonic development of some organisms
bronch (Gk.) windpipe	*bronchitis,* inflammation of the bronchial tubes of the lungs
calor (L.) heat	*calorie,* the amount of heat required to raise the temperature of 1 gram of water 1° C
card (Gk.) the heart	*cardiac,* pertaining to the heart

cephal (Gk.) the head	*cephalic,* pertaining to the head
chlor (Gk.) green	*chlorophyll,* a photosynthetic pigment found in green plants
chondr (Gk.) cartilage	*chondrichthyes,* fish having an endoskeleton of cartilage
chrom (Gk.) color	*chromoplast,* a plant organelle containing certain colored pigments
cide (L.) kill	*insecticide,* a chemical that kills insects
coel (Gk.) hollow	*coelenterate,* an invertebrate having a hollow gut or intestine
cyan (Gk.) dark blue	*hemocyanin,* a dark-blue pigment
cyt (Gk.) a cell	*cytology,* the study of cells
de (L.) down, away	*decomposer,* an organism that breaks down dead organic material
dendr (Gk.) a tree	*dendrite,* a branching nerve cell fiber
derm (Gk.) skin	*endoderm,* the innermost layer of skin
di (Gk.) double	*dicotyledon,* a plant embryo having two embryonic leaves
dis (Gk.) apart, double	*dissect,* to cut apart
dys (Gk.) bad, difficult	*dyspepsia,* poor or bad digestion
e (L.) without, from	*enucleate*, without a nucleus
eco (Gk.) house, abode	*ecology*, the study of the interrelationships within the "house" of nature
ecto (Gk.) outside, outer	*ectoparasite*, a parasite that exists on the outside of the body of the host
en (Gk.) in, into	*engulf*, to surround and take into (the body)
end (Gk.) within, inner	*endoskeleton*, a skeleton within the body
epi (Gk.) upon, over	*epinephrine*, a hormone secreted by the adrenal glands, one of which is situated on each kidney
eu (Gk.) good, true	*eukaryotic*, pertaining to a cell having a "true" nucleus
exo (Gk.) outside, without	*exoskeleton*, a skeleton covering the outside of the body
extra (L.) outside, more	*extracellular*, outside a cell
fer (L.) to bear, carry	*conifer*, a tree that bears cones
fibr (L.) fiber	*myofibril*, a microscopic muscle fiber
gam (Gk.) marriage	*gamete,* a reproductive cell
gast (Gk.) stomach	*gastric,* pertaining to the stomach
gen (Gk.) to bear, produce	*genetics*, the science of heredity
gluc, *glyc* (Gk.) sweet, sugary	*glycolysis*, the splitting of a sugar molecule
gymn (Gk.) naked, bare	*gymnosperm*, a type of tree that produces "naked" seeds, that is, seeds not enclosed within female tissues
hem (Gk.) blood	*hematology*, the scientific study of blood
heter (Gk.) other, different	*heterozygous*, having two different alleles for a specific gene
hom (Gk.) alike, same	*homology*, similarity of organisms resulting from a common evolutionary origin

hydr (Gk.) water	*hydrosphere*, that area of the biosphere consisting of water
hyper (Gk.) over, above	*hypertension*, high blood pressure
hypo (Gk.) under, beneath	*hypotonic*, solute concentration less than that of another solution
ichthy (Gk.) a fish	*ichthyology*, the scientific study of fish
infra (L.) below, beneath	*infracostal*, below the ribs
itis (Gk.) inflammation	*neuritis*, inflammation of a nerve
kary (Gk.) nucleus	*karyokinesis*, division of a cell nucleus
kilo (Gk.) one thousand	*kilocalorie*, one thousand calories
laryn (Gk.) voice box, larynx	*laryngeal*, pertaining to the larynx
leuc (Gk.) white	*leucocyte*, a white blood cell
log (Gk.) discourse, study of	*physiology*, the scientific study of the functioning of living organisms
lys (Gk.) a loosening	*hydrolysis*, the breaking down of a substance by the addition of water
macr (Gk.) large	*macroscopic*, large enough to be seen with the unaided eye
meg (Gk.) large, great	*megaspore*, a large female spore
mer (Gk.) a part	*centromere*, the structure that joins sister chromatids of a duplicated chromosome
mes (Gk.) the middle	*mesophyll*, the layer of cells between the upper and lower epidermis of leaves
micr (Gk.) small	*micrometer*, a unit of length equal to one millionth of a meter
mill (L.) one thousand	*milliliter*, a unit of volume equal to one thousandth of a liter
morph (Gk.) form	*morphology*, the scientific study of the form of organisms
myo (Gk.) muscle	*myocardium*, the muscle of the heart
nephr (Gk.) the kidney	*nephritis*, inflammation of a kidney
neuro (Gk.) nerve	*neuron*, a nerve cell.
oid (Gk.) like, form	*mucoid*, resembling, or like, mucus
oma (Gk.) a tumor	*carcinoma*, a cancerous tumor
oö (Gk.) an egg	*oocyte*, a female reproductive cell
osis (L.) a condition of	*cirrhosis*, a diseased condition of the liver
ost (Gk.) bone	*osteoblast*, a young bone cell
par (Gk.) beside, beyond, near	*parathyroid*, an endocrine gland located near the thyroid gland
path (Gk.) disease	*pathogenic*, disease-causing
ped (L.) a foot; (Gk.) a child	*biped*, having two feet; *pediatrics*, the medical specialty concerned with the treatment of children
peri (Gk.) around	*perinuclear*, around the nucleus
phag (Gk.) eat	*phagocyte*, a cell that engulfs and destroys foreign substances
plasm (Gk.) something molded	*cytoplasm*, the semiliquid cell contents lying within the cell membrane
pod (Gk.) a foot	*podiatry*, the medical specialty concerned with the treatment of feet

poly (Gk.) many, much — *polysome,* a group of ribosomes

pro (Gk.) before, in front of — *prophase,* the first stage of cell division

sacchar (Gk.) sugar — *disaccharide,* a carbohydrate composed of two simple sugars

soma (Gk.) body — *somatic,* pertaining to the body

spor (Gk.) seed — *sporophyte,* a plant generation that produces spores

stas, *stat* (Gk.) standing — *homeostasis,* staying the same

stom (Gk.) mouth — *stomate,* a microscopic opening on the surface of leaves

sub (L.) under, below — *subcutaneous,* under the skin

sym, *syn* (Gk.) with, together — *symbiosis,* living together

tax (Gk.) arrangement — *chemotaxis,* the orientation of an organism to a chemical stimulus

thromb (Gk.) clot — *thrombus,* a stationary blood clot

tom Gk.) cut — *anatomy,* the scientific study of body structure; literally, "to cut up"

trop (Gk.) turn, change — *tropism,* a turning response in some plants

troph (Gk.) food, nourish — *heterotroph,* an organism that requires food from an outside source

vas (L.) vessel, duct — *vascular,* pertaining to vessels

vor (L.) eat, devour — *carnivore,* an animal that feeds on flesh

Bibliography

Barnes, R.D. *Invertebrate Zoology,* 3rd ed. Philadelphia, PA: W.B. Saunders Co., 1974.

Black, J.G. *Microbiology: Principles and Applications,* 3rd ed. Upper Saddle River, NJ: Prentice-Hall, 1996.

Bloom, W., and D.W. Fawcett. A *Textbook of Histology,* 12th ed. Philadelphia, PA: W.B. Saunders Co., 1995.

Darwin, C. *On the Origin of Species by Means of Natural Selection or The Preservation of Favored Races in the Struggle for Life.* New York: Washington Square Press, 1963.

DeDuve, C. *A Guided Tour of the Living Cell.* New York: *Scientific American* Books, 1984.

Kormandy, E.J. *Concepts of Ecology,* 4th ed. Upper Saddle River, NJ: Prentice-Hall, Inc., 1996.

Matthews, C., and K. van Holde. *Biochemistry,* 2nd ed. Redwood City, CA: Benjamin/Cummings, 1996.

Martini, F. *Fundamentals of Anatomy and Physiology,* 2nd ed. Englewood Cliffs, NJ: Prentice-Hall, Inc., 1992.

Masterson, W.L., E.J. Slowinski, and C.L. Stanitski. *Chemical Principles,* 5th ed. Philadelphia, PA: Saunders College Publishing, 1981.

Nebel, B.J., and R.T. Wright. *Environmental Science: The Way the World Works,* 5th ed. Upper Saddle River, NJ: Prentice-Hall, Inc., 1996.

Poirier, F.E. *Understanding Human Evolution,* 3rd ed. Englewood Cliffs, NJ: Prentice-Hall, Inc., 1993.

Raven, P., R. Evert, and S. Eichhorn. *Biology of Plants,* 4th ed. New York: Worth. 1986.

Romer, A.S., and T.S. Parsons. *The Vertebrate Body,* 6th ed. Philadelphia, PA: Saunders, 1986.

Schlesinger, A.B. *Explaining Life.* New York: McGraw-Hill, Inc. 1994.

Seeley, R.R., T.D. Stephens, and P. Tate. *Anatomy and Physiology,* 2nd ed. St. Louis, MO: Mosby, 1992.

Stine, G.J. *The New Human Genetics.* Dubuque, IA: Wm. C. Brown, 1989.

Tizard, I. *Immunology: An Introduction,* 4th ed. Philadelphia, PA: Saunders, 1995.

Weaver, R.F., and P.W. Hendrick. *Genetics.* Dubuque, IA: Wm. C. Brown, 1989.

Wilkins, M. *Plantwatching: How Plants Remember, Tell Time, Form Relationships and More.* London: Roxby Reference Books Limited, 1988.

Wolfe, S. *An Introduction to Cellular and Molecular Biology.* Belmont, CA: Wadsworth, 1995.

Index